Traffic Engineering

Traffic Engineering

Editor: Dominic Brady

MURPHY & MOORE
www.murphy-moorepublishing.com

Published by Murphy & Moore Publishing
1 Rockefeller Plaza,
New York City, NY 10020, USA
www.murphy-moorepublishing.com

Traffic Engineering
Edited by Dominic Brady

International Standard Book Number: 978-1-63987-537-5 (Hardback)

Cataloging-in-Publication Data

Traffic engineering / edited by Dominic Brady.
p. cm.
Includes bibliographical references and index.
ISBN 978-1-63987-537-5
1. Traffic engineering. 2. Transportation engineering. 3. Highway engineering. I. Brady, Dominic.
HE333 .T73 2022
388.413 12--dc23

Contents

Preface

Traffic engineering is a field of civil engineering which uses engineering techniques for achieving the safe and efficient movement of passengers and goods. It is focused on safe and efficient traffic flow by using innovation in road geometry, traffic lights and traffic signs, and through the design of sidewalks, crosswalks, road surface markings and cycling infrastructure. Traffic flow management can be long-term or short-term. Traffic problems can largely be addressed by developing intelligent transportation systems, such as by integrating sensors to measure traffic flows, as well as interconnected and automated guidance systems to manage traffic. An important area of traffic engineering is highway safety engineering, in which factors such as road user psychology, vehicle dynamics and human factors engineering are taken into account to reduce the incidence of crashes. This book provides comprehensive insights into the field of traffic engineering. It traces the progress of this field and highlights some of its key concepts and practices. Coherent flow of topics, student-friendly language and extensive use of examples make this book an invaluable source of knowledge.

The information contained in this book is the result of intensive hard work done by researchers in this field. All due efforts have been made to make this book serve as a complete guiding source for students and researchers. The topics in this book have been comprehensively explained to help readers understand the growing trends in the field.

I would like to thank the entire group of writers who made sincere efforts in this book and my family who supported me in my efforts of working on this book. I take this opportunity to thank all those who have been a guiding force throughout my life.

Editor

1

A State-of-the-Art Review on Empirical Data Collection for External Governed Pedestrians Complex Movement

Xiaomeng Shi ,[1,2,3] Zhirui Ye ,[1,2,3] Nirajan Shiwakoti,[4] and Offer Grembek[5]

[1]*Jiangsu Key Laboratory of Urban ITS, Southeast University, China*
[2]*Jiangsu Province Collaborative Innovation Center of Modern Urban Traffic Technologies, China*
[3]*School of Transportation, Southeast University, 2 Dongnandaxue Rd, Nanjing, Jiangsu 211189, China*
[4]*School of Engineering, RMIT University, Carlton, Melbourne, VIC 3053, Australia*
[5]*Safe Transportation Research & Education Center, Institute of Transportation Studies, UC Berkeley, 2614 Dwight Way, Berkeley, CA 94720-7374, USA*

Correspondence should be addressed to Zhirui Ye; yezhirui@seu.edu.cn

Academic Editor: Shamsunnahar Yasmin

Complex movement patterns of pedestrian traffic, ranging from unidirectional to multidirectional flows, are frequently observed in major public infrastructure such as transport hubs. These multidirectional movements can result in increased number of conflicts, thereby influencing the mobility and safety of pedestrian facilities. Therefore, empirical data collection on pedestrians' complex movement has been on the rise in the past two decades. Although there are several reviews of mathematical simulation models for pedestrian traffic in the existing literature, a detailed review examining the challenges and opportunities on empirical studies on the pedestrians complex movements is limited in the literature. The overall aim of this study is to present a systematic review on the empirical data collection for uni- and multidirectional crowd complex movements. We first categorized the complex movements of pedestrian crowd into two general categories, namely, external governed movements and internal driven movements based on the interactions with the infrastructure and among pedestrians, respectively. Further, considering the hierarchy of movement complexity, we decomposed the externally governed movements of pedestrian traffic into several unique movement patterns including straight line, turning, egress and ingress, opposing, weaving, merging, diverging, and random flows. Analysis of the literature showed that empirical data were highly rich in straight line and egress flow while medium rich in turning, merging, weaving, and opposing flows, but poor in ingress, diverging, and random flows. We put emphasis on the need for the future global collaborative efforts on data sharing for the complex crowd movements.

1. Introduction

Pedestrian traffic is an essential mode in a multimode and multilevel transportation system. In contrast with vehicular traffic, pedestrians tend to change their speed and direction more frequently resulting in complex interactions with other pedestrians, transport modes, and the traffic infrastructures. These interactions can be generally categorized into four types: Pedestrian to Infrastructure (P2I), Pedestrian to Pedestrian (P2P), Pedestrian to Motor Vehicle (P2V), and Pedestrian to Nonmotorized Transportation (P2N) [1]. In transport hubs (e.g., inside train station) or mass gathering (e.g., a concert), the interaction types for pedestrian traffic are typically limited to P2I and P2P. The simple maneuver could be pedestrians walking relatively straight in one direction at uniform speed (unidirection movement) to their destination. However, as the direction of movements and speed changes (e.g., multidirection movements), this can lead to complex interactions and competitiveness among pedestrians [2]. The movements of pedestrians to a large extent are governed by external factors such as infrastructure constraints (e.g., people at bottlenecks are forced to slow down as a result of space competitions) and maintaining proximity from other moving objects (e.g., a pedestrian has to change his/her walking direction for collision avoidance purpose). As a result, we observe pedestrian crowd movement ranging from unidirectional to multidirectional movement at major public infrastructure such as train stations, sports stadiums, and other outdoor or indoor public gatherings. These multidirection movements can result in increased number of conflicts resulting in delays and congestion [3]. In addition, the additional delays due to conflicts can increase egress time during emergency evacuation thereby impacting the efficiency of the evacuation process [4]. Further, the additional delays can cause impatience among escapee resulting in pushing behavior which can lead to trampling and stampede [5]. Recent studies have discovered that the small architectural modifications in an escape area can have large potential effects in terms of outflow and safety of individuals [3, 6–11]. Therefore, pedestrians crowd movements and interactions with the escape area or infrastructure is a subject of greater research interests among researchers working in the field of transport engineering, infrastructure engineering, and architecture. As such, these complex multidirection movements have been studied extensively in the past [2, 3, 9, 12, 13]. Both microscopic and macroscopic approaches have been proposed to understand the mechanism of the pedestrian complex movements [3].

Recent advancements in computer processing power have prompted the researchers to collect and analyze the empirical data to a finer grain for model calibration and validation. Therefore, empirical data collections through controlled laboratory experiments have been on the rise in the past decade [2]. Although there are several reviews of mathematical simulation models for pedestrians' traffic in the existing literature, only few studies have reviewed the state of the art of empirical data for pedestrian traffic [2, 14]. Even those limited studies have examined the empirical data collection for pedestrian traffic from broader perspective rather than exploring the detailed debates, challenges, and opportunities for multidirection complex movements. To the authors' knowledge, there is no detailed and systematic review on empirical data collection for pedestrians' complex traffic movement in the literature.

Therefore, the overall aim of this paper is to present a systematic review on the empirical data collection for uni- and multidirection complex movements. Considering the length and scope of this paper, we restrict the review to only uni- and multidirection movements governed by external factor by considering only P2I and P2P interactions. Please note that pedestrian traffic can also be governed by internal interactions among pedestrians which can lead to some self-motivated movement behaviors such as overtaking [15], self-slowing [16], queuing [17], and grouping [18], as well as crowds' self-organization phenomena such as 'faster-is-slower' effect [5], "lane formation" [9, 19], "zipper effects" [20, 21], "herding" [22], and 'freezing-by-heating' [23]. These emergent behaviors are driven by the individual's self-consciousness or the internal interactions within the crowd. These movement types have been identified as the underlying mechanisms of several crowd disasters [24–33]. For example, under severely congested situation, the individual body's movements are largely restricted by the surrounding pedestrians and any secondary increase of flow could result in the forward and backward compression waves leading to stampede accidents [24, 27].

The exclusive review of this internally governed behavior is not within the scope of this review. However, we have occasionally mentioned these phenomena when examining externally governed movement patterns at instances where the internally governed movements also played an important role in the efficiency of pedestrians walking operations. Also, this review is not intended to review the mathematical models for pedestrian traffic or the results obtained from simulation models on complex movements. Interested readers on simulation models for pedestrian traffic can look into the review work by Duives et al. (2013) [34]. This is necessary to balance the breadth and the depth of the review article.

Our paper is structured as follows. Section 2 describes the methodology of this literature review paper. Section 3 gives the classification methods of the complex movement pattern. Section 4 presents the discussion based on the reviews. Finally in Section 5, the summary of this review article is presented.

2. Methodology

We used Boolean searches of databases to obtain literatures from electronic database including ISI Web of Science, Scopus, and Google Scholar. Other online social platforms such as Researchgate and Mendeley as well as well-known conference proceedings in this field such as the Pedestrian and Evacuation Dynamics (PED), Traffic and Granular Flows (TGF), and Transportation Research Board (TRB) were also used as the search database. We first checked the literature review papers related to pedestrian traffic topic, especially multidirection movements, published in recent years. The literature search was then further supplemented by relevant publications in the reference lists of those recent publications.

The time period for the review is based on the past two decades (from 1998 to 2018) as it is the time-frame where we witnessed a significant advancement made to the data collection and modeling of pedestrian traffic. Only papers published in English are considered in this review as they are widely accessible to the researchers. We only reviewed the articles published in journal, conference, and book or book chapters. Therefore data from the government reports, industry related journals, or other sources are not considered. If the reviewed conference papers were extended and published in journals afterwards, we considered only the journal paper as the primary source and vice versa.

Reference management software, Mendeley Desktop, was adopted for the collection of literature, organization of the document, and generation of the cited list. This tool significantly improved the efficiency of writing the review paper.

3. Complex Movements of Pedestrian Flow

Complex movements are the aggregates of individual and crowd collective motions. To decompose the collective movement into critical movement behaviors, many researchers have studied the crowd motion behaviors both from pedestrian traffic system and other biological systems [35, 36].

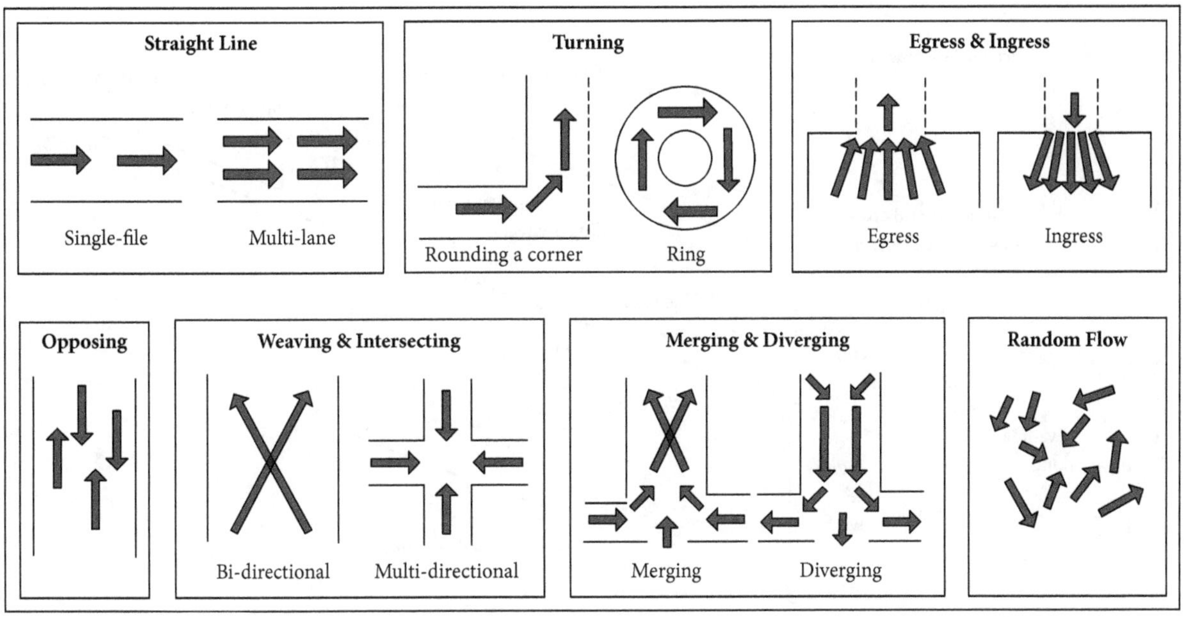

FIGURE 1: Conceptual diagrams of external governed complex movements.

However, a generic classification of pedestrian crowd behaviors is still missing from the literature [34, 35]. To the authors' knowledge, a review on the hierarchy of the crowd motion behaviors with regard to the movement complexity is missing in the literature.

Duives et al. (2013) proposed a classification of crowd motion cases based on the flow patterns [34]. They decomposed the crowd movements into eight motion based cases belonging to two categories: unidirectional flow (straight flow, rounding a corner, entering, and exiting) and multi-directional flow (bidirectional flow, 2-strips crossing flows, >2-strips crossing flows, and random crossing flows without focal point). They stated that the mix of these cases could be able to cover the entire range of complex movements. In addition, they also summarized six types of self-organization phenomena from the literature: lane formation, 'stop-and-go' waves, turbulent flows, herding, the zipper effect, and the 'faster-is-slower' effect. However, the focus of that paper is the assessment of crowd simulation models, rather than the empirical evidences under each crowd motion phenomenon.

Based on the classification of interrupted and uninterrupted flows in vehicular traffic [37] and considering the predominant driving forces of complex interactions, we classify complex movements into two categories:

(i) *Externally Governed Movement.* Movements are strongly restricted/affected by external factors such as infrastructure constraints (e.g., merging pedestrian crowd flow at T-intersection [38]). Pedestrians have to change their desired walking manner due to the competition of space and time. Therefore, under this category, movements are aggregated into patterns which have strong relationships with the geometrics of architectural configurations. In addition, the interpersonal interactions are also predominately affected by the external environments but not the internal factors.

(ii) *Internally Driven Movement.* Movements are spontaneously organized within the crowd in an observable manner under appropriate situations. Under this category, the formulations of crowd movement patterns are less influenced by external factors. And the interpersonal interactions are usually driven by pedestrian's willingness to behave his/her own manner or follow others' behaviors (e.g., lane formation within a pedestrian crowd [19])

In this paper, we build our review based on the eight classifications as stated by Duives et al. (2013) and discussed earlier. Additionally, we have extended and considered other complex movements that have been investigated theoretically or empirically in the literature such as turning, weaving, merging, and diverging movements as shown in Figure 1.

Figure 2 shows the classification of complex movements that shows the category of pedestrians' movements under external governed and internal driven movements. The external governed movements include unidirectional and multi-directional flows while internal driven movements include individual and crowd motion behaviors and self-organized phenomena. A dotted line representing the interface between internal driven and external governed movements is also drawn demonstrating that there could be overlap between these two types of movements (e.g., a merging flow may display 'faster-is-slower effect' or "lane formation"). As stated earlier in the introduction section, the scope of this paper is to review the externally governed pedestrian crowd movement. Figure 2 displays the conceptual diagram of these flow patterns as shown in the Figure 1 with a stepwise hierarchy. Starting with a simple straight unidirection movement, the pedestrian crowd movement can get complex as we move

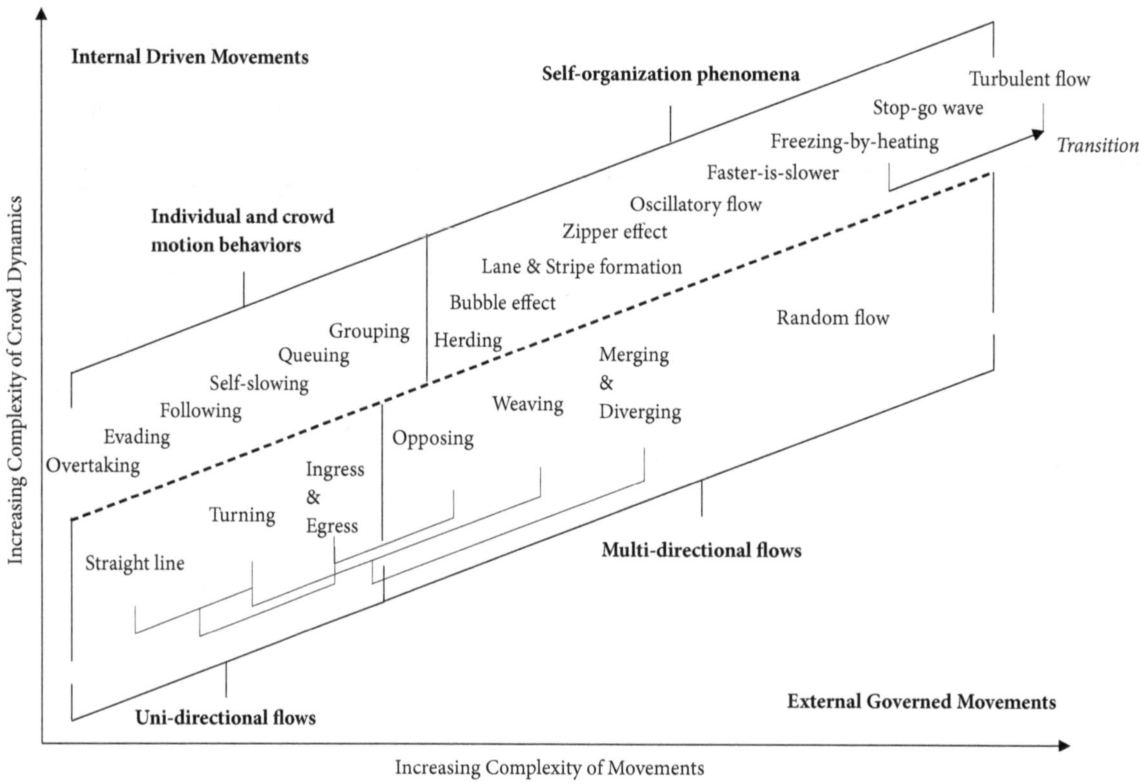

FIGURE 2: Classification of complex movements.

towards opposing, weaving, merging, diverging, and random flows. Please note that random flow can be the combinations of the above movement patterns. The greatest challenge for studying random flow is the uncertainty of conflicts. Therefore, the prediction of walking directions and potential conflicts is at the heart of this category.

As shown in Figure 2, the crowd dynamics becomes complex as the complexity of movement increases. Additionally, the figure also shows the connections and overlapping among these complex interactions.

Externally governed movement usually exhibits observable flow patterns as it usually occurs at fixed locations. In this category, we have put the externally governed movements into eight subcategories: straight line, turning, egress and ingress, opposing, weaving, merging, diverging, and random flows. The first three flow patterns (straight line, turning, and egress and ingress) belong to unidirectional flow, as the desired flow directions are united. Meanwhile, the others are affiliated with multidirectional flow, as the desired flow directions are varied (e.g., opposing, weaving, merging, and diverging). Please note that multidirectional movement like crossing can be a part of opposing flow or weaving flow.

Internal driven movement is more complicated because it can be observed only under certain conditions such as location and flow status. By expanding the summary of individual and crowd motion cases, which are primarily internal driven as highlighted in [39], we developed a comprehensive classification of the internal driven movement as shown in Figure 2. We first listed the self-motivated individual motion behaviors including overtaking [15], evading [40], following [41], and self-slowing [16], as well as the crowd motion behaviors, i.e., queuing [17] and grouping [42]. Then, we categorized 10 types of self-organization phenomena including herding [43], bubble effect [44], lane formation [45], stripe formation [9], zipper effect [20], oscillatory flow [46], faster-is-slower [5], freezing-by-heating [23], stop-go waves [47], and turbulent flow [48].

The above classification method is not restricted to a type of facilities, but is extended to the operations of pedestrians flow on those facilities. In addition, it should be noted that crowd dynamics are affected by both internal and external factors simultaneously, the movements in one category are not absolutely independence with the other category. Under some circumstances, the governing mechanisms of flow might be transformed to another category. Therefore, the overlapping among those interactions occurs as shown in Figure 2. Also, please note that the logical flow for each subcategory may vary. For example, some empirical studies contain more than one crowd phenomena, such as both unidirectional and multidirectional flows. In such case we separated the key findings but introduced the common experiment setups in their first belonging category.

In the following subsections, we describe the empirical studies for each external governed complex movement. Finally, a summary table highlighting the contributions and recommendations from those studies are presented at the end.

3.1. Straight Line. Straight line walking, as the basic form of walking, refers to the kind of unidirectional movement (i.e., without any direction change). It has been used as a benchmark for pedestrian dynamics studies. Datasets for straight line walking can be seen in several well-recognized handbooks in transportation and fire engineering fields as well as many early studies on pedestrian traffic [49–55]. Some early year's empirical data were not well documented as of missing details due to the limitation of the traditional data collection techniques. In addition, large discrepancy existed in these studies for the insufficient influence factor consideration and uncontrollability. In this section, we highlight the recent empirical data collected through more advanced techniques [48, 56–59].

As shown in Figure 1, straight line walking can be segmented into single-file and multilanes movements. In single-file movement, pedestrian has no lateral interaction with other pedestrians; i.e., the interactions are purely longitudinal. Meanwhile, multilanes straight line movement requires the consideration of the horizontal oscillations from the other walking streamlines.

Single-file movement experiments were carried out with participants from different cultural background. It is worth mentioning that despite that many experiment setups in straight line walking are loop corridors/passageways, most of them have not considered the turning movements at corners and restricted the measuring areas to the straight parts. Such oval-shaped loop corridors are also different from the settings of ring-shaped corridors where pedestrian kept turning during their movements [57, 60].

Seyfried et al. (2005) set up experiments in a loop corridor with up to 34 German participants [57]. Through manual video analysis, velocity-density relationships for single-file movement were determined and compared with the movement on 2D plane. They found little difference between single-file movement and 2D plane movement in terms of density range and concluded that the internal friction and other lateral interventions had little influence on density-speed relations. Later, Liu et al. (2009) replicated Seyfried et al. (2005)'s setup and performed a similar single-file movement experiment involving Chinese participants [61]. Through the trajectories derived from automatic image processing, velocity, density, amplitude, and frequency of lateral oscillation, time headway values and distributions were calculated. They also observed that the lateral oscillations change with velocity or density. As density increased, time headway amplified while spacing reduced. After comparing the results with Seyfried et al. (2005), they found that their results exhibited higher speed under same density.

In the same year, Chattaraj et al. (2009) also carried out the same experiments in India to compare the fundamental diagrams of straight line movement with those from Seyfried et al. (2005) in Germany [62]. They found significant cultural difference between the two groups. For example, German group tend to have more minimum personal space than Indian group while German group was more sensitive to the increase in density than Indian group. In addition, they also discovered that the length of corridor have no influence on the fundamental diagram. Apart from the effect of corridor length, boundary condition effects on fundamental diagrams were also empirically examined. Zhang et al. (2015) set up the similar corridors with Seyfried et al. (2015) with closed and open boundaries [63]. The influence boundary conditions on flow properties including the fundamental diagrams and density transition over time were investigated. They found that boundary condition has an effect on the flow properties and the effect was exceedingly high at jamming state. Under same density level, the velocities were lower under closed boundary condition. Cao et al. (2016) also set up the similar single-file experiments with 80 young students and 47 aged people in China [64]. It was discovered that jams tend to occur in mixed group due to the variation of maneuverability and self-adaptive abilities.

Recently, Seyfried's team further continued their researches on single-file movement by considering the stepping behaviors of pedestrian locomotion. Adopting the similar setups, Wang et al. (2018) organized 196 participants to walk in the oval passageway at different densities [65]. Step length, step frequency, swaying amplitude, and step synchronization were measured from trajectories. They found that power functions had better goodness-of-fit than linear function in terms of describing the relations between step length and frequency and speed. In addition, the relations between swaying amplitude and speed could be segmented into two regimes. Likewise, Cao et al. (2018) further analyzed the experiment data in Cao et al. (2016) with the similar analysis of stepping behavior with Wang et al. (2018) [66]. They discovered additional findings on age effects on stepping behavior, e.g., young people tend to control the lateral movement better than the old people. They also fitted the relations between step frequency and speed and found a quadratic fit in three groups of experiments which were different than those observed with Wang et al. (2018). Moreover, they observed no significant effect of height and gender on the fundamental diagram.

In terms of multilanes movements, Daamen and Hoogendoorn (2003) performed a series of controlled experiments with 60-80 human participants [56]. Their experiments included straight line walking scenarios with different desired speed and density levels (Experiment 1-3). Flow characteristics such as density and speed distribution, travel time, and fundamental diagrams were calculated from the extracted video trajectories. Likewise, Helbing et al. (2005) organized about 100 college students to walk through a straight passageway bounded with desks and chairs [9]. Bottleneck was created by moving the desk to narrow the passageway along the walking line. Mean flow and time headways were counted from videos. Results were compared between the corridor setups with and without bottleneck and they found significant capacity drop in bottleneck situation under unidirectional cases.

Similarly, Asano et al. (2007) conducted crowd experiments in a campus with 94 students [12]. Flow demand (30, 94) and controllability (with/without instruct to walk) were considered in straight line scenario. Multilanes walking features with different desired speed level in each stream were also investigated. Trajectories were extracted from video and headway-speed relationship for straight flow was calculated for different cases. Likewise, Zhang et al. (2011) performed a series of large-scale experiments in well-designed corridors with up to 350 people [58]. Controlling the number of pedestrians, width of corridor, width of entrance, and exits, 28 runs of straight line experiments were performed. From the individual trajectories extracted using a software called PeTrack [67], the influences of density measurement methods and corridor width on fundamental diagrams were analyzed. It was found that, for density less than 3.5 m^{-2}, measurements method had minor influence and the specific flow was independent of corridor width as the density level was not highly congested.

More recently, Seyfried's team reported their new large-scale experiments results in Cao et al. (2017) [13]. About 2000 participants were involved in this series of experiments, including multilanes straight line movements in a corridor with a width of 5 m. Fundamental diagrams for different density measurements were presented and the estimated specific flow for unidirectional flow was 1.4 Ped/m/s at density level of 1.5 Ped/m^2. Likewise, Sharifi et al. (2017) conducted a series of laboratory experiments in a circuit corridor that contained straight line passageway [68, 69]. Heterogeneity of crowd dynamics was investigated through the involvement of 189 normal people and 42 disabled people [70]. Trajectories of participants were extracted through automatic video processing and the macroscopic and microscopic crowd motion parameters were calculated. Capacities and level-of-service for straight line facilities with disabled people were further analyzed through the modeling of time headway [71].

Several field studies of straight line movement on different types of walking facilities were carried out [72–74]. To calibrate and validate Legion's simulation model, Berrou et al. (2007) collected 4762 valid samples of pedestrians' walking speeds from 18 hours of video recorded by 23 cameras in New York City Transit Station. The researchers compared their results (mean speed 1.50m/s with standard deviation of 0.21m/s) with several published free speed statistics. Similarly, to derive pedestrian's walking speed distribution on a long stairway, Kretz et al. (2008) carried out a field observation at a stadium in Germany [72]. 485 pedestrians were observed and divided into three categories according to the degree of influence by surrounding people. Results showed that the maximum horizontal walking speeds distribution exhibited 0.4m-0.5m/s depending on density level. They also compared the speed distribution with a short stairway case and found the speed in a long stairway could be the half of that on a short stairway. They also discovered the gap of lack of a universal scaling factor for the speed on stairs relying on the length of stairway. Regarding high density situation in multilanes straight flow, Zhang et al. (2013) collected a field data of dense crowd during a mass gathering event through the combined use of camera and active infrared counter [73]. Flow rate and velocity of the crowd were measured and the fundamental diagrams were painted and compared with previous studies. Three additional flow operational indices including flow rate/velocity transition, velocity field, and speed map were calculated. Results suggested that the capacity for the street was estimated between 1.73 and 1.98 /m/s under normal situation and the density should be controlled under 5 /m^2.

Apart from human subjects, several researchers also investigated the straight line movement through animal experiments and examined their relevancy to pedestrian crowd movement. John et al. (2009) reported the collective dynamics of straight line movements of ants on trails creating a nature-like experiments [75]. They observed no overtaking maneuvers in ants traffic trails but noticed the ants platoons that could be forecasted by simple models. It was observed that the flow increased monotonically with the density and no jam branch could be observed. Sharp contrast was found in ants traffic as compared with vehicular traffic and other transport modes where overtaking and congestion are frequently observed. John et al. (2009)'s findings were observed under nonstressed situations. To examine the straight line movement of stressed ants, Wang and Song (2016) made a passageway (1cm wide and 20 cm long) to measure the stressed ants' movements with different body size [76]. Fundamental diagrams were presented for both small and large ants. They found no evidence of jamming in stressed ants trails which was in line with John et al. (2009)'s findings. Also, they observed that speed seemed to be constant with density. Likewise, they also noted the differences of ants traffic behavior comparing to vehicular and pedestrian traffic in terms of fundamental diagrams.

3.2. Turning. Turning movement can be regarded as the response to the requirement of changing walking direction. It is usually associated with walking maneuver around corners or at exits, angled corridors, or ring corridor scenarios [77]. For single turning movement, the number of walking streamlines does not change before and after turning process and turning flow does not interact with flows in other direction. Therefore, turning movement can be also categorized into unidirectional flow.

It is worth mentioning that turning movement in this review has different meaning than those gait-level of works such as body-turning behavior as the turning movement should show direction change in the general walking streamlines [78].

Literatures documented several simulation studies on turning movements [79–82]. In addition, empirical evidences could be found with both human subjects and biological entities [81, 83–85].

Regarding turning movements around the corners and exits, Yanagisawa et al. (2009) organized 18 participants to perform an experiment with a special attention on the turning locomotion at bottlenecks [83]. They found the theoretical relations of turning angle on the parameters of floor field model under different cases. They compared the theoretical results with empirical data and found the complete agreements. In addition, obstacle effects were also examined by placing a column with 20 cm diameter in front of the exit and the increment of outflow from the obstacle was observed. Apart from the straight line experiments, Zhang et al. (2012) also carried out turning experiments at the cornered corridor with 2.4m width [84]. Fundamental diagram of turning flows at a corner was observed. With regard to turning movements around stair corners, Burghardt et al. (2013) organized experiments to observe the turning movements on stairways [86]. They presented the fundamental diagrams and conducted profiles analysis of density, velocity, and specific flow at stairways corner and compared their results with four planning handbooks. They found that the turning bends could be the potential constraints for safety issue so that stairs without bend should be preferred by designers.

In terms of turning movements at angled-corridor, Dias et al. (2014) performed controlled experiments to obtain the microscopic characteristics of 16 individuals walkers in angle-changing corridors (0°, 45°, 60°, 90°, 135°, and 180°) under three different desired speed levels (normal walking, faster walking, and slow running) [77, 87]. They found the speed drop at a fixed turning region and the dimensions of such region was independent with turning angle but was dependent with desired speed levels. It was observed that the speed around the corner dropped with the increment of turning angle. Later, Dias et al. (2015) extended their previous turning experiments with up to 55 participants [88]. It was found that the longitudinal spacing between pedestrians tends to increase with the increased speed. The collected data were further utilized to develop an optimal trajectories algorithm and to calibrate a cellular automation model [79, 89]. Similarly, Sharifi et al. (2014) and Sharifi et al. (2017)'s experiments also included two angled corridors (right and oblique angle) as well as a spiral stairs with right corners [68, 90]. Walking speed distributions and time-space diagrams for both homogenous and heterogeneous flows were analyzed from the calculation of extracted trajectories. Their results exhibited a correspondence with Dias et al. (2014) as the speed reduction increased with the rise of turning angle [87]. Recently, Seyfried's team reported the results of two sets of experiments in Sieben et al. (2017) [91]. One experiment set a right-angle corridor with two narrow entrances and more than 270 people were asked to enter the corridor. Density bursts and fluctuations were observed at turning sections. Apart from turning angle, the curve radius of turning region was also considered in a recent experiment in China. Sun et al. (2017) studied the combined effects of turning angle and bend radius on the operational characteristics of pedestrians [92]. Results indicated that, under different flow volumes, the cumulative density and average speed of pedestrian flow were affected by both angle and radius effect. Angle's effect was more significant than radius, and the influence was maximized in right-angle case.

In terms of ring-shaped corridor scenarios, Jelić et al. (2012) performed single-file walking experiments with up to 28 pedestrians in a ring corridor with inner radius of 2 m and outer radius of 4.5m [93]. They derived their fundamental diagrams and instantaneous velocity and distance headway from both global and local measurements and compared the density-velocity relations with Seyfried et al. (2005) [57]. Their results showed that three linear regimes could be discovered from the velocity-spatial-headway relations which could be useful for further modeling. Likewise, Moussaïd et al. (2012) organized 119 people in France and adopted the same ring corridor as Jelić et al. (2012) with an advanced thinking of bidirectional flow design and allowance of overtaking maneuver [93, 94]. They found the transitions between organized traffic state and disorganized stated and the unstable dynamics arose from the speed variations. Later in China, Kuang et al. (2015) performed a simplified single-file experiments with up to 26 students walking in a circular path with open boundary [95]. From video observation, they discovered the velocity variation, step fluctuation, and the delay from 'stop-and-go' waves.

To investigate the flow states under hypercongested situation, Jin et al. (2017) organized up to 278 college students to walk in a ring corridor with an inner width of 2 m and outer width of 4 m [60]. As a combined consideration of Jelić et al. (2012) and Moussaïd et al. (2012), both unidirectional and bidirectional flows were considered and the state transitions were described before and after lane formation process [93, 94]. Also, the fundamental diagrams for both flows were calculated and compared with previous works. Recently, Rahman et al. (2017) established similar experiments with Dias et al. (2014) considering the impact of different turning angles (60°, 90°, and 135°) on walking velocity [87, 96]. Results showed that instead of a linear decrement of speed with angle as discovered in Dias et al. (2014), 90° corridor exhibited the lowest mean speed of the participants. They also found that pedestrians preferred walking at the inner side of the corridor due to the shorter path.

Animal-based turning experiment and its relevancy to pedestrian crowd under panic conditions started from Shiwakoti et al. (2011) [7, 80]. The experiments described the egress of panicking ants from a squared and a circular rooms with an attention on the turning movement effects on egress dynamics. Results showed that turning movements could have negative effects on ants' outflows. The findings were also compared with a real-life in-store stampede where turning movement occurred at the door and the results showed a consistency between humans and ants in terms of the negative effects of turning movements. They also found that the placement of an obstacle at the exits could increase the flow. Later, Dias et al. (2012) set up a similar experiment with panicking ants egressing from a squared chamber [81]. They also found the flow reduction arose from turning movement that the flow was 20% lower than straight movement. Dias et al. (2013) further set up a more complex environments with turning, merging, and weaving configurations [97]. Escape rates under four angles of turning corridors (30°, 45°, 60°, and 90°) were compared with straight corridor and the flow reduction percentages were -8%, -20%,-16%, and -27%, respectively.

3.3. Egress and Ingress. Egress refers to the movement process of or exiting/outflow of a space while ingress means the opposite action, i.e., entering/inflow of a space. This two processes are also named as "entering and exiting", "inflow and outflow", or "inlet and outlet" by some other studies. As shown in Figure 2, egress and ingress process can be interpreted as the combination of straight line and turning movements. Since there is no direction change during egress and ingress process, the two types of movements also belong to the unidirectional flows. Egress and ingress movements often occur at fixed bottleneck sites such as exit and entrance. Space competition resulting from bottleneck geometric restriction is the main cause for the flow reduction.

Given the imperative role of egress process for both daily activities and emergency evacuations, during the past two decades, crowd egress has been one of the most popular topics among all the complex movements. Literatures are quite rich in egress experiments using both humans and animals under normal and emergency situations. Based on the past studies in this topic, the geometric of bottlenecks can be divided into two general categories, i.e., channel or corridor bottleneck and door bottleneck.

In addition to the measurements of the macroscopic and microscopic flow characteristics, the top two research topics for egress and ingress movements are the influence of bottleneck width on capacity, and the performance of obstacles near the exits. Specific flow rate of exit segment is often considered as the benchmark for comparison analysis among different measuring results.

Regarding channel/corridor bottleneck, Daamen and Hoogendoorn (2003)'s experiments series included the egress of crowd into a channel bottleneck with a width of 1 m and 2 m, respectively [56]. They further presented the detailed analysis results in Hoogendoorn and Daamen (2005) [20]. The microscopic flow characteristics for the two bottlenecks were analyzed and compared to the straight corridor without bottleneck. Based on the analysis, the researchers found the self-organized phenomena including the "lane formation" and the "zipper" effect.

With respect to door bottleneck, Helbing et al. (2005) conducted egress experiments under both regular and panic-like escape situations [9]. They examined the effect of placing an obstacle near the exit on the outflow and found the obstacle placement could avoid the clogging effect and increase the flow under both competitive intensities. Likewise, in Japan, Nagai et al. (2006) examined the effect of different movement maneuvers (walking and crawling) on the egress flow through a door [98]. They found the increase in the mean escape time with the number of walker as well as crawlers. It was observed that the mean flow increased with density and saturated at the capacity of the door. Later in China, Zhang et al. (2008) organized 60 students to egress a classroom with fixed exit width of 1.1m [99]. Based on video observation, they calculated the distribution of premovement times, velocity variations, continuous outflow at exit, dislocable queue, and monopolizing exit. They discovered that evacuation times displayed a normal distribution and arrival time and escape order showed linear dependence with similar slope. It was also noticed that coordination among escapers is beneficial to the outflow.

To examine the quantitative relations between the width of bottleneck and the specific flow, Kretz et al. (2006) organized 94 participants (40, 80 persons cases) to walk through a door of varying widths ranging from 40 cm to 120 cm with a stepwise increment of 20 cm [100]. Time headway and specific flow were analyzed for different scenarios. Results showed that the specific flow declined with the increase of width when the door only allowed one person to pass and remained a constant value at larger width. They also compared the flow results with other studies. Similarly, Seyfried et al. (2009) performed controlled experiments controlling the channel width from 0.8m to 1.2m with a stepwise increment of 0.1m [21]. They measured the specific flow under different density levels (20, 40, and 60 pedestrians). Individual velocities, local densities, and time gaps for different width of the corridors were analyzed from the trajectories. They discovered that bottleneck capacity increased almost linearly with width and congestion might occur below maximum capacity.

Likewise, to develop an automatic trajectory extraction software for pedestrians, Boltes et al. (2010) organized 250 pedestrians to walk in two types of bottleneck corridors [67]. Microscopic characteristics such as the spatial and temporal variations at bottlenecks were analyzed later by Liddle et al. (2010) and Liddle et al. (2011) [101, 102]. They stated the reasons behind the flow fluctuations at bottlenecks as a physical effect, i.e., a new stepping function manner that could be only performed at bottleneck.

Later in China, Song et al. (2011) and Tian et al. (2012) carried out similar experiments with Seyfried et al. (2009) considering exit corridor width from 0.5m to 1.4m with a stepwise length of 0.1m [103, 104]. Flow features such as average speeds, group speeds, density-speed-flow relations, and time headway distributions were analyzed. They also compared their specific flow results with Kretz et al. (2006) and found an agreement before 0.7m and large disagreement after 0.7m. In contrast with the previous focus on narrow bottlenecks, Liao et al. (2014) adopted a wide door with up to 5 m width to study the egress flow characteristics [105]. Density-velocity relationships inside the bottleneck were found to be independent with door width. In addition, a linear dependency was discovered between the flow and the bottleneck width.

The optimal design of exit areas also attracted empirical researchers. Sun et al. (2017) conducted controlled experiments in a subway station environment to examine the funnel shape bottleneck design on pedestrian flow operation [106]. Testing different angles of funnel shape, the optimal funnel angle was found between 46° and 65° under all flow conditions. This result was in agreement with the previous empirical evidences on turning angle effects, as the funnel shaped exit design could reduce turning angle and conflicts at exit regions [10].

Thereafter, more influence variables were considered to study the egress capacity at bottleneck especially with regard to the heterogeneity flow. Daamen and Hoogendoorn (2010 and 2012) conducted a series of experiments to examine the door capacities considering different door width (from 50 cm to 275 cm), population composition, stress level, and the presence of door [107, 108]. After a series of statistical tests, except for stress level, all the other parameters passed the significance tests. Later, Garcimartín et al. (2014) and Garcimartín et al. (2016) conducted a series of egress experiments considering door width (0.69m, 0.75m) and competitiveness (low, moderate, and high) [109, 110]. They discovered that the time headways displayed a heavy-tailed distributions while the burst sizes decayed exponential distribution.

Bode et al. (2015) involved 12 pedestrians to egress from a room with 6 exits considering the social group effects on egress time [111]. Their results suggested that the presence of social groups increased egress times. Further, no clear time variation was found between individual and group movements. Similarly, von Krüchten et al. (2016) and von Krüchten and Schadschneider (2017) arranged 32-46 participants (dividing into social groups with 4, 6, and 8 people) to egress a narrow door with widths of 0.8m and 1.2m [112, 113]. Macroscopic and microscopic features of the egress flows were analyzed. They discovered that coordinated social grouping, e.g., queuing at exit and moving in a compact manner, could beneficial to the outflow by reducing the conflicts. This finding also indicated that cooperative compact movements had different effects on the flow operation than the bursts release after clogging. Similarly, Nicolas et al. (2016 and 2017) examined the effects of individual's selfish and selfishness behaviors on the global flow characteristics by organizing 80 people to egress through a narrow doorway with a 72 cm width [114, 115]. They found that flow increased monotonically with density and the selfish fraction. It was an interesting finding that "selfishness does not mean coordination" because evading behavior could also lag the flow. A tradeoff between the number of evading and overtaking behaviors at the bottleneck that could maximize the flow should be further studied.

Concerning the ingress movement from the outside area into a confined space, only a few of empirical studies could be found. Ezaki et al. (2016) conducted controlled experiments employing a single-file stream of pedestrians successively ingress into a confined room with closed boundary [116]. Collective features, e.g., location and density distribution and individual characteristics such as the choice of location, were analyzed from video recordings. The revealed pattern such as people preferred to select location near corner indicated the significance of psychological and anticipative factors around an individual. The same year in China, Liu et al. (2016) drafted 40 students to reproduce the ingress and egress process in a rectangular room with a 1 m entrance and 1.5m exit [117]. It was discovered that the ingress order had a significant influence on pedestrian's location distribution in the steady state and the egress order. They also found that inactive pedestrians had a negative effect on the movement as they impeded the flow. Another experiments in Sieben et al. (2017) aimed at reproducing the entrance process of music events [91]. More than 270 people rounding a semicircle room were asked to enter into two narrow exits. It was found that there was a sharp increase of density from 3.8 to 8 Ped/m^2 within 10s.

Regarding field data, Cepolina and Tyler (2005) observed the ingress and egress process of metro passengers flowing into the entrance of escalator in London underground station. They discovered the temporal transition of inflow and outflow from the video. In addition, they established the inflow and outflow relationships at a bottleneck with a width of 80 cm. Apart from the individual walking speed in straight movement, Berrou et al. (2007) also observed the ingress process of passengers entering a train station from a high view [74]. Simulation parameters required by crowd simulation software Legion [118], such as the estimate arrival rate and preferred speed, exit flow, and densities, were extracted from the videos and adopted to set up simulation scenarios. Results showed that the simulated flows and densities were in line with empirical data.

Apart from the empirical investigation of human subjects, animals experiment approaches in studying egress movements are also found in the literature.

Most of the animal experiments were carried out using ants. The experiment setups for ants egress experiments were developed to replicate the similar set up for human experiments, i.e., circular chamber [8, 119, 120], squared chamber [121–123], and squared chamber with oblique corner near exits [124–126]. The common parameter measured is the escape flow at exits. Animal models have been the popular choice to study stressful condition like panic behavior as ethical considerations prevent creating a panic-like situation in pedestrian crowd situation [7].

Altshuler et al. (2005) established a circular chamber with two symmetrically located exits to examine the symmetry breaking phenomena in panicking ants [119]. It was discovered that under panic state, the use of two exits were asymmetrical while the use of exits displayed symmetric under normal situation. Later, Shiwakoti et al. (2009), Shiwakoti et al. (2010), and Shiwakoti et al. (2011) conducted ants experiments to study the effect of exit location and obstacle placement on outflow of panicking ants [7, 120, 127]. It was found that corner exits tend to have higher flow compared to middle exits in squared chambers. Also, the placement of obstacle could in general increase the flow as compared with column free situation. Burd et al. (2010) followed Shiwakoti et al. (2009)'s experiments to study ants' nested behaviors in normal state [8]. They also found the obstacle could increase the egress flow of normal ants. Soria et al. (2012) found the 'faster-is-slower' effect in ants traffic under various stress levels [124]. Their results also exhibited no selfish traffic behavior among ants. Likewise, Boari et al. (2013) observed the consistence finding in terms of the lack of selfish behavior [125]. However, 'faster-is-slower' effect was not found in their result. Later, Parisi et al. (2015) further analyzed Soria et al. (2012)'s data and carried out a discussion with human traffic rules [128].

Recently in China, Wang et al. (2015) used a single-exit chamber with different exit widths to perform ants egress experiments under stressed situations [121]. They found that the average flow was independent linearly on the exit width which was in contrast with human behaviors. Wang et al. (2016) further built a two-exit chamber to study ants egress through exits with the change of exit location and spacing [122]. It was found that the most efficient exit layout was the longest spacing of the two exits and this finding was correspondent with Shiwakoti et al. (2013) [123]. Symmetric breaking was also observed among stressed ants along with Altshuler et al. (2005). They also discovered the different performance of Social Force Model with respect to describing humans and ants movement in positions analysis, density map, velocity direction, and trajectories.

Another insect that has been used to study crowd collective movement is woodlice. Sobhani et al. (2014) and Sobhani et al. (2015) placed 120 panicking woodlice in a squared chamber with a 1 cm-width middle exit [129, 130]. Densities in near exit region and the flow of the exits were measured based on video analysis and then the relationship between jam density and exit capacity and the fundamental diagrams for near exit region were discovered. They found that the fundamental diagrams for animals in panic status had some variations with the normal situation. And as the rise of stress level among woodlice, the number of blockage near the exit increased.

Apart from ants and woodlice experiments, several researchers used mice to study the egress movement under stressed conditions. Saloma et al. (2003) carried out an experiment with 60 mice egressing from a water pool onto a dry land through door bottleneck with different widths and separations [131]. It was observed that mice behavior was to some extent similar to the simulated results from pedestrian models. Self-organized queuing could be observed at the width that could only allow one mice to pass at the same time. As the door width increased, the self-organized pattern was broken due to the space competition. Saloma et al. (2015) further analyzed their experiments data from the aspects of individual training and crowded degree [132]. They found that prior training of individual mouse could increase the escape efficiency compared with the untrained situation. The training effects were more significant under more crowded situation.

Later in China, Lin et al. (2016) purchased a group of 95 female mice with a uniform body size to perform experiments in a dry rectangular ground with 2 cm width middle exit [133]. Different number of Joss-sticks was burned into smoke to stimulate the mice escape at various level of panic or stress. It was observed that egress time and the number of clogs increased with the levels of stress. It was concluded that the selfish competitive behavior displayed by mice was the main cause for the increase of egress time. Lin et al. (2017) further improved their experiment setting with a consideration of obstacle placement distance and the partition of near exit regions [134]. It was observed that the placement of obstacle could increase the flow by a maximum of 36% and 26% for regular exit region and partitioned exit region, respectively. The same group of researchers further analyzed the effects of exit width and position on mice egress movements. Chen et al. (2017) found that enough exit width enabling two mice pass the exit side by side could avoid 'faster-is-slower' effect [135]. Further, Chen et al. (2018) found that 'faster-is-slower' effect was less likely to occur at corner exits than middle exits [136]. In another mice egress experiments, Oh and Park (2017) examined the influence of the exit angle on egress dynamics with a group of 50 mice [137]. They found that, in general, mean velocity and total egress time decreased with angle.

Regarding larger animals, Zuriguel et al. (2014) studied the egress dynamics by observing the feeding process of sheep flock passing through a narrow fence with in a farm [138]. They found that the time headway of sheep flock exhibited a power-law distribution and the burst sizes obeyed an exponential decay. To further investigate the door width and obstacle effects, Garcimartín et al. (2015) compared the effect of 77 cm and 94 cm doors in with and without obstacle situations [139]. By examining the mean egress time per animal for the four scenarios, it could be found that the most efficient setup was the 94 cm door with obstacle and the least case was the 77 cm door without obstacle. Zuriguel et al. (2016) further studied the effect of obstacle position on the egress flow [140]. They found obstacle position had an influence on the sheep flow and 80 cm was the best distance for the 77 cm door.

3.4. Opposing. Multidirectional movements are more complex compared with unidirectional movements due to the increase possibilities of interactions. The simplest case of multidirectional flows is usually called 'bidirectional' flow where movements are only operated in two directions.

Opposing movement refers to the situation when two streams of people coming from two opposed directions confront at a confined region. It can be also named as counterflow adopted by many studies [141–145]. Because there can only be two direction of flows in opposing movements, in this case, opposing can be regarded as a special scenario of bidirectional flows (or head-on weaving) where the two streamlines interact with an angle of 180°.

Opposing movements can be frequently observed in the daily life such as the boarding and alighting process in a crowded metro station, the movements on bridge during special events [146, 147]. In addition, opposing movements have been identified as the cause of several crowd disasters around the global occurring from the early 1980s to the late 2010s [3, 32].

Given its imperative role in practice, due amount of research efforts has been put into the study of opposing movement which include many empirical evidences.

Several Japanese researchers conducted opposing experiments with an initial purpose to validate the lattice gas model proposed by Tajima and Nagatani (2001) [148]. Isobe et al. (2004) organized up to 70 people to walk from two opposed sides of a rectangular corridor [141]. They presented the evolutions of arrival time and mean velocity as the increase of density. They found that, as the rise of density, the arrival time increased but the mean velocity decreased. Nagai et al. (2005) performed similar experiments with both walkers and crawlers in the opposing case [142]. The arrival time of crawler was dependent on the initial location and the mean arrival time increased with density. In addition to the straight line scenario, Asano et al. (2007) also included an opposing scenario. They illustrated the headway-speed relationships of opposing flows and found that mean speed tend to decrease in shorter headways.

In Europe, controlled experiments of opposing movements were first conducted by Daamen and Hoogendoorn (2003) [56]. Their experiment series also included the case of opposing movements. They applied the similar configuration of their previous straight line and egress scenarios including the desired speed and density levels. However, they only provided the qualitative analysis results for the bottleneck scenarios. Likewise, Helbing et al. (2005)'s experiments sets also contained the opposing movements in a corridor with a short and a long bottlenecks [9]. They presented the flow values and time gaps distributions for the two opposing cases and results showed that long bottleneck was worse than short bottleneck in terms of the effects on opposing movements. Later, Kretz et al. (2006) carried out an opposing experiment with 67 participants in a 2 m width corridor [143]. They calculated the passing times, speeds, and flows and found that the sum of flow in opposing flow was larger than the straight line flow. They also investigated the lane formation and symmetric breaking in opposing flows and presented the frequency of number of lanes in the corridor.

Moussaïd et al. (2009) performed laboratory experiments (40 participants) and field observations (2670 samples) including opposing movement [149]. In their findings, individual's side preference in the evading maneuvers occurring in opposing movement was interpreted as a behavioral coordination behavior as a result of cultural differences. In Dutch, Versluis (2010) conducted a series of laboratory experiments considering age, body size, gender, free speed, travel purpose, maneuverability, pedestrian number, and predicted postencroachment time [150]. They investigated the lateral and longitudinal evasion of pedestrians with a mean value of 0.26m and 0.11m/s for opposing case. Back to Germany, Seyfried's team also collected datasets for opposing movement. Zhang et al. (2012) separated the opposing flows into stable separated lanes and dynamical multilanes flows with balanced and unbalanced flow ratio based on different ordering levels. They painted the fundamental diagrams for each flow types with different corridor width and compared the results with their previous straight line movement data [151]. Results indicated that ordering levels and density measurement methods did not significantly affect the fundamental diagrams. They also compared the maximum flow values of bidirectional data (1.5 Ped/m/s) with their previous unidirectional data (2.0 Ped/m/s). In addition to unidirectional flow, Cao et al. (2017) also reported the experiment results of opposing movements in a corridor with a width of 4 m [13]. Under the density level of 1.5Ped/m^2, the estimated specific flow value for opposing movements was 1.1Ped/m/s. This value was observed to be significantly lower than the unidirectional movement.

To investigate the impact of social group and flow ratio on opposing movement, Gorrini et al. (2016) organized controlled walking experiments in Japan [42]. Their results suggested that the rise of flow ratio had negative impact on the speed of pedestrians. Walking in groups could slow the speed because of the difficulty in moving coordination among group members. In contrast, coordination effects were also found to be positive in opposing movement. To explore the experimental evidence in the "gridlock" effect existing in CA models when simulating the opposing movement, Xue et al. (2017) conducted comparative experiments in both discrete and continuous space [152]. Through the time step analysis of pedestrians' position as well as the global density and specific flow, gridlock was not observed even under high density opposing flow. This result demonstrated the remarkable collaboration behaviors of pedestrian group movement especially at congested situation.

Recently, opposing movements were also investigated in animal systems. Wang et al. (2018) observed the bidirectional movement characteristics for nested ants with loaded and unloaded status through a less competitive experiment [153]. They measured the speed for unloaded and loaded ants, distance headway distribution over speed, and the spatial-temporal analysis of encounter behaviors through trajectories. Results showed that ants appeared to be equally sensitive to distance headway in both uni- and bidirectional flows. Head-on encounter behaviors with the following ants could reduce movement efficiency. Unloaded ants tend to spend more time in communicating with encountering ants than the loaded ones. However, quantitative comparison with vehicular and pedestrian traffic was not conducted.

3.5. Weaving and Intersecting. In opposing flows, the interacting angle of the two streamlines remains a constant value of 180°. Past studies have highlighted that oblique interacting angle of different streamlines can also have a significant impact on the crowd movements [12, 87, 150].

In this study, weaving movement refers to the situation when two streams of pedestrian flows arrive at a fixed region (such as an intersection) at the same time and they can only keep their walking direction by cross over the other steam (please refer Figure 1). It is a commonly observed type of multidirectional movement as more than two numbers of streams may involve in the interactions. Weaving has several alternative forms such as intersecting and crossing, and if the number of involving streams is only 2, it can be also categorized into bidirectional movement.

The first empirical study on weaving movement for pedestrians was found in Daamen and Hoogendoorn (2003) [56]. However, no empirical analysis results for opposing flows were presented in the study. Later, Helbing et al. (2005)'s experiment sets also contained a right-angle weaving scenario with two stripes of flows [9]. Compared to their straight line data, they found the stripe formation phenomenon as there was a significant irregularity and clusters in the passing time. Asano et al. (2007) further analyzed the speed-density relationships (Speed=A+B∗Density) under the impact of interacting angle [12]. They found that parameter A was significantly influenced by several angles (0°, 45°, and 90°) while parameter B was not influenced by the angles except for the right-angle case. Apart from the opposing movement reviewed in the above section, Versluis (2010) also considered oblique interacting angle including 45°, 90°, and 135°. It was found that the angle exhibited a positive effect on mean interaction point and showed a negative effect on mean longitudinal evasion. Similarly, Plaue and Chen (2011) organized two groups of people (54:46) to participant a right-angle weaving experiment [154]. They applied a crowd density estimation method to analyze the trajectories extracted from video. It was found that the nearest-neighbor kernel density method could provide a reasonable estimate of the density of human crowds. Back to Seyfried's new experiments, Cao et al. (2017) also reported the results of weaving scenarios including the right-angle bidirectional and four-directional weaving flows [13]. Similarly, at the density level of 1.5Ped/m^2, the estimated specific flow for bidirectional and four-directional weaving flows was 1.2m/s and 1.1m/s.

In Hong Kong, Wong et al. (2010) performed controlled experiments considering five intersecting angles (0°, 45°, 90°, 135°, and 180°) [155]. The aggregate empirical data were used to calibrate a model and the results showed highly significance. Adopting the calibrated model, they compared the mean speed transitions with reference stream under different density ratio and total densities. The researchers found that as the rise of angle, the speed for reference stream decreased. Later, Wu and Lu (2013) conducted weaving experiments considering the dimension of passageway and flow ratio [156]. They defined a pedestrian weaving zone (PWZ) and discovered the flow characteristics within PWZ. They found that the flow values and flow ratios were more concerned with PWZ instead of the entire walking region. They also established a utility function to assess the PWZ operations based on three novel indices, i.e., weaving intensity, trajectory offset ratio, and density distribution factor of weaving points. Likewise, Sun et al. (2014) performed controlled experiments with 50 students to understand the mechanism of weaving movement at congested situations. Along with the flow characteristics of weaving movements, they also found that weaving tend to occur within an area which was in line with Wu and Lu (2013). Later, Lian et al. (2015) performed a four-directional intersecting experiments with 364 young Chinese students [157]. They examined the effects of side preference on lane formation and the placement of obstacle on the flow stability. It was discovered that lane formation process could be faster when students selected right-hand lane in the corridor. In addition, the placement of obstacle in the middle of the intersection could stabilize the intersecting flows. Sun et al. (2017) and Sun et al. (2018) further investigated the impact of intersecting angles and obstacle effects on intersecting movements [149, 158]. It was found that the effects of angles on flow speed vary at different flow conditions. But the effects of the size of the circular obstacle were monotonous with the speed increment and acceleration dispersion.

Animal-based interesting studies could be only found in the ants experiments conducted by Dias et al. (2013) [97]. In addition to the turning configuration, they also set up a junction corridor with right angle that enabled the bidirectional weaving movements. They discovered the transitions between the two weaving streams; i.e., one moving stream could block the other stream.

3.6. Merging and Diverging. Merging movement can be regarded as the combination of turning and weaving movement. It can be frequently observed at angled corridors/passageways such as T-junction or the stair-floor interface of a high-rise building. Moreover, merging configurations have also been identified as the causes for several stampede accidents including the famous Love Parade disaster [3, 32]. Due to the multiplication of the interactions among turning and weaving movements, merging movements are more complex.

Empirical evidences on merging movements were not sufficient before the year of 2014 [4, 58, 159, 160]. There was a burst of empirical studies on merging behaviors in the most recent four years [3, 161–168].

Boyce et al. (2012) examined the merging behaviors at stair-floor interfaces by performing evacuation drills with 581 participants in three buildings. They noted that the location of stair-floor connection points and the population composition could have a potential influence on merging flow patterns. The same year, Ma et al. (2012) conducted a similar evacuation drills in ultra-high-rise building [169]. They also observed that the merging behaviors at stair-floor interface could impede the flow of evacuees. To replicate the emptying process of mass gathering events in a stadium, Burghardt et al. (2013) performed controlled experiments in the exit of a football stadium where pedestrians moved from spectator seats upon receiving the signal and successively formed three stable streamlines merging into the outlet doors [86]. They only presented the fundamental diagrams of the downstream stairs; details on the merging characteristics were not analyzed. More recently, Huo et al. (2016) performed evacuation drills with 73 participants in a 9-floor high-rise building [161]. They discovered the negative effect of merging behavior on total evacuation time and speed through quantitatively comparison.

With respect to the merging flows on ground surface, Zhang et al. (2011 and 2012) performed merging experiments at T-junctions and compared the fundamental diagrams of merging with straight walking [58, 84]. After deriving the fundamental diagrams of merging flows, they found an increase in speed within after merging areas compared to before merging regions. Shiwakoti et al. (2015) studied the impact of merging angle and desired speed on the microscopic characteristics of merging flows in before and after merging areas. They found that, as the rise of merging angle, the reduction in speed increased. Similarly, Aghabayk et al. (2014 and 2015) also investigated the crowd merging behaviors considering merging angles and desired speed with asymmetric design of corridor [162, 163]. They found that the arrival and departure flow increased with speed and time headways decreased with speed. It was observed that merging angle was also influential to flow and time headways which was in consistent with the findings from Shiwakoti et al. (2015).

Later, Shahhoseini et al. (2016) conducted ants experiments to further understand the merging angle effects (60°, 90°, and 120°) [164]. They found the negative effects of merging angle in ant trials in panic situation. Further, Shahhoseini et al. (2017) combined the factors of merging angle and symmetrical configuration and designed 5 types of merging corridors to perform laboratory experiments with human subjects [170]. It was discovered that the merging behaviors of ants and humans displayed similarly effects arising from differences in architectural configurations.

Same year in China, Lian et al. (2017) established a merging configuration of branched flow merged into main flow with a further consideration of changing the width of branched corridor (0.8m, 1.6m, and 2.4m) [165]. It was observed that the merging ratios had a significant influence on the operations of upstream and downstream of merging areas for pedestrians flow. In the same year, Cuesta et al. (2017) performed merging experiments in a mock-up tunnel with 77 participants to reproduce the merging process of passengers exiting from the train and pedestrians walking on the platform [171]. They used the distribution of instantaneous specific flow to represent the flow dynamics. It was found that merging section had negative effects on the outflow of people at rail tunnel in terms of specific flow.

As illustrated in Figure 2, diverging movement can be defined as a reverse process of merging movement, where a united stream of pedestrians separated into multiple streams with different destinations. Different from the fluctuating flows near a bottleneck, the flow states are stable in both before and after diverging.

A few empirical studies on exit choice behaviors could be categorized into diverging movements. Haghani et al. (2016) and Haghani and Sarvi (2017) performed exit choice experiments considering various exit number (2, 3, and 4) and exit position and exit width (50cm, 100 cm) [172, 173]. Instead of movement characteristics, their focuses were in the decision-making aspects. In another exit choice experiments, Wagoum et al. (2017) reported the flow features, e.g., flow, density relationships of the participants at exits. It was found that under normal situation individual's choice of exit only was dependent on the shortest distance while under higher density situation the load balancing for two exits could be observed.

3.7. Random Flow. Researchers have tried to reproduce the random flow situations by conducting experiments with multiple crossing flows. Dyer et al. (2008) set up a circle with sixteen equally angle-divided directions on the edge [174]. People were asked to stand inside at the center of the circle and individuals would be informed with a direction to walk outside the circle. They tested the role of prior information of walking direction and the conflicting directional information on the group directional decision-making. However, the flow characteristics of the experiments were not analyzed as the researchers were more interested on the psychological aspects of the participants. However, it is to be noted that there is a greater need of empirical data to predict the walking directions and potential conflicts due to random flow.

Researchers from computer vision field have also contributed to the empirical understanding of random flow. Zhou et al. (2012) analyzed the collective behaviors a crowd of pedestrians in a train station walking in random flow directions [175]. Using a mixture model of dynamic pedestrian agents, the future trajectories of pedestrians were predicted based on the observed past trajectories. Also, they classified the random walkers into various collective motion patterns and visualized them in different clusters.

3.8. Summary. Table 1 shows the summary of key findings and recommendations from the literature that conducted empirical studies on the externally governed complex multidirectional pedestrian traffic movements. The table shows the source of the literature, the movement type, the type of infrastructure considered (walking facility, bottleneck, and obstacle), subjects considered (pedestrians, animals), controllability of the variables (controlled laboratory experiments, field observation), and competitiveness among pedestrians (normal, emergency/evacuation drills, and panic). Further, the table provides a summary of the key findings, limitations, and recommendations for future studies and further use of the data generated from a particular literature in other studies. The filled dot in the table shows what approach has been followed in a particular literature while empty cell shows missing approach in a particular literature.

From the table, it is clear that initial studies have focused mainly on the empirical studies of unidirectional pedestrian flow, mostly at the corridor or at the bottleneck like door. Recent empirical studies have focused more on the complex multidirectional movements like weaving, crossing, and merging/diverging. Further, most of the studies are conducted under normal walking conditions. There have been increasing interests on the pedestrians' multidirectional movements under emergency or panicked conditions in recent time. As such, there has been rise in the study of animal models, particularly their relevancy to pedestrian traffic under emergency conditions.

In the next section, we present a discussion on the implications of these findings and directions for future research.

4. Discussions

Empirical data of complex pedestrian crowd movements are playing increasing imperative role in the verification and validation of crowd simulation models, design and planning of pedestrian walking facilities, scheme and evaluation of evacuation plans, development of image processing and visualization tools, coding of game engines, and trajectory planning for robots as well as the understanding of the nature of collective traffic phenomena in biology systems. Therefore, as observed from the literature review, there has been surge on empirical data collection and analysis over the past two decades. From the synthesis of the findings from these literatures, we have identified some opportunities and challenges as described in the following sections.

4.1. Anticipation of Future Empirical Data Requirements. To evaluate the current work and identify the research gaps based on the reviewed empirical studies, we utilize the data in Table 1 based on the number of studies and their belonging to each category. The summary of the statistics is presented in Figure 3.

Figure 3(a) summarizes the number of studies conducted over three time periods (2003-2007, 2008-2012, and 2013-2018) under each movement type. From the general trend of column height showing in Figure 3(a), we can notice that egress and straight line movements are the most popular topics among empirical studies while diverging and random flows are less studied through empirical approach. Also, from the distribution of studies, we can observe that earlier studies (2003-2007) in unidirectional flow have large proportion of straight line and egress experiments while multidirectional flow contains study mostly on the opposing and weaving movements. Meanwhile in the recent decade (2008-2018), researchers not only continue to explore the empirical evidences in straight line and egress movements, but also started to collect empirical data on turning, merging, and ingress movements. The emergence of increased studies on these movement patterns may arise as a result of several recent crowd disasters that involved turning and merging movements [32]. It can be further noticed that the dynamic features of diverging and random flows still require further research attentions.

Table 1: Summary of empirical results on externally governed complex pedestrian traffic movements.

Ref.	Source	Movement Type	Infrastructure			Subject		Controllability		Competitiveness			Summaries		
			Facility Type	Bottleneck	Obstacle	Animal	Pedestrian	Laboratory	Field	Normal	Emergency	Panic	Key Findings	Limitations & Future Recommendations	Further Data Usage
[56]	Daamen and Hoogendoorn (2003)	Straight line; Egress; Opposing; Intersecting	Narrow corridor; Large hallway ground	●			●	●		●			(i) Demonstrate a detailed process for designing controlled experiments (ii) Define and calculate the microscopic and macroscopic characteristics from trajectories for straight, opposing and intersecting flows (iii) Find different pedestrian walking features with and without bottleneck	(i) Only present the analysis of bottleneck experiments data (ii) Data analysis methods are not comprehensive	(i) Bottleneck experiments data were further analyzed [20, 176, 177]
[131]	Saloma et al. (2003)	Egress (Mice)	Water pool	●		●		●				●	(i) Burst sizes of mice egress through the exit exhibit exponential or power-law distribution depending on exit size (ii) Self-organized queuing behavior can be observed when the door width only enables one mice to pass (iii) The queue will be broken with the increase of door width	(i) Whether the findings from the mice can be applicable of humans should be further studies by comparing with empirical data	(i) Further analysis of the experiments data considering the effect of prior individual training [132]
[141]	Isobe et al. (2004)	Opposing	Narrow channel				●	●		●			(i) Experiment data can help improve the lattice gas model by considering the front watching effect and back step (ii) Jamming transition is not found in experiment	(i) The experiment data have not been used for the analysis of microscopic flow features	(i) –
[142]	Nagai et al. (2005)	Opposing	Narrow corridor				●	●			●		(i) The arrival time of crawler depends highly on the initial location (ii) Mean arrival time increases with density	(i) It is interesting to know how the vision condition among crawlers affect the opposing movement	(i) –

TABLE 1: Continued.

Ref.	Source	Movement Type	Infrastructure			Subject		Controllability		Competitiveness			Summaries		
			Facility Type	Bottleneck	Obstacle	Animal	Pedestrian	Laboratory	Field	Normal	Emergency	Panic	Key Findings	Limitations & Future Recommendations	Further Data Usage
[57]	Seyfried et al. (2005)	Straight line	Narrow passageway				●	●		●			(i) Determine the fundamental diagram for single-file movements (ii) Single-file and 2D movements have agreements on velocity-density relations (iii) Required length and velocities indicated a linear relation	(i) Lack of automatic trajectories extraction (ii) Participants did not behave like real life walking in experiments	(i) Across cultural comparison of fundamental diagrams [61, 62]
[178]	Cepolina & Tyler (2005)	Ingress; Egress	Narrow corridor	●			●		●	●			(i) Capacity drop at bottleneck arises from a geometric issue and a time effect (ii) Discover the relationships between average inflow and outflow	(i) The data comparison is not reasonable due to different setups	(i) –
[9]	Helbing et al. (2005)	Straight line; Opposing; Intersecting; Egress; Counteracting	Narrow passageway	●	●		●	●	●	●	●		(i) Boundary conditions influence capacity of facilitates and the headway distribution of pedestrians (ii) The proper use of obstacle can stabilize flow patterns and increase flow fluidity (iii) For opposing flows, long bottleneck is less efficient than short bottleneck (iv) For intersecting flows, lane formation is the result of a significant irregularity and clusters in the passing time (v) Several design solutions are proposed based on simulations	(i) Additional empirical data are needed to verify the proposed design solutions (ii) More psychological factors of pedestrians should be considered	(i) –

Table 1: Continued.

Ref.	Source	Movement Type	Infrastructure			Subject		Controllability		Competitiveness			Summaries		
			Facility Type	Bottleneck	Obstacle	Animal	Pedestrian	Laboratory	Field	Normal	Emergency	Panic	Key Findings	Limitations & Future Recommendations	Further Data Usage
[119]	Altshuler et al. (2005)	Egress (Ants)	Chamber	●		●		●		●		●	(i) Symmetric breaking can be found in panicking ants	(i) The ants' data can be used for comparison with the symmetric breaking phenomena in humans	(i)–
[100]	Kretz et al. (2006)	Egress	Width-changing door	●			●	●		●			(i) Specific flow decrease with the increase of width when the width is under 100cm and keeps a constant when the width exceeds 100cm	(i) Further experiments should consider the heterogeneous group	(i) Data have been used to compare with ants experiments [7]
[143]	Kretz et al. (2006)	Opposing	Corridor				●	●		●			(i) Total flow is always larger in opposing movement than straight line movement (ii) The number of lanes in opposing flow can be as much as 4	(i) Repeated experiments should be carried out considering elderly people, different corridor width and shape, side preference etc.	(i) –
[98]	Nagai et al. (2006)	Egress	Room	●			●	●			●		(i) Channel capacity is different for walkers and crawlers (75 vs. 35) (ii) Mean escape time increases with the number of walkers and crawlers (iii) Mean flow increases with density till reaching the capacity or clog occurs (iv) Lattice gas model can reproduce the egress of walker and crawlers well in a correspondence with empirical data	(i) Detailed gait characteristics of walker and crawlers should be analyzed by introducing more advanced computer vision techniques	(i) –

Table 1: Continued.

Ref.	Source	Movement Type	Infrastructure			Subject		Controllability		Competitiveness			Summaries		
			Facility Type	Bottleneck	Obstacle	Animal	Pedestrian	Laboratory	Field	Normal	Emergency	Panic	Key Findings	Limitations & Future Recommendations	Further Data Usage
[12]	Asano et al. (2007)	Straight line; Opposing; Intersecting	Restricted passageway				●	●		●			(i) Crossing angle may affect speed-density parameters (ii) Distance headway affects mean speed	(i) Require considering heterogeneity in crowds	(i) Development of a combined micro-meso-model [179]
[74]	Berrou et al. (2007)	Straight line; Ingress; Opposing	Passageway; door	●			●		●	●			(i) Strong context-dependencies require to be considered into the modeling of crowd motion pattern	(i) The capability of samples to represent the population are not examined	(i) –
[174]	Dyer et al. (2008)	Random flow	Circle room				●	●		●			(i) Small directionally informed pedestrian can guide the group to targeted directions without verbal communication	(i) Flow characteristics for the different directional flows are not analyzed	(i) Further analysis on three new questions [180]
[72]	Kretz et al. (2008)	Straight line	Stairs				●		●	●			(i) 485 individuals' walking speed on a long stairway (ii) Mean upward walking speed on a long stairway can be half of that on a short stairway	(i) No universal scaling factor for speed on stairways in dependence of the stairway length	(i) –
[99]	Zhang et al. (2008)	Egress	Room	●			●	●		●			(i) Evacuation times display a normal distribution (ii) Arrival time and escape order shows linear dependence with similar slope (iii) Coordination among escapers is beneficial to the outflow	(i) Assumptions cannot be tested under panic situations	(i) –
[62]	Chattaraj et al. (2009)	Straight line	Narrow passageway				●		●	●			(i) Free flow speed keeps the same between Indian and German participants (ii) German tend to have more minimum personal space than Indian (iii) German are more sensitive to the increase in density than Indian (iv) The length of corridor does not have influence on the fundamental diagram	(i) Lack of a method that can measure the dependence of fundamental diagram on security distance	(i) –

TABLE 1: Continued.

Ref.	Source	Movement Type	Infrastructure			Subject		Controllability		Competitiveness			Summaries		
			Facility Type	Bottleneck	Obstacle	Animal	Pedestrian	Laboratory	Field	Normal	Emergency	Panic	Key Findings	Limitations & Future Recommendations	Further Data Usage
[75]	John et al. (2009)	Straight line (Ants)	Nature path			●		●	●	●			(i) No overtaking behavior in straight movement ants trails (ii) No jammed branch for ant traffic in straight movement (iii) No hysteresis and synchronized flow in ant traffic	(i) Lack of the exploration of analogy between ants traffic system and other traffic systems such as vehicular flow and pedestrian flow	(i) –
[61]	Liu et al. (2009)	Straight line	Narrow passageway				●		●	●			(i) Lateral oscillation remains 5.5 cm at free flow (ii) With the increase of density, the velocity drops, the lateral oscillation rises and the time headway surges	(i) No clear explanation on the differences with Seyfried et al. (2015)'s results	(i) –
[181]	Moussaïd et al. (2009)	Straight line; Opposing	Narrow corridor		●		●	●		●			(i) Individual evading behaviors for straight and opposing situations are different (ii) Side preference can be interpreted as a coordination behavior or a cultural bias	(i) Successive interaction with other people should be investigated	(i) –
[21]	Seyfried et al. (2009)	Egress	Narrow corridor	●			●	●		●			(i) Bottleneck capacity rises with the widen of width (ii) Jam can occur at below max. capacity situation (iii) Zipper effect starts to be observed at the bottleneck width of 0.7m (iv) Absolute values and figures from a lab experiment cannot simply be adopted for design usage.	(i) Panic and emergency situation have not been considered	(i) –
[120]	Shiwakoti et al. (2009)	Egress (Ants)	Chamber	●	●	●		●				●	(i) Placement of obstacle near the exit can increase the flow in general cases (ii) Corner exits tend to have higher flow comparing to middle exits in squared chambers	(i) Lack of human empirical data on obstacle performance and exit location effect	(i) Development of an animal-based microscopic model [7, 127]. (ii) Study the effect of obstacle on egress dynamics [123].

TABLE 1: Continued.

Ref.	Source	Movement Type	Infrastructure			Subject		Controllability		Competitiveness			Summaries		
			Facility Type	Bottleneck	Obstacle	Animal	Pedestrian	Laboratory	Field	Normal	Emergency	Panic	Key Findings	Limitations & Future Recommendations	Further Data Usage
[83]	Yanagisawa et al. (2009)	Turning; Egress	Wide-changing bottleneck	●	●		●	●		●			(i) The introduction of turning and conflicts functions to floor field model agree well with empirical data (ii) Outflow depends on the position of obstacle near the exits (iii) About four pedestrians are trying to exit at the same time when the exit width is 50cm	(i) The floor field model is not intelligent enough to capture the lane formation process as empirical data indicate	(i) –
[67]	Boltes et al. (2010)	Egress	Bottleneck corridor	●			●	●		●			(i) Automatic trajectories extraction can be realized in PeTrack with high accuracy in space and time based on markers	(i) The recognition approach needs to be replaced for marker-less tracking	(i) Analysis of the bottleneck flow patterns [101, 102].
[8]	Burd et al. (2010)	Egress (Ants)	Chamber	●	●	●		●		●			(i) Obstacle can increase the ants' egress flow under normal nested situations	(i) Comparison of non-panic data with panic data is required	(i) –
[107]	Daamen and Hoogendoorn (2010)	Egress	Narrow door	●			●	●		●	●		(i) Capacity increases with the rise of stress level at 220cm width and the finding reverses at 110cm width (ii) Population with more children has the highest capacity (mean value of 3.31P/m/s) (iii) Population with 5% disabled people has the lowest capacity (mean value of 2.02P/m/s) (iv) The open door leads to more interactions between participants resulting in reduce in speed and outflow	(i) Latent factors on pedestrian dynamics should be analyzed	(i) Data were used to calibrate several pedestrian simulation models [182]
[150]	Versluis (2010)	Straight line; Intersecting; Opposing	Passageway in a hall				●	●		●			(i) 88% movements are related lateral and longitudinal interaction (ii) Approach side has no effect on the overtaking side (iii) Pedestrians prefer larger lateral evasion in opposing condition and larger longitudinal evasion in intersecting condition (iv) As the rise of interacting angle, mean interaction point increases and mean longitudinal evasion decreases	(i) Variable selections, measurement and experimental setups should be improved	(i) Collison avoidance behavior was analyzed [183]

Table 1: Continued.

Ref.	Source	Movement Type	Infrastructure			Subject		Controllability		Competitiveness			Summaries		
			Facility Type	Bottleneck	Obstacle	Animal	Pedestrian	Laboratory	Field	Normal	Emergency	Panic	Key Findings	Limitations & Future Recommendations	Further Data Usage
[155]	Wong et al. (2010)	Intersecting	Passageway in a stadium				●	●		●			(i) Conflict-induced interactions increase with the angle (ii) People are less willing to follow the front pedestrian to reduce conflicts (iii) The capacity for intersecting flow grows with the increasingly imbalance of pedestrian density ratio	(i) Cultural factors should be considered in pedestrian models (ii) Field observation data should be collected for intersecting flows	(i) Further analysis on the look-ahead behavior of pedestrians [184] (ii) Using a Bayesian inference method to improve the current model [185]
[154]	Plaue et al. (2011)	Intersecting	Entrance area of a building				●	●		●			(i) Nearest-neighbour kernel density method is suitable for the density estimation of human crowds	(i) The flow characteristics have not been analyzed enough	(i) –
[80]	Shiwakoti et al. (2011)	Turning; Egressing (Humans & Ants)	Circular chamber; Doorway	●	●	●	●	●	●			●	(i) Turning movements create push and delay in egress (ii) Column at the exit can increase the flow of panicking ants (iii) Observation of the in-store stampede shows consistency with ants experiments	(i) Need a scaling model to find the analogism between animals and humans quantitatively	(i) Develop an animal-based model to simulate uni-directional crowd movement [7]
[103]	Song et al. (2011)	Egress	Wide-changing corridor	●			●	●		●			(i) Compare with Kretz et al. (2006), densities outside the bottleneck region is the same, densities are higher and velocities are smaller inside the bottleneck	(i) The explanation for the difference between the compared researches are not fully explained	(i) Further analyze of experiments data [104]
[58]	Zhang et al. (2011)	Straight line; Merging	Wide-changing corridor				●	●		●			(i) Use four density measurements methods and automatic video processing (ii) Specific flow is independent of corridor width under less congested situation without bottleneck (iii) Fundamental diagrams for different shapes of facilities are not comparable	(i) Influence of crowd heterogeneity on the shape of fundamental diagrams has not been discussed	(i) –

Table 1: Continued.

Ref.	Source	Movement Type	Infrastructure			Subject		Controllability		Competitiveness			Summaries		
			Facility Type	Bottleneck	Obstacle	Animal	Pedestrian	Laboratory	Field	Normal	Emergency	Panic	Key Findings	Limitations & Future Recommendations	Further Data Usage
[159]	Boyce et al. (2012)	Merging	Stair-floor interface				●	●			●		(i) Location of stair-floor connections have influences on merging flows (ii) Population composition affects the merging patterns	(i) Lack of real panic data on merging flows	(i) –
[81]	Dias et al. (2012)	Straight line; Turning (Ants)	Chamber	●		●		●				●	(i) Right angled path are 20% less efficient than straight path	(i) Design solutions are not proposed based on the results	(i) –
[93]	Jelić et al. (2012)	Turning	Ring corridor				●	●		●			(i) Provide a measurement method for instantaneous fundamental diagrams (ii) Velocity-distance headway relations can be used to segment three linear regimes	(i) The overall cross-cultural comparison is weak	(i) –
[169]	Ma et al. (2012)	Merging	Ultra high-rise building				●	●			●		(i) Merging behavior can affect the flow of evacuees (ii) Overtaking behavior can be frequently observed on the refugee floor	(i) It should be a more significant contribution if the individuals' entire evacuation process can be tracked	(i) –
[94]	Moussaïd et al. (2012)	Turning; Opposing	Ring corridor				●	●	●	●			(i) Find the transition between organized and disorganized states in pedestrian traffic (ii) Unstable dynamics is the result of speed variations among pedestrians	(i) The underlying reason for the speed variations has not been explored	(i) –
[124]	Soria et al. (2012)	Egress (Ants)	Chamber			●		●		●		●	(i) Selfish behavior does not exist in ants traffic (ii) Discover the faster-is-slower effect in escaping ants under different stress levels	(i) Lack of discussion and comparison with human empirical data	(i) Faster-is-slower effects of ants are revisited and compared with human crowds [128]

TABLE 1: Continued.

Ref.	Source	Movement Type	Infrastructure			Subject		Controllability		Competitiveness			Summaries		
			Facility Type	Bottleneck	Obstacle	Animal	Pedestrian	Laboratory	Field	Normal	Emergency	Panic	Key Findings	Limitations & Future Recommendations	Further Data Usage
[151]	Zhang et al. (2012)	Opposing; Intersecting	Corridor				●	●		●			(i) Bi-directional flow is classified into stable separated lanes (SSL), dynamical multi-lanes (DML) flows with balanced and unbalanced flow ratio (BFR, UFR) (ii) Density measurement methods do not affect fundamental diagrams significantly (iii) Different ordering flows do not affect fundamental diagrams significantly	(i) Experiments under hyper-congested situation should be further conducted	(i) –
[84]	Zhang et al. (2012)	Turning; Merging	Corridor				●	●		●			(i) Present Voronoi diagrams for speed and density distribution over space for turning and merging flows (i) Pedestrians tend to accelerate their speed after merging	(i) No further analysis on turning movements considering locomotion characteristics	(i) Same data were used for developing PeTrack [186]
[125]	Boari et al. (2013)	Egress (Ants)	Chamber	●		●		●		●	●	●	(i) Ants' egress efficiency is independent of hurry degree (ii) Selfish egress behavior is not found in ants traffic (iii) Faster-is-slower effect is not observed	(i) The further underlying reasons for the results difference are not discussed well	(i) –
[86]	Burghardt et al. (2013)	Turning	Stairs				●	●		●			(i) Maximum specific flow values are up to 1.25m/s for upward movement and 1.22m/s for downward movement (ii) Stairs without bends are preferred	(i) Stairs evacuation data on turning bend section should be compared	(i) –
[97]	Dias et al. (2013)	Turning; Merging; Intersecting (Ants)	Complex chamber	●			●		●	●			(i) Turning angle affects escape flow significantly (ii) Clogging transitions can be found in intersecting flows	(i) Effects of intersecting and merging should be further investigated	(i) –

Table 1: Continued.

Ref.	Source	Movement Type	Infrastructure			Subject		Controllability		Competitiveness			Summaries		
			Facility Type	Bottleneck	Obstacle	Animal	Pedestrian	Laboratory	Field	Normal	Emergency	Panic	Key Findings	Limitations & Future Recommendations	Further Data Usage
[156]	Wu and Lu (2013)	Weaving	Corridor				●	●		●			(i) The effects of weaving are more significant within a region (ii) Flow values and flow ratios are significant with the operation of weaving area	(i) Safety indices for pedestrian weaving zones can be further established	(i) Level-of-service for pedestrian weaving zone were analyzed [187]
[73]	Zhang et al. (2013)	Straight line	Long street				●		●	●			(i) Combine use of camera and active infrared counter (ii) A complete dataset of flow rate and velocity for high density straight flow (iii) Find the boundary effect of uni-directional dense laminar flow (iv) Multi-directional flow data cannot be directly applied to uni-directional flow	(i) Measurements cannot reach individual trajectory level (ii) Personal and crowd collective factors are not investigated (iii) Data availability and usability are unknown	(i) –
[163]	Aghabayk et al. (2014)	Merging	Corridor				●	●		●	●		(i) Crowd arrival and departure flow increase with the rise of desired speed (ii) Time headways decrease with speed (iii) Merging angle affects flow and headways	(i) Lack of detailed analysis within the merging regions	(i) More detailed analysis in a journal paper [162]
[77]	Dias et al. (2014)	Turning	Corridor		●		●	●		●	●		(i) Speed tends to reduce at a fixed region around turning point (ii) The speed reduction increases with the rise of merging angle	(i) Sample size is not enough (ii) High density situation should be investigated	(i) Develop the optimal trajectories method for turning manoeuvres [89]
[105]	Liao et al. (2014)	Egressing	Wide door	●			●	●		●			(i) The density and speed inside the bottleneck are independent on the width of door (ii) A linear dependency is observed between flow and bottleneck width	(i) The flow and density are calculated using the trajectories under both stable and unstable state	(i) Trajectories were further utilized to test a method that can measure the steady flow state [188]

Table 1: Continued.

Ref.	Source	Movement Type	Infrastructure			Subject		Controllability		Competitiveness			Summaries		
			Facility Type	Bottleneck	Obstacle	Animal	Pedestrian	Laboratory	Field	Normal	Emergency	Panic	Key Findings	Limitations & Future Recommendations	Further Data Usage
[129]	Sobhani et al. (2014)	Egressing (woodlice)	Chamber	●		●		●				●	(i) Fundamental diagrams near exit regions for animal flow have differences in normal and panic situations. (ii) Stress level of woodlice has positive relation with the number of blockages near the exit	(i) Quantitative comparison between animals and humans should be conducted to test their hypothesis	(i) More detailed analysis with the same datasets [130]
[189]	Sun et al. (2014)	Opposing; Weaving	Corridor		●		●	●		●			(i) Weaving conflicts tends to be centralized within a stable region (ii) Lane changing behaviors are modelled in weaving movements	(i) The asymmetric of weaving area location may arise from the organization of experiments	(i) Data were further analyzed to reproduce subway corridor flow [190]
[109]	Garcimartín et al. (2014)	Egress	Room	●			●	●		●	●		(i) Time headways display a heavy-tailed distribution (ii) Burst sizes fit an exponential function	(i) Steady state flows should be measured to fit the model	(i) Extended analysis were conducted [110]
[138]	Zuriguel et al. (2014)	Egress (Sheep)	Width-changing fence	●	●	●		●			●		(i) Time headways for sheep passing the bottleneck displays a power-law distribution (ii) The burst sizes for sheep flock obey an exponential decay (iii) Universal laws that can describe the flow and clogs of different particle systems are established	(i) Require more real life empirical data on human egress considering the effect of obstacle size and position	(i) Use the sheep data to study the effect of door width and obstacle placement on sheep flow [139] (ii) Effect of obstacle position on flow and clogging patterns were further studied [140]
[111]	Bode et al. (2015)	Egress	Room	●			●	●		●			(i) Social grouping behavior has negative effects on egress efficiency (ii) No clear time differences can be found between individual and group movements in front of the exits	(i) Samples might not be enough	(i) –

Table 1: Continued.

Ref.	Source	Movement Type	Infrastructure			Subject		Controllability		Competitiveness			Summaries		
			Facility Type	Bottleneck	Obstacle	Animal	Pedestrian	Laboratory	Field	Normal	Emergency	Panic	Key Findings	Limitations & Future Recommendations	Further Data Usage
[88]	Dias et al. (2015)	Turning	Angle-changing corridor				●	●		●	●		(i) Longitudinal spacing between pedestrians tends to rise as speed increases (ii) Speed levels affect flow-density and speed-spacing relationships	(i) Require further investigation on the combined effects under complex environments	(i) Data were utilized to calibrate a cellular automation model [79]
[157]	Lian et al. (2015)	Intersecting	Corridor		●		●	●		●			(i) Among Chinese students, walking in the right-hand side can accelerate the lane formation process (ii) Obstacle can help to stable intersecting flows	(i) The effect of smoothing the corners of the intersection can be a future work	(i) –
[95]	Kuang et al. (2015)	Turning	Ring path				●	●		●			(i) Find the speed variation, step fluctuation and delay from Stop-and-Go waves in single-file ring movement	(i) Insides of flow characteristics are not presented	(i) –
[166]	Shiwakoti et al. (2015)	Merging	Corridor				●	●		●	●		(i) Pedestrian tends to reduce velocity in merging zone (ii) The speed reduction is significantly with merging angle and desired speed (iii) The speed reduction can impede the flow of out stream corridor	(i) Experiments with more participants should be conducted (ii) Effect of blocked vision should be studied	(i) –
[121]	Wang et al. (2015)	Egress (Ants)	Chamber	●		●		●		●	●		(i) Time headway displayed an exponential decay in ants traffic (ii) Mean flow is independent linearly on the exit width (iii) Exit width affects the time headway (iv) Ant's group size affects the flow at certain exit width	(i) Accuracy of data extraction should be improved	(i) –

Table 1: Continued.

Ref.	Source	Movement Type	Infrastructure			Subject		Controllability		Competitiveness			Summaries		
			Facility Type	Bottleneck	Obstacle	Animal	Pedestrian	Laboratory	Field	Normal	Emergency	Panic	Key Findings	Limitations & Future Recommendations	Further Data Usage
[63]	Zhang et al. (2015)	Straight line	Narrow corridor				●	●		●			(i) Boundary has an effect on fundamental diagram (ii) Higher maximal specific flow can be detected under open boundary condition (iii) Stop-and-go waves can be observed under closed boundary condition	(i) Lack of empirical data of high density situations for this setup	(i) –
[116]	Ezaki et al. (2016)	Ingress	Room	●			●	●		●			(i) Individual prefers to select location near corner and wall (ii) Flow avoidance, boundary preference, distance cost and angle cost are modelled	(i) Design and guidance solutions for ingress can be further presented	(i) –
[42]	Gorrini et al. (2016)	Opposing	Corridor				●	●		●			(i) Higher flow ratio has negative influence on speed (ii) Walking in group is less efficient than individual walker	(i) The impact of coordination movement should be further investigated	(i) –
[161]	Huo et al. (2016)	Merging	Stair-floor interface				●	●			●		(i) Merging movement at stair landings decrease the speed and increase the total evacuation time	(i) More detailed analysis on the interaction mechanisms on the stair-floor interfaces are required	(i) –
[133]	Lin et al. (2016)	Egress (Mice)	Room	●	●	●		●		●	●	●	(i) Burning joss-sticks can be used to stimulate mice to different stress levels (ii) Egress time and the number of clogs increase with levels of stimulus (iii) Faster-is-slower effect is observed	(i) The quantitative relationships between the stress levels of the mice and the burning of joss-sticks should be presented	(i) The effect of obstacle was further analyzed [134] (ii) The effect of exit position was further analyzed [136] (iii) The effect of bottleneck width was further analyzed [135]

Table 1: Continued.

Ref.	Source	Movement Type	Infrastructure			Subject		Controllability		Competitiveness			Summaries		
			Facility Type	Bottleneck	Obstacle	Animal	Pedestrian	Laboratory	Field	Normal	Emergency	Panic	Key Findings	Limitations & Future Recommendations	Further Data Usage
[191]	Liu et al. (2016)	Ingress; Egress	Room	●	●		●	●		●	●		(i) Ingress order has significant influence on the location of people in steady state and the egress order (ii) Crowd distribution in steady state is not uniform (iii) Inactive person may affect inflow and outflow process	(i) The method to detect steady state is questionable	(i) –
[117]	Liu et al. (2016)	Ingress	Room	●			●	●		●			(i) Ingress order is significant with individual's location distribution in the steady state and the egress order. (ii) Inactive pedestrians contribute a negative effect on the movement	(i) The effects of inhomogeneous composition are not necessarily negative	(i) –
[114]	Nicolas et al. (2016)	Egress	Room	●			●	●		●			(i) Flow increases monotonically with density (ii) Selfishness behavior may impede the egress flow at bottleneck (iii) A generalized zipper effect is observed (iv) Faster-is-slower effect is not observed	(i) Latent parameter effects should be further analyzed	(i) A more comprehensive journal paper [115]
[3]	Shi et al. (2016)	Merging	Corridor				●	●		●	●		(i) Merging angle and flow direction affect flow characteristics such as flow and time headway (ii) Blocked vision effects can be found at merging section (iii) Branched merging flow is less safe than opposed merging flow	(i) Safety indices for merging section should be established (ii) Level-of-service for merging region should be evaluated	(i) Further analysis of right angle merging case [38] (ii) Further analysis of vision condition [192]

Table 1: Continued.

Ref.	Source	Movement Type	Infrastructure			Subject		Controllability		Competitiveness			Summaries		
			Facility Type	Bottleneck	Obstacle	Animal	Pedestrian	Laboratory	Field	Normal	Emergency	Panic	Key Findings	Limitations & Future Recommendations	Further Data Usage
[164]	Shahhoseini et al. (2016)	Merging (Ants)	Angle-changing chamber	●		●		●				●	(i) Merging layouts can create stop-and-go waves (ii) Merging angle affects traffic flow characteristics	(i) Comparison analysis with human data are needed	(i) Compare ants data with human experiments with similar setups [193]
[92]	Sun et al. (2016)	Turning	Angle-changing corridor				●	●		●			(i) Smoothed turning curve point design based on the Beijing subway corridor angle (ii) Flow characteristics are significant with different angles and radii (iii) Right angle corridor has large impact on the cumulative density and mean speed	(i) The boundary materials of the corridor are not solid	(i) –
[113]	von Krüchten and Schad-schneider (2016)	Egress	Square room	●			●	●		●			(i) Queuing at the exit can shorten egress time comparing with board distribution (ii) Moving as a compact crowd can increase the egress flow	(i) Differences between ordered compactly movement and the bursts after clogs are not discussed	(i) More detailed analysis were published in a journal [112]
[76]	Wang and Song (2016)	Straight line; Egress (Ants)	Restricted passageway	●		●		●			●		(i) No jam among stressed ants trails (ii) Speed seems to be constant with density	(i) Effect of stressfulness is not quantified	(i) –
[122]	Wang et al. (2016)	Egress (Ants)	Chamber	●		●		●		●		●	(i) Stressed ants exhibit symmetry breaking (ii) Escape flow reaches the peak when the spacing of two exits is largest (iii) Social force model has different performance in terms of describing the traffic rules of ants and humans	(i) Comparison with human empirical data is encouraged	(i) –

Table 1: Continued.

Ref.	Source	Movement Type	Infrastructure			Subject		Controllability		Competitiveness			Summaries		
			Facility Type	Bottleneck	Obstacle	Animal	Pedestrian	Laboratory	Field	Normal	Emergency	Panic	Key Findings	Limitations & Future Recommendations	Further Data Usage
[13]	Cao et al. (2017)	Straight line; Opposing; Intersecting	Corridor				●	●		●			(i) The difference in the effects of measurements of fundamental diagrams gradually appears as the increase of density level (ii) No difference is found between bi-directional and four-directional flows in fundamental diagrams under same density level	(i) Microscopic properties should be further analyzed for uni-, bi-, four-directional flows	(i) –
[171]	Cuesta et al. (2017)	Merging	Tunnel	●			●	●		●			(i) Merging impedes the evacuation efficiency in tunnel (ii) Merging decreases the rail car flow and the walkway flow (iii) As the increase of height difference between rail car exit and walkway, the more walkway flow dominates the flow (iv) No gender effects can be found in deference behavior	(i) Experiments should be conducted in real tunnel for more realism	(i) –
[165]	Lian et al. (2017)	Merging	Corridor	●			●	●		●			(i) Lane formation can be found in main flow (ii) Main and branched flow can be mutual bottleneck when the flow saturated at the 2.4m width branched corridor (iii) Speed in main channel decreases as the approaching of merging areas in 1.6m and 2.4m corridors	(i) More statistical approaches should be applied to verify the effects of different variables	(i) –
[60]	Jin et al. (2017)	Turning	Ring corridor				●	●		●			(i) Uni-directional flow experiment exhibits four traffic states: free flow, congested, over-congested and hyper congested state (ii) Transition from stopped state and moving state can be observed under hyper-congested situation (iii) Flow rates in over-capacity situation are almost constant (iv) Three types of lane formations could be found in bi-directional flows	(i) The segmentation of steady flow states is done by manual observation	(i) –

Table 1: Continued.

Ref.	Source	Movement Type	Infrastructure			Subject		Controllability		Competitiveness			Summaries		
			Facility Type	Bottleneck	Obstacle	Animal	Pedestrian	Laboratory	Field	Normal	Emergency	Panic	Key Findings	Limitations & Future Recommendations	Further Data Usage
[137]	Oh and Park (2017)	Egress (Mice)	Angle-changing chamber	●		●		●		●			(i) Mice do not often follow the ideal route (ii) Total mean velocity, total egress time decrease with the exit angle	(i) Empirical validation with human crowd data should be considered	(i) –
[96]	Rahman et al. (2017)	Turning	Angle-changing corridor				●	●		●			(i) L-shaped corridor has the lowest average speed compared with 60° and 135° (ii) Participant prefers to walk in the inner side of the corridor due to the shorter walking distance	(i) More advanced statistical methods should be applied to examine the mixed variable effects	(i) –
[170]	Shahhoseini et al. (2017)	Merging	Corridor				●	●		●	●		(i) Asymmetric setups can create more delays compared to symmetric setups (ii) Asymmetric setups also can create more imbalance between merging streams	(i) Further analysis within the merging sections should be conducted	(i) Find the faster-is-slower effects in merging sections [167] (ii) Compare the data with ants experiments [193]
[68]	Sharifi et al. (2017)	Straight line; Turning	Loop corridor	●			●	●		●			(i) Homogenous flow experience a 75% capacity drop at bottleneck (ii) Disable people need more space to keep their desired speed (iii) Larger spacing is kept within the groups with wheelchair users	(i) Gait-level features of disabled people and normal people should be a further research direction	(i) Analysis of the data with a focus on speed distribution [90] and time headway [71]
[91]	Sieben et al. (2017)	Ingress; Turning	Semicircle & right-angle corridor	●			●	●		●			(i) Spatial structures have influence on entrance dynamics in terms of density, waiting time and speed transition (ii) Constriction effects are found in both structures	(i) Experiments with single exit and dual exits can be further conducted	(i) –

Table 1: Continued.

Ref.	Source	Movement Type	Infrastructure			Subject		Controllability		Competitiveness			Summaries		
			Facility Type	Bottleneck	Obstacle	Animal	Pedestrian	Laboratory	Field	Normal	Emergency	Panic	Key Findings	Limitations & Future Recommendations	Further Data Usage
[152]	Xue et al. (2017)	Opposing	Ground with grid				●	●		●			(i) Comparative experiments to examine the gridlock effect in CA model (ii) Gridlock is not spotted even under high density opposing flow	(i) Different competitiveness levels should be considered	(i) –
[17]	Wagoum et al. (2017)	Diverging; Turning	Corridor	●			●	●		●			(i) Under normal situation people choice exit only concerned with shortest path (ii) Under high density situation, the load balancing for two exits can be detected	(i) Experiment data under emergency situations should be used for comparison	(i) Further development of a route choice model [194]
[66]	Cao et al. (2018)	Straight line	Narrow corridor				●	●		●			(i) Young people tend to control the lateral movement better than the aged people (ii) Step length increases as the rise of velocity and they exhibit a quadratic relation for three groups (iii) Gender and height effects on fundamental diagrams are not found	(i) More statistical analysis should be applied to test the significance among the relations	(i) Study the heterogeneous crowd considering age composition [64]
[158]	Sun et al. (2017)	Egress	Narrow corridor	●			●	●		●			(i) Use of funnel shape exit design with different angles (ii) The optimal funnel angle is between 46° to 65°	(i) The width and length of corridor are fixed (ii) Extra obstacle effects are not considered	(i) –

Table 1: Continued.

Ref.	Source	Movement Type	Infrastructure			Subject		Controllability		Competitiveness			Summaries		
			Facility Type	Bottleneck	Obstacle	Animal	Pedestrian	Laboratory	Field	Normal	Emergency	Panic	Key Findings	Limitations & Future Recommendations	Further Data Usage
[158]	Sun et al. (2017)	Intersecting	Ground		●		●	●		●			(i) Effects of 5 types of intersecting angle and obstacle implement under three flow conditions are investigated (ii) The influences of angles on flow operation speed vary at different flow conditions	(i) The effects of obstacle shapes and positions are not considered	(i) Obstacle effects in intersecting roundabout region is further studied [149]
[65]	Wang et al. (2018)	Straight line	Narrow passageway				●	●		●			(i) Effects of height on step length and duration change with density (ii) Step length & frequency-speed relations can be better described by power function (iii) Swaying amplitude-speed relations can be segmented into two regimes	(i) Unintended factors affecting the stepping characteristics should be eliminated	(i) –
[153]	Wang et al. (2018)	Straight line; Opposing	Restricted passageway	●		●		●		●			(i) Ants show unified sensitivity to distance headway in both uni- and bi-directional flow (ii) Head-on encounter and evading behaviors can be frequently observed (iii) Ants spend more time for encounter communication than humans (iv) Following behavior of ants can improve the flow of opposing movement	(i) Quantitative comparison with empirical data in bi-directional vehicular traffic and pedestrian traffic is encouraged	(i) –

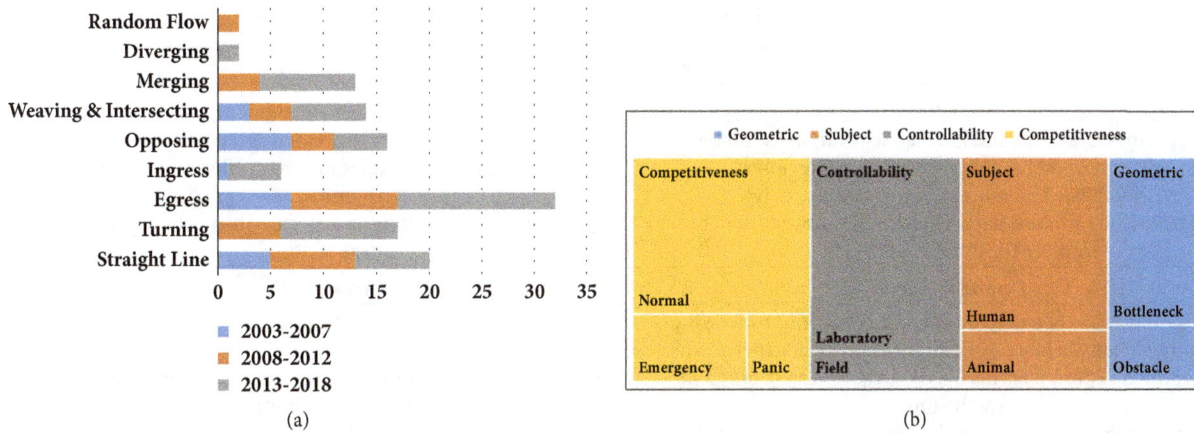

FIGURE 3: Summary statistics from Table 1. (a) Histogram diagram illustrating the distributions of the quantity and proportion of empirical studies on different multidirectional movements over different time periods. (b) Tree map to compare the proportion of the empirical study that focused on competitiveness, controllability, subjects, and geometric conditions.

In addition to the flow-based summary, we also present the comparison of contextual-based statistics of the reviewed studies in a set of nested rectangles as shown in Figure 3(b). The hierarchy of data generated from Table 1 in terms of the geometric (bottleneck or obstacle), investigating subject (human or animal), controllability (laboratory experiment or field observation), and competitiveness (normal or emergency or panic status) is presented in Figure 3(b).

Overall, laboratory experiments occupy the predominant proportion compared with field observations. In addition, the majority of empirical evidences are in human subjects while it is minor in biological entities. Regarding the study on competitive behavior, studies are mostly conducted under normal status but less in emergency and panic situations. Therefore, in future, empirical studies that can improve our understanding on mechanisms of multidirectional movements under emergency situations need to be investigated.

4.2. SWOT Analysis for Empirical Data Collection Approaches. During the review process, we find that the data collection methods can have substantial influences on the obtained results. To help researchers select the suitable approach to satisfy their data requirements, assessments of different data collection approaches in terms of their underlying internal advantages and disadvantages as well as the potential external factors that can exploit or obstruct the advantages are needed.

Therefore, we perform comparative SWOT (Strengths, Weaknesses, Opportunities, and Threats) analysis for different data collection methods to identify the positive and negative prospects and classifying them into internal (Strengths and Weaknesses) and external (Opportunities and Threats) factors [195].

We conducted a first set of SWOT analysis to examine the variations of controllability of experiments between laboratory experiments versus field observations.

The utmost advantage for laboratory experiments is the strong controllability towards investigating targets that satisfied the straightforward requirements. Researchers can set up the predesigned walking environment, select the satisfying participants, choose the intended data measurement techniques, or even define the manners of movements by giving instructions or pressing forces to the participants. Through such purpose-oriented approach, the specific influencing factors of the crowd movements can be examined one by one in a flexible manner and in detail. Therefore, empirical data from laboratory experiments should be the suitable materials for the development of microscopic models. However, the benefits of the well controllability for laboratory experiments may have to tradeoff with the risk of not completely representing the real-life situation. For instance, differences between the pedestrian movements during emergency escape and laboratory experiments have been quantitatively compared in Shiwakoti (2016) [196]. It is suggested that the physiological aspects of pedestrian should be further examined such as the effect of peoples' consciousness of their involvement in an experiment on their performed behaviors [197]. Moreover, not all the factors are controllable especially in animal experiments. This is perhaps the right reason for the absence of opposing experiments using biological entities. Further, due to the simplifications of the experiment setups, the effects of latent uncontrolled variables can also affect the global results and such effects might be amplified due to the control of certain variables. Therefore, in the future, in-depth statistical analysis should be conducted to examine the interactions of manifest and latent variables of pedestrian dynamics by using more comprehensive empirical datasets. Nevertheless, the advancements made on data collection and analysis techniques as well as optimal experimental design methods may help increase the size of the database, improve the data accuracy of experiments, and reduce the time and labour costs of the researchers.

With regard to field observation, the major pros for this approach are the authenticity of data quality and the relative lower baseline requirement to carry out the field observation. Field data can represent the real-life circumstance giving no other effects such as the subjects' awareness of being monitored. In addition, it can be easily implemented, e.g., placing a camera at a fixed location or even using the

Table 2: SWOT analysis for laboratory experiments versus field observations.

Laboratory Experiments	
Strengths	**Weaknesses**
(i) Specific dependent and independent variables can be controlled and examined one by one flexibly (ii) Repeated measures can be realized easily (iii) Save a lot of time comparing with field observation to capture the intended samples	(i) Discrepancy exists between real life situation (ii) Uncontrolled latent variables may have effects on the global dynamics (iii) Contextual limitations on the experiments targets and purposes
Opportunities	**Threats**
(i) The application of emerging data collection and processing methods can attract more researchers to perform experiments (ii) Optimal experiments design can help lower the cost and save the time (iii) A comprehensive empirical database can add on the values of current experimental data (iv) Advanced statistical approaches might improve the performance of empirical data in verification and validation processes	(i) Organization and performance of experiments require a lot of money and labour (ii) Data integration is hard due to the barriers among disciplines
Field Observations	
Strengths	**Weaknesses**
(i) Ground truth data (ii) Labour cost effective and easy to carry out (iii) Large quantity of existing datasets	(i) Take long time to obtain adequate samples (ii) Not easy to capture specific behavior compared with laboratory experiments (iii) Restrictions of access into certain areas or contexts
Opportunities	**Threats**
(i) Smart city can provide vast data source for collecting pedestrian movement behaviors (ii) Big data solutions might be able to reduce the need of data quality (iii) Advancement in image processing industry offer the chance to collect large scale field data	(i) Complex patterns and behaviors are difficult to recognize and capture from field data (ii) Privacy & security issues should be highlighted as the potential large scale in data volume (iii) Minimize the effects of the placement of cameras or UAVs on the investigating subjects

exiting videos shot by surveillance devices. However, there are limitations of field observation as well. Frist of all, not all the areas or contexts are permitted to conduct field observation research because of privacy and security reasons. This point is particularly suitable to the unmanned aircraft vehicle (UAV) as many countries have banned UAVs and drones in their restricted airspaces. Second, it usually takes a long time period or requires switching various sites to capture the intended contexts and obtain the adequate samples. Fortunately, the emerging of smart cities and big data analytics provide opportunities for large field data collection from multiple sources. Recent technological advances in the imaging-sensing industry and the machine learning research community have made previously inaccessible data largely available. The imaging industry enables high-resolution (1080p) CCTV cameras to perform crowd monitoring. Therefore, real time field data collection and short term analysis of pedestrian crowd have been implemented in many cities [198]. Meanwhile, the ubiquity of smartphones has also offered various opportunities for collecting pedestrian traffic data in the real-life fields [199]. Consequently, the relating issues on personal privacy and public information security should also be the threats for this approach.

The highlights of the SWOT analysis for laboratory experiments and field observations are summarized in Table 2.

We further conducted a second set of SWOT analysis between animal-based approach versus human-based approach.

The foremost strength for animal-based approach is identified as the possibility of conducting real panic experiments using biological entities. Further, animal behaviors can be more easily captured through laboratory experiments or field observations with lower cost than human approaches. Although there exists obvious discrepancy between animal movement and human motion, the empirical data in panicking animals are also more 'life-like' than many mathematical models [36, 196]. There is opportunity to use the animal models to further improve our understanding of collective motion. For example, the communication and coordination mechanisms of ants can be further examined to develop the methods for controlling human crowd [127].

In terms of human-based approach, apart from the most apparent advantage of ground truth data, we also highlight the advantage of providing instructions to pedestrians as the secondary pro in conducting experiments with human subjects. However, human data show large contextual

TABLE 3: SWOT analysis for animal-based approaches versus human-based approaches.

Animal-Based Approaches	
Strengths	**Weaknesses**
(i) Real panic experiments can be performed (ii) Relative lower cost than human experiments (iii) Can be more life-like than mathematical models	(i) Taxonomic differences in animals with humans (ii) Less convincing to some crowd researchers in terms of contextual realism
Opportunities	**Threats**
(i) Biological intelligence can be beneficial to mankind e.g. ants show more coordination in panic escape than humans (ii) Animal behaviors can be used as the education materials for children to augment their safety awareness in terms of evacuation (iii) Resolving optimal design issue in panic escape	(i) The universal law for the scaling of different species of biological entities with various body sizes (ii) Further analysis of the behavioural similarities and dissimilarities between animals and humans (iii) The selection of organisms for different experimental design (iv) Conduct experiments using panicking animals may arise debates from bestiarian
Human-Based Approaches	
Strengths	**Weaknesses**
(i) Ground truth data of real pedestrians (ii) Participants are more instructive than animals	(i) Panic experiments cannot be performed due to ethical and safety reasons (ii) Contextual dependency
Opportunities	**Threats**
(i) Emerging virtual interactive techniques such as AR and VR can help examine human behaviors in more extreme scenarios	(i) Risk free/immigration experimental design methods (ii) Physiological aspects of humans during collective movements should be further exploited

dependency arising from both internal and external factors such as heterogeneity and behavioral uncertainty. Therefore, future empirical databases should be enriched with more psychological aspects [196]. In addition, one major con for human approach is the prohibition of performing dangerous panic experiments due to the ethical and safety reasons. However, the emerge of virtual interactive techniques such as augmented reality (AR) and virtual reality (VR) can help examine human behaviors in more extreme scenarios including the panic escape [200]. Also, researchers should also explore risk free or risk immigration experimental design methods to extend the research scopes to extreme behaviors.

The summaries of the SWOT analysis for animal-based approaches versus human-based approaches are presented in Table 3.

4.3. Future Global Collaborative Efforts on Empirical Data Collection. In the future, in order to accelerate the path of crowd dynamics research, a comprehensive open-source empirical database on root of the different external complex movement patterns and internal influence variables should be established. To realize this goal, interdiscipline collaborations should be encouraged. Some of the researchers from various disciplines have made their datasets available to the public such as open videos and trajectories datasets of controlled laboratory experiments [201], pedestrian crossing behaviors on street [202], field datasets of pedestrian's walking behaviors in train station, school forum or gallery [203–205], and collective motion databases for computer vision analysis [206]. All the above datasets contain pedestrian trajectories and some of the researchers have also uploaded the original videos.

Therefore, mutual assistance and cooperation should be encouraged between research groups across disciplines and across regions around the world. A data-sharing platform should be established so that all researchers working in this field can contribute their datasets.

During our review study, we noticed that many empirical data are not well recorded and have little possibilities for further usage due to the limitations of experiments setups or measuring methods. Therefore, a common standard guidance on performing controlled experiments and field observations should be developed. Further, automatic image processing tools that can restore the trajectories from the published papers can be developed in future to make full use of the existing empirical studies.

Further, researchers working in this field should come together to establish a series of quantitative standardization methods to normalize the empirical data [207]. Also, a comprehensive guideline and standard for performing controlled laboratory experiments are required.

5. Conclusions

Collective motion of pedestrian traffic is sophisticated under complex environments. To better understand the underlying mechanism of collective crowd movements, a lot of empirical studies were carried out through controlled laboratory experiments and field observations with humans subjects and non-human biological entities under various scenarios. However, a comprehensive review of the detailed classification of these empirical works from the aspects of complex movements was absent in the literature.

To this end, in this review article, we first categorized the complex movements of pedestrian crowd into two general categories, i.e., external governed movements and internal driven movements based on the governing factors of the motion pattern. Further, considering the hierarchy of movement complexity, we decomposed the external governed movements of pedestrian traffic into several motion based patterns including straight line, turning, egress and ingress, opposing, weaving, merging, diverging, and random flows. Then, we reviewed the related literatures under each motion based pattern with a focus on the complex movement characteristics. Further, a series of evaluations were performed towards each reviewed empirical study in terms of the geometrics setups (whether contained bottleneck or obstacle), investigating subjects (humans or animals), controllability (laboratory experiments or field observations), and competitiveness (under normal, emergency, or panic situations). Moreover, the key findings, limitations, or future recommendations and the further usage of the empirical data in other studies were summarized in a table.

Summary statistics from the aggregated table showed that empirical data were highly rich in straight line, egress flow, and medium rich in turning, merging, weaving, and opposing, but poor in ingress, diverging, and random flows. Researchers were more concerned with human subjects than animal subjects and laboratory experiments were more preferred than field observations. Empirical data were mostly collected under normal situations but less in emergency and panic conditions. Likewise, the studies on bottleneck effects were more popular than the obstacle effects.

We further presented the comparative SWOT analysis for different data collection approaches that included two comparisons: laboratory experiments versus field observation and animal-based approaches versus human-based approaches. At last, we put emphasis on the need for the future global collaborative efforts on data sharing and developing guideline for performing controlled laboratory experiments for the pedestrians complex movements.

Conflicts of Interest

The authors declare that they have no conflicts of interest.

Acknowledgments

This research is sponsored by National Science Foundation of China (no. 71701108) and National Science Foundation of Zhejiang Province (no. LQ17E080007). The third coauthor would like to acknowledge the funding received from ARC Linkage Project LP120200361 for his contribution in this paper.

References

[1] X. Shi, *Empirical Investigation on Pedestrian Complex Movement Characteristics and Interaction Behaviors [Ph.D. thesis]*, Southeast University, 2018.

[2] X. Shi, Z. Ye, N. Shiwakoti, and Z. Li, "A Review of Experimental Studies on Complex Pedestrian Movement Behaviors," in *Proceedings of the 15th COTA International Conference of Transportation Professionals, CICTP 2015*, pp. 1081–1096, American Society of Civil Engineers, Reston, VA, USA, 2015.

[3] X. Shi, Z. Ye, N. Shiwakoti, D. Tang, C. Wang, and W. Wang, "Empirical investigation on safety constraints of merging pedestrian crowd through macroscopic and microscopic analysis," *Accident Analysis & Prevention*, vol. 95, pp. 405–416, 2016.

[4] E. R. Galea, G. Sharp, and P. J. Lawrence, "Investigating the representation of merging behavior at the floor—stairs interface in computer simulations of multi-floor building evacuations," *Journal of Fire Protection Engineering*, vol. 18, no. 8, pp. 291–316, 2008.

[5] D. Helbing, I. Farkas, and T. Vicsek, "Simulating dynamical features of escape panic," *Nature*, vol. 407, no. 6803, pp. 487–490, 2000.

[6] R. Escobar and A. De La Rosa, "Architectural design for the survival optimization of panicking fleeing victims," in *Advances in Artificial Life*, pp. 97–106, 2003.

[7] N. Shiwakoti, M. Sarvi, G. Rose, and M. Burd, "Animal dynamics based approach for modeling pedestrian crowd egress under panic conditions," *Procedia - Social and Behavioral Sciences*, vol. 17, pp. 438–461, 2011.

[8] M. Burd, N. Shiwakoti, M. Sarvi, and G. Rose, "Nest architecture and traffic flow: large potential effects from small structural features," *Ecological Entomology*, pp. no–no, 2010.

[9] D. Helbing, L. Buzna, A. Johansson, and T. Werner, "Self-organized pedestrian crowd dynamics: experiments, simulations, and design solutions," *Transportation Science*, vol. 39, no. 1, pp. 1–24, 2005.

[10] N. Shiwakoti, M. Sarvi, and M. Burd, "Using non-human biological entities to understand pedestrian crowd behaviour under emergency conditions," *Safety Science*, vol. 66, pp. 1–8, 2014.

[11] N. Shiwakoti, X. Shi, Z. Ye, Y. Liu, and J. Lin, "A comparative study of pedestrian crowd flow at middle and corner exits," in *Proceedings of the Australasian Transport Research Forum 2016*, 2016, http://atrf.info/papers/2016/files/ATRF2016_Full_papers_resubmission_147.pdf.

[12] M. Asano, M. Kuwahara, and S. Tanaka, "Multi-directional pedestrian flow model based on empirical data," in *Proceedings of the 11th World Conference on Transport Research*, 2007, http://trid.trb.org/view.aspx?id=877133.

[13] S. Cao, A. Seyfried, J. Zhang, S. Holl, and W. Song, "Fundamental diagrams for multidirectional pedestrian flows," *Journal of Statistical Mechanics: Theory and Experiment*, vol. 2017, no. 3, p. 033404, 2017.

[14] M. Haghani and M. Sarvi, "Crowd behaviour and motion: Empirical methods," *Transportation Research Part B: Methodological*, vol. 107, pp. 253–294, 2018.

[15] J. K. K. Yuen and E. W. M. Lee, "The effect of overtaking behavior on unidirectional pedestrian flow," *Safety Science*, vol. 50, no. 8, pp. 1704–1714, 2012.

[16] Y. Zhao, T. Lu, M. Li, and L. Tian, "The self-slowing behavioral mechanism of pedestrians under normal and emergency conditions," *Physics Letters A*, vol. 381, no. 37, pp. 3149–3160, 2017.

[17] A. U. Kemloh Wagoum, A. Tordeux, and W. Liao, "Understanding human queuing behaviour at exits: an empirical study," *Royal Society Open Science*, vol. 4, 2017.

[18] M. Moussaïd, N. Perozo, S. Garnier, D. Helbing, and G. Theraulaz, "The walking behaviour of pedestrian social groups and its impact on crowd dynamics," *PLoS ONE*, vol. 5, no. 4, Article ID e10047, 2010.

[19] S. P. Hoogendoorn and W. Daamen, "Self-organization in walker experiments," in *Traffic Granul. Flow '03*, p. 11, 2004.

[20] S. P. Hoogendoorn and W. Daamen, "Pedestrian behavior at bottlenecks," *Transportation Science*, vol. 39, no. 2, pp. 147–159, 2005.

[21] A. Seyfried, O. Passon, B. Steffen, M. Boltes, T. Rupprecht, and W. Klingsch, "New insights into pedestrian flow through bottlenecks," *Transportation Science*, vol. 43, no. 3, pp. 395–406, 2009.

[22] D. J. Low, "Following the crowd," *Nature*, vol. 407, no. 6803, pp. 465-466, 2000.

[23] D. Helbing, I. J. Farkas, and T. Vicsek, "Freezing by heating in a driven mesoscopic system," *Physical Review Letters*, vol. 84, no. 6, pp. 1240–1243, 2000.

[24] D. Helbing, A. Johansson, and H. Z. Al-Abideen, "Crowd turbulence: the physics of crowd disasters," 2007, https://arxiv.org/abs/0708.3339.

[25] X. Yang, Z. Wu, and Y. Li, "Difference between real-life escape panic and mimic exercises in simulated situation with implications to the statistical physics models of emergency evacuation: The 2008 Wenchuan earthquake," *Physica A: Statistical Mechanics and its Applications*, vol. 390, no. 12, pp. 2375–2380, 2011.

[26] B. Krausz and C. Bauckhage, "Loveparade 2010: Automatic video analysis of a crowd disaster," *Computer Vision and Image Understanding*, vol. 116, no. 3, pp. 307–319, 2012.

[27] D. Helbing and P. Mukerji, "Crowd disasters as systemic failures: analysis of the Love Parade disaster," *EPJ Data Science*, vol. 1, no. 1, 2012.

[28] M. D'Orazio, L. Spalazzi, E. Quagliarini, and G. Bernardini, "Agent-based model for earthquake pedestrians' evacuation in urban outdoor scenarios: Behavioural patterns definition and evacuation paths choice," *Safety Science*, vol. 62, pp. 450–465, 2014.

[29] Z. Gu, Z. Liu, N. Shiwakoti, and M. Yang, "Video-based analysis of school students' emergency evacuation behavior in earthquakes," *International Journal of Disaster Risk Reduction*, vol. 18, pp. 1–11, 2016.

[30] G. Bernardini, E. Quagliarini, and M. D'Orazio, "Towards creating a combined database for earthquake pedestrians' evacuation models," *Safety Science*, vol. 82, pp. 77–94, 2016.

[31] H. Gayathri, P. M. Aparna, and A. Verma, "A review of studies on understanding crowd dynamics in the context of crowd safety in mass religious gatherings," *International Journal of Disaster Risk Reduction*, vol. 25, pp. 82–91, 2017.

[32] G. K. Still, "Crowd Disasters, Crowd Saf. Risk Anal," 2018, http://www.gkstill.com/CV/ExpertWitness/CrowdDisasters.html.

[33] J. J. Fruin, "The causes and prevention of crowd disasters," in *Engineering for Crowd Safety*, pp. 1–10, 1993.

[34] D. C. Duives, W. Daamen, and S. P. Hoogendoorn, "State-of-the-art crowd motion simulation models," *Transportation Research Part C: Emerging Technologies*, vol. 37, pp. 193–209, 2013.

[35] V. J. Kok, M. K. Lim, and C. S. Chan, "Crowd behavior analysis: A review where physics meets biology," *Neurocomputing*, vol. 177, pp. 342–362, 2016.

[36] T. Vicsek and A. Zafeiris, "Collective motion," *Physics Reports*, vol. 517, no. 3, pp. 71–140, 2012.

[37] R. P. Roess, E. S. Prassas, and W. R. Mcshane, *Traffic Engineering*, Pearson/Prentice Hall, 2011.

[38] N. Shiwakoti, X. Shi, Z. Ye, and W. Wang, "Empirical study on pedestrian crowd behaviour in right angled junction," in *Proceedings of the Australasian Transport Research Forum*, p. 12p, 2015.

[39] X. Chen, M. Treiber, V. Kanagaraj, and H. Li, "Social force models for pedestrian traffic – state of the art," *Transport Reviews*, vol. 38, no. 5, pp. 625–653, 2018.

[40] X. Jia, H. Yue, X. Tian, and H. Yin, "Simulation of pedestrian flow with evading and surpassing behavior in a walking passageway," *Simulation*, vol. 93, no. 12, pp. 1013–1035, 2017.

[41] M. Haghani and M. Sarvi, "Following the crowd or avoiding it? Empirical investigation of imitative behaviour in emergency escape of human crowds," *Animal Behaviour*, vol. 124, pp. 47–56, 2017.

[42] L. Crociani, A. Gorrini, K. Nishinari, and S. Bandini, "Social Groups and Pedestrian Crowds: Experiment on Dyads in a Counter Flow Scenario," in *Proceedings of the Proceedings of the 8th International Conference on Pedestrian and Evacuation Dynamics (PED2016)*, Hefei, China, 2016.

[43] M. Kremer and L. G. Debo, "Herding in a Queue: A Laboratory Experiment," *SSRN Electronic Journal*, pp. 12–28, 2013.

[44] G. K. Still, *Crowd Dynamics*, University of Warwick, 2000.

[45] S. Hoogendoorn and W. Daamen, "Self-organization in pedestrian flow," in *Traffic Granul. Flow ' 03*, pp. 373–382.

[46] D. Helbing, P. Molnár, I. J. Farkas, and K. Bolay, "Self-organizing pedestrian movement," *Environment and Planning B: Planning and Design*, vol. 28, no. 3, pp. 361–383, 2001.

[47] D. Helbing, "Traffic and related self-driven many-particle systems," *Reviews of Modern Physics*, vol. 73, no. 4, pp. 1067–1141, 2001.

[48] D. Helbing, A. Johansson, and H. Z. Al-Abideen, "Dynamics of crowd disasters: An Empirical Study," *Physical Review E: Statistical, Nonlinear, and Soft Matter Physics*, vol. 75, no. 4, Article ID 046109, 2007.

[49] J. J. Fruin, *Pedestrian Planning and Design*, vol. 77, Elevator World, Inc, 1971.

[50] *Transportation Research Board, Highway Capacity Manual*, National Research Council, Washington, DC, USA, 2010.

[51] A. Gupta and N. Pundir, "Pedestrian flow characteristics studies: a review," *Transport Reviews*, vol. 35, no. 4, pp. 445–465, 2015.

[52] U. Weidmann, *Transporttechnik der Fussgänger, Transporttechnische Eigenschaften des Fussgängerverkehrs (Literturauswertung)*, 1993.

[53] V. M. Predtechenskii and A. I. Milinskii, *Planning for Foot Traffic in Buildings (translated from the Russian)*, 1969.

[54] R. A. Klein, *SFPE Handbook of Fire Protection Engineering*, 5th edition, 2016.

[55] W. H. K. Lam, J. F. Morrall, and H. Ho, "Pedestrian flow characteristics in Hong Kong," *Transportation Research Record Journal of the Transportation Research Board*, vol. 1995, pp. 56–62, 1995.

[56] W. Daamen and S. Hoogendoorn, "Controlled experiments to derive walking behaviour," *European Journal of Transport and Infrastructure Research*, vol. 3, pp. 39–59, 2003.

[57] A. Seyfried, B. Steffen, W. Klingsch, and M. Boltes, "The fundamental diagram of pedestrian movement revisited," *Journal of Statistical Mechanics: Theory and Experiment*, no. 10, Article ID P10002, pp. 41–53, 2005.

[58] J. Zhang, W. Klingsch, A. Schadschneider, and A. Seyfried, "Transitions in pedestrian fundamental diagrams of straight corridors and T-junctions," *Journal of Statistical Mechanics: Theory and Experiment*, vol. 2011, no. 6, Article ID P06004, 2011.

[59] J. Zhang and A. Seyfried, "Empirical Characteristics of Different Types of Pedestrian Streams," *Procedia Engineering*, vol. 62, pp. 655–662, 2013.

[60] C.-J. Jin, R. Jiang, S. C. Wong et al., "Large-scale pedestrian flow experiments under high-density conditions," 2017, https://arxiv.org/abs/1710.10263.

[61] X. Liu, W. Song, and J. Zhang, "Extraction and quantitative analysis of microscopic evacuation characteristics based on digital image processing," *Physica A: Statistical Mechanics and its Applications*, vol. 388, no. 13, pp. 2717–2726, 2009.

[62] U. Chattaraj, A. Seyfried, and P. Chakroborty, "Comparison of pedestrian fundamental diagram across cultures," *Advances in Complex Systems (ACS)*, vol. 12, no. 3, pp. 393–405, 2009.

[63] J. Zhang, A. Tordeux, and A. Seyfried, "Effects of boundary conditions on single-file pedestrian flow," in *Proceedings of the 9th International Conference on Cellular Automata, ACRI 2014*, J. Wąs, G. C. Sirakoulis, and S. Bandini, Eds., pp. 462–469, Springer International Publishing, Cham, Switzerland, 2015.

[64] S. Cao, J. Zhang, D. Salden, J. Ma, C. Shi, and R. Zhang, "Pedestrian dynamics in single-file movement of crowd with different age compositions," *Physical Review E: Statistical, Nonlinear, and Soft Matter Physics*, vol. 94, no. 1, 2016.

[65] J. Wang, M. Boltes, A. Seyfried, J. Zhang, V. Ziemer, and W. Weng, "Linking pedestrian flow characteristics with stepping locomotion," *Physica A: Statistical Mechanics and its Applications*, vol. 500, pp. 106–120, 2018.

[66] S. Cao, J. Zhang, W. Song, C. Shi, and R. Zhang, "The stepping behavior analysis of pedestrians from different age groups via a single-file experiment," *Journal of Statistical Mechanics: Theory and Experiment*, vol. 2018, no. 3, p. 033402, 2018.

[67] M. Boltes, A. Seyfried, B. Steffen, and A. Schadschneider, "Automatic Extraction of Pedestrian Trajectories from Video Recordings," in *Pedestrian and Evacuation Dynamics*, W. W. F. Klingsch, C. Rogsch, A. Schadschneider, and M. Schreckenberg, Eds., pp. 43–54, Springer, Berlin, Germany, 2010.

[68] M. S. Sharifi, K. Christensen, A. Chen, D. Stuart, Y. S. Kim, and Y. Chen, "A large-scale controlled experiment on pedestrian walking behavior involving individuals with disabilities," *Travel Behaviour and Society*, vol. 8, pp. 14–25, 2017.

[69] M. S. Sharifi, *Analysis and modeling of pedestrian walking behaviors involving individuals with disabilities*, Utah State University, 2016, https://digitalcommons.usu.edu/etd/4959.

[70] M. S. Sharifi, D. Stuart, K. Christensen, and A. Chen, "Traffic Flow Characteristics of Heterogeneous Pedestrian Stream Involving Individuals with Disabilities," *Transportation Research Record*, vol. 2537, pp. 111–125, 2015.

[71] M. S. Sharifi, D. Stuart, K. Christensen, and A. Chen, "Time Headway Modeling and Capacity Analysis of Pedestrian Facilities Involving Individuals with Disabilities," *Transportation Research Record*, vol. 2553, pp. 41–51, 2016.

[72] T. Kretz, A. Grünebohm, A. Kessel, H. Klüpfel, T. Meyer-König, and M. Schreckenberg, "Upstairs walking speed distributions on a long stairway," *Safety Science*, vol. 46, no. 1, pp. 72–78, 2008.

[73] X. L. Zhang, W. G. Weng, H. Y. Yuan, and J. G. Chen, "Empirical study of a unidirectional dense crowd during a real mass event," *Physica A: Statistical Mechanics and its Applications*, vol. 392, no. 12, pp. 2781–2791, 2013.

[74] J. L. Berrou, J. Beecham, P. Quaglia, M. A. Kagarlis, and A. Gerodimos, "Calibration and validation of the Legion simulation model using empirical data," in *Pedestrian and Evacuation Dynamics*, pp. 167–181, 2007.

[75] A. John, A. Schadschneider, D. Chowdhury, and K. Nishinari, "Trafficlike collective movement of ants on trails: absence of a jammed phase," *Physical Review Letters*, vol. 102, no. 10, pp. 1–4, 2009.

[76] S. Wang and W. Song, "Experimental Study of Ant Movement in a Straight Passageway under Stress Conditions," *Journal of Insect Behavior*, vol. 29, no. 6, pp. 735–743, 2016.

[77] C. Dias, O. Ejtemai, M. Sarvi, and M. Burd, "Exploring Pedestrian Walking through Angled Corridors," *Transportation Research Procedia*, vol. 2, pp. 19–25, 2014.

[78] C. Jin, R. Jiang, J. Yin, L. Dong, and D. Li, "Simulating bi-directional pedestrian flow in a cellular automaton model considering the body-turning behavior," *Physica A: Statistical Mechanics and its Applications*, vol. 482, pp. 666–681, 2017.

[79] C. Dias and R. Lovreglio, "Calibrating cellular automaton models for pedestrians walking through corners," *Physics Letters A*, vol. 382, no. 19, pp. 1255–1261, 2018.

[80] N. Shiwakoti, M. Sarvi, G. Rose, and M. Burd, "Consequence of Turning Movements in Pedestrian Crowds during Emergency Egress," *Transportation Research Record*, vol. 2234, no. 1, pp. 97–104, 2011.

[81] C. Dias, M. Sarvi, N. Shiwakoti, and M. Burd, "Turning Angle Effect on Emergency Egress: Experimental Evidence and Pedestrian Crowd Simulation," *Transportation Research Record*, vol. 2312, no. 1, pp. 120–127, 2012.

[82] A. Matsui, T. Mashiko, and T. Nagatani, "Traffic flow of mobile objects through obstacles: Turning and translational objects," *Physica A: Statistical Mechanics and its Applications*, vol. 388, no. 2-3, pp. 157–173, 2009.

[83] D. Yanagisawa, A. Kimura, A. Tomoeda et al., "Introduction of frictional and turning function for pedestrian outflow with an obstacle," *Physical Review E: Statistical, Nonlinear, and Soft Matter Physics*, vol. 80, no. 3, Article ID 036110, 2009.

[84] J. Zhang, W. Klingsch, T. Rupprecht, A. Schadschneider, and A. Seyfried, *Empirical Study of Turning And Merging of Pedestrian Streams in T-Junction*, 2011.

[85] N. Shiwakoti, M. Sarvi, C. Dias, and M. Burd, "Understanding crowd panic at turning and intersection through model organisms," in *Pedestrian and Evacuation Dynamics*, pp. 1175–1183, Springer International Publishing, Cham, Switzerland, 2014.

[86] S. Burghardt, A. Seyfried, and W. Klingsch, "Performance of stairs—fundamental diagram and topographical measurements," *Transportation Research Part C: Emerging Technologies*, vol. 37, pp. 268–278, 2013.

[87] C. Dias, O. Ejtemai, M. Sarvi, and N. Shiwakoti, "Pedestrian Walking Characteristics through Angled Corridors," *Transportation Research Record*, vol. 2421, no. 1, pp. 41–50, 2014.

[88] C. Dias, M. Sarvi, O. Ejtemai, and M. Burd, "Elevated Desired Speed and Change in Desired Direction," *Transportation Research Record*, vol. 2490, pp. 65–75, 2015.

[89] C. Dias, M. Sarvi, and M. Asano, "Optimal trajectories for constrained pedestrian turning manoeuvres," in *Proceedings of the 8th International Conference on Pedestrian and Evacuation Dynamics*, pp. 1–9, Hefei, China, 2016.

[90] M. S. Sharifi, D. Stuart, K. Christensen, A. Chen, Y. S. Kim, and Y. Chen, "Analysis of walking speeds involving individuals with disabilities in different indoor walking environments," *Journal of Urban Planning and Development*, vol. 142, no. 1, 2014.

[91] A. Sieben, J. Schumann, A. Seyfried, and D. R. Chialvo, "Collective phenomena in crowds—Where pedestrian dynamics need social psychology," *PLoS ONE*, vol. 12, no. 6, p. e0177328, 2017.

[92] L. Sun, L. Cui, S. Qiu, L. Yao, and J. Rong, "Impact on pedestrian flow of bends in passenger access tunnels," *Proceedings of the Institution of Civil Engineers - Transport*, pp. 1–10, 2017.

[93] A. Jelić, C. Appert-Rolland, S. Lemercier, and J. Pettré, "Properties of pedestrians walking in line: Fundamental diagrams," *Physical Review E: Statistical, Nonlinear, and Soft Matter Physics*, vol. 85, no. 3, 2012.

[94] M. Moussaïd, E. G. Guillot, M. Moreau et al., "Traffic instabilities in self-organized pedestrian crowds," *PLoS Computational Biology*, vol. 8, no. 3, Article ID e1002442, 2012.

[95] H. Kuang, M.-J. Cai, X.-L. Li, and T. Song, "Asymmetric effect on single-file dense pedestrian flow," *International Journal of Modern Physics C*, vol. 26, no. 6, 1550064, 13 pages, 2015.

[96] N. A. Rahman, N. A. Alias, and S. H. M. Adenan, "Empirical investigation of trajectories and desired walking velocity of pedestrian walking through angled-corridor," in *AIP Conference Proceedings*, AIP Publishing LLC, 2017.

[97] C. Dias, M. Sarvi, N. Shiwakoti, O. Ejtemai, and M. Burd, "Investigating collective escape behaviours in complex situations," *Safety Science*, vol. 60, pp. 87–94, 2013.

[98] R. Nagai, M. Fukamachi, and T. Nagatani, "Evacuation of crawlers and walkers from corridor through an exit," *Physica A: Statistical Mechanics and its Applications*, vol. 367, pp. 449–460, 2006.

[99] J. Zhang, W. Song, and X. Xu, "Experiment and multi-grid modeling of evacuation from a classroom," *Physica A: Statistical Mechanics and its Applications*, vol. 387, no. 23, pp. 5901–5909, 2008.

[100] T. Kretz, A. Grünebohm, and M. Schreckenberg, "Experimental study of pedestrian flow through a bottleneck," *Journal of Statistical Mechanics: Theory and Experiment*, vol. 2006, no. 10, pp. P10014–P10014, 2006.

[101] J. Liddle, A. Seyfried, B. Steffen et al., "Microscopic insights into pedestrian motion through a bottleneck, resolving spatial and temporal variations," 2011, https://arxiv.org/abs/1105.1532.

[102] J. Liddle, A. Seyfried, and B. Steffen, "Analysis of bottleneck motion using Voronoi diagrams," in *Pedestrian and Evacuation Dynamics*, pp. 833–836, 2010.

[103] W. Song, J. Zhang, and A. Seyfried, "Experimental Study of Pedestrian Flow in the Channel through Bottleneck," in *Pedestrian and Evacuation Dynamics*, R. D. Peacock, E. D. Kuligowski, and J. D. Averill, Eds., pp. 875–879, Springer, Boston, MA, USA, 2011.

[104] W. Tian, W. Song, J. Ma, Z. Fang, A. Seyfried, and J. Liddle, "Experimental study of pedestrian behaviors in a corridor based on digital image processing," *Fire Safety Journal*, vol. 47, pp. 8–15, 2012.

[105] W. Liao, A. Seyfried, J. Zhang, M. Boltes, X. Zheng, and Y. Zhao, "Experimental Study on Pedestrian Flow through Wide Bottleneck," *Transportation Research Procedia*, vol. 2, pp. 26–33, 2014.

[106] L. Sun, W. Luo, L. Yao, S. Qiu, and J. Rong, "A comparative study of funnel shape bottlenecks in subway stations," *Transportation Research Part A Policy & Practice*, vol. 98, pp. 14–27, 2017.

[107] W. Daamen and S. Hoogendoorn, "Capacity of doors during evacuation conditions," *Procedia Engineering*, vol. 3, pp. 53–66, 2010.

[108] W. Daamen and S. P. Hoogendoorn, "Emergency Door Capacity: Influence of Door Width, Population Composition and Stress Level," *Fire Technology*, vol. 48, no. 1, pp. 55–71, 2012.

[109] A. Garcimartín, I. Zuriguel, J. Pastor, C. Martín-Gómez, and D. Parisi, "Experimental Evidence of the "Faster Is Slower" Effect," *Transportation Research Procedia*, vol. 2, pp. 760–767, 2014.

[110] A. Garcimartín, D. R. Parisi, J. M. Pastor, C. Martín-Gómez, and I. Zuriguel, "Flow of pedestrians through narrow doors with different competitiveness," *Journal of Statistical Mechanics: Theory and Experiment*, no. 4, pp. 1–16, 2016.

[111] N. W. Bode, S. Holl, W. Mehner, A. Seyfried, and G. Xiao, "Disentangling the Impact of Social Groups on Response Times and Movement Dynamics in Evacuations," *PLoS ONE*, vol. 10, no. 3, p. e0121227, 2015.

[112] C. von Krüchten and A. Schadschneider, "Empirical study on social groups in pedestrian evacuation dynamics," *Physica A: Statistical Mechanics and its Applications*, vol. 475, pp. 129–141, 2017.

[113] C. von Krüchten, F. Müller, A. Svachiy, O. Wohak, and A. Schadschneider, "Empirical study of the influence of social groups in evacuation scenarios," in *Traffic and Granular Flow '15*, pp. 65–72, Springer International Publishing, Cham, Switzerland.

[114] A. Nicolas, S. Bouzat, and M. Kuperman, "Influence of selfish and polite behaviours on a pedestrian evacuation through a narrow exit: A quantitative characterisation," in *Proceedings of the 8th International Conference on Pedestrian and Evacuation Dynamics*, Hefei, China, 2016.

[115] A. Nicolas, S. Bouzat, and M. N. Kuperman, "Pedestrian flows through a narrow doorway: Effect of individual behaviours on the global flow and microscopic dynamics," *Transportation Research Part B: Methodological*, vol. 99, pp. 30–43, 2017.

[116] T. Ezaki, K. Ohtsuka, M. Chraibi et al., "Inflow Process of Pedestrians to a Confined Space," *Collective Dynamics*, vol. 1, 2016.

[117] X. Liu, W. Song, L. Fu, and Z. Fang, "Experimental study of pedestrian inflow in a room with a separate entrance and exit," *Physica A: Statistical Mechanics and its Applications*, vol. 442, pp. 224–238, 2016.

[118] Legion Software — Science in Motion, Legion Software, 2018, http://www.legion.com/legion-software.

[119] E. Altshuler, O. Ramos, Y. Núñez, J. Fernández, A. J. Batista-Leyva, and C. Noda, "Symmetry breaking in escaping ants," *The American Naturalist*, vol. 166, no. 6, pp. 643–649, 2005.

[120] N. Shiwakoti, M. Sarvi, G. Rose, and M. Burd, "Enhancing the Safety of Pedestrians during Emergency Egress," *Transportation Research Record*, vol. 2137, no. 1, pp. 31–37, 2009.

[121] S. Wang, W. Lv, W. Song, and D. R. Chialvo, "Behavior of Ants Escaping from a Single-Exit Room," *PLoS ONE*, vol. 10, no. 6, p. e0131784, 2015.

[122] S. Wang, S. Cao, Q. Wang, L. Lian, and W. Song, "Effect of exit locations on ants escaping a two-exit room stressed with repellent," *Physica A: Statistical Mechanics and its Applications*, vol. 457, pp. 239–254, 2016.

[123] N. Shiwakoti and M. Sarvi, "Enhancing the panic escape of crowd through architectural design," *Transportation Research Part C: Emerging Technologies*, vol. 37, pp. 260–267, 2013.

[124] S. Soria, R. Josens, and D. Parisi, "Experimental evidence of the "Faster is Slower" effect in the evacuation of ants," *Safety Science*, vol. 50, no. 7, pp. 1584–1588, 2012.

[125] S. Boari, R. Josens, D. R. Parisi, and J. A. Marshall, "Efficient Egress of Escaping Ants Stressed with Temperature," *PLoS ONE*, vol. 8, no. 11, p. e81082, 2013.

[126] J. M. Pastor, A. Garcimartín, P. A. Gago et al., "Experimental proof of faster-is-slower in systems of frictional particles flowing through constrictions," *Physical Review E: Statistical, Nonlinear, and Soft Matter Physics*, vol. 92, no. 6, 2015.

[127] N. Shiwakoti, M. Sarvi, G. Rose, and M. Burd, "Biologically Inspired Modeling Approach for Collective Pedestrian Dynamics under Emergency Conditions," *Transportation Research Record*, vol. 2196, no. 1, pp. 176–184, 2010.

[128] D. R. Parisi, S. A. Soria, and R. Josens, "Faster-is-slower effect in escaping ants revisited: Ants do not behave like humans," *Safety Science*, vol. 72, pp. 274–282, 2015.

[129] A. Sobhani, M. Sarvi, D. Duives, O. Ejtemai, K. Aghabayk, and S. Hoogendoorn, "Exploring the relationship of exit flow and jam density in panic scenarios using animal dynamics," *Transportation Research Procedia*, pp. 745–751, 2014.

[130] A. Sobhani, M. Sarvi, D. Duives, O. Ejtemai, K. Aghabayk, and S. Hoogendoorn, "Exploring the Relationship of Exit Flow and Jam Density in Panic Scenarios Using Animal Dynamics," *Transportation Research Procedia*, vol. 2, pp. 745–751, 2015.

[131] C. Saloma, G. J. Perez, G. Tapang, M. Lim, and C. Palmes-Saloma, "Self-organized queuing and scale-free behavior in real escape panic," *Proceedings of the National Academy of Sciences of the United States of America*, vol. 100, no. 21, pp. 11947–11952, 2003.

[132] C. Saloma, G. J. Perez, C. A. Gavile, J. J. Ick-Joson, and C. Palmes-Saloma, "Prior individual training and self-organized queuing during group emergency escape of mice from water pool," *PLoS ONE*, vol. 10, no. 2, 2015.

[133] P. Lin, J. Ma, T. Liu, T. Ran, Y. Si, and T. Li, "An experimental study of the "faster-is-slower" effect using mice under panic," *Physica A: Statistical Mechanics and its Applications*, vol. 452, pp. 157–166, 2016.

[134] P. Lin, J. Ma, T. Y. Liu et al., "An experimental study of the impact of an obstacle on the escape efficiency by using mice under high competition," *Physica A: Statistical Mechanics and its Applications*, vol. 482, pp. 228–242, 2017.

[135] Y. Zhang, J. Ma, Y. Si et al., "Required width of exit to avoid the faster-is-slower effect in highly competitive evacuation," *Chinese Physics B*, vol. 26, no. 8, p. 084504, 2017.

[136] J. M. Chen, P. Lin, F. Y. Wu, D. L. Gao, and G. Y. Wang, "Revisit the faster-is-slower effect for an exit at a corner," *Journal of Statistical Mechanics: Theory and Experiment*, vol. 2018, no. 2, p. 023404, 2018.

[137] H. Oh and J. Park, "Main factor causing "faster-is-slower" phenomenon during evacuation: rodent experiment and simulation," *Scientific Reports*, vol. 7, no. 1, 2017.

[138] I. Zuriguel, D. R. Parisi, R. C. Hidalgo et al., "Clogging transition of many-particle systems flowing through bottlenecks," *Scientific Reports*, vol. 4, no. 1, 2014.

[139] A. Garcimartín, J. M. Pastor, L. M. Ferrer, J. J. Ramos, C. Martín-Gómez, and I. Zuriguel, "Flow and clogging of a sheep herd passing through a bottleneck," *Physical Review E: Statistical, Nonlinear, and Soft Matter Physics*, vol. 91, no. 2, 2015.

[140] I. Zuriguel, J. Olivares, J. M. Pastor et al., "Effect of obstacle position in the flow of sheep through a narrow door," *Physical Review E: Statistical, Nonlinear, and Soft Matter Physics*, vol. 94, no. 3, 2016.

[141] M. Isobe, T. Adachi, and T. Nagatani, "Experiment and simulation of pedestrian counter flow," *Physica A: Statistical Mechanics and its Applications*, vol. 336, no. 3-4, pp. 638–650, 2004.

[142] R. Nagai, M. Fukamachi, and T. Nagatani, "Experiment and simulation for counterflow of people going on all fours," *Physica A: Statistical Mechanics and its Applications*, vol. 358, no. 2-4, pp. 516–528, 2005.

[143] T. Kretz, A. Grünebohm, M. Kaufman, F. Mazur, and M. Schreckenberg, "Experimental study of pedestrian counterflow in a corridor," *Journal of Statistical Mechanics: Theory and Experiment*, no. 10, Article ID P10001, 2006.

[144] Z. Yang, Y. Zhang, and O. Grembek, "Combining traffic efficiency and traffic safety in countermeasure selection to improve pedestrian safety at two-way stop controlled intersections," *Transportation Research Part A: Policy and Practice*, vol. 91, pp. 286–301, 2016.

[145] M. Muramatsu, T. Irie, and T. Nagatani, "Jamming transition in pedestrian counter flow," *Physica A: Statistical Mechanics and its Applications*, vol. 267, no. 3-4, pp. 487–498, 1999.

[146] Q. Zhang, B. Han, and D. Li, "Modeling and simulation of passenger alighting and boarding movement in Beijing metro stations," *Transportation Research Part C: Emerging Technologies*, vol. 16, no. 5, pp. 635–649, 2008.

[147] S. Seriani and R. Fernandez, "Pedestrian traffic management of boarding and alighting in metro stations," *Transportation Research Part C: Emerging Technologies*, vol. 53, pp. 76–92, 2015.

[148] Y. Tajima and T. Nagatani, "Scaling behavior of crowd flow outside a hall," *Physica A: Statistical Mechanics and its Applications*, vol. 292, no. 1-4, pp. 545–554, 2001.

[149] L. Sun, S. Hao, Q. Gong, S. Qiu, and Y. Chen, "Pedestrian roundabout improvement strategy in subway stations," in *Proceedings of the Institution of Civil Engineers - Transport*, vol. 171, pp. 20–29, 2017.

[150] D. Versluis, *Microscopic Interaction Behavior between Individual Pedestrians*, 2010.

[151] J. Zhang, W. Klingsch, A. Schadschneider, and A. Seyfried, "Ordering in bidirectional pedestrian flows and its influence on the fundamental diagram," *Journal of Statistical Mechanics: Theory and Experiment*, vol. 2012, Article ID P02002, 2012.

[152] S. Xue, R. Jiang, B. Jia, Z. Wang, and X. Zhang, "Pedestrian counter flow in discrete space and time: experiment and its implication for CA modelling," *Transportmetrica B: Transport Dynamics*, pp. 1–16, 2017.

[153] Q. Wang, W. Song, J. Zhang, and S. Lo, "Bi-directional movement characteristics of Camponotus japonicus ants during nest relocation," *Journal of Experimental Biology*, 2018.

[154] M. Plaue, M. Chen, G. Bärwolff, and H. Schwandt, "Trajectory Extraction and Density Analysis of Intersecting Pedestrian Flows from Video Recordings," in *Photogrammetric Image Analysis*, vol. 6952 of *Lecture Notes in Computer Science*, pp. 285–296, Springer, Berlin, Germany, 2011.

[155] S. C. Wong, W. L. Leung, S. H. Chan et al., "Bidirectional pedestrian stream model with oblique intersecting angle," *Journal of Transportation Engineering*, vol. 136, no. 3, pp. 234–242, 2010.

[156] J. Wu and S. Lu, "Feature Analysis and Operation Evaluation of Pedestrian Weaving Zone," *Transportation Research Record*, vol. 2393, no. 1, pp. 66–74, 2013.

[157] L. Lian, X. Mai, W. Song, Y. K. Kit Richard, X. Wei, and J. Ma, "An experimental study on four-directional intersecting pedestrian flows," *Journal of Statistical Mechanics: Theory and Experiment*, vol. 2015, Article ID P08024, 2015.

[158] L. Sun, Q. Gong, S. Hao, C. Wang, and Y. Chen, "Experimental Study of Oblique Pedestrian Streams," *PROMET - TrafficTransportation*, vol. 30, pp. 151-10, 2018.

[159] K. E. Boyce, D. A. Purser, and T. J. Shields, "Experimental studies to investigate merging behaviour in a staircase," *Fire and Materials*, vol. 36, no. 5-6, pp. 383–398, 2012.

[160] X. Shi and W. Wang, *Characteristics of Merging Pedestrian Flows at Congested Situation*, 2015.

[161] F. Huo, W. Song, L. Chen, C. Liu, and K. M. Liew, "Experimental study on characteristics of pedestrian evacuation on stairs in a high-rise building," *Safety Science*, vol. 86, pp. 165–173, 2016.

[162] K. Aghabayk, M. Sarvi, O. Ejtemai, and A. Sobhani, "Impacts of Different Angles and Speeds on Behavior of Pedestrian Crowd Merging," *Transportation Research Record*, vol. 2490, pp. 76–83, 2015.

[163] K. Aghabayk, O. Ejtemai, M. Sarvi, and A. Sobhani, "Understanding Pedestrian Crowd Merging Behavior," *Transportation Research Procedia*, vol. 2, pp. 768–773, 2014.

[164] Z. Shahhoseini, M. Sarvi, and M. Saberi, "Insights toward characteristics of merging streams of pedestrian crowds based on experiments with panicked ants," *Transportation Research Record*, vol. 2561, pp. 81–88, 2016.

[165] L. Lian, X. Mai, W. Song, Y. K. K. Richard, Y. Rui, and S. Jin, "Pedestrian merging behavior analysis: An experimental study," *Fire Safety Journal*, vol. 91, pp. 918–925, 2017.

[166] N. Shiwakoti, Y. Gong, X. Shi, and Z. Ye, "Examining influence of merging architectural features on pedestrian crowd movement," *Safety Science*, vol. 75, pp. 15–22, 2015.

[167] Z. Shahhoseini, M. Sarvi, and M. Saberi, "Pedestrian crowd dynamics in merging sections: Revisiting the "faster-is-slower" phenomenon," *Physica A: Statistical Mechanics and its Applications*, vol. 491, pp. 101–111, 2018.

[168] M. Chen, J. Wang, Y. Zhi, and L. Gao, "Effect of the Ratio of the Branch Inflow to the Total Inflow on Evacuation Efficiency of Pedestrians Merging at T-junctions," *Procedia Engineering*, vol. 211, pp. 70–77, 2018.

[169] J. Ma, W. Song, W. Tian, S. Lo, and G. Liao, "Experimental study on an ultra high-rise building evacuation in China," *Safety Science*, vol. 50, no. 8, pp. 1665–1674, 2012.

[170] Z. Shahhoseini, M. Sarvi, M. Saberi, and M. Haghani, "Pedestrian Crowd Dynamics Observed at Merging Sections," *Transportation Research Record*, vol. 2622, pp. 48–57, 2017.

[171] A. Cuesta, O. Abreu, A. Balboa, and D. Alvear, "An experimental data-set on merging flows in rail tunnel evacuation," *Tunnelling and Underground Space Technology*, vol. 70, pp. 155–165, 2017.

[172] M. Haghani, M. Sarvi, Z. Shahhoseini, M. Boltes, and J. Jing, "How Simple Hypothetical-Choice Experiments Can Be Utilized to Learn Humans' Navigational Escape Decisions in Emergencies," *PLoS ONE*, vol. 11, no. 11, p. e0166908, 2016.

[173] M. Haghani and M. Sarvi, "Stated and revealed exit choices of pedestrian crowd evacuees," *Transportation Research Part B: Methodological*, vol. 95, pp. 238–259, 2017.

[174] J. R. G. Dyer, C. C. Ioannou, L. J. Morrell et al., "Consensus decision making in human crowds," *Animal Behaviour*, vol. 75, no. 2, pp. 461–470, 2008.

[175] B. Zhou, X. Wang, and X. Tang, "Understanding collective crowd behaviors: Learning a Mixture model of Dynamic pedestrian-Agents," in *Proceedings of the 2012 IEEE Conference on Computer Vision and Pattern Recognition (CVPR)*, pp. 2871–2878, Providence, RI, USA, June 2012.

[176] M. Campanella, S. Hoogendoorn, and W. Daamen, "Improving the Nomad microscopic walker model," *IFAC Proceedings Volumes*, vol. 42, no. 15, pp. 12–18, 2009.

[177] T. Robin, G. Antonini, M. Bierlaire, and J. Cruz, "Specification, estimation and validation of a pedestrian walking behavior model," *Transportation Research Part B: Methodological*, vol. 43, no. 1, pp. 36–56, 2009.

[178] E. Cepolina and N. Tyler, "Understanding Capacity Drop for designing pedestrian environments, Walk 21 Everyday Walk. Cult," (2005), http://discovery.ucl.ac.uk/1412/1/Cepolina_Tyler_paper_Walk21.pdf.

[179] M. Asano, T. Iryo, and M. Kuwahara, "Microscopic pedestrian simulation model combined with a tactical model for route choice behaviour," *Transportation Research Part C: Emerging Technologies*, vol. 18, no. 6, pp. 842–855, 2010.

[180] J. R. G. Dyer, A. Johansson, D. Helbing, I. D. Couzin, and J. Krause, "Leadership, consensus decision making and collective behaviour in humans," *Philosophical Transactions of the Royal Society B: Biological Sciences*, vol. 364, no. 1518, pp. 781–789, 2009.

[181] M. Moussaïd, D. Helbing, S. Garnier, A. Johansson, M. Combe, and G. Theraulaz, "Experimental study of the behavioural mechanisms underlying self-organization in human crowds," *Proceedings of the Royal Society B Biological Science*, vol. 276, no. 1668, pp. 2755–2762, 2009.

[182] W. Daamen and S. Hoogendoorn, "Calibration of pedestrian simulation model for emergency doors by pedestrian type," *Transportation Research Record*, vol. 2316, pp. 69–75, 2012.

[183] D. R. Parisi, P. A. Negri, and L. Bruno, "Experimental characterization of collision avoidance in pedestrian dynamics," *Physical Review E: Statistical, Nonlinear, and Soft Matter Physics*, vol. 94, no. 2, 2016.

[184] R. Guo, S. C. Wong, Y. Xia, H. Huang, W. H. Lam, and K. Choi, "Empirical Evidence for the Look-Ahead Behavior of Pedestrians in Bi-directional Flows," *Chinese Physics Letters*, vol. 29, no. 6, p. 068901, 2012.

[185] S. Xie, S. C. Wong, W. H. K. Lam, and A. Chen, "Development of a bidirectional pedestrian stream model with an oblique intersecting angle," *Journal of Transportation Engineering*, vol. 139, no. 7, pp. 678–685, 2013.

[186] M. Boltes, J. Zhang, A. Seyfried, and B. Steffen, "T-junction: Experiments, trajectory collection, and analysis," in *Proceedings of the IEEE International Conference on Computer Vision Workshops (ICCV '11)*, pp. 158–165, Barcelona, Spain, November 2011.

[187] J. Wu, S. Lu, Y. Wang, and X. Liu, "Level of Service Evaluation Method of Pedestrian Weaving Area," in *Proceedings of the 96th Annual Meeting on Transportation Research Board*, pp. 1–16, Washington, DC, USA, 2017.

[188] W. Liao, A. Tordeux, A. Seyfried et al., "Measuring the steady state of pedestrian flow in bottleneck experiments," in *Physica A: Statistical Mechanics and its Applications*, pp. 1–13, 2016.

[189] L. Sun, Z. Yang, J. Rong, and X. Liu, "Study on the Weaving Behavior of High Density Bidirectional Pedestrian Flow," *Mathematical Problems in Engineering*, vol. 2014, Article ID 765659, 9 pages, 2014.

[190] J. Qiao, L. Sun, X. Liu, and J. Rong, "Reducing the impact of speed dispersion on subway corridor flow," *Applied Ergonomics*, vol. 65, pp. 362–368, 2017.

[191] Y. Liu, X. Shi, Z. Ye, N. Shiwakoti, and J. Lin, "Controlled experiments to examine different exit designs on crowd evacuation dynamics," in *CICTP 2016: Green and Multimodal Transportation and Logistics*, pp. 779–790, American Society of Civil Engineers, Reston, VA, USA, 2016.

[192] N. Shiwakoti, X. Shi, Z. Ye, Y. Gong, and W. Wang, "Empirical Investigation on merging crowd behaviour with and without

blocked vision," in *Proceedings of the World Conference on Transport Research*, Shanghai, China, 2016.

[193] Z. Shahhoseini, M. Sarvi, and E. Ito, "Collective movements of pedestrians: How we can learn from simple experiments with non-human (ant) crowds," *PLoS ONE*, vol. 12, no. 8, p. e0182913, 2017.

[194] W. Liao, A. U. Kemloh Wagoum, and N. W. Bode, "Route choice in pedestrians: determinants for initial choices and revising decisions," *Journal of the Royal Society Interface*, vol. 14, no. 127, p. 20160684, 2017.

[195] G. Warnaby, *Marketing Strategy*, JMark Management LLC, 1999, http://ezproxy.library.capella.edu/login?url=http://search.ebscohost.com/login.aspx?direct=true&db=bth&AN=5659377&site=ehost-live&scope=site.

[196] N. Shiwakoti, "Understanding differences in emergency escape and experimental pedestrian crowd egress through quantitative comparison," *International Journal of Disaster Risk Reduction*, vol. 20, pp. 129–137, 2016.

[197] S. Gwynne, E. Kuligowski, K. Boyce et al., "Enhancing egress drills: Preparation and assessment of evacuee performance," *Fire and Materials*.

[198] A. Leeson, P. Alvarez, and S. Ghosh, "Understanding How Big Data and Crowd Movements will Shape the Cities of Tomorrow," in *Proceedings of the European Transport Conference*, Frankfurt, Germany, 2014, http://abstracts.aetransport.org/paper/index/id/4296/confid/19.

[199] X. Shi, X. Zhao, Z. Ye, N. Shiwakoti, and J. Lu, "Estimating pedestrian walking characteristics by use of smartphone sensing: an experimental study," in *Proceedings of the 17th COTA conference International Conference of Transportation Professionals (CICTP2017)*, American Society of Civil Engineers, 2017.

[200] R. Lovreglio, V. Gonzalez, and Z. Feng, "Prototyping virtual reality serious games for building earthquake preparedness: the auckland city hospital case study," 2018, https://arxiv.org/abs/1802.09119.

[201] Civil-Security-and-Traffic, Jülich-Supercomuting-Center, and Forschungszentrum-Jülich, *Data Archive of Experimental Data from Studies about Pedestrian Dynamics*, 2018, http://ped.fz-juelich.de/db/.

[202] Pablo Augusto Negri, GSdataset-PANKit - Traffic Intersection Video Sequence DataSet, 2017, http://pablonegri.free.fr/Downloads/GSdataset-PANKit.htm.

[203] CSAI, Vittorio Emanuele II Gallery Dataset, An Annot. Video about Pedestr. Gr. Dyn. Proxemic Behav, 2013.

[204] B. Zhou, *Grand Central Station Dataset, Train Stn. Dataset*, 2012, http://www.ee.cuhk.edu.hk/~xgwang/grandcentral.html.

[205] B. Majecka, *Edinburgh Informatics Forum Pedestrian Database*, 2010, http://homepages.inf.ed.ac.uk/rbf/FORUMTRACKING/.

[206] B. Zhou, *Collective Motion Database*, 2013, http://mmlab.ie.cuhk.edu.hk/projects/collectiveness/dataset.htm.

[207] W. Song and L. Fu, "Verification and validation methods," in *Evacuation Model*, pp. 81–102, 2016.

2

Theoretical Comparison of the Effects of Different Traffic Conditions on Urban Road Traffic Noise

Mohammad Maghrour Zefreh and Adam Torok

Budapest University of Technology and Economics, Department of Transport Technology and Economics, Stoczek St. 2, Building ST, 1111 Budapest, Hungary

Correspondence should be addressed to Adam Torok; torok.adam@mail.bme.hu

Guest Editor: Vladislav Zitricky

Road traffic noise is one of the most relevant sources in the environmental noise pollution of the urban areas where dynamics of the traffic flow are much more complicated than uninterrupted traffic flows. It is evident that different traffic conditions would play the role in the urban traffic flow considering the dynamic nature of the traffic flow on one hand and presence of traffic lights, roundabouts, etc. on the other hand. The main aim of the current paper is to investigate the effect of different traffic conditions on urban road traffic noise. To do so, different traffic conditions have been theoretically generated by the Monte Carlo Simulation technique following the distribution of traffic speed in the urban roads. The "ASJ RTN-Model" has been considered as a base road traffic noise prediction model which would deal with different traffic conditions including steady and nonsteady traffic flow that would cover the urban traffic flow conditions properly. Having generated the vehicles speeds in different traffic conditions, the emitted noise (L_{WA}) and subsequently the noise level at receiver (L_A) were estimated by "ASJ RTN-Model." Having estimated L_{WA} and L_A for each and every vehicle in each traffic condition and taking the concept of transient noise into account, the single event sound exposure levels (*SEL*) in different traffic conditions are calculated and compared to each other. The results showed that decelerated traffic flow had the lowest contribution, compared to congestion, accelerated flow, free flow, oversaturated congestion, and undersaturated flow by 16%, 14%, 12%, 12%, and 10%, respectively. Moreover, the distribution of emitted noise and noise level at receiver were compared in different traffic conditions. The results showed that traffic congestion had considerably the maximum peak compared to other traffic conditions which would highlight the importance of the range of generated noise in different traffic conditions.

1. Introduction

Highly industrialized living style of modern societies has produced a dramatic impact on the environment [1]. Nowadays, the environment noise and emission [2] have become a worldwide problem and the environmental impact control is one of the most important problems in the urban areas [3]. It is reasonable to affirm that, in urban areas, road traffic noise is the most relevant source [4] since airports are usually placed outside the downtowns and railways are usually designed to move out from the center of the cities [5] and rarely cross the residential districts [6]. Furthermore, in urban areas, the concentration of both road networks and city dwellers makes of traffic flow the main culprit of noise annoyance, which has to be precisely estimated and treated through relevant road traffic mitigation [7]. Although the road noise is not usually loud enough to cause hearing problems, continuous exposure to unacceptable noise levels can create the adverse effects on health [8]. Moreover, environmental noise exposure is associated with annoyance [9, 10], sleep disturbance [11], cognitive ability in schoolchildren [12], and health impacts, especially cardiovascular conditions [13]. Exposure to environmental noise is pervasive and increasing in terms of road traffic noise and the reduction of the night-time quiet period [14]. Traffic-related noise is one of the major environmental impacts of roadways [15] and is said to account for over 1 million healthy years of life lost annually to ill health and may lead to a disease burden that is second only in magnitude to that from air pollution [16]. According to the WHO, the environment noise costs societies 0.2%-2.0% of the gross domestic product [17].

Therefore, urban road traffic emission and noise should be included in the economic evaluation of an urban road traffic policy [18] since different traffic conditions, as consequence of these policies, might have a considerable contribution in the urban road traffic noise emission which would have a significant impact on the gross domestic product of the societies. Noise prediction is one of the vital tools for city planners for noise abatement and control [19]. Traffic management can be a very effective policy to fight against urban traffic noise [20, 21] which is increasingly implemented in European [22] noise reduction projects. By taking a wide look at the literature, mostly, noise impacts of traffic management policies have been measured/investigated by conducting the on-field studies [23]. The current research instead seeks to provide a theoretical comparison environment in which the average impact of different traffic conditions on road traffic noise emission would be compared. Generally, the estimation of the noise emissions generated by road traffic requires both a noise emission model that captures the impact of vehicle kinematics, namely, speed and acceleration, on sound power levels, and an accurate estimation of the kinematic variables of interest in the area under study [24]. There are various models developed recently in order to evaluate the road traffic noise emissions. For instance, the Federal Highway Administration Traffic Noise Model (FHWA TNM) of USA [25] predicts the hourly equivalent sound level of the ith class of vehicles based on the reference energy mean emission level of the ith class of vehicles which refers to the maximum sound level emitted by a vehicle pass-by at a reference distance of 15 (m) through a series of adjustments as follows.

$$L_{eq}(h)_i = \left(\overline{L_0}\right)_{Ei} + \Delta_{f(i)} + \Delta_d + \Delta_l + \Delta_s \quad (1)$$

where

$L_{eq}(h)_i$ is hourly equivalent sound level ($dB(A)$)

$(\overline{L_0})_{Ei}$ is reference energy mean emission level ($dB(A)$)

$\Delta_{f(i)}$ is traffic flow adjustment

Δ_d is distance adjustment

Δ_l is length of the roadway adjustment

Δ_s is shielding adjustment

The CoRTN (Calculation of Road Traffic Noise) model of the United Kingdom estimates the A-weighted sound pressure level that is exceeded for 10% of the measurement period at a reference distance of 10 (m) away from the nearside carriageway edge as follows [26].

$$L_{10} = 42.2 + 10\log_{10} q + \Delta_f + \Delta_g + \Delta_p + \Delta_d + \Delta_s + \Delta_a + \Delta_r \quad (2)$$

where

L_{10} is A-weighted sound pressure level that is exceeded for 10% of the measurement period (dB(A))

q is total hourly flow (veh/hr) calculated at a reference distance of 10 (m) at an average traffic speed of 75 (km/h)

Δ_f is traffic flow adjustment

Δ_g is gradient adjustment

Δ_p is pavement type adjustment

Δ_d is distance adjustment

Δ_s is shielding adjustment

Δ_a is angle of view adjustment

Δ_r is reflection adjustment

The HARMONOISE (Harmonised Accurate and Reliable Methods for the EU Directive on the Assessment and Management of Environmental Noise) model [27] of the European Union is defined based on two sound power values for different vehicle categories as follows.

$$L_{WR}(f) = a_R(f) + b_R(f)\log\left(\frac{V}{V_{ref}}\right) \quad (3)$$

$$L_{WP}(f) = a_P(f) + b_P(f)\log\left(\frac{V - V_{ref}}{V_{ref}}\right) \quad (4)$$

where

$L_{WR}(f)$ is rolling noise generated by tire-road interaction and aerodynamic drag (dB(A))

$L_{WP}(f)$ is propulsion noise generated by the power-train and the exhaust (dB(A))

V_{ref} is 70 (km/h)

$a_R(f)$, $b_R(f)$, $a_P(f)$, and $b_P(f)$ are given coefficients in $1/3^{rd}$ octave bands in frequency range 25 to 10 kHz

The ASJ RTN-Model (Acoustical Society of Japan Road Traffic Noise prediction model) considers a single vehicle as an omnidirectional point source passing along the road under consideration and predicts the noise emitted by each vehicle in different categories as a function of the vehicle speed in different traffic conditions (steady and nonsteady traffic flow, acceleration, and deceleration running condition) as follows [28].

$$L_{WA} = a + b\log(V) + c \quad (5)$$

where

L_{WA} is A-weighted sound power level (noise emitted by each vehicle) (dB(A))

V is vehicle speed (km/h)

a and b are regression coefficients

c is correction term for road conditions (pavement type and road gradient)

Furthermore, noise propagation from the ith source position to the prediction point will be predicted as follows.

$$L_{A,i} = L_{WA,i} - 8 - 20\log r_i + \Delta L_{cor,i} \quad (6)$$

where

$L_{A,i}$ is A-weighted sound pressure level (noise level at receiver) (dB(A))

r_i is direct distance from the ith source position to the prediction point (m)

$\Delta L_{cor,i}$ is correction for diffraction, ground effect, and atmospheric absorption (dB(A))

The main aim of the current research is to investigate the contribution of different traffic conditions in noise emission of the urban road traffic flow by preparing a theoretical comparison in which each traffic condition would generate road traffic noise independently. In the current study, different traffic conditions have been represented by the distribution of traffic speed in that traffic condition obtained by on-field data collection. It is worth to mention that the role of speed distribution on road traffic emission [29] and noise has been investigated in the literature. For instance, the effects of speed distributions on the Harmonoise model predictions have been investigated by [30]. They assumed when traffic is freely moving the speed distribution of a given category of vehicle approximates to a normal distribution. It is a very general assumption and they did not consider the situation in which the traffic flow freely moves but it is close to the capacity. They examined the errors in noise prediction which would result if the mean speed was used for prediction purposes rather than the actual speed distribution. They pointed out that it is likely that the assumptions concerning a normal distribution are not so robust due to the potential periods of congested traffic so that the estimates of standard deviation could be misleading. Also, due to the lack of data at the individual level, they have roughly assumed some other distributions for the other simple traffic conditions. The main difference of this paper with our paper is that we have investigated the effect of different traffic conditions based on the actual (fitted to empirical data) speed distributions adapted to the empirical traffic speed data obtained by the on-site traffic data collection in the city of Budapest, Hungary. In 2013, Iannone et al. [31] studied the influence of speed distribution on road traffic noise prediction. In their study, they investigated the effect of speed distribution (different speed distributions were simulated following the same average speed, 70 (km/h) and standard deviation, 10 (km/h)) on road traffic noise prediction using the "Micro to Macro" approach that adopts the single vehicle noise emission. The random extraction of the speed allows running the model also when single vehicle speed data are not available. In the current paper, the effects of different traffic conditions (different speed distributions and different speed specifications obtained by empirical data collection in city of Budapest) on urban road traffic noise were theoretically compared to each other using the Japanese reference noise model which is calibrated for different traffic conditions.

A majority of the researches in dynamic road traffic noise prediction topic considered the microscopic traffic flow characteristics (speed, acceleration, etc. of each and every single vehicles) instead of the macroscopic traffic flow characteristics (average parameters of the entire traffic flow) as the main element of the road traffic dynamics for noise prediction [24, 32, 33]. In the current paper, different traffic conditions have been theoretically generated based on the distribution of vehicles speed on that traffic condition. Furthermore, the reference road traffic noise model of Japan, the ASJ RTN-Model, has been selected as a base road traffic noise prediction model due to the fact that it is calibrated for both steady and nonsteady traffic flow (adapted for urban condition). Moreover, this model is also calibrated for acceleration and deceleration running conditions which would be appropriate in predicting the road traffic noise emissions close to the intersections. Apart from these, the base of the model is directly dependent on the speed of vehicles which is the considered characteristic of different traffic conditions in this research.

The rest of the paper is organized as follows. In Section 2, at first, different traffic conditions have been theoretically generated according to the distribution of vehicles speed in different traffic conditions and then the generated road traffic noise in each traffic condition is estimated using the ASJ RTN-Model. The estimated road traffic noises in different traffic conditions have been compared and discussed in terms of magnitude and continuity in Section 3. Finally, the paper is concluded in Section 4.

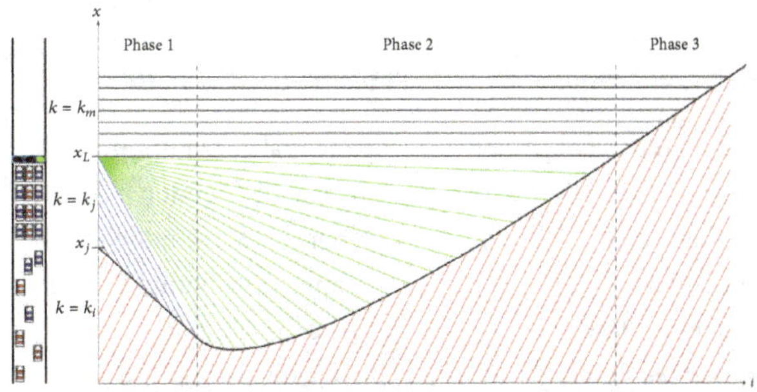

FIGURE 1: Characteristic diagram for traffic moving at a traffic light.

2. Methodology

The main aim of the current research is to compare the effects of different traffic conditions on urban road traffic noise. To do so, a field measurement campaign (manual video recording) has been done by the Stipendium Hungaricum (2016-2017/MSc) transportation engineering students in the two-lane Villnyi street and Karolina street intersection in Budapest, XI district, Hungary, during the entire day. Six different traffic conditions (scenarios) have been defined based on the recorded videos. The defined scenarios are theoretically explained here using a theoretical representation of a characteristic diagram for traffic moving at a traffic light when it turns to green as shown in Figure 1. The condition in which the traffic congestion would not be disappeared in a cycle (exceeds the capacity) is considered as oversaturated congestion. This is the region $x_j \leq x < x_L$ in Figure 1 where density is close to Jam density (k_j). The undersaturated flow is considered when the traffic is close to the capacity (and will be discharged in a cycle) in the region $x < x_j$ where traffic is flowing with the density lower than optimal density ($k_i < k_m$), as the flow is not unimpeded. The free

TABLE 1: Specifications of traffic speed in different traffic conditions.

Traffic condition	Fitted speed distribution	Minimum speed (km/h)	Maximum speed (km/h)	Average speed (km/h)	SD of speed (km/h)
Over-saturated congestion	Exponential distribution	1	15.5	10.37	2.74
Under saturated flow	Normal distribution	25	55	38.14	6.46
Free flow	Log-normal distribution	35	55	47.18	5.5
Congestion	Gamma distribution	10	25	14.53	3.91
Accelerated flow	Beta distribution	1	45	27.41	13.15
Decelerated flow	Chi-square distribution	1	35	18.76	11.82

flow traffic is considered when there are a few numbers of vehicles in the street (much lower than the capacity) in the region $x < x_j$ where traffic is flowing with the density lower than optimal density ($k_i < k_m$), as the flow is not unimpeded. The deceleration process is considered in the condition where traffic flow is getting closer to x_j (where the shockwave, black curve, travels backward through the traffic in phase 1 shown in Figure 1). When the traffic light turns green, vehicles are able to leave the light entirely unimpeded, so the density would be equal to optimal density ($k = k_m$) and obviously flow would be in its maximum. This condition is considered as the acceleration process in which the shockwave slows down and starts to move back towards the traffic light (phase 2). The condition in which traffic light turns to green but the intersection is not completely empty yet is considered as congestion.

Having collected the speed of vehicles in different traffic conditions, the distribution of speed in these traffic conditions was estimated by the Maximum Likelihood Estimation method as follows.

$$L(\theta) = \prod_{i=1}^{n} f(x_i \mid \theta) \tag{7}$$

where

x_i is observed traffic speed (km/h)

$f(\cdot \mid \theta)$ is density function of the candidate parametric distribution

Having estimated different candidate parametric distributions for traffic speed in different traffic conditions based on (7), the candidate distributions were compared to each other by three goodness-of-fit tests (Kolmogorov-Smirnov, Cramer-von-Mises, and Anderson-Darling) in order to find best-fitted speed distribution in each traffic condition as follows.

$$KS: \sup \left| F_n(x) - F(x) \right| \tag{8}$$

$$CvM: \int_{-\infty}^{\infty} \left(F_n(x) - F(x)\right)^2 dx \tag{9}$$

$$AD: n \int_{-\infty}^{\infty} \frac{\left(F_n(x) - F(x)\right)^2}{F(x)\left(1 - F(x)\right)} dx \tag{10}$$

where

F_n is empirical cumulative distribution function of the vehicles speeds

F_x is fitted theoretical parametric distribution

Apart from these three goodness-of-fit tests, two classical penalized criteria (Akaike and Bayesian information criteria) based on the log-likelihood were further considered to tackle the overfitting problems as follows.

$$AIC: 2k - 2\ln\left(\widehat{L}\right) \tag{11}$$

$$BIC: \ln(n)\,k - 2\ln\left(\widehat{L}\right) \tag{12}$$

where

k is number of estimated parameters in the model

$\widehat{L}$ is maximum value of the likelihood function for the model

n is number of observations

The best-fitted distributions to the traffic speed in different traffic conditions along with the statistical specifications of the traffic speed in different traffic conditions are shown in Table 1.

Taking Table 1 into account, one would find out that the fitted distributions are all unimodal distributions. It should be highlighted that, in the current study, a homogeneous traffic flow (passenger cars) has been considered in both data collection and simulation phases. Therefore, single-mode distributions (fitted unimodal distribution) have been considered to simulate traffic speed in different traffic conditions since the literature suggests considering bimodal and multimodal distributions while there is a heterogeneity in traffic flow [34]. There is a lot of research that has examined distribution models for motorized vehicle speed data in uncongested traffic condition. On one hand, [35, 36] found that, for lightly trafficked two-lane roads where most vehicles are traveling freely, car speeds measured in time are approximately normally distributed. Also, Minh et al. mentioned that speed distribution would follow normal distribution on the urban roads, 2005 [37]. From the other hand, log-normal distribution has been considered for modeling speed data [38, 39] since it offers the advantage that the same functional form is retained when the time speed distribution is transformed into a space-speed distribution and avoids the theoretical difficulty of the negative speeds given by the infinite tails of

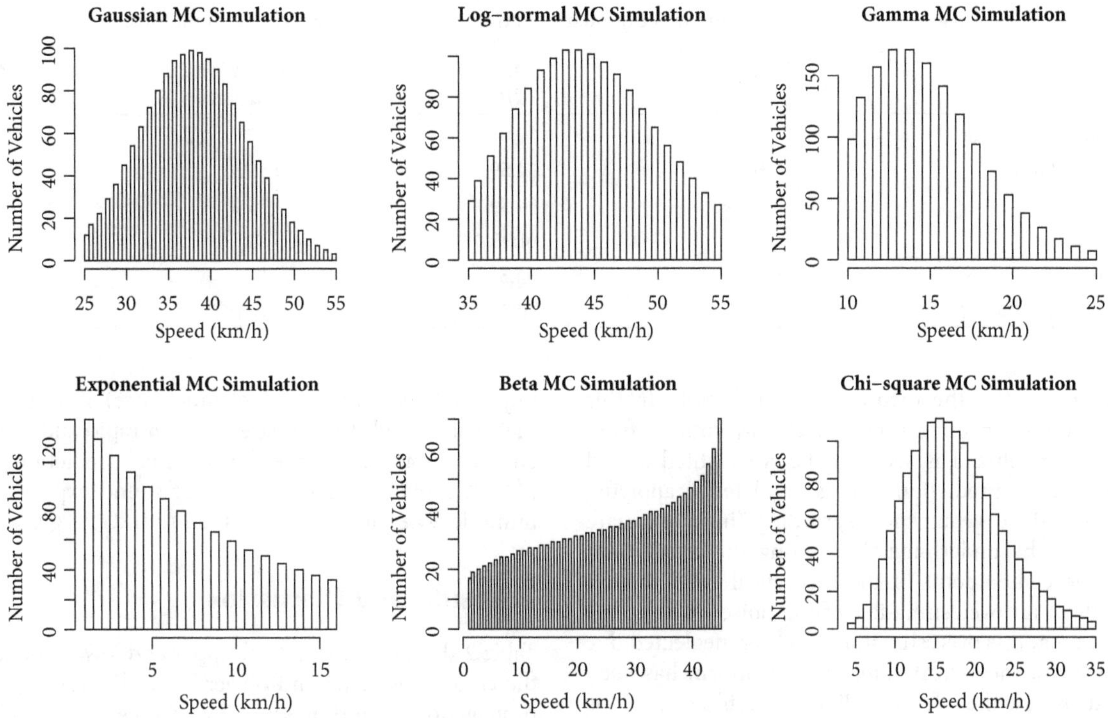

FIGURE 2: Random speed generation by Monte Carlo (MC) Simulation.

the normal distribution. In regard to log-normal distribution, Haight and Mosher [40] considered log-normal distribution since the speed data could be well represented by log-normal distribution in that study. Furthermore, [41, 42] suggested considering log-normal speed distribution for traffic speed simulation. Considering the fundamental relationship of the traffic flow, in the current study, the log-normal distribution is fitted better to the collected traffic speed in free flow condition (the condition in which much lower vehicles than the capacity were available in the roadway) and normal distribution is fitted better to the traffic speed in undersaturated flow condition (close to capacity condition).

2.1. Simulation of Different Traffic Conditions. In order to study the effect of different traffic conditions on road traffic noise emission, these traffic conditions were theoretically simulated by Monte Carlo (MC) simulation technique using the specifications of traffic speed in different traffic conditions shown in Table 1. It should be highlighted that the number of vehicles was considered to be constant (1600 vehicles) in each traffic condition in order to release the effect of traffic flow on further road traffic noise estimation. The randomly generated vehicles speeds which would represent the speed of vehicles in different traffic conditions are shown in Figure 2.

2.2. Road Traffic Noise Estimation. Having simulated the traffic speed in different traffic conditions theoretically, the A-weighted sound power level, L_{WA}, which would represent noise emitted by each single vehicle was calculated based on (5) in which the entire traffic flow has been considered to be in the same classification (passenger cars) with the same engine rotational speed, the same engine load, the same road pavement type, and the same road geometry in order to release the effect of determinants other than traffic condition on emission estimation.

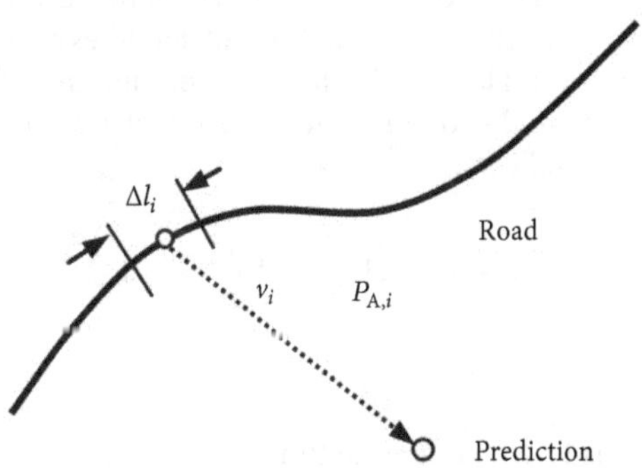

FIGURE 3: Sound propagation from a sound source to a prediction point [28].

The calculated L_{WA} for each vehicle is further considered as the basis of sound propagation as suggested by [28] and shown schematically in Figure 3 in which $P_{A,i}$ is the A-weighted sound power, $\Delta l_i(m)$ is the length of theoretical road, and v_i (km/h) is the running speed of vehicle.

Hence, the A-weighted sound pressure level, L_A, which would represent the noise level at receiver was calculated based on (6) considering the same distance from the source position to the prediction point (20 (m)), and the same $\Delta L_{cor,i}$ for the entire traffic conditions in order to release the effect of determinants other than traffic condition on sound propagation estimation. Moreover, it is worth to mention that a representative point (fixed point source) has been set

TABLE 2: Theoretical upper and lower boundaries of traffic noise in different traffic conditions.

Traffic condition	Min L_A $(dB(A))$	Max L_A $(dB(A))$	Min L_{WA} $(dB(A))$	Max L_{WA} $(dB(A))$
Over-saturated congestion	47.97	60.02	82	94.04
Under saturated flow	54.31	64.59	88.33	98.61
Free flow	58.7	64.59	92.72	98.61
Congestion	57.97	61.95	92	95.97
Accelerated flow	47.97	64.51	82	98.53
Decelerated flow	42.38	58.7	76.4	92.72

at the center point of the theoretical road section. In this case, the A-weighted sound power level L_{WA} emitted from the source (the vehicle) is set, and the A-weighted sound pressure level at the prediction point is calculated by applying a formula based on geometrical spreading (inverse-square law). It should be underlined that setting the fixed point source at the center point of the theoretical road section might result in an overestimation of the noise level. In the current research, this overestimation has been neglected due to the fact that a theoretical comparison platform has been provided such that the effects of different traffic conditions on urban road traffic noise have been theoretically compared to each other.

Having calculated the A-weighted sound power level and A-weighted sound pressure level received at the receiver from each and every vehicle, the single event sound exposure level (a constant sound level which has the same amount of energy in 1 second as the original noise level) of the vehicles is calculated as follows:

$$L_{AE} = 10\log\left(\frac{1}{T_0}\sum_i 10^{L_{A,i}/10}.\Delta t_i\right) \quad (13)$$

where

T_0 is 1 s (the reference time)

$L_{A,i}$ is A-weighted sound pressure level in the ith section (dB(A))

Δt_i is the time when the sound source exists in the ith section (s)

The calculated sound exposure level of the entire traffic flow in each traffic condition is summed up according to [28], in order to calculate the total theoretical sound energy emitted from the traffic flow in each traffic condition. Since the point source has been considered to be fixed in the current study, choosing the sound exposure level parameter would help to (to a certain extent) overcome the overestimation of energy. This is due to the fact that we consider the energy in 1 second, with little variations. In general, considering the sound exposure level for calculating the total sound energy is more preferred for the situations we are facing the transient noise (e.g., an aircraft fly-over or a vehicle drive-by). These situations often result in wide variations from background noise to maximum level and if only the maximum level is reported, information on the duration of the noise (an important feature for rating annoyance) is lost. This also makes it difficult to compare between rapid and slow events and to combine different events for noise prediction purposes [43]. It should be highlighted that sound exposure level is numerically equivalent to the total sound energy.

3. Results and Discussion

Table 2 shows the theoretical upper and lower boundaries of the emitted noise (sound power level (dB(A)), L_{WA}) and its propagation (sound pressure level (dB(A)), L_A) considering 20 (m) distance from the source position in each traffic condition. Having estimated the noise emitted by running vehicles in different traffic conditions, L_{WA} $(dB(A))$, the noise level at receiver considering 20 (m) distance, L_A $(dB(A))$, is calculated based on (6). The theoretical relationship between speed ranges and sound pressure level in different traffic conditions were compared to each other in Figure 4 considering the same number of vehicles (1600 vehicles) in each scenario. By taking a wide look at Figure 4, one can simply find out that the vehicles in free flow condition have emitted sound pressure level in higher ranges compared to the other traffic conditions. It seems to be logical since the range of traffic speed in free flow condition is the highest among the other conditions (35 (km/h) to 55 (km/h)) based on Table 1. Also, taking the decelerated traffic flow into account, it is clear that the vehicles emitted sound pressure in a wider range compared to the other traffic conditions. It should be highlighted that, considering the speed-sound pressure level relationship of decelerated flow in Figure 4, the vehicles with the speeds lower than 10 (km/h) have emitted the same sound pressure level as the vehicles with the speed of 10 (km/h). This is due to the fact that, in deceleration running condition, the sound power level at speed of 10 (km/h) is applied at speeds of less than 10 (km/h) according to [28].

Figure 4 suggests the need for considering the dynamics of traffic flow in economic investigations of traffic/transportation policies. For instance, although the vehicles in free flow condition would have the higher sound pressure level compared to the other traffic conditions, they would pass the roadway faster than the vehicles in congestion or any other traffic conditions defined in Figure 1. Therefore, the overall generated single event sound exposure level in different traffic conditions has been considered for a theoretical comparison platform. Since the vehicles in decelerated traffic flow emitted sound pressure in a wider range compared to the

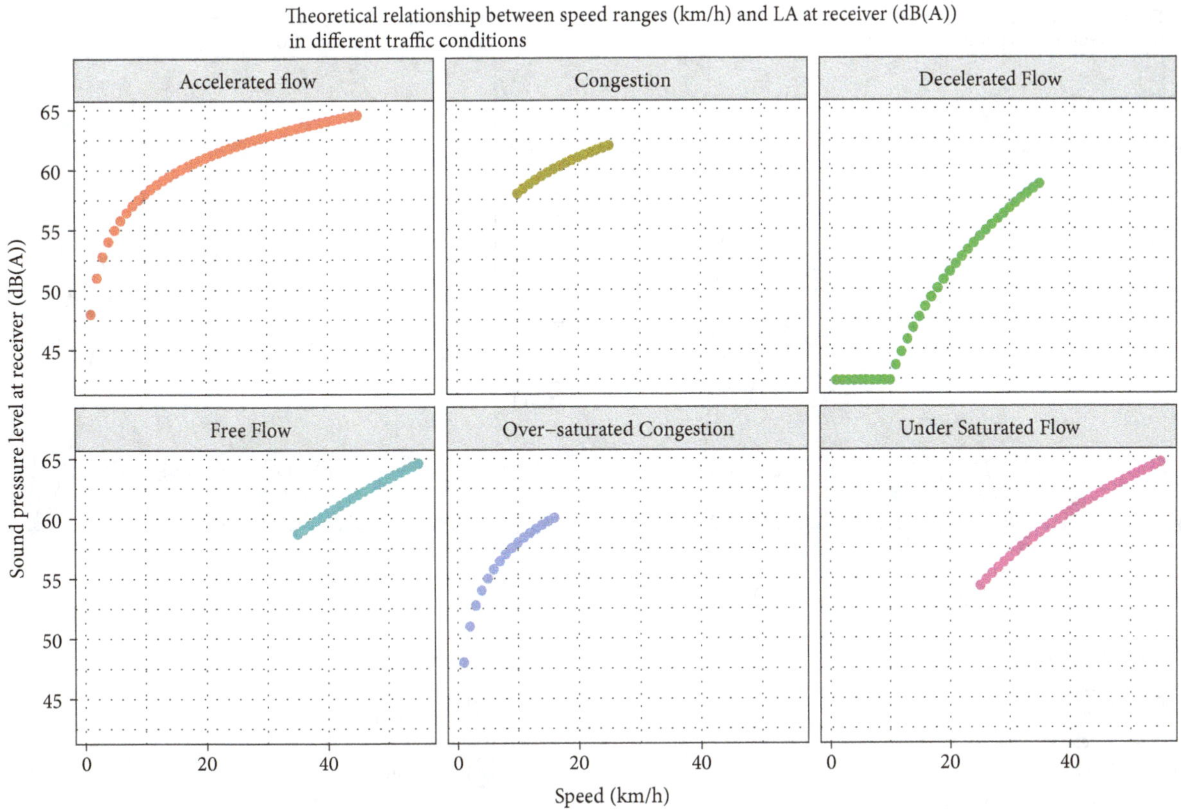

FIGURE 4: Theoretical relationship between speed ranges and the generated sound pressure in different traffic conditions.

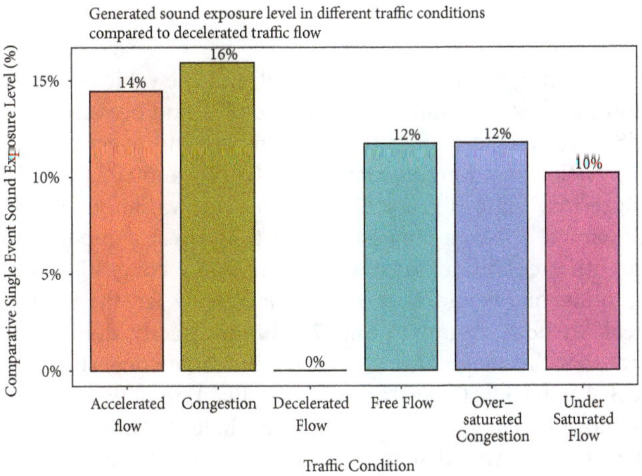

FIGURE 5: Comparing generated single event sound exposure levels in different traffic conditions.

other traffic conditions, the generated sound exposure level in each traffic condition has been compared to the generated sound exposure level in decelerated traffic flow in Figure 5.

Figure 5 clearly shows the fact that the vehicles in decelerated traffic flow would generate considerably lower sound energy compared to the other traffic conditions. Furthermore, comparing Figures 4 and 5 implies the fact that the dynamics of traffic flow which would generate different traffic conditions are of great importance in road traffic noise assessment. For example, although the free flow condition had the highest range of emitted sound pressure level, it has a lower sound exposure level compared to congestion and accelerated flow. Taking (13) into account, one would clearly understand that, apart from the emitted sound pressure level (L_A), the time when sound source exists in the roadway is of great importance in sound energy (sound exposure level) calculations. Therefore, compared to decelerated flow, traffic congestion, accelerated traffic flow, free flow, and undersaturated flow would theoretically generate more sound energy by 16%, 14%, 12%, and 10%, respectively.

It should be noted that the generated sound energy in decelerated flow condition is lower by 12% even compared to oversaturated congestion where cars have lower speed range (1 (km/h) to 15.5 (km/h)) compared to decelerated flow (1 (km/h) to 35 (km/h)) according to Table 1. This result might also confirm the suggestion of not just considering the speed ranges in different traffic conditions but also the dynamic nature of the traffic flow might have a considerable impact on the generated sound energy in traffic flow.

According to WHO, continuous exposure to unacceptable noise levels can create the adverse effects on health [8]. Moreover, the research evidence shows that the physiological effect of transportation noise on human sleep may depend more on the level and number of noise events in traffic streams than on energy equivalent measures [44, 45]. Therefore, not just the generated sound energy in road traffic is of great importance but also the distribution (continuity) of road traffic noise would be important. Hence, apart from the

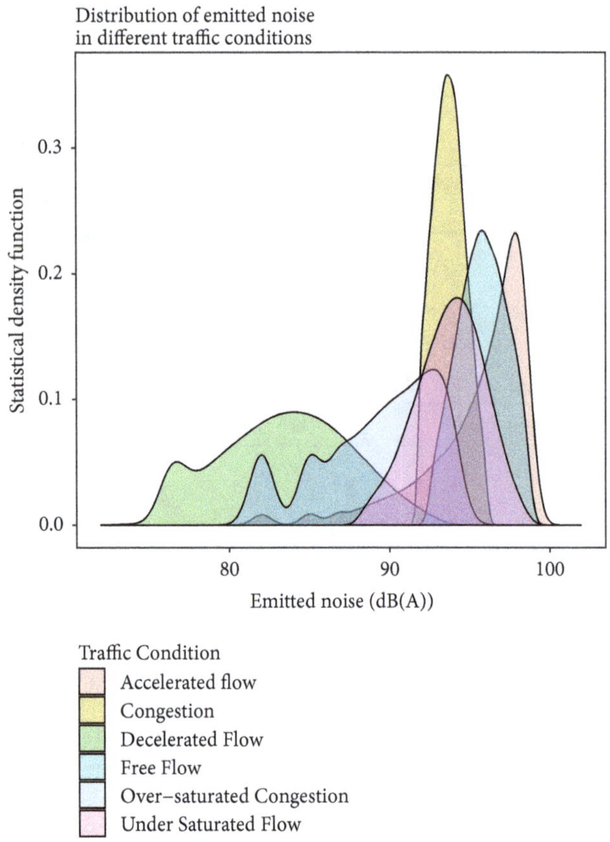

FIGURE 6: Comparing the distributions of emitted noise in different traffic conditions.

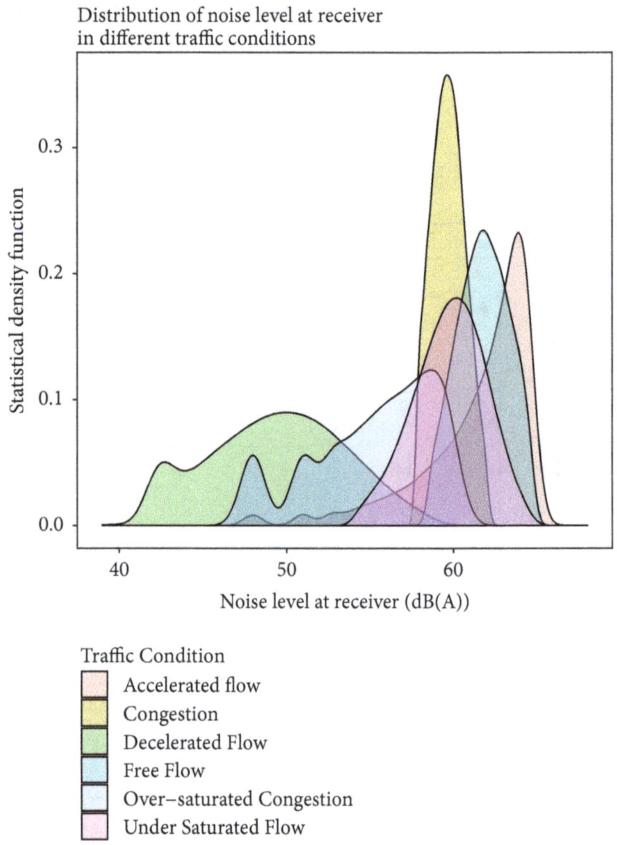

FIGURE 7: Comparing the distributions of noise level at receiver in different traffic conditions.

comparison of road traffic noise in different traffic conditions in terms of generated sound energy (Figure 5), they were compared in terms of continuity of road traffic noise by the means of density plots as shown in Figures 6 and 7, respectively. Most of the researches in this field suggested that by decreasing the speed rate the road traffic noise will be decreased. For instance, Ellebjerg proposed that noise abatement can be obtained through speed reduction [46]; furthermore Robertson et al. said that traffic calming policy ensures noise reduction since it forbids strong accelerations [47]. In fact, these statements are both true in terms of the generated sound pressure level such that the vehicles with the lower speeds would generate lower sound pressure level (see Figure 4 for illustration).

But the other parameter that needs to be considered in the evaluations is the number of noise events that are exceeded from the maximum thresholds since the physiological effect of transportation noise on human sleep may depend more on the level and number of noise events in traffic streams than on energy equivalent measures [44, 45]. Based on the abovementioned and considering Figures 6 and 7 which would clearly show the importance of the range of L_{WA} and L_A, it is evident that different traffic conditions, even with the same generated sound exposure level, might have different effects on their ambience. For instance, comparing Figures 5 and 7 implies the fact that the free flow condition and the oversaturated condition would generate the same amount of sound energy (sound exposure level) but the vehicles in free flow condition produce higher range of sound pressure level (58.7 (dB(A)) to 64.59 (dB(A))) compared to oversaturated congestion that ranges from 47.97 (dB(A)) to 60.02 (dB(A)). Moreover, Figures 6 and 7 show that the variations of the generated noise in congestion are the lowest compared to the other condition. Also, taking the traffic congestion into account, one would find out that it has, by far, the highest peak in both Figures 6 and 7. This is exactly due to the fact that the range of L_{WA} and L_A in traffic congestion (92 (dB(A)) to 95.97 (dB(A)) and 57.97 (dB(A)) to 61.95 (dB(A))) is shorter than the others based on Table 2. Therefore, the density plots would suggest paying attention to not just the maximum possible road traffic noise which may be caused by vehicle characteristics (speed, acceleration, etc.), but also the possible range of traffic noise which may be caused by nature of traffic condition in evaluation of the consequence of urban road traffic policies.

4. Conclusions

The rapid urbanization throughout the world extends the noise pollution, especially those induced by road traffic in the urban districts. Environmental noise exposure is associated with annoyance [9, 10], sleep disturbance [11], cognitive ability in schoolchildren [12], and health impacts, especially cardiovascular conditions [13]. In order to tackle these issues caused by road traffic, different traffic and transportation policies might be applied that would generate different traffic conditions. Hence, the effects of different traffic conditions on urban road traffic noise have been theoretically compared in this paper. The results show that when the traffic flow is in deceleration process it would theoretically generate lower sound energy compared to congestion, accelerated flow, free flow, and undersaturated flow by 16%, 14%, 12%, and 10%, respectively (see Figure 5). It should be noted that the generated sound energy in decelerated flow condition is lower by 12% even compared to oversaturated congestion where cars have lower speed range (1 (km/h) to 15.5 (km/h)) compared to decelerated flow (1 (km/h) to 35 (km/h)) according to Table 1. Furthermore, by taking a wide look at Figure 5, one would find out that free flow and oversaturated congestion would theoretically generate equal sound energy (sound exposure level) whereas they have different traffic speed ranges according to Table 1. In fact, these results would suggest that the emitted road traffic noise is the production of both traffic speed and dynamic nature of the traffic flow (pay attention to the range of the traffic noise in Table 2). Most of the researches in this field suggested that by decreasing the speed rate the road traffic noise will be decreased. For instance, Ellebjerg proposed that noise abatement can be obtained through speed reduction [46]; furthermore Robertson et al. said that traffic calming policy ensures noise reduction since it forbids strong accelerations [47].

Based on the abovementioned and considering Figures 6 and 7 which would clearly show the importance of the range of L_{WA} and L_A, it is evident that different traffic conditions, even with the same generated sound exposure level, might have different effects on their ambience. For instance, comparing Figures 5 and 7 implies the fact that the free flow condition and the oversaturated condition would generate the same amount of sound energy (sound exposure level) but the vehicles in free flow condition produce higher range of sound pressure level (58.7 (dB(A)) to 64.59 (dB(A))) compared to oversaturated congestion that ranges from 47.97 (dB(A)) to 60.02 (dB(A)). Moreover, Figures 6 and 7 show that the variation of the generated noise in congestion is the lowest compared to the other condition. Lastly, comparing the generated noises in different traffic conditions by density plots revealed the fact that not only is the generated sound exposure level in road traffic noise important but also the range of the generated noise in different traffic conditions is of significant importance. In this paper, the results showed that different traffic conditions might theoretically generate the same amount of noise (considering the equal number of vehicles) but in different ranges. Therefore, it might be concluded that the traffic condition that would generate the noise in the higher ranges (considering the distribution of generated noise in different traffic conditions) might have more negative effect on its ambience compared to the one that generates noise in lower ranges.

In the end, it should be highlighted that this research attempts to study neither the influence of junctions nor the influence of different driving patterns on noise emissions. But instead, it seeks to compare the average influences of different traffic conditions on urban road traffic noise in a theoretical environment. The scope for the future study is to calibrate the range of emitted noise in different traffic conditions in a real practice and build up a comparison environment based on the collected on-site values for further incorporation in the economic investigation of the road traffic policies.

Conflicts of Interest

The authors declare that there are no conflicts of interest regarding the publication of this paper.

Acknowledgments

The research reported in this paper was supported by the Higher Education Excellence Program of the Ministry of Human Capacities in the frame of Artificial Intelligence research area of Budapest University of Technology and Economics (BME FIKP-MI/FM). Moreover, the authors are grateful for the support of HAS (Hungarian Academy of Science) for providing the János BÓLYAI Scholarship. The first author further gratefully acknowledges the support of Stipendium Hungaricum of Tempus Public Foundation.

References

[1] J. B. Zavada, B. Abramovic', and Š. Denis, "A strategic model of sustainable mobility in the city of zagreb and its surrounding area," *International Journal for Traffic & Transport Engineering*, vol. 7, no. 4, 2017.

[2] I. Ivković, O. Čokorilo, and S. Kaplanović, "The estimation of GHG emission costs in road and air transport sector: Case study of Serbia," *Transport*, vol. 33, no. 1, pp. 260–267, 2018.

[3] K. Jadaan, H. Khreis, and Á. Török, "Exposure to traffic-related air pollution and the onset of childhood asthma: a review of the literature and the assement methods used," *Periodica Polytechnica Transportation Engineering*, vol. 46, no. 1, p. 21, 2017.

[4] B. S.-O. S.-C. Li, "The issues of measuring the exterior and interior noise of road vehicles," *Energy Intensity and GHG Production of Chosen Propulsions Used in Road Transport*, vol. 19, p. 50, 2017.

[5] B. Abramović, "Infrastructure access charges," in *Sustainable Rail Transport*, Lecture Notes in Mobility, pp. 45–58, Springer International Publishing, Cham, 2018.

[6] C. Guarnaccia, "Advanced tools for traffic noise modelling and prediction," *WSEAS Transactions on Systems*, vol. 12, no. 2, pp. 121–130, 2013.

[7] I. Varga, "A congestion detection based traffic control for signalized intersection," *Periodica Polytechnica Civil Engineering*, vol. 62, no. 2, p. 398, 2015.

[8] WHO, "Night noise guidelines for Europe," *World Health Organization-Regional Office for Europe*, 2009.

[9] W. Babisch, G. Pershagen, J. Selander et al., "Noise annoyance - A modifier of the association between noise level and cardiovascular health?" *Science of the Total Environment*, vol. 452-453, pp. 50–57, 2013.

[10] C. Clark, R. Crombie, J. Head, I. Van Kamp, E. Van Kempen, and S. A. Stansfeld, "Does traffic-related air pollution explain associations of aircraft and road traffic noise exposure on children's health and cognition? A secondary analysis of the United Kingdom sample from the RANCH project," *American Journal of Epidemiology*, vol. 176, no. 4, pp. 327–337, 2012.

[11] WHO-JRC, "Burden of disease from environmental noise: Quantification of healthy life years lost in Europe," OCLC: 779684347, 2011.

[12] C. Clark, J. Head, and S. A. Stansfeld, "Longitudinal effects of aircraft noise exposure on children's health and cognition: A six-year follow-up of the UK RANCH cohort," *Journal of Environmental Psychology*, vol. 35, pp. 1–9, 2013.

[13] A. L. Hansell, M. Blangiardo, L. Fortunato et al., "Aircraft noise and cardiovascular disease near Heathrow airport in London: Small area study," *BMJ*, vol. 347, no. 7928, 2013.

[14] M. S. Hammer, T. K. Swinburn, and R. L. Neitzel, "Environmental noise pollution in the United States: Developing an effective public health response," *Environmental Health Perspectives*, vol. 122, no. 2, pp. 115–119, 2014.

[15] P. Pamanikabud and P. Vivitjinda, "Noise prediction for highways in Thailand," *Transportation Research Part D: Transport and Environment*, vol. 7, no. 6, pp. 441–449, 2002.

[16] J. Gulliver, D. Morley, D. Vienneau et al., "Development of an open-source road traffic noise model for exposure assessment," *Environmental Modeling and Software*, vol. 74, pp. 183–193, 2015.

[17] World Health Organization, "Prevention of noise-induced hearing loss," Report of a WHO-PDH Informal Consultation, Geneva, 1997.

[18] M. M. Zefreh, F. Meszaros, R. Junevičius, and A. Torok, "Economic investigation of a public transport support policy: a case study at Budapest," *Promet-Traffic & Transportation*, vol. 29, no. 1, pp. 77–84, 2017.

[19] N. Garg and S. Maji, "A critical review of principal traffic noise models: strategies and implications," *Environmental Impact Assessment Review*, vol. 46, pp. 68–81, 2014.

[20] L. Ellebjerg and H. Bendtsen, "Possibilities of noise reduction through road traffic flow management," in *Proceedings of the INTER-NOISE and NOISE-CON Congress and Conference Proceedings*, pp. 2601–2609, 2007.

[21] H. Bendtsen, J. Haberl, U. Sandberg, G. Watts, and E. Pucher, *Traffic Management and Noise Reducing Pavements-Recommendations on Additional Noise Reducing Measures (Silvia, Sustainable Road Surfaces for Traffic Noise Control)*, Roskilde, Denmark, 2005.

[22] A. Rust, J. Affenzeller, and J. Hinton, "Plans for future european research to reduce transport noise," in *Proceedings of the INTER-NOISE and NOISE-CON Congress and Conference Proceedings*, pp. 2048–2057, Institute of Noise Control Engineering, 2007.

[23] V. Desarnaulds, M. Gilbert, and A. Carvalho, "Noise reduction by urban traffic management," in *Proceedings of the ICA*, 2004.

[24] A. Can and P. Aumond, "Estimation of road traffic noise emissions: The influence of speed and acceleration," *Transportation Research Part D: Transport and Environment*, vol. 58, pp. 155–171, 2018.

[25] M. T. Barry and A. J. Reagan, "Fhwa highway traffic noise prediction model," Tech. Rep., 1978.

[26] UK Department of Transport, *Calculation of Road Traffic Noise*, 1988.

[27] G. Watts, "Harmonoise models for predicting road traffic noise," *Acoustics Bulletin*, vol. 30, no. 5, pp. 19–25, 2005.

[28] S. Sakamoto, "Road traffic noise prediction model "ASJ RTN-Model 2013": Report of the Research Committee on Road Traffic Noise," *Acoustical Science and Technology*, vol. 36, no. 2, pp. 49–108, 2015.

[29] A. Török and M. M. Zefreh, "Assessing the Need for Applying Multimodal Speed Distribution in Road Transport Macro Emission Estimation," *Acta Technica Jaurinensis*, vol. 9, no. 2, pp. 118–127, 2016.

[30] G. Watts, D. van Maercke, H. van Leeuwen et al., "Effects of speed distributions on the harmonnoise model predictions," in *Proceedings of the INTER-NOISE and NOISE-CON Congress and Conference Proceedings*, vol. volume, pp. 2530–2536, Institute of Noise Control Engineering, 2004.

[31] G. Iannone, C. Guarnaccia, and J. Quartieri, "Speed distribution influence in road traffic noise prediction," *Environmental Engineering and Management Journal*, vol. 12, no. 3, pp. 493–501, 2013.

[32] F. Li, Y. Lin, M. Cai, and C. Du, "Dynamic simulation and characteristics analysis of traffic noise at roundabout and signalized intersections," *Applied Acoustics*, vol. 121, pp. 14–24, 2017.

[33] L. Estévez-Mauriz and J. Forssén, "Dynamic traffic noise assessment tool: A comparative study between a roundabout and a signalised intersection," *Applied Acoustics*, vol. 130, pp. 71–86, 2018.

[34] P. P. Dey, S. Chandra, and S. Gangopadhaya, "Speed distribution curves under mixed traffic conditions," *Journal of Transportation Engineering*, vol. 132, no. 6, pp. 475–481, 2006.

[35] H. J. Leong, "The distribution and trend of free speeds on two lane two way rural highways in new south wales," in *Proceeding of the 4th Australian Road Research Board (ARRB) Conference*, vol. 4, 1968.

[36] McLean. J. r., "Observed speed distributions and rural road traffic operations," in *Proceedings of the Australian Road Research Board Conference*, vol. 9, 1979.

[37] C. Minh, K. Sano, and S. Matsumoto, "Characteristics of passing and paired riding maneuvers of motorcycle," *Journal of the Eastern Asia Society for Transportation Studies*, vol. 6, pp. 186–197, 2005.

[38] Y. Wang, W. Dong, L. Zhang et al., "Speed modeling and travel time estimation based on truncated normal and lognormal distributions," *Transportation Research Record*, no. 2315, pp. 66–72, 2012.

[39] D. L. Georlough and M. J. Huber, "Traffic flow theory: A monograph, special report 165," *Transportation Research Board*, 1975.

[40] A. F. Haight and W. W. Mosher, "A practical method for improving the accuracy of vehicular speed distribution measurements," *Highway Research Board Bulletin*, vol. 341, pp. 92–116, 1962.

[41] M. Elhenawy, H. Rakha, and H. Chen, "An automatic traffic congestion identification algorithm based on mixture of linear regressions," in *Smart Cities, Green Technologies, and Intelligent Transport Systems*, vol. 738 of *Communications in Computer and Information Science*, pp. 242–256, Springer International Publishing, Cham, 2017.

[42] M. Elhenawy, H. Rakha, and H. Chen, "A unified real-time automatic congestion identification model considering weather

and roadway visibility conditions," in *Proceedings of the 2nd International Conference on Vehicle Technology and Intelligent Transport Systems, VEHITS 2016*, pp. 39–48, Italy, April 2016.

[43] P. Bernard, *Leq, SEL: When? Why? How?* Brüel & Kjær, 1975.

[44] B. Griefahn, A. Marks, and S. Robens, "Noise emitted from road, rail and air traffic and their effects on sleep," *Journal of Sound and Vibration*, vol. 295, no. 1-2, pp. 129–140, 2006.

[45] S. Pirrera, E. De Valck, and R. Cluydts, "Nocturnal road traffic noise: A review on its assessment and consequences on sleep and health," *Environment International*, vol. 36, no. 5, pp. 492–498, 2010.

[46] L. Ellebjerg, "Effectiveness and benefits of traffic flow measures on noise control," *Deliverable H. D*, vol. 1, 2007.

[47] S. Robertson, W. Heather, and G. Marsden, *Ulf Sandberg, and Ulf Hammerstrom. The effects of speed on noise, vibration and emissions from vehicles [Master, thesis]*, Public Master, 1998.

3

An Evaluation Method of Urban Public Transport Facilities Resource Supply based on Accessibility

Chao Sun, [1,2] **Xiaohong Chen,** [1] **H. Michael Zhang,** [1,3] **and Ze Huang** [2]

[1] *The Key Laboratory of Road and Traffic Engineering, Ministry of Education, Tongji University, Shanghai 201804, China*
[2] *Shenzhen Urban Transport Planning Center Co., Ltd., Shenzhen Key Laboratory for Traffic Information and Traffic Engineering, Shenzhen 518021, China*
[3] *Department of Civil and Environmental Engineering, University of California Davis, California, CA 95616, USA*

Correspondence should be addressed to Chao Sun; 649167196@qq.com

Academic Editor: Jaeyoung Lee

The two indices of the supply level of urban public transport facilities resource and the public transport-private cars traveling time ratio are used to evaluate the public transport accessibility in urban networks. An evaluation method for urban public transport facilities resource supply is proposed based on accessibility by taking Shenzhen as the research area and by using the London PTAL evaluation system. The method takes into account the walking time outside the public transit station and the spatial layout of public transit stations, reflects regional development opportunity, and can reflect regional differences regarding public transport facilities resources in various spatial scales. This paper evaluates the relative service competiveness of public transports and private cars in the entire travel of individuals based on the public transport-private cars travelling time ratio, which lays a foundation to judge the service level of public transport and to make a purpose-specific transportation policy and public transport development scheme.

1. Introduction

Accessibility has been applied in the fields such as urban and land use planning, transportation management, and public facility location analysis. The earliest research on accessibility is the classical location theory. Reilly defined accessibility as the index to calculate transport cost in 1931 [1–3], and Hansen indicated that accessibility presented the possibility of interaction among various nodes of the transportation network in 1959 [4, 5]. As people have deeper and deeper understanding of the concept of accessibility, both foreign and domestic scholars give explanation on accessibility from several aspects such as time, space, sociology, and psychology, and many classical accessibility calculation models have been created, but they are not internationally recognized [6]; from the point of urban area and land use, accessibility can be used to calculate space among different areas and to analyze the convenience to reach other places from current area; from the point of transportation system, accessibility can be used to evaluate the difficultly level of traveling by vehicles; from the point of individuals, accessibility reflects the possibility and effectiveness of people's trip [7] and it can be used to measure the degree of freedom of individuals of traveling by vehicles in traffic environment and ecological environment. Currently, foreign and domestic research methods for urban public transport accessibility can be divided into two categories, i.e., the method based on classical space obstacle model [8, 9], taking time distance as the measurement index and calculating the cost of passengers by taking public transport facilities, the method reflecting the level of accessibility based on indexes such as line coverage and density from the point of infrastructure availability [9, 10]. There are two problems in current methods; i.e., the "non-door-to-door" feature of public transport facility makes time outside the station become the key factor impacting the accessibility of public transport facilities, while most public transport accessibility evaluation method does not consider the walking time outside the station because it is difficult to create a pedestrian network, or straight-line distance is used to replace actual walking distance [11, 12], resulting in a certain discrepancy between the evaluation results and the actual situation. According to the General Investigation on Basic Data of

Shenzhen Passenger Transport Planning, the walking time needed to reach a conventional public transport accounts for 33% of the total travelling time [13], so slow-moving traffic such as walking is the most important connection method to achieve door-to-door service of public transport. Some studies about public transport accessibility considered actual walking time outside the station, which is usually represented by the shortest walking time of the public transit stop at the starting point or the end point [14–20], which include walking from the point of origin to the appropriate public transport station ("access time") and walking from the last public transit station to the final destination ("egress time"). The second problem is that such methods fail to evaluate the accessibility from the point of individuals by considering the selection or opportunities of individuals.

In the new cosmopolis plan, the transport for London promotes public transport accessibility to urban strategic level and the Public Transport Accessibility Levels (PTAL for short) proposed by it take into account the difference in the distribution of public facilities resource and the level of public service enjoyed by individuals in various areas [21]; New York transportation planning emphasizes the fairness of transportation system and measures the public transport accessibility based on the population and number of posts covered by public transport taken by individuals under certain time constraints [22]. The London PTAL accessibility evaluation method analyzes the level of service enjoyed by individuals under certain time-space constraints and reflects the real accessibility of urban public transport facilities resource from the point of passengers' demand. Currently, the PTAL evaluation method on urban public transport facilities resource supply and public transport service level has not been applied yet in China. Nowadays, urban center has the highest public transport supply level but also heavy congestion, mainly because people making transportation policy only pay attention to the balance of facility supply but fail to realize fairness from the point of users.

In general, travel time, travel expenses, travel comfort, and safety are important factors affecting the competitiveness of private cars and public transport. Service quality is perceived as an important determinant of users' travel demand [23]. Yet the measurement of service quality remains a challenging and important research area with practical implications for service providers [24]. Service quality is difficult to measure by quantitative comprehensive evaluation method [25]. Existing research usually selects one or more quantifiable evaluation indicators for comparison the competitiveness of public transport and private cars. For example, a methodological analysis framework was constructed to quantify operational performance measures that enable the comparison of the different travel modes. This analysis framework was then applied to two cities—Auckland, New Zealand, and Paris, France—to assess the overall performance metrics of PC, PT, and CB, such as travel costs, travel time, and fuel consumption. This comparison sheds light on the differences between the travel modes, their viability, and their competitiveness [26, 27]. Some studies use accessibility to compare the relative competitiveness between public and private cars; spatial accessibility comparison [28–32] and traffic isochronism [33, 34] are two important methods. Few researches aim to compare the travel time between public transport and private cars. Travel time is one of the key indicators to reflect accessibility, and it is also the most intuitive indicator of service level. The travel time ratio is used to define the relative competitiveness between public transport and the private car [14, 35, 36]. The purpose is to provide a service that can compete with the private car by improving the resources of the public transport facility.

Therefore, this paper proposes an evaluation method for urban public transport facilities resource supply based on accessibility by taking Shenzhen as the research area and by using the London PTAL evaluation system. The method takes into account the time a passenger needs to reach a public transit stop and the spatial layout of public transit stops, reflects regional development opportunity, and can reflect regional differences regarding public transport facilities resources in various spatial scales. This paper evaluates the relative service competiveness of public transports and private cars in the entire travel of individuals based on the public transport-private cars travelling time ratio by using the evaluation system of Singapore land transport planning [35], which lays a foundation to judge the service level of public transport and to make a purpose-specific transportation policy and public transport development scheme.

2. Evaluation Method of Urban Public Transport Resource Supply

2.1. Technical Route. The public transportation methods of this paper include bus, subway, and railway. This paper proposes an accessibility-based evaluation method for urban public transport facilities resource supply by using London PTAL public transport evaluation method and proposes the relative service competiveness of public transport and private cars in the entire travel of individuals based on the public transport-private cars travelling time ratio by using the evaluation system of Singapore land transport planning. The main technical route is as follows: obtaining basic data, using big data acquisition technology during research, and knowing the actual time passengers need to reach a bus stop and public transit network data based on network map. The isochron from the demand point is drawn by using ArcGIS space analysis technology based on the walking time and concept of isochron. Walking range and covered stops are determined based on the maximum acceptable walking time of passengers. Then the accessibility of demand point can be achieved by calculating total accessibility of stations in the walking range of all demand points by taking into account station type, number of routes, weight, and walking walk-to-public transit-stop time. Finally public transport-private cars travelling time ratio is calculated based on the network map and public transport-private cars travelling time to reflect the relative service competitiveness of public transport and private cars and to make it possible to evaluate the public transport accessibility of main areas. For the technical route, please see Figure 1.

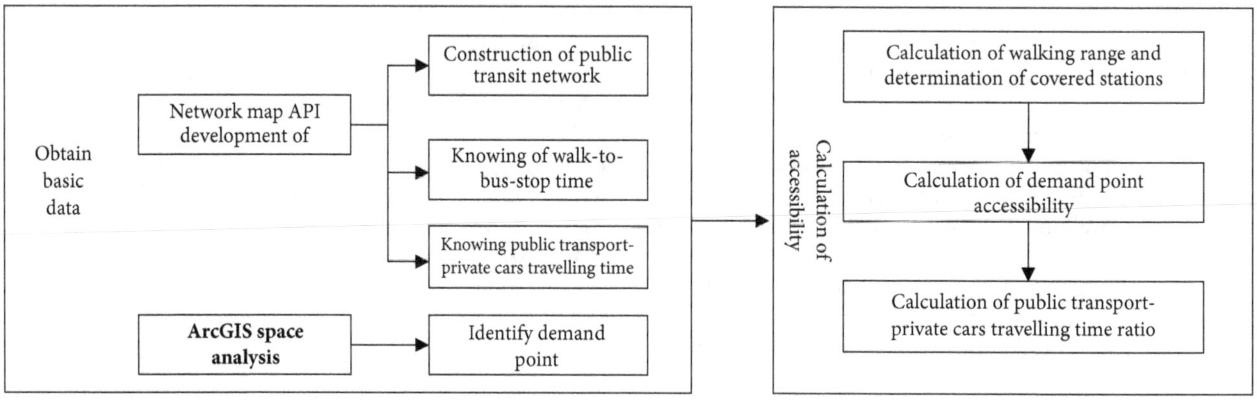

FIGURE 1: Technology roadmap.

2.2. Data Acquisition

2.2.1. Demand Point. Traditional accessibility research generally takes small traffic area or fixed grid as analysis unit, and the evaluation result focuses more on macrospace and cannot reflect microtraffic characteristics. According to relevant regulations, spaces between public transport stations are different in different areas, which is 300-500m in urban center and larger in suburban areas. To reflect the accessibility of any point in a city as well as spatial layout of public transport stations, the research uses GIS space analysis tool to insert sample points in the research area every 200m, which covers essentially all public transport stations, where one sample point represents a demand point, and there are 50,549 demand points in total. The demand points distribution can be seen in Figure 2. Demand points will be taken as center points, and public transport stations covered in a walking range will be the basic unit to measure the public transport accessibility. The selection of demand point can be adjusted based on spatial scale, the shorter the distance between demand points is, the higher the calculation accuracy will be.

2.2.2. Public Transit Network. Urban public transit network is always changing and being adjusted, so it is difficult to get real-time and latest basic data on the network. Online maps such as Baidu, Google, and Amap provide rich geographic information, and their data updating is fast and coverage is extensive. The research takes Baidu map as the example to develop a urban public transit network capture system by using Baidu development interface and doing secondary development of ArcGIS Engine, which has successfully captured Shenzhen public transport and transit network data, which mainly include public transport station data (station name as well as longitude and latitude) and route data. This system can be applied to all domestic cities which are covered by network maps to obtain true and comprehensive public transit network data.

The captured Shenzhen public transit network contains 47,858 stations and 904 routes, and the number of routes publicized by Shenzhen transport and communications committee in January 2016 was 908, so the two numbers are almost the same. All data are integrated on the GIS platform for the convenience of later visualization and analytical calculation. For data of Shenzhen public transit network, please see Figure 3.

2.2.3. Walking Time. Traditional walking time calculation is usually straight-line distance divided by walking speed, which is less than actual time. Baidu map contains actual walking route network which can be used to obtain real walking time and distance, so walking time of passengers to the public transport station is calculated with Baidu map Direction API which is a bus, drive, and walking query and retrieval interface in the form of http and then returning retrieved data in format xml or json. The process can be seen in Figure 4. Baidu map interface can be used to obtain walking time to the station to the accuracy of second.

2.3. Model Building. Model calculation mainly considers two factors: ① the type of public transit stations in a certain walking range of the demand point. For example, large junction station, transfer station, or common stop station. ② The number and characteristics of public transit stations which can be reached from the demand point in certain walking time.

2.3.1. Calculation of Walking Range. In the ideal and accessible plane space, the walking range is determined by taking the initial point as the center and walking distance in certain time as the radius. In the actual road network, however, constraints caused by the road network and impedimental buildings are also main factors determining the walking range. The real walking time between demand point and various stations can be obtained based on network map, and the walking time to the points in other locations in the area can be calculated according to the walking time to the adjacent stations. Here we use Inverse Distance Weighted (IDW) method to realize spatial interpolation and rendering of walking space.

IDW calculates the value of sample point by taking the distance between insertion point and sample point as the weight; the smaller the distance between the sample point and insertion point is, the larger the impact on interpolating point will be; otherwise the impact will be smaller. When it is assumed that there are several metro stations ($D(x_i, y_i)(i = 1, 2, \cdots, m)$) available in an area, the accessibility of such

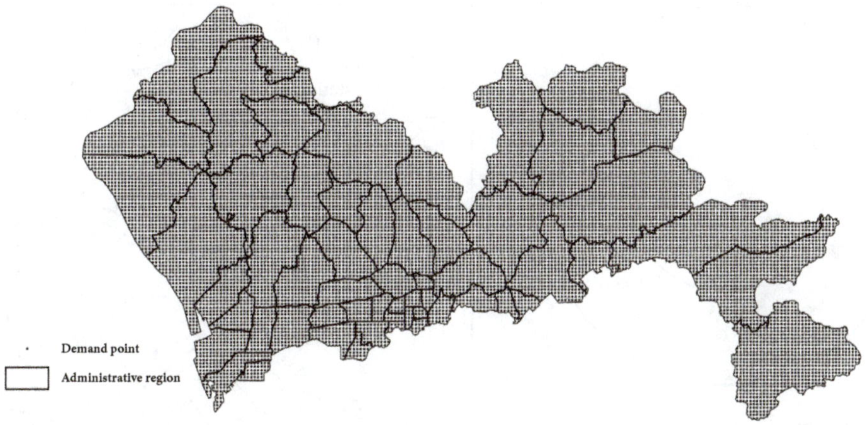

FIGURE 2: Demand points distribution.

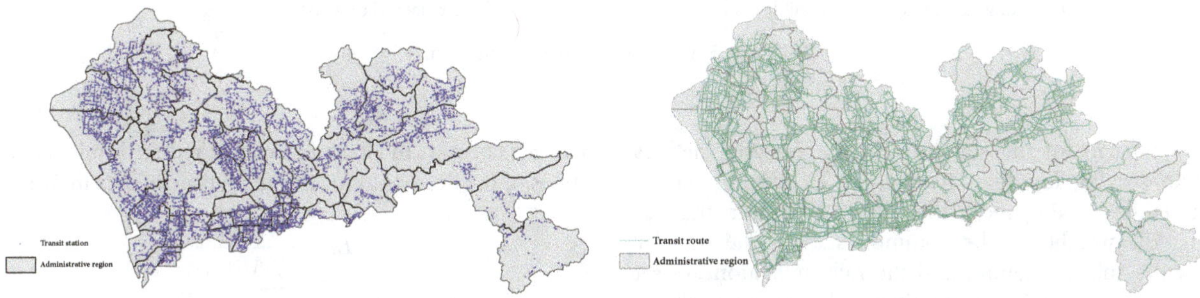

(a) Public transport station (b) Public transport route

FIGURE 3: Shenzhen public transportation network diagram.

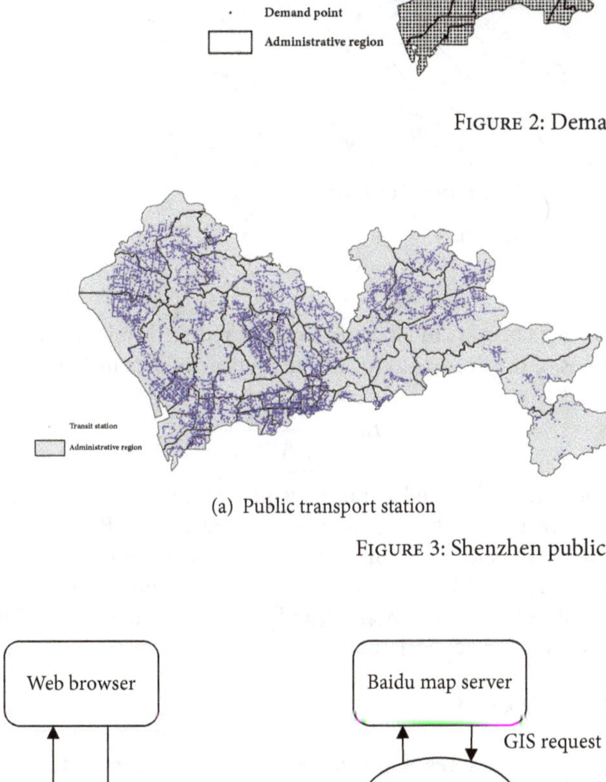

FIGURE 4: Process of direction API to obtain data.

stations will be A_i. If stations with equal A_i value are connected with each other, a smooth curve cannot be obtained since the number of sample points is insufficient. Smooth curve can be obtained with IDW method, and the equation is as follows.

$$A(Z) = \frac{\left(\sum_{i=1}^{m}\left(Z_i / [d_i(x,y)]^n\right)\right)}{\left(\sum_{i=1}^{m}\left(1 / [d_i(x,y)]^n\right)\right)} \quad (1)$$

where $d_i(x,y) = \sqrt{(x-x_i)^2 + (y-y_i)^2}$, $d_i(x,y)$ is the distance between insertion point $A(x,y)$ and station $D(x_i, y_i)$; m is the number of stations in the area; $A(Z)$ is the accessibility of insertion point; Z_i is the accessibility of station i; in the weight coefficient $w_i(x,y) = 1/[d_i(x,y)]^n$, n controls the degree of reduction of weight coefficient when the distance between insertion point and surrounding sample points is increased. Generally speaking, people will feel easy if they can arrive at the destination point or traffic interest point in a walking time of 10 minutes, and most people will be tired if the walking time exceeds 10 minutes [12]. Therefore, this paper decides 10 minutes is the maximum acceptable walking time. Stations covered by the demand point in different time limits can be obtained by using the spatial interpolation, as shown in Figure 5.

2.3.2. Calculation of Accessibility. Different types of stations are of different grades, can provide different service levels, and have different levels of attraction to passengers; the research indicates that station weight, walking time, and the number of lines are key factors affecting station accessibility, and the equation to calculate station accessibility is as follows:

$$W_i = \partial_m \times \frac{L_i}{T_w} \quad (2)$$

where W_i is the attraction of station i, ∂_m is the weight of class m station (common station, transfer station, subway station, and railway station), L_i is the total number of lines passing station i, and t_w is the walking time of individuals to public transport station. Station weight is determined by

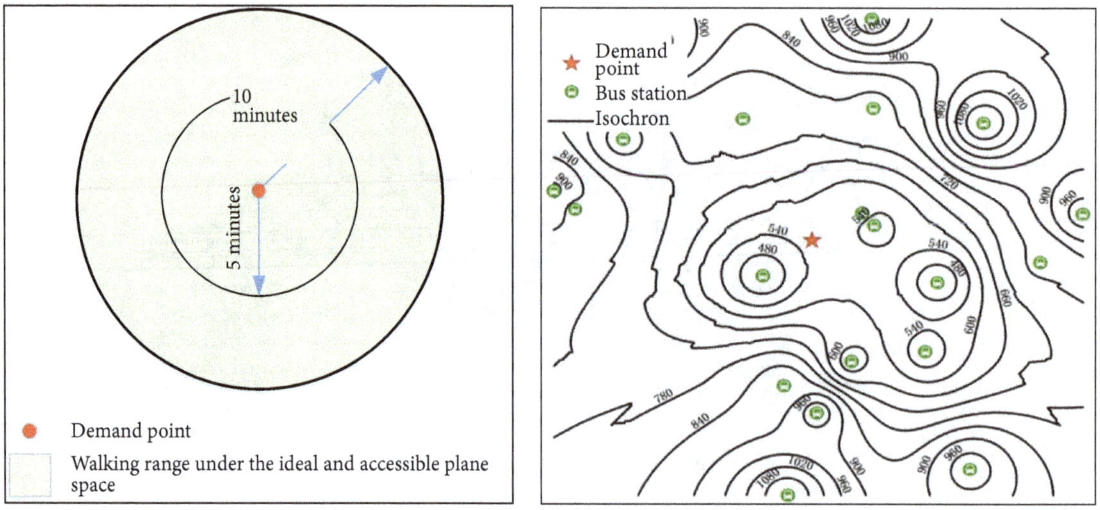

(a) Walking range under the accessible plane

(b) Actual walking distance isochron

Figure 5: Walking distance calculation.

transfer capacity of the station, and the research defines a station serving less than 10 lines as a common station and a station serving more than 10 stations as a transfer station. The weights of the common station, the transfer station, the subway station, and the railway station are set to 0.1, 0.2, 0.3, and 0.5, respectively. The equation shows that the station accessibility is in direct proportion to station weight and the number of public transport lines, and the accessibility is inversely proportional to walking time to the station.

Therefore, the accessibility of passengers from the demand points equals the sum of accessibility of all bus stops in the walking range:

$$A_i = W_i \times R(o,d)_j = \sum_{m}^{k} \sum_{j}^{n} \partial_m \times L_i \times \frac{R(o,d)_j}{T_w} \quad (3)$$

where $R(o,d)_j$ can be used to judge whether station j is in the walking range of demand point i, and only the bus stations in the walking range of demand point need to be considered if the walks start from the demand point.

2.3.3. Calculation of Public Transport-Private Cars Travelling Time Ratio. The public transport-private cars travelling time ratio equals the ratio of public transport travelling time and private cars travelling time from the initial point A and end point B, which is an important indicator to reflect the accessibility competitiveness of public transport compared to private cars. Success of public transportation in international cities shows that public transport can be competitive against cars if the total travelling time of public transport is limited to 1.5 times of that of cars. The overall Singapore land traffic planning indicates that the door-to-door travelling time of public transport in Singapore will be reduced to less than 1.5 times of the door-to-door travelling time of cars in 2020 [35].

$$E_i = \frac{T(AB)_{PTi}}{T(AB)_{cari}} \quad (4)$$

where E_i is the public transport-private cars travelling time ratio from initial point A to end point B, $T(AB)_{PTi}$ is the public transport travelling time from initial point A to end point B, $T(AB)_{cari}$ is the private cars travelling time from initial point A to end point B, and this paper obtains the index based on the network map.

3. Case Study

Calculate the accessibility value from different demand points to bus stations under certain time-space limits with accessibility calculation model by taking Shenzhen as the example when above statistics processing is finished, and obtain the spatial distribution of public transport accessibility by using GIS analysis tools. The public transport accessibility, public transport, and private car travel time of this article refers to off-peak.

3.1. Evaluation on Public Transport Facilities Resource Supply Level of the City. The 500m coverage rate of Shenzhen public transport has reached 97%, but Shenzhen's public transport accessibility renderings (see Figure 6) calculated with the method proposed in this paper show that there are obvious regional differences in the distribution of Shenzhen's public transport facilities resource supply, which has the following three features: firstly, the overall Shenzhen public transport accessibility structure is highly identical to the railway line network and public transport corridor form; secondly, the public transport accessibility inside the original special zone is obviously superior to that outside the special zone, Luohu District, and Futian railway network in the original special zone is extensive, and its public transport accessibility is

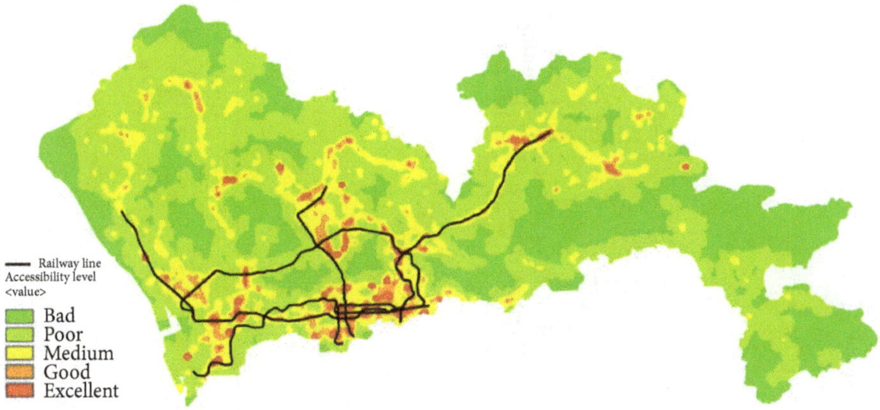

FIGURE 6: Shenzhen public transport accessibility evaluation.

TABLE 1: The correspondence between accessibility index and accessibility level.

Accessibility Level	Accessibility Index	Color
Excellent	>25.01	
Good	15.01-25.00	
Medium	10.01-15.00	
Poor	5.01-10.00	
Bad	0.01-5.00	

higher than that of the Nanshan District; areas with high public transport accessibility outside the original special zone are along railway and some public transport corridors, and the public transport accessibility of most areas is low; thirdly, local areas with high accessibility are near railway and public transport transfer stations as well as traditional commercial area and job centers, etc. The former includes Futian station, Shenzhen North Railway Station, and Buji Railway Station, and the latter includes Luohu Golden Triangle and the convention and exhibition center. The correspondence between accessibility index and accessibility level can be seen in Table 1.

3.2. Evaluation on the Public Transport Accessibility of Key Areas. In terms of accessibility, the average public transport travelling time from key areas such as Luohu District, Huaqiang North Commercial Area, Futian CBD, Che Kung Temple, five job centers in the science park, Longcheng Square in Longgang District, Longhua Bantian, the center of Pingshan District and the center of Guangming District to other places of the city is 118-183 minutes, and the average private cars travelling time is 58-69 minutes, so the overall accessibility of private cars is higher than that of public transport. Shenzhen key areas public transport accessibility can be seen in Figure 7.

In terms of public transport-private cars travelling time ratio, the average ratio from key areas such as Luohu District, Huaqiang North Commercial Area, Futian CBD, Che Kung Temple, five job centers in the science park, Longcheng Square in Longgang District, Longhua Bantian, the center of Pingshan District, and the center of Guangming District to other places of the city is 1.79-3.09, which is above 1.5, so the public transport service competitiveness should be improved further. For areas with a public transport-private cars travelling time ratio of more than 1.5, Luohu District, Huaqiang North Commercial Area and Futian CBD are larger than other areas, and their public transport competitiveness in the 1-hour public transport coverage area is also higher than that of other areas; in addition, the public transport service level of areas outside the original special zone, especially Guangming District and Pingshan District, shall be improved greatly. Shenzhen key areas public transport-private cars travelling time ratio can be seen in Figure 8.

Above accessibility research on key areas of the city can be used to analyze public transport travelling time, private cars travelling time, and the public transport-private cars travelling time ratio, which is the basis to assess public transport facilities resource supply level and select proper travel mode. Shenzhen key areas public transport accessibility analysis can be seen in Table 2.

3.3. Further Discussion of Application. The accessibility calculation method proposed in this paper can directly reveal the spatial distribution characteristics of public transport facilities resource supply and can be used to

Table 2: Shenzhen key areas public transport accessibility analysis.

Calculation Index	Area								
	Luohu Commercial Center	Huaqiang North Commercial Area in Futian District	Futian CBD	Futian Che Kung Temple	Nanshan Science Park	Huawei in Longhua Bantian	Longcheng Square in Longgang District	The Center of Pingshan District	The Center of Guangming District
Average public transport travelling time (min)	127	121	117	118	134	143	129	169	183
Average private cars travelling time (min)	59	69	60	66	67	61	58	69	63
Average public transport-private cars travelling time ratio	2.18	1.79	1.97	1.92	2.11	2.42	2.39	2.77	3.09

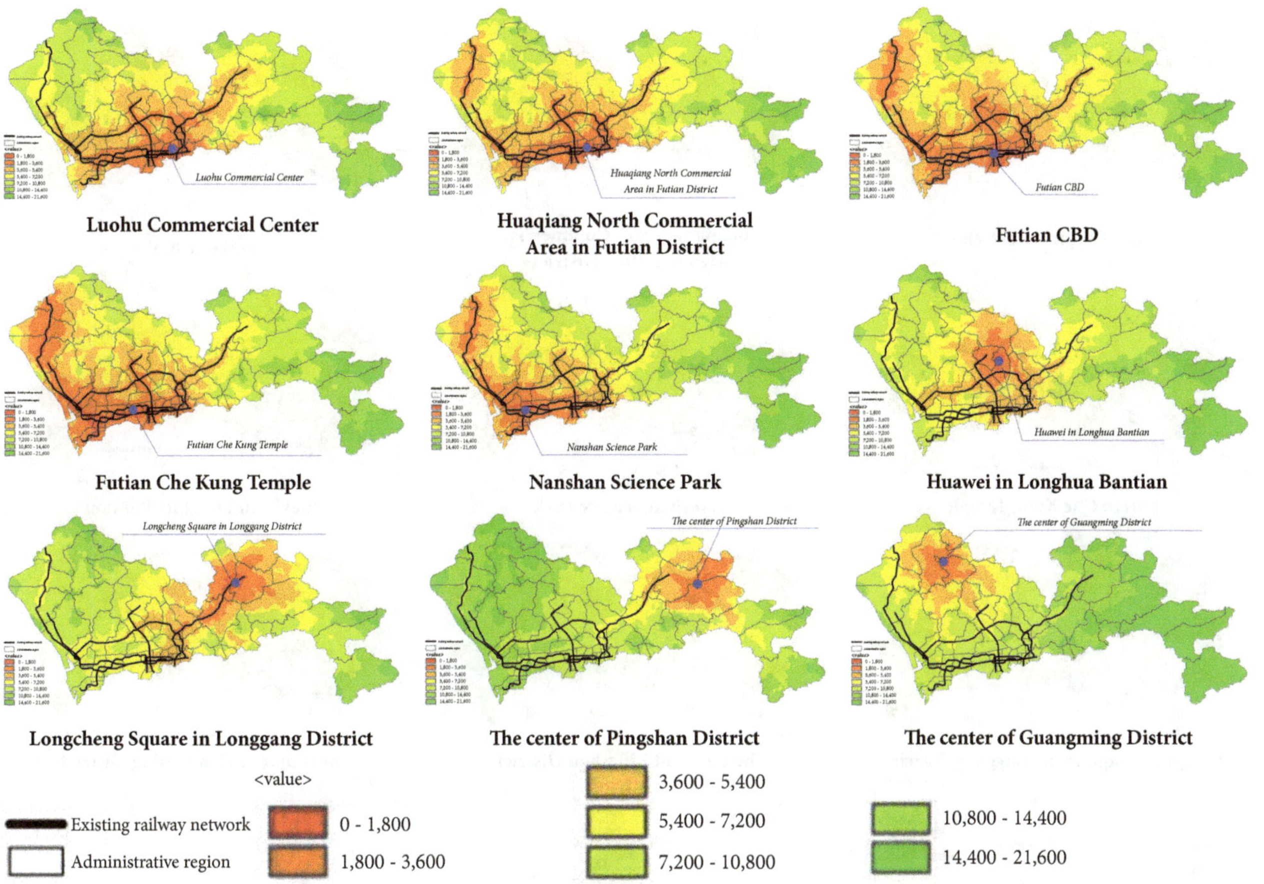

FIGURE 7: Shenzhen key areas public transport accessibility (unit: seconds).

effectively analyze the arrangement of spatial elements and preparation of spatial development policy. The method is highly useful in the preparation of urban public transport policy and traffic demand management of key areas.

3.3.1. Preparation of Public Transport Development Policy of the City Based on Population Density. Public transport resource distribution reflects the difference of regional development opportunity to some extent; the public transport accessibility evaluation method proposed by this paper is combined with population density and structure distribution, which can provide decision-making support for balanced development of public transport. Figure 9(a) shows that Shenzhen population is evenly distributed; both central urban area and peripheral streets such as Fuyong, Buji, Longcheng, and Bantian have high population density. Figure 6 shows that public transport resource in peripheral areas is not evenly distributed; the public transport accessibility level of some highly populated areas is low, which shall draw more attention. Of course, more attention shall be paid to the difference of population structure and public transport demand in addition to population density. From Figure 9(b) we can see that large number of people with junior high school degree and lower degree (most of them are peasant-workers) are distributed in areas outside the Airport-He'ao expressway outside the original special zone; the public transport demand of such areas is greatly different from that of areas inside the Airport-He'ao expressway, and a differentiated public transport development policy shall be made.

3.3.2. Preparation of Demand Management Policy Based on Traffic Operation State. The traffic demand management policy can be made based on public transport accessibility level of the area and the actual traffic operation state. According to the 2015 Shenzhen frequently congested road diagram, roads such as Luohu District Binhe Road, Shennan East Road, and Aiguo Road are heavily congested (see Figure 10). Shenzhen public transport accessibility calculation result shows that the public transport accessibility level of Luohu District, especially the Luohu Golden Triangle Central Business District, is very high. It is possible to carry out congestion charging and increase parking charge in some areas of Luohu District to ease traffic jam and improve traffic operation state. But in the science park with severe traffic jam and livelihood issues, the substitutability of public transport shall be evaluated before traffic demand management policy is prepared, in order to prevent deceasing the overall residents benefit level, because current public transport accessibility level of this area is low.

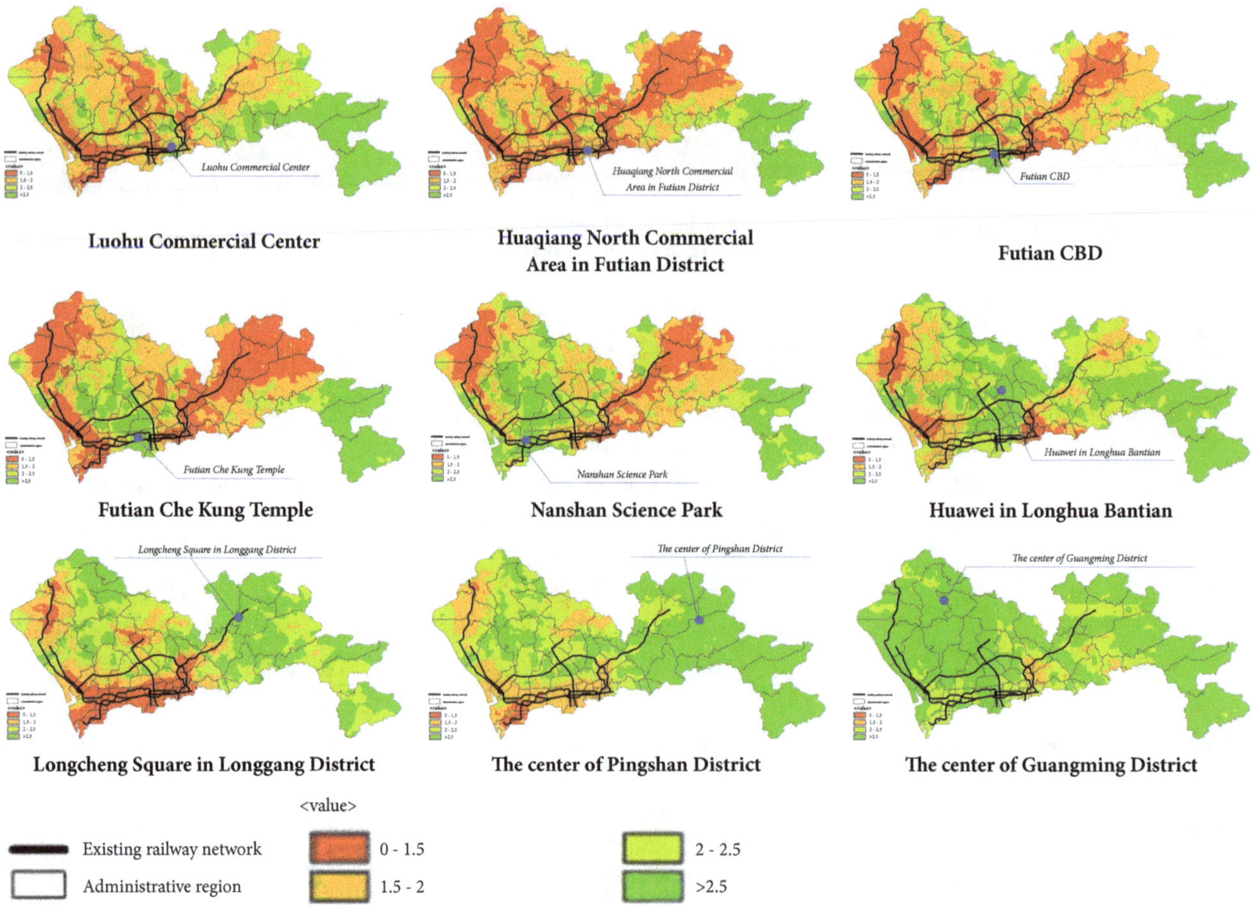

FIGURE 8: Shenzhen key areas public transport-private cars travelling time ratio.

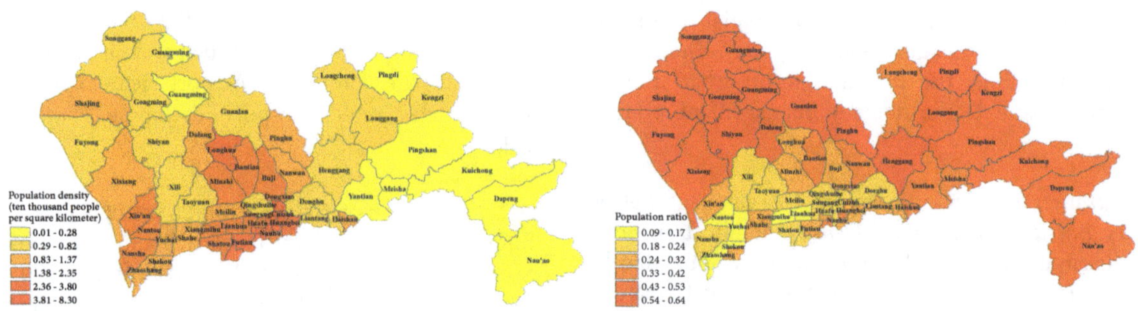

(a) Shenzhen Population Density Distribution Diagram

(b) Proportion of people with junior high school degree and lower degree

FIGURE 9: Shenzhen population distribution.

4. Conclusion

This paper considers walking time of passengers to stations as well as other factors such as property, class, and spatial distribution characteristics of station and proposes an urban public transport resource supply evaluation method based on the accessibility. Equity is an important principle of the sustainable urban traffic system. The method proposed in this paper essentially reflects the distribution of urban public transport facilities resource supply, evaluates the equity of urban public transport facilities resource distribution based on the index of accessibility, calculates the public transport-private cars relative service competitiveness based on the index of public transport-private cars travelling time ratio, and provides conditions for the public transport accessibility evaluation of key areas. The proposed method is applied by taking Shenzhen as the example. The result shows that the method can truly reflect the spatial distribution characteristics of Shenzhen public transport facilities resource and can provide basis to assess travel mode in main areas, prepare

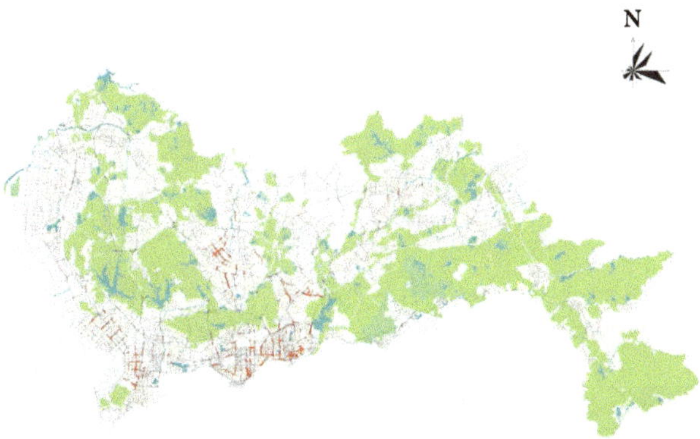

FIGURE 10: 2015 Shenzhen recurrent congested road map.

public transport development policy for the whole city, and make demand management policy for key areas. Further research shall consider more factors which may impact the accessibility, such as the departure frequency of the line and setting of station weight parameters, to establish a better calculation model. In addition, this paper only applies the method to the city and districts; subsequent research shall apply the method to more fields and further verify reliability and practicability of the method.

Conflicts of Interest

The authors declare that they have no conflicts of interest.

Acknowledgments

This work is funded by Natural Science Foundation of China, no. 71641005, Project name: Urban Transportation Theory System Research Focusing on the Organization and Operation of Urban Transportation Network.

References

[1] J. Chen, F. Lu, and X. C. Cheng, "Advance in Accessibility Evaluation Approaches and Applications," *Process in Geography*, vol. 26, no. 5, pp. 100–110, 2007.

[2] H. P. Li and L. Y. Lu, "Review and Prospection of Accessibility Research," *Process in Geography*, vol. 24, no. 3, pp. 69–78, 2005.

[3] X. T. Liu, "General Description of Spatial Accessibility," *Urban Transport of China*, vol. 5, no. 6, pp. 36–43, 2007.

[4] M. X. Zhang, *Research of Public Transport Accessibility on GIS-A Case Study of Luzhou. Southwest Jiaotong University*, Research of Public Transport Accessibility on GIS-A Case Study of Luzhou, Ed., Southwest Jiaotong University, 2011.

[5] W. G. Hansen, "How accessibility shapes land use," *Journal of the American Institute of Planners*, vol. 25, no. 2, pp. 73–76, 1959.

[6] W. Nie and C. F. Shao, "Research on Accessibility Calculating Method of Regional Transportation," *Technology & Economy in Areas of Communications*, vol. 10, no. 4, pp. 85–87, 2008.

[7] A. Páez, D. M. Scott, and C. Morency, "Measuring accessibility: Positive and normative implementations of various accessibility indicators," *Journal of Transport Geography*, vol. 25, pp. 141–153, 2012.

[8] T. S. Shen, *Spatial-Temporal Evolution of Urban Public Transportation Accessibility-A Case Study in Nanjing Metropolitan Area*, Southeast University, 2014.

[9] S. X. Cao, Y. X. Huang, and Z. Dong, "Transit Accessibility and Characteristics of Residents Transportation Based on GIS," *Journal of South China Normal University (Natural Science Edition)*, vol. 45, no. 5, pp. 98–105, 2013.

[10] Y. Y. Chen, Y. P. Wei, and H. J. Lai, "Regional Public Transportation Accessibility Calculation Method Based on GIS," *Journal of Transportation Systems Engineering and Information Technology*, vol. 2015, no. 2, pp. 61–67, 2016.

[11] O. Manout, P. Bonnel, and L. Bouzouina, "Transit accessibility: A new definition of transit connectors," *Transportation Research Part A: Policy and Practice*, vol. 113, pp. 88–100, 2018.

[12] D. M. Li, "Study on Walking Accessibility and subway station service range," in *Proceedings of the Urban Planning Under Ecological Civilization Perspective: 2008 China Urban Planning Symposium*, Dalian Press, 2008.

[13] Shenzhen Urban Planning and Land Development Research Centre. Shenzhen Resident Trip Characteristics 2014.

[14] M. Salonen and T. Toivonen, "Modelling travel time in urban networks: Comparable measures for private car and public transport," *Journal of Transport Geography*, vol. 31, pp. 143–153, 2013.

[15] I. Benenson, K. Martens, Y. Rofé, and A. Kwartler, "Public transport versus private car GIS-based estimation of accessibility applied to the Tel Aviv metropolitan area," *Annals of Regional Science*, vol. 47, no. 3, pp. 499–515, 2011.

[16] T. L. Lei and R. L. Church, "Mapping transit-based access: Integrating GIS, routes and schedules," *International Journal of Geographical Information Science*, vol. 24, no. 2, pp. 283–304, 2010.

[17] S. Liu and X. Zhu, "Accessibility analyst: an integrated GIS tool for accessibility analysis in urban transportation planning," *Environment and Planning B: Planning and Design*, vol. 31, no. 1, pp. 105–124, 2004.

[18] W. Chen, "Methodological study on bus station transfer system evaluation based on accessibility of all modes of transportation," *Jiangsu Urban Planning*, vol. 2014, no. 12, pp. 34–37, 2014.

[19] S. Hou and H. Jiang, "An analysis on accessibility of hospitals in Changchun based on urban public transportation," *Geographical Research*, vol. 33, no. 5, pp. 915–925, 2014.

[20] Y. Chen and Y. Li, "Index of public transport accessibility based on GIS," *Geomatics & Spatial Information Technology*, vol. 34, no. 2, pp. 177–179, 2011.

[21] B. M. Wu and J. P. Hine, "A PTAL approach to measuring changes in bus service accessibility," *Transport Policy*, vol. 10, no. 4, pp. 307–320, 2003.

[22] The City of New York, "One New York: The Plan for a Strong and Just City," 2015, http://www.nyc.gov/html/onenyc/downloads/pdf/publications/OneNYC.pdf.

[23] P. Prioni and D. Hensher, "Measuring Service Quality in Scheduled Bus Services," *Journal of Public Transportation*, vol. 3, no. 2, pp. 51–74, 2001.

[24] D. A. Hensher, P. Stopher, and P. Bullock, "Service quality—developing a service quality index in the provision of commercial bus contracts," *Transportation Research Part A: Policy and Practice*, vol. 37, no. 6, pp. 499–517, 2003.

[25] G. Beirão and J. A. Sarsfield Cabral, "Understanding attitudes towards public transport and private car: A qualitative study," *Transport Policy*, vol. 14, no. 6, pp. 478–489, 2007.

[26] R. Woods and J. Masthoff, "A comparison of car driving, public transport and cycling experiences in three European cities," *Transportation Research Part A: Policy and Practice*, vol. 103, pp. 211–222, 2017.

[27] T. Liu, A. A. Ceder, and R. Bologna, "Commuting by customized bus: a comparative analysis with private car and conventional public transport in two cities," *Journal of Public Transportation*, vol. 19, no. 2, p. 4, 2016.

[28] C. M. Burns and A. D. Inglis, "Measuring food access in Melbourne: Access to healthy and fast foods by car, bus and foot in an urban municipality in Melbourne," *Health & Place*, vol. 13, no. 4, pp. 877–885, 2007.

[29] S. Hou Y, *An analysis on accessibility of urban parks based on urban public transportation and private car–A case study of*, Nanjing University, Changchun, China, 2011.

[30] Y. Zhang T, *Study of traffic accessibility impact in applying trip generation rate method*, Hebei University of Technology, 2013.

[31] T. Yang, *Research on transportation equity in GuangzhouCompare between public and private transportation*, GuangZhou University, 2012.

[32] A. S. Huang, F. W. Li, and Y. D. Yang, "Transport Accessibility of Community Health Service Based on the Point-of-Interests (POIs) Data: A Case Study of Shanghai," *China transportation review*, vol. 39, no. 5, pp. 80–84, 2017.

[33] C. Duan, *Research on the accessibility of car travel in Beijing*, Beijing Jiaotong University, Beijing, China, 2017.

[34] F. L. Deng and H. S. Tan, "Comparative study on accessibility of urban central areas based on traffic isochron–Taking 15 key cities as examples," in *Planning for 60 Years: Achievements and Challenges - Proceedings of 2016 China Urban Planning Annual Conference*, Building Industry Press, 2016.

[35] Singapore's Land Transport Authority, *Singapore's Land Transport MasterPlan*, Singapore's Land Transport Authority, Singapore, 2013.

[36] G. Hitge and M. Vanderschuren, "Comparison of travel time between private car and public transport in Cape Town," *Journal of the South African Institution of Civil Engineering*, vol. 57, no. 3, pp. 35–43, 2015.

4

Traffic State Estimation using Connected Vehicles and Stationary Detectors

Ellen F. Grumert[1,2] **and Andreas Tapani**[1]

[1] *Swedish National Road and Transport Research Institute (VTI), 581 95 Linköping, Sweden*
[2] *Department of Science and Technology, Linköping University, 601 74 Norrköping, Sweden*

Correspondence should be addressed to Ellen F. Grumert; ellen.grumert@vti.se

Academic Editor: Fernando García

Real-time traffic state estimation is of importance for efficient traffic management. This is especially the case for traffic management systems that require fast detection of changes in the traffic conditions in order to apply an effective control measure. In this paper, we propose a method for estimating the traffic state and speed and density, by using connected vehicles combined with stationary detectors. The aim is to allow fast and accurate estimation of changes in the traffic conditions. The proposed method does only require information about the speed and the position of connected vehicles and can make use of sparsely located stationary detectors to limit the dependence on the infrastructure equipment. An evaluation of the proposed method is carried out by microscopic traffic simulation. The traffic state estimated using the proposed method is compared to the true simulated traffic state. Further, the density estimates are compared to density estimates from one detector-based method, one combined method, and one connected-vehicle-based method. The results of the study show that the proposed method is a promising alternative for estimating the traffic state in traffic management applications.

1. Introduction

Density, speed, and flow are important measures for describing the characteristics of the traffic on a road segment. The density of traffic is defined as the number of vehicles located on a road segment. The speed can be defined either as the time mean speed, which is the mean speed of all vehicles passing a specific location within a given time interval, or as space mean speed, which is the mean speed of all vehicles travelling over a road segment at a certain point in time. Finally, the flow is defined as the number of vehicles passing a specific location within a given time interval. The three measures are commonly referred to as the traffic state. The traffic state is of special interest for traffic management systems based on automatic control. Examples of such systems are variable speed limit systems and ramp metering. The purpose of the traffic management system is to improve the traffic conditions during, for example, congested periods and incidents. The systems should be able to detect abrupt changes in the traffic state, such as lower speeds and flows and higher densities, and use this information as input to apply a suitable control strategy. Thus, representative real-time estimation of the traffic state is of importance.

The traditional way to measure and estimate the traffic state is by the use of stationary equipment, for example, loop detectors and radars. Due to recent development in vehicle technology, different types of connected vehicles are being introduced and the expectation is that in 2020 75% of newly produced vehicles will be equipped with technology that enables the possibility to connect to the surroundings [1]. The connected vehicles facilitate communication between vehicles and between vehicles and the infrastructure. This allows for frequent updates of individual vehicle measures such as their speed and position. Hence, it is possible to use connected vehicles in combination with stationary detectors, or as a standalone data source, to estimate the traffic state. This can also result in improved spatial estimates instead of the traditionally used point estimates.

In this study, we propose a method for estimating the traffic state based on vehicle-to-infrastructure communication. The required information is speed and positioning measurements from connected vehicles in combination with counts from stationary detectors. By assuming that the connected vehicles have the same distribution of speed as regular vehicles, the speed is estimated as an average of the speeds of the connected vehicles. The only connected vehicle data needed to estimate the density is information about the current road segment of the connected vehicles. This makes the method robust with respect to errors in the positioning data. Each connected vehicle continuously communicates its location, which is used to estimate the total number of connected vehicles on a specific road segment. Further, the number of connected vehicles passing stationary detectors is, together with the total number of passing vehicles, used to estimate the penetration rate of connected vehicles. Thereby, the total number of vehicles located on a segment can be estimated.

Our hypotheses are that the proposed method will result in (1) density and speed estimates that can capture the current traffic conditions on the road, (2) a possibility to use more sparsely placed stationary detectors without considerably reducing the performance of the density estimation, given that the share of connected vehicles is assumed to be approximately the same over the road stretch, and (3) precise and fast detection of changes in the traffic state, especially for higher penetration rates of connected vehicles. In real applications, the traffic state estimation is one component of a larger model system, often including both data assimilation and fusion techniques. The aim of this paper is to find a straightforward approach to estimate the density and speed by the use of connected vehicles and investigate how well the traffic state estimation can capture the actual traffic situation as a first step towards using the proposed method as such a component in traffic management applications.

To study these hypotheses, the proposed method is evaluated by the use of microscopic traffic simulation. The density estimates are compared to one detector-based method, one combined method, and one connected-vehicle-based method. The comparisons are done in order to investigate if the proposed method gives estimates that are comparable to estimates of existing methods. A simulation scenario with an incident is analyzed to study how well the proposed method can capture abrupt changes in the traffic state. Two different distances between the detectors are applied to examine how the method performs with sparsely placed detectors. To isolate the effects related to the method, we use a simple design of the road network and assume that no measurement errors exist in detector and connected vehicle data.

The remainder of the paper is organized as follows. In Section 2, an overview of traffic state estimation methods is given with focus on density estimation. The proposed method for estimating the traffic state based on connected vehicles in combination with stationary detectors is presented in Section 3. The simulation setup and the evaluation method are described in Section 4, including an overview of the methods used for comparison. In Section 5, the performance of the proposed method is presented and compared to other methods. Finally, conclusions from the study and directions for further research are discussed in Section 6.

2. Traffic State Estimation Using Connected Vehicles

The traditional way to estimate the traffic state is by the use of stationary detectors such as loop detectors and radar detectors, as described by Kurkjian et al. [2], Coifman [3], and Singh and Li [4]. This is limiting the estimation to specific points in space, and the conditions in between detectors remain unknown. Hence, the estimation will be a good representation of the traffic state on the road section only under steady-state conditions, that is, when there is no change in the traffic conditions in space and time. This is usually not the case, particularly not for bottlenecks or during incidents, and therefore the density estimated with these methods will most probably deviate from the true density on the section. Hence, to give enough information about the traffic conditions on a longer road section, the detector-based method does require densely placed detectors (see, e.g., the method proposed by Singh and Li [4]). Data assimilation and fusion techniques including a traffic model are common methods to get a picture of the traffic state also in between the detectors. A number of studies using different underlying traffic models and different filtering approaches exist in the literature (see, e.g., Kurkjian et al. [2], Muñoz et al. [5], Wang and Papageorgiou [6], Mihaylova et al. [7], Singh and Li [4], and Duret et al. [8]). Methods have also been proposed, where no underlying model and no filtering are needed. See, for example, Coifman [3], where reidentification of vehicles and the vehicle conservation law is used to estimate the density between two detectors. Darwish and Bakar [9] conclude that methods using different types of stationary detectors can estimate the traffic state accurately but they are often expensive to install and maintain and are limited to small areas. Also, the information is usually transmitted with delay, since it has to be processed through a traffic information center.

Lately, when more data sources have become available, connected vehicles have been used as input to the filtering approaches in order to update the modeled traffic state. See, for example, Herrera and Bayen [10], Work et al. [11], Yuan et al. [12], Seo et al. [13], Astarita et al. [14], and Bekiaris-Liberis et al. [15]. Other traffic state estimation methods making use of connected vehicle data without an underlying traffic model are presented by Herring et al. [16], Herrera et al. [17], Van Lint and Hoogendoorn [18], Qiu et al. [19], Ma et al. [20], Bhaskar et al. [21], Zhang et al. [22], Seo et al. [23], and Montero et al. [24].

When speed measurements from connected vehicles are available, the speed can be estimated by calculating an average of the speeds of the connected vehicles. This requires that the connected vehicles have the same distribution of speeds as regular vehicles, similar to what has been done in the works of Astarita et al. [14] and Bekiaris-Liberis et al. [15]. Otherwise, the speed estimate would be biased towards the average speed of the connected vehicles.

The density estimate using connected vehicles requires some more calculations. One way of estimating the density is by using connected vehicles together with traditional stationary detectors, here referred to as combined methods. For the combined methods, a weighted estimate based on both traditional detector measurements and connected vehicle measurements of, for example, speed, travel time, and/or location is used. Examples are the methods presented by Astarita et al. [14], Qiu et al. [19], Ma et al. [20], Bhaskar et al. [21], Zhang et al. [22], and Bekiaris-Liberis et al. [15]. The method by Qiu et al. [19], which was later extended by Ma et al. [20], detects the number of vehicles located within a segment. The density estimate is calculated by counting the number of vehicles that have passed the detector upstream of the segment at the times when a connected vehicle enters and exits a segment. According to Qiu et al. [19], the accuracy of the density estimates is better than when only stationary detector data is used to estimate density. Zhang et al. [22] use probe vehicle data and detector stations in order to estimate the space mean speed, and not density, on a road stretch. Astarita et al. [14] and Bekiaris-Liberis et al. [15] develop macroscopic cell transmission type models for the dynamics of the percentage of connected vehicles along the considered road. It is assumed that the connected vehicles move with the same average speed as the nonconnected vehicles, and hence no modeling of the speed dynamics is needed. In the work of Astarita et al. [14], the density is estimated by the percentage of connected vehicles based on counts of connected vehicles moving from one segment to another in the network and inflows measured at the ramps and at the boundaries of the network. Similarly, Bekiaris-Liberis et al. [15] use the penetration rate of connected vehicles, together with measurements of speed from the connected vehicles, the boundary flow, and the ramp flows measured through stationary detectors, to estimate the density. Also the combined methods do often require densely placed detectors to get good estimates. The methods are often based on retrospective measurements, such as travel time at an earlier point in time (see, e.g., Qiu et al. [19] and Ma et al. [20]). As a result, the density estimate might not reflect the current situation.

For the connected-vehicle-based methods, the connected vehicles are used to capture the surrounding traffic conditions at every point in space, and the methods are therefore not limited to fixed locations. Recent studies by Seo et al. [13, 23] investigate how the gap to a leading vehicle can be used to estimate the density on a road segment. The same method has been applied to an urban area in Montero et al. [24]. Further, the method is extended to also include measurements of the gap to the following vehicle. Another method is making use of vehicle spacings and speed as input data for estimating density [12]. This method requires a numerical model of the relationship between the traffic states to describe the changes in speed, flow, and density. Finally, Seo and Kusakabe [25] propose a method based on the number of vehicles located in between two connected vehicles to estimate the traffic conditions on the road. For methods using only connected vehicle data, the measurements used to estimate density are local, only including the connected vehicle and its surroundings, which might not necessarily reflect the density on a larger section of the road. Also, the methods often require identification of current lane, speed of the vehicle, distance to vehicle in front, and so forth.

To conclude, detector-based density estimation techniques make use of measurements from stationary detectors, usually consisting of flow and speed, to estimate density. However, the density estimates using stationary detectors are based on point estimates and are therefore limited due to the fact that the conditions in between detectors are not known. Therefore, the methods require densely spaced detectors to give density estimates that correspond well with the traffic conditions on the road. The detector-based method can be improved by including connected vehicle data. For combined methods, the need for continuous updates from connected vehicles can be limited; that is, it is enough with low frequency data communicated from the connected vehicles. However, the combined methods do usually still require densely spaced detectors and the information is sometimes based on retrospective connected vehicle measurements. The density estimates using only connected vehicle data are based on local density estimates including precise estimates of the density surrounding the connected vehicle. The connected vehicle density estimates are transmitted for further processing and converted to a density estimate of a larger area by including estimates from many connected vehicles. Hence, a representative density estimate can often only be reached with a high connected vehicle penetration rate or for a high flow level. Further, the connected vehicles are assumed to be able to continuously transmit information about their location, speed, gap to proceeding vehicle, and so forth.

3. A Method for Estimating the Traffic State by Using Connected Vehicle and Detector Data

We propose a combined method for estimating the traffic state on the road. The connected vehicle data is based on vehicle-to-infrastructure communication. The method is straightforward and it is possible to estimate the speed and density accurately based on limited information from the connected vehicles. Measurements from sparsely placed stationary detectors can be used without reducing the performance of the estimates. The purpose of the method is to get fast and representative traffic state estimates that can also be used to identify changes in the traffic conditions. Before introducing the method, a few essential assumptions are given:

(i) The connected vehicles are able to report their position, including information about their current road segment and speed, with a frequency of 1 Hz.

(ii) The stationary detectors are able to count and report the total number of vehicles, N, and the total number of connected vehicles, M, passing the detector within a given aggregation time period, T.

(iii) The connected vehicles are assumed to have the same distribution of speeds as the nonconnected vehicles.

Hence, neither the equipment used for communication of information for the connected vehicle nor the type of stationary detector (radar, Bluetooth, loop, etc.) is defined and

TABLE 1: Required measurements for estimating density based on CCV method, SD method, CC method, and GAP method.

Method	Connected vehicle data			Stationary detector data		
	Time at detector	Position	Measurement time stamp	Count of vehicles	Count of connected vehicles	Speed
CCV		X		X	X	
SD				X		X
CC	X			X		
GAP		X	X			

may vary as long as they fulfil the requirements presented above.

The traffic state estimation consists of two parts, a speed estimate and a density estimate for each segment on the considered road stretch. The speed estimates are based on simple calculations. It is assumed that the connected vehicles have the same distribution of speeds as the nonconnected vehicles. By communication of the individual speed of each connected vehicle i located on segment k at time t, the average speed of connected vehicles, $v_k^i(t)$, can be calculated. Then, the average speeds at time t are averaged over the aggregation time period, T, to get the final speed estimate,

$$\widehat{v}_k(T) = \frac{1}{T}\sum_{t=1}^{T}\frac{1}{n_k}\sum_{i=1}^{n_k} v_k^i(t), \tag{1}$$

where n_k is the total number of connected vehicles on road segment k.

The density estimates make use of the position of each connected vehicle to get the total number of connected vehicles at each road segment, k, and for each time step, t. The average number of connected vehicles, $\overline{PV}_k(T)$, in segment k and for the aggregation time period, T, is used to estimate the total density. The stationary detector data is used to estimate the penetration rate on segment k as the number of connected vehicles, M_k, divided by the total number of vehicles, N_k. The penetration rate at the detector station located just upstream of segment k is used as input for the density estimate at segment k. The density estimate becomes

$$\widehat{\rho}_k(T, X_k) = \frac{\overline{PV}_k(T)}{X_k}\frac{1}{(M_k/N_k)}. \tag{2}$$

The new speed and density estimates are becoming available after the latest aggregation time period T, based on the measurements within the same aggregation time period and, hence, the estimates are varying with time. The temporal indices in the density estimates have been suppressed to increase readability. The method is hereafter referred to as the Count Connected Vehicle (CCV) method. By assuming that the penetration rate is constant over a longer road section, detectors can be sparsely placed in order to have as little requirements on detectors as possible. In this case, each estimate of the penetration rate is applied to many segments before a new estimate of the penetration rate becomes available.

The information from the connected vehicles and the detectors is collected at each time step and communicated to a central unit, where it is being processed, resulting in time-dependent speed and density estimates. Finally, the estimates are aggregated over the aggregation time period T. Figure 1 gives an illustration of the process. The local units are the individual connected vehicles and the detectors. The central unit can, for example, be a roadside unit or a traffic management center used for further processing of data.

4. Evaluation Method

The analysis is divided into three parts. First, CCV is evaluated with respect to the aggregation time period, T, where the aggregation time period is defined as the time interval over which the traffic state is estimated. The performance of the traffic state estimation is examined for four different time periods. This will indicate for which aggregation time periods the method gives traffic state estimates that are representative for the traffic conditions on the road. Both density and speed estimates are considered. Second, the density estimates of the CCV are compared to similar methods: one method using only stationary detector data, one combined method, and one method using only connected vehicle data. The comparison is used to evaluate if the density estimates of the proposed method are comparable to those of detector-based, combined, and connected-vehicle-based methods found in the literature. Finally, it is investigated how well CCV manages to detect changes in the traffic states at different distances between detectors. Two cases are included with detectors placed 500 and 2500 meters away from each other. The results are compared to another combined method. This section describes the methods used for comparison, the simulation setup in a microscopic traffic simulation environment, including the choice of parameters, and the performance indicators used for evaluation.

4.1. Comparison of the Density Estimates. The density estimates are somewhat more complex to calculate compared to the speed estimates. Hence, methods with the same level of complexity are chosen for comparison. However, the methods require different types of data to estimate the density and include one stationary-detector-based method, SD, one combined method, CC, [19, 20], and one connected-vehicle-based method, GAP [23]. The required measurements for the different density estimation methods are summarized in Table 1. The measurements are divided into stationary-detector-based measurements and connected-vehicle-based measurements. The methods used for comparison with CCV are described below.

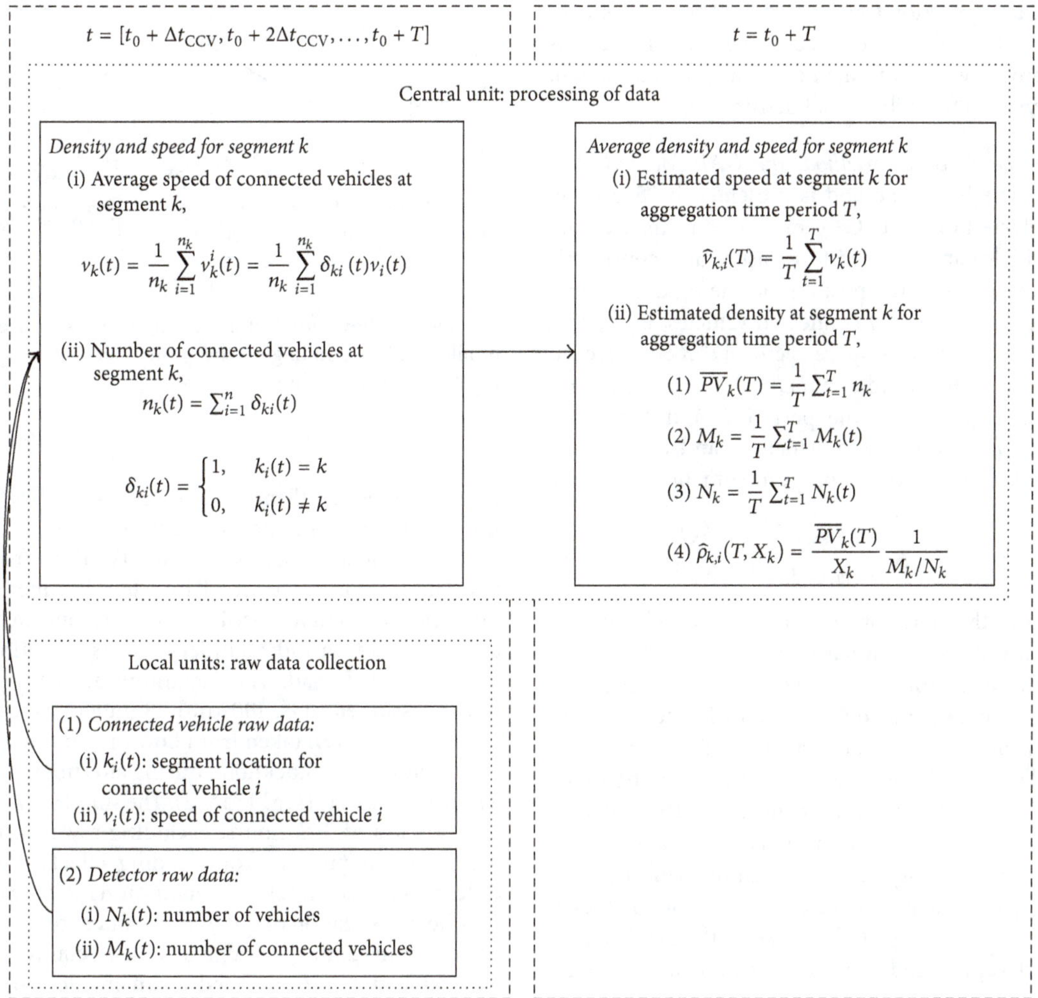

FIGURE 1: The communication flow between the local units (connected vehicles and detectors) and the central unit for the aggregation time period T.

A Detector-Based Method: the Stationary Detector (SD) Method. The detector-based method is using speed and flow measurements to estimate the density by the fundamental relationship

$$\overline{\rho}_k(T) = \frac{\overline{F}_k}{\overline{v}_k}, \quad (3)$$

where $\overline{F}_k$ and $\overline{v}_k$ are the mean flow and harmonic mean speed, respectively, detected at a detector station located just upstream of segment k and for the aggregation time period T. The method is hereafter referred to as stationary detector (SD) method. An alternative to using flow and speed measurements is to use the detector occupancy, that is, the amount of time a detector is occupied by vehicles. The occupancy can then be translated to density based on an estimate of the vehicle length. However, also by using occupancy, the resulting density will be represented only at a specific point. Further, the length of the vehicles has to be estimated or assumed.

A Combined Method: the Cumulative Count (CC) Method. The combined method is presented by Qiu et al. [19] and later extended by Ma et al. [20] and here is referred to as the Cumulative Count (CC) method. The method detects the number of vehicles located within segment k. Let $C(t_k^m)$ be the number of vehicles that have been detected up until time t_k^m, where t_k^m is the time connected vehicle m enters segment k. The density estimate for connected vehicle m is given by counting the number of vehicles, $C(t_k^m) - C(t_{k+1}^m)$, which have passed the detector upstream of the segment at the time when a connected vehicle m enters, t_k^m, and exits, t_{k+1}^m, the segment. The resulting density estimate becomes

$$\tilde{\rho}_k(T, X_k) = \frac{1}{P_k}\sum_{m=1}^{P_k}\frac{C(t_k^m) - C(t_{k+1}^m)}{X_k}. \quad (4)$$

Here, the total number of connected vehicles, P_k, which exit segment k within the aggregation time period T is used to get an average density estimate. This means that the travel time for each connected vehicle within the segment can be longer than the aggregation time period. Further, no vehicles are assumed to overtake the connected vehicle. This means that, on a segment with more than one lane, overtaking will result in a deviation from the actual density.

A Connected-Vehicle-Based Method: the GAP Method. The connected-vehicle-based method is presented by Seo et al. [23] and is referred to as the GAP method. In this method, connected vehicles are assumed to measure and communicate the gap to its leader, the position on the road, and time of the measurement. A set of connected vehicles, $\mathbf{P}_k(T, X_k)$, are located within the time-space region $[T, X_k]$. For each connected vehicle m, the total time, $tt_k^m(T, X_k)$, spent on segment k in aggregation time period T and the time-space gap, $|g_k^m(T, X_k)|$, between vehicle m and its leader is communicated. The density estimate on segment k becomes

$$\breve{\rho}_k(T, X_k) = \frac{\sum_{m \in \mathbf{P}_k(T,X_k)} tt_k^m(T, X_k)}{\sum_{m \in \mathbf{P}_k(T,X_k)} |g_k^m(T, X_k)|}. \quad (5)$$

This means that the trajectory of a connected vehicle, and its leader, within the segment has to be recorded in order to estimate the density. It should be noted that the measurement range of the gap between a connected vehicle and its leader has to be considered. If the gap is assumed to be measured through local on-board equipment, longer gaps might not be included due to a limited measurement range. This will lead larger gaps to be excluded from the density estimations, and as a result, the uncertainty in the estimations will increase. Further, the gap behind the connected vehicle is not included, which is of importance at platooning, since the gap between the connected vehicle and the vehicle behind can be long and maybe even not in the same segment. Since this gap is excluded, the density estimate becomes higher than the actual density on the road. Also, the frequency at which the connected vehicle data is transmitted is important. Between each time of communication for the connected vehicles, a linear relation in measured data is assumed. This can result in discretization errors that are larger at lower transmission frequencies.

4.2. Simulation Setup. The proposed method, as well as two of the comparison methods, requires identification and communication of information from connected vehicles. Microscopic traffic simulators describe individual vehicles in the traffic stream, allowing for gathering of information from single vehicles within the simulation. Thus, microscopic traffic simulation is suitable for analysis of the proposed method. In this study, we use the open-source microscopic traffic simulation tool SUMO (version 0.27.1) [26, 27]. SUMO is multimodal, space continuous, and time discrete. The car-following model used to model vehicle interactions is a further development of the work by Krauß [28] and is based on the calculation of a safe speed, compared to the approach of Gipps [29]. The lane-changing model is rule-based. The core model in SUMO is further described by Krajzewicz [30]. The connected vehicle and stationary detector data are accessed during the simulation through SUMO's Traffic Control Interface (TraCI). Python scripts [31] are used to implement the traffic state estimation methods.

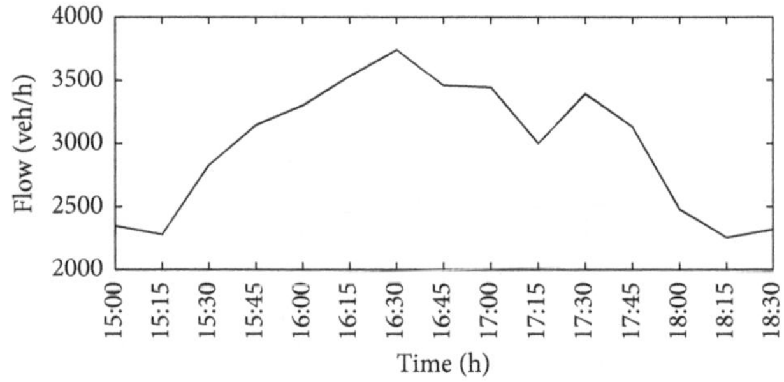

FIGURE 2: Inflow profile for the simulated scenario. The inflow profile is corresponding to the peak hours on a two-lane urban motorway in Stockholm.

The simulated scenario consists of a one-directional two-lane motorway, divided into ten 500-meter segments. Further, a segment for loading of vehicles and an end segment are included to avoid boundary effects, resulting in a 6 km long simulated road. The maximum allowed speed on the road is assumed to be 100 km/h. Simulations are performed with a flow pattern taken from flow measurements on a two-lane motorway in Stockholm during afternoon peak hours on a normal weekday (see Figure 2). The simulation is performed for a period of 3.75 hours, excluding a warm-up period of 5 minutes to prevent from loading effects. Further, eight different connected vehicle penetration rates are investigated. For the investigation of different distances between detectors, abrupt changes in the traffic state are required. Changes in the traffic state are modeled as an incident by letting one vehicle halt on a road segment for ten minutes after one hour. This is resulting in a temporary drop in capacity, which is considerably changing the traffic conditions.

4.3. Vehicle Parameters. The data used for calibration is collected through stationary radar detectors from four different locations on a two-lane urban motorway in Stockholm. The measurement from each detector consists of speed and flow measurements averaged over 15 minutes from three days in April 2016, with a typical pattern and no larger incidents reported. The flow profile is presented in Figure 2. The speed measurements, averaged over the three days, at the different detector locations and for different time instants are given in Figure 3(a). However, the original road stretch includes some on- and off-ramps, which have been excluded in this simulation study. The reason for this is to isolate the effects of the method by using a simple simulation scenario before trying more complex scenarios. Additionally, vehicle data at free flow conditions for four different vehicle classes are available for calibration of the desired speed distribution. The composition of vehicles and the speed distribution for each vehicle class are given in Table 2.

The calibration of a SUMO model of the two-lane motorway in Stockholm has resulted in calibrated vehicle parameters that are applied for this study. Figure 3 gives an overview of the speed at the four detectors and at

TABLE 2: Free flow speed distribution for the different vehicle classes.

Class	Allowed speed limit (km/h)	Vehicle length (m)	Mean speed (km/h)	Std of speed (km/h)	Composition (%)
1	110	0–8	109	13.13	91
2	90	8–12.5	92	10.33	5
3	90	12.5–24	90	8.35	3
4	80	24–36.5	88	5.55	1

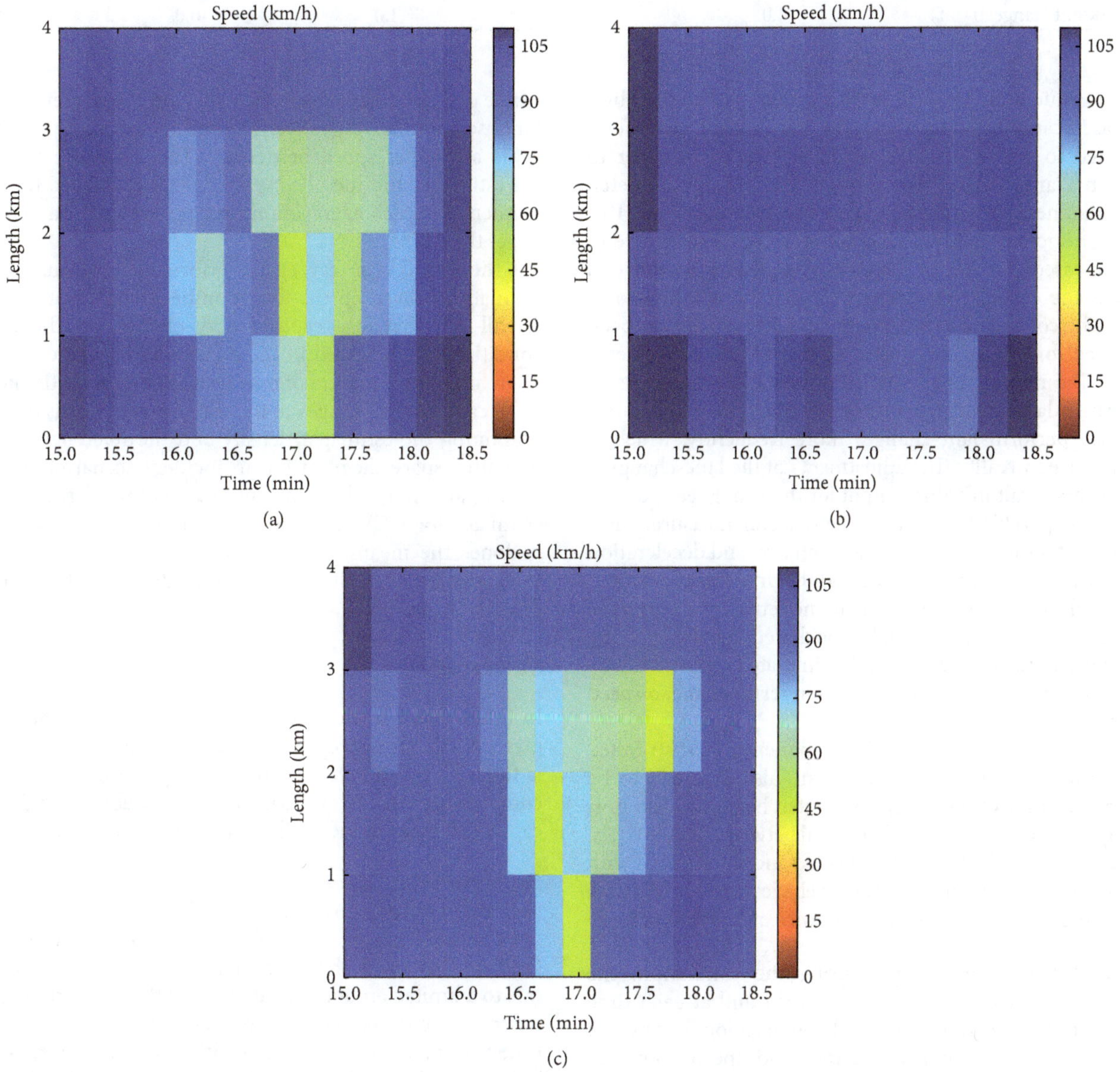

FIGURE 3: The mean speed for the given detector measurements (a), the uncalibrated scenario (b), and the calibrated scenario (c). The detectors are represented on y-axis and the time instants are represented on x-axis (15 min intervals). The mean speed is ranging from 0 km/h (red) to 110 km/h (blue).

15 min intervals for the measurements (a), the uncalibrated simulated measurements in SUMO (b), and the calibrated measurements in SUMO (c). The speed ranges from 0 km/h (red) to 110 km/h (blue). The default parameters and the calibrated parameters in SUMO are given in Table 3. The lane-changing parameters, lcCooperative and lcSpeedGain, have been changed substantially compared to the default values in SUMO. The reason for this is that by using the default values the throughput becomes much higher than observed on the motorway in Stockholm (see Figure 3(b)). Here, lcCooperative parameter controls the degree of cooperation with other vehicles when performing a lane-change.

TABLE 3: Vehicle parameters based on the calibrated scenario.

Parameter	Default	Class 1	Class 2	Class 3	Class 4
Accel. ability (m/s^2)	2.6	0.8	0.8	0.6	0.6
Decel. ability (m/s^2)	4.5	8.0	8.0	6.0	6.0
Imperfection (range: 0-1)	0.5	0.8	0.8	0.8	0.8
Reaction time (s)	1.0	0.85	0.85	0.85	0.85
lcCooperative (range: 0-1)	1.0	0.4	0.4	0.4	0.4
lcSpeedGain (range: 0–inf)	1.0	45.0	45.0	45.0	45.0
lcKeepRight (range: 0–inf)	1.0	1.0	1.0	10.0	10.0

A lower value results in decreased cooperation and thereby a decreased capacity on the road. The lcSpeedGain parameter is related to the willingness to increase speed in order to perform a lane-change. Hence, by increasing this parameter, the willingness to perform a lane-change is increased. This results in increased interaction between the vehicles due to the increased number of lane-changes, which in the end reduces the capacity on the road. Also, the willingness to keep right, controlled by the parameter lcKeepRight, has been increased for trucks and buses, corresponding to classes 3 and 4. The reason for this is that when the willingness to perform a lane-change is increased, the trucks and buses will also perform lane-changes more frequently, which is not the case in reality. The adjustments of the lane-changing parameters result in a throughput for the simulated scenario that is comparable to the available detector measurements. In the car-following model, the acceleration and deceleration abilities, as well as reaction time and the driver imperfection, have been adjusted to correspond to the actual capacity on the road. From Figure 3(c), it can be concluded that, by applying the final calibrated parameters, the simulated scenario is able to reproduce the measured scenario in terms of mean speed levels.

Vehicles are generated with exponentially distributed headways. The connected vehicles are also assumed to be generated with exponentially distributed headways and uniformly distributed in the total flow and between vehicle types. The connected vehicle data is collected and transmitted with a frequency of 1 Hz. The connected vehicles are able to detect a vehicle in front at a maximum distance of 1500 meters.

4.4. Performance Indicators. During the simulation, the number of vehicles within each segment and at each time step is counted and averaged over the aggregation time period to get the "true" simulated density and speed, hereafter referred to as the reference density. The reference density is compared to the density estimated using the proposed method, CCV, and the methods used for comparison (SD, CC, and GAP). The true simulated space mean speed is calculated by averaging of the speeds of all vehicles located within a segment for the considered aggregation time period. The comparison of both density and speed estimates is done with respect to the Root Mean Square Error (RMSE) for the total number of observations N:

$$\text{RMSE} = \sqrt{\frac{\sum_{n=1}^{N} (\text{est}_n - \text{obs}_n)^2}{N}}, \quad (6)$$

where est_n and obs_n are the estimate and the reference value for observation n, respectively. The estimate and the reference value are either speed or density. The difference between the estimated and the observed values is calculated for each segment and each aggregation time period and summarized to get the RMSE.

The RMSE is an aggregate performance measure of the estimations based on the total number of observations and it will not therefore capture how changes in the traffic conditions are reflected in the density and speed estimates. Therefore, an evaluation of how well the density estimates of CCV capture the changes in the traffic conditions is done by examining the estimated density and the reference density in a time-space diagram for an incident scenario. Further, a comparison of the time-space diagram of the density estimates for CCV and the other combined method, CC, is done. The means and standard errors of the means are calculated based on 10 replications of the simulation for the different scenarios.

5. Results

In this section, results from the simulation experiments are given. First, the aggregation time period is examined. This is followed by an investigation of the performance of CCV compared to SD, CC, and GAP. Finally, the ability for CCV to capture changes in the traffic conditions is presented.

5.1. Aggregation Time Period for CCV. The aggregation time period has to be chosen carefully. For a short aggregation time period, changes in the density can be discovered fast. On the other hand, the estimations become more uncertain due to a limited amount of data to base the estimation upon, especially for lower penetration rates. For a large aggregation time period, the traffic state estimation becomes more stable, but there is also a risk to smooth out changes and thereby miss useful and relevant information. The performance of CCV for the aggregation time periods of 15, 30, 60, and 120 seconds is examined by comparing the estimated density and speed to the reference density and speed, as explained in Section 4.4.

At low penetration rates and for short aggregation time periods, it is often not possible to estimate the traffic state due to no connected vehicle measurements to base the estimation on. In this case, the missing estimates are excluded from the RMSE and the resulting RMSE becomes uncertain. To give an indication of the amount of missing estimates at different penetration rates, the mean percentage of missing

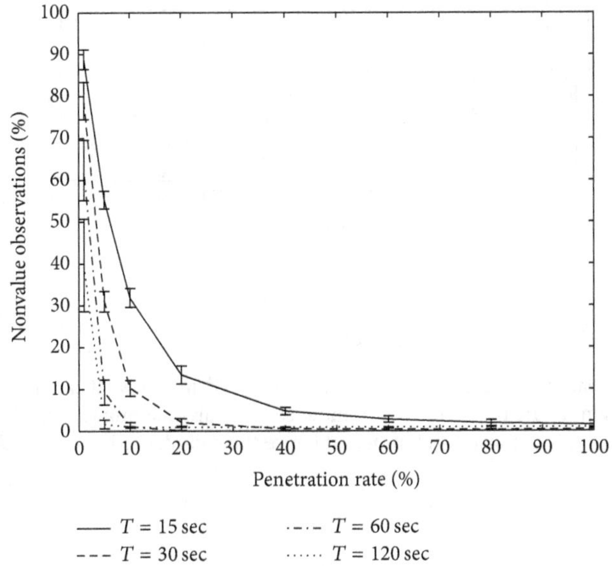

FIGURE 4: Mean percentage of missing estimates per segment for CCV at different penetration rates. The error bars are 95% confidence intervals, assuming normally distributed missing number of estimations over the simulation runs.

estimates per segment is given in Figure 4. The mean RMSE for speed and density for different penetration rates is shown in Figure 5.

From Figure 5(a), it is observed that the aggregation time period has limited effect on the RMSE of speed. Further, the RMSE of speed is highest for low penetration rates and decreases for higher penetration rates. However, this is the case only when it can be assumed that the connected vehicles have the same speed distribution as nonconnected vehicles. If the distribution of speed for the connected vehicles deviates from the distribution of speed of nonconnected vehicles, the proposed method for estimating the speed will be biased towards the speed of the connected vehicles. This will result in a higher or lower speed estimate depending on whether the connected vehicles are assumed to drive faster or slower than nonconnected vehicles.

Further, it is observed that the aggregation time period will have a great effect on the density estimates. As can be seen in Figure 4, a low penetration rate of connected vehicles will result in a large percentage of missing measurements. As a result, no, or only uncertain, traffic state estimates are available. This does explain the increase in RMSE with increasing penetration rates observed in Figure 5(b) for low penetration rates and for the aggregation time periods of 15–60 seconds. Since nonvalue estimates are excluded from the resulting RMSE at low penetration rates, an increase in penetration rate leads more observations to become available and more, but uncertain, estimates contribute to a larger RMSE. However, at penetration rates above 5–20%, depending on the aggregation time period, enough measurements are included in the estimation to give a reliable result and the RMSE is starting to decrease with increased penetration rate of connected vehicles. When the penetration rate is 100%, the RMSE of both speed and density are, as expected, zero due to the exclusion of measurement errors.

As a conclusion, the RMSE of speed and density are smallest for the aggregation time period of 120 seconds. However, by using such a large aggregation time period, the estimates become smoothed over a longer period. This might result in the fact that important phenomena in the traffic conditions are detected late or are even not detected at all. In order to reduce the uncertainty in the estimations and at the same time be able to capture the changes in the traffic flow, which is important for identifying changes in the traffic conditions, an aggregation time period of 60 seconds is therefore chosen for further investigations.

5.2. Performance of CCV. The density estimates of the CCV are compared to the density estimates of GAP, CC, and SD in order to investigate how the proposed method performs compared to existing methods. The mean RMSE of density for the four methods and using different penetration rates are presented in Figure 6. The aggregation time period is set to 60 seconds based on the results presented in Section 5.1.

As concluded in Section 5.1, the CCV method does not show a clear relationship between penetration rate and performance for penetration rates below 10% as a result of the missing or limited measurements to base the estimations on. Hence, for a penetration rate of connected vehicles of 1 and 5%, the RMSE of density for the CCV is not trustworthy. However, since the CCV performs poorly at low penetration rates as shown in Figure 6, it can be concluded that the penetration rate must be higher than 10% to give reliable estimates.

The density estimates based on GAP, CC, and CCV are, as expected, improved with an increased penetration rate. The CCV gives density estimates comparable to GAP at a penetration rate of 20%. It is first at a penetration rate of 40–50%; the accuracy in the estimates of CCV is comparable to estimates of CC and SD, although after a penetration rate of 40%, the improvements in the density estimates are rapid with increased penetration rate. When approaching a penetration rate of 100%, the estimated density is close to the reference density, unlike the other methods where a larger difference still exists. The reason for this is that, for CCV, a penetration rate of 100% means identifying all vehicles located within the segment and this is expected to be the same as the reference density over the same aggregation time period.

The other combined method, CC, has a larger RMSE of density at higher penetration rates. One reason for this might be the fact that it is assumed that no vehicles are overtaking the connected vehicle. Therefore, when the connected vehicles arrive at the downstream detector, all vehicles arriving at the upstream detector after the connected vehicle are assumed to still be within the segment. But since the segment consists of two lanes and the vehicles within the simulation have different desired speed, overtaking will occur, especially for connected vehicles with a low desired speed.

By using only connected vehicles, as is done in GAP, the local nature of the method does not seem to capture the traffic conditions in a larger area during inhomogeneous

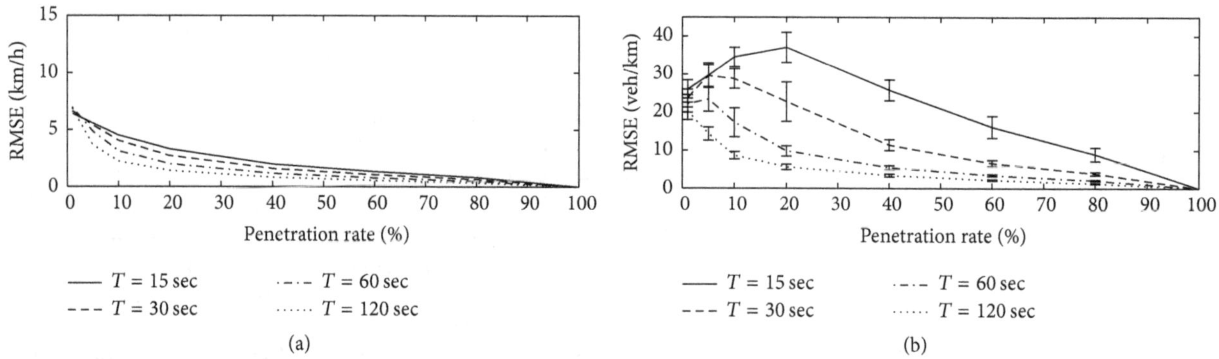

Figure 5: Mean RMSE for the estimated speed (a) and density (b) compared to the reference density and speed for CCV at different penetration rates and for different aggregation time periods. The error bars in (b) are based on 95% confidence intervals, assuming normally distributed RMSE over the simulation runs. The standard error of mean in (a) is at most 0.44 km/h, and hence the confidence intervals are excluded from the figure for greater legibility.

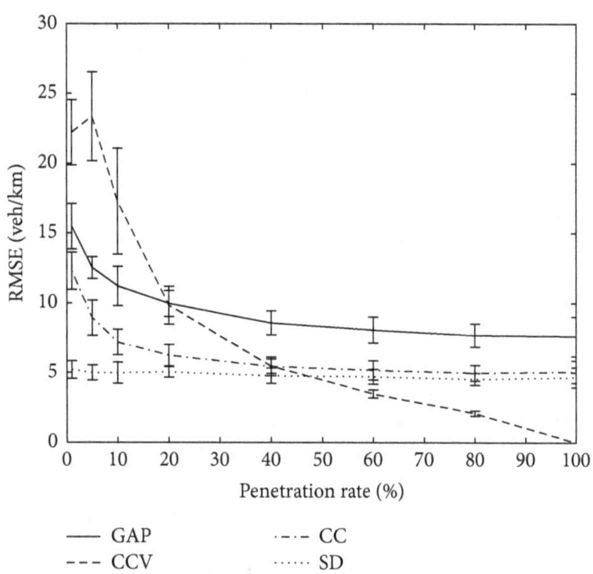

Figure 6: Mean RMSE of the estimated density compared to the reference density for the different methods and using different penetration rates. The aggregation time period is 60 seconds. The error bars are 95% confidence intervals, assuming normally distributed missing number of estimations over the simulation runs.

traffic conditions where the speed distribution of the vehicles becomes wider. By only considering the gap to the vehicle in front, the formation of platoons, with small gaps for the vehicles within the platoons and larger gaps in between platoons, will decrease the performance and there will be a bias towards gaps within the platoons. This is especially observed at medium flow levels where the left lane is mostly used for overtaking. In this case, a leader is sometimes missing as a result of the limited measurement range or the time-space distance behind the platoon is unknown in the overtaking lane where the traffic flow is more inhomogeneous. This is why the RMSE for the GAP method does not decrease as much as the other methods at higher penetration rates.

Finally, it is concluded that the density estimates using SD are, as expected, independent of the penetration rate. The difference between the true density and the density estimates for SD is due to the fact that the density estimates for SD are based on measurements at specific locations and not measurements for the whole segment.

As a conclusion, the SD is stable and gives reliable estimates of the density independent of the connected vehicle penetration rate and flow level. For the combined methods, the CC is preferable at lower penetration rates, while the CCV is more accurate at higher penetration rates. Results from GAP show that using only local measurements from connected vehicles gives a lower accuracy in the density estimates under inhomogeneous traffic conditions as a result of the inability to measure the gap behind vehicles and the limited measurement range. Since the accuracy of the density estimates is higher for the combined methods and the detector-based method, it is preferable to include stationary detectors when estimating the density.

5.3. Evaluation of Effects of Different Distances between Detectors. One of the advantages of CCV is that it can capture changes in the traffic conditions in between detectors, meaning that even though the detectors are sparsely placed, smaller segments in between the detectors can be considered for estimating the density. However, this is based on the assumption that the connected vehicle penetration rate is approximately constant in between detectors. For this reason, CCV is compared to the other combined method, CC, to investigate how the ability to detect changes in the traffic conditions, in this case modeled as an incident, is affected by sparsely located detectors for the two methods. Since CC has to include the whole stretch between two detectors as one segment, it is expected that the resulting performance is improved for the CCV.

Two time-space diagrams of density are presented in Figure 7. The distances between detectors are 500 and 2500 meters, respectively. An aggregation time period of 60 seconds is used based on the investigation in Section 5.1. A penetration rate of 40% is used, since CCV and CC are

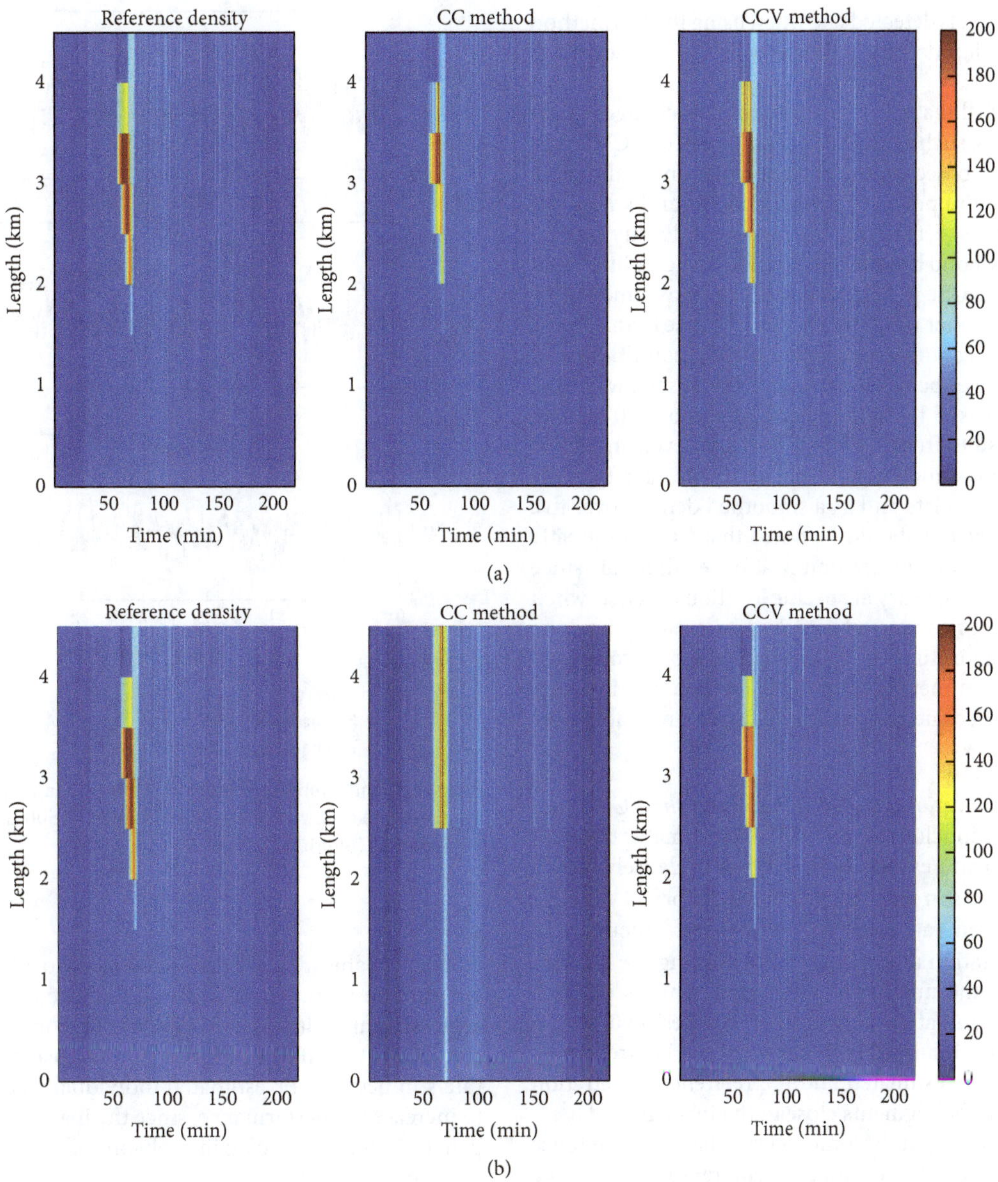

FIGURE 7: Reference density and density estimates using CC and CCV in time (min) and space (km) for a simulated incident scenario. The aggregation time period is 60 sec and the detector interval is 500 meters and 2500 meters in (a) and (b), respectively. The color map shows density (veh/km).

comparable in accuracy at this level. The presented results are means over 10 simulation runs. As can be seen in Figure 7(a), both CCV and CC manage to capture the tail of the incident with a distance between the detectors of 500 meters. However, both methods seem to overestimate the density at the incident and underestimate the density in the tail of the incident. An explanation to the overestimation for the CC is that vehicles might overtake the connected vehicle at an incident and pass to the next segment before the connected vehicle has exited the considered lane. This will result in a too large density estimate. Further, the connected vehicle data is sometimes based on retrospective information and information smoothed out over a longer time period than the aggregation time period, which will reduce the performance of CC especially during congested conditions, such as an incident. One reason for the deviation in the estimation for CCV might be due to an uncertain estimate of the penetration rates caused by the low speed levels at the detectors. This is further investigated below.

With more sparsely placed detectors, as in Figure 7(b), the same behaviour is seen for CCV, since the segment length is still 500 meters and the penetration rate in this example is approximately the same over the whole stretch. However, for CC, the distance between detectors and thereby the distance over which the density is estimated is considerably increased, which will result in decreased accuracy of the density estimates. Further, a consequence of the more sparsely located detectors is an even longer time delay in the estimates. As a result, the incident is detected late when using the CC method and as a smoothed density estimate over the five segments without detectors.

It is concluded that CCV seems to capture changes in the traffic conditions, such as an incident. A benefit of CCV is that the estimates are based on the current situation on the road. This is becoming important during incidents and where the longer travel times seem to cause delays in the density estimation for CC. A second benefit is that, under the condition that the connected vehicle penetration rate can be assumed to be constant over a longer road stretch, as is the case in this study, CCV manages to capture changes in the traffic conditions also for longer distances between detectors. This is seen when the distance is increased to 2500 meters instead of 500 meters, whereas the results from CC are affected by even increased delay, as well as smoothing, resulting in the fact that the incident is detected late and as a smoothed density over five segments. However, it should be noted that CC does greatly limit the dependence of communication equipment, since reporting is only necessary at specific locations. Hence, when communication of connected vehicle data can only be done at specific points in time, or at specific locations, combined density estimation methods not dependent on continuously transmission of connected vehicle data are preferable over CCV.

The Effect of Penetration Rates at Incidents for the CCV. From Section 5.3, it is concluded that CCV overestimates the density at incidents and especially in the case of densely placed detectors. One reason for this could be an incorrect estimate of the penetration rate close to the incident. Therefore, a reference penetration rate, gathered by counting the number of vehicles and the number of connected vehicles on each segment in the simulation, is compared to the estimates from the CCV for the simulated incident scenario. Figure 8 gives an overview of the estimated and the reference penetration rate over time at the segments close to the incident.

From the figure, it becomes clear that the reference penetration rate and the estimated penetration rate deviate more during the incident and for the affected segments. This confirms that the estimate of the penetration rate is the reason for the uncertain estimates. It seems like, in order for the model to perform well under low-to-medium penetration levels, the traffic condition within a segment has to be homogeneous. Hence, if the segment can be subdivided to one congested and one uncongested traffic state, the penetration rate between those states might differ, resulting in an inaccurate estimate of the penetration rate and as a result an inaccurate estimate of the density. Further, the estimates of the penetration rate at detector intervals of 500 meters are more unstable close to the incident. This is probably due to the fact that the congestion is moving upstream, resulting in slow moving vehicles and a more uncertain number of connected vehicles passing the detectors, which creates local variations in the connected vehicle distribution in space. This corresponds well with the larger error in the density estimates at the incident for more densely placed detectors, which can be seen in Figure 7(a). For that reason, a more sophisticated method for estimating the penetration rate

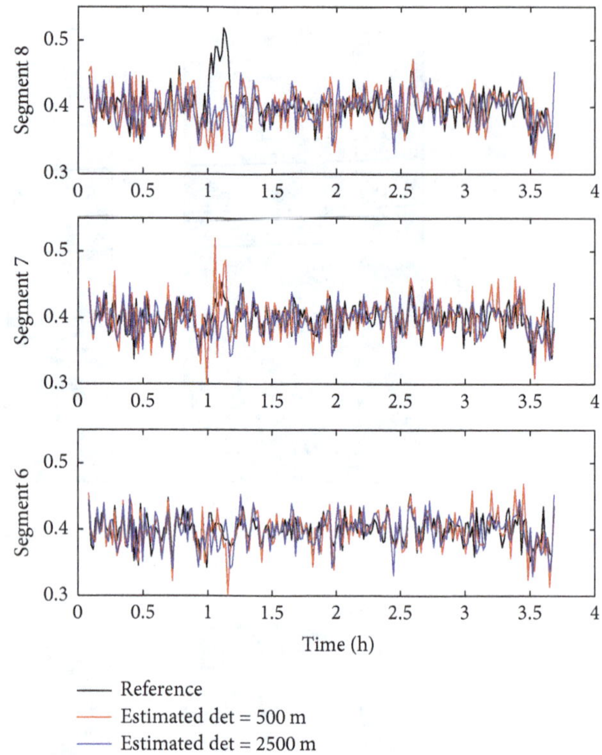

FIGURE 8: The reference (black) and the estimated penetration rate for segments 6–8 with a detector distance of 500 meters (red) and 2500 meters (blue).

would probably increase the performance of CCV even at low penetration rates. Also the division of the road into smaller segments can reduce the possibility of having different traffic conditions within one segment. In this case, the penetration rate still needs to be estimated individually at each segment to increase the performance, since the inaccurately estimated penetration rate is the main problem.

6. Conclusions

We propose the Count Connected Vehicle (CCV) method for estimating the traffic state using connected vehicles in combination with stationary detectors. The method provides a straightforward approach to estimate the speed and density based on vehicle-to-infrastructure communication. The purpose of the method is to get fast and accurate traffic state estimations that can be used to detect changes in the traffic conditions and at the same time limit the dependence on detailed measurements communicated from the connected vehicles using as few stationary detectors as possible. The only measurements required from the stationary detectors are the number of passing vehicles and the number of passing connected vehicles. Hence, the complexity of the stationary detectors is limited. Moreover, if the penetration rate is assumed to be approximately the same over a longer road stretch, the stationary detectors can be sparsely placed. The limited dependence on measurements from the connected vehicles is due to the fact that the requirement in precision in position data can be low. The only requirement is that the current segment be correctly reported. Hence, the measurement errors for estimating the number of connected vehicles located on a segment are limited to the boundaries of each segment.

The method is evaluated by the means of microscopic traffic simulation. The speed and density estimates of the proposed method are compared to the true simulated values. Further, the density estimates are compared to the density estimates using one detector-based method, one combined method, and one connected-vehicle-based method. The results of the study show that the proposed method is a promising alternative for accurately estimating density and speed on the road, especially at medium-to-high penetration rates of connected vehicles. Since the method is based on real-time positioning data from connected vehicles, it can capture abrupt changes in the traffic conditions, such as incidents. The traffic state can be estimated instantaneously given that there are at least a few connected vehicles at the segment. Hence, the aggregation time period can be short. This makes the method useful for traffic management purposes. Note that the penetration rate can be based on previous knowledge, that is, a moving average, during short aggregation time periods, to increase the accuracy in the estimation of the connected vehicle penetration rate.

Many interesting topics are identified for future research. First, a low connected vehicle penetration rate and a short aggregation time period are concluded to result in few or sometimes missing density estimates. Hence, by including data assimilation techniques, such as the ones described by Evensen [32], Wang and Papageorgiou [6], Antoniou et al. [33], and Seo et al. [13], the performance of the proposed method can be improved. Second, the penetration rate becomes inaccurate during inhomogeneous traffic conditions within a segment. By using a more advanced method to estimate the penetration rate, the accuracy of the method can be improved. For example, the dynamics of the penetration rate can be modeled using a macroscopic model as in Astarita et al. [14] and Bekiaris-Liberis et al. [15]. Third, the effect on the traffic state estimation for more complex designs of the road network and inclusion of measurements errors has to be further investigated. Finally, the use of the proposed method in traffic management systems is also a topic for future research.

Conflicts of Interest

The authors declare that there are no conflicts of interest regarding the publication of this paper.

Acknowledgments

The authors would like to acknowledge the Swedish Transportation Administration (Trafikverket) for the financial support.

References

[1] R. Coppola and M. Morisio, "Connected car: technologies, issues, future trends," *ACM Computing Surveys*, vol. 49, no. 3, article 46, 2016.

[2] A. Kurkjian, S. B. Gershwin, P. K. Houpt, A. S. Willsky, E. Y. Chow, and C. S. Greene, "Estimation of roadway traffic density on freeways using presence detector data," *Transportation Science*, vol. 14, no. 3, pp. 232–261, 1980.

[3] B. Coifman, "Estimating density and lane inflow on a freeway segment," *Transportation Research Part A: Policy and Practice*, vol. 37, no. 8, pp. 689–701, 2003.

[4] K. Singh and B. Li, "Estimation of traffic densities for multilane roadways using a markov model approach," *IEEE Transactions on Industrial Electronics*, vol. 59, no. 11, pp. 4369–4376, 2012.

[5] L. Muñoz, X. Sun, R. Horowitz, and L. Alvarez, "Traffic density estimation with the cell transmission model," in *Proceedings of the American Control Conference*, vol. 5, pp. 3750–3755, IEEE, June 2003.

[6] Y. Wang and M. Papageorgiou, "Real-time freeway traffic state estimation based on extended Kalman filter: a general approach," *Transportation Research Part B: Methodological*, vol. 39, no. 2, pp. 141–167, 2005.

[7] L. Mihaylova, R. Boel, and A. Hegyi, "Freeway traffic estimation within particle filtering framework," *Automatica*, vol. 43, no. 2, pp. 290–300, 2007.

[8] A. Duret, L. Leclercq, and N.-E. El Faouzi, "Data assimilation using a mesoscopic Lighthill-Whitham-Richards model and loop detector data: methodology and large-scale network application," *Transportation Research Record*, vol. 2560, pp. 26–35, 2016.

[9] T. Darwish and K. A. Bakar, "Traffic density estimation in vehicular ad hoc networks: a review," *Ad Hoc Networks*, vol. 24, pp. 337–351, 2015.

[10] J. C. Herrera and A. M. Bayen, *Traffic Flow Reconstruction Using Mobile Sensors And Loop Detector Data*, University of California Transportation Center, 2007.

[11] D. B. Work, S. Blandin, O.-P. Tossavainen, B. Piccoli, and A. M. Bayen, "A traffic model for velocity data assimilation," *Applied Mathematics Research eXpress*, vol. 2010, no. 1, pp. 1–35, 2010.

[12] Y. Yuan, H. Van Lint, F. Van Wageningen-Kessels, and S. Hoogendoorn, "Network-wide traffic state estimation using loop detector and floating car data," *Journal of Intelligent Transportation Systems: Technology, Planning, and Operations*, vol. 18, no. 1, pp. 41–50, 2014.

[13] T. Seo, T. Kusakabe, and Y. Asakura, "Traffic state estimation with the advanced probe vehicles using data assimilation," in *Proceedings of the 18th IEEE International Conference on Intelligent Transportation Systems, ITSC 2015*, pp. 824–830, Spain, September 2015.

[14] V. Astarita, R. L. Bertini, S. d'Elia, and G. Guido, "Motorway traffic parameter estimation from mobile phone counts," *European Journal of Operational Research*, vol. 175, no. 3, pp. 1435–1446, 2006.

[15] N. Bekiaris-Liberis, C. Roncoli, and M. Papageorgiou, "Highway traffic state estimation with mixed connected and conventional vehicles," *IEEE Transactions on Intelligent Transportation Systems*, vol. 17, no. 12, pp. 3484–3497, 2016.

[16] R. Herring, A. Hofleitner, P. Abbeel, and A. Bayen, "Estimating arterial traffic conditions using sparse probe data," in *Proceedings of the 13th International IEEE Conference on Intelligent Transportation Systems, ITSC 2010*, pp. 929–936, Portugal, September 2010.

[17] J. C. Herrera, D. B. Work, R. Herring, X. Ban, Q. Jacobson, and A. M. Bayen, "Evaluation of traffic data obtained via GPS-enabled mobile phones: the mobile century field experiment,"

Transportation Research, Part C: Emerging Technologies, vol. 18, no. 4, pp. 568–583, 2010.

[18] J. W. C. Van Lint and S. P. Hoogendoorn, "A robust and efficient method for fusing heterogeneous data from traffic sensors on freeways," *Computer-Aided Civil and Infrastructure Engineering*, vol. 25, no. 8, pp. 596–612, 2010.

[19] T. Z. Qiu, X.-Y. Lu, A. H. F. Chow, and S. E. Shladover, "Estimation of freeway traffic density with loop detector and probe vehicle data," *Transportation Research Record*, no. 2178, pp. 21–29, 2010.

[20] Y. Ma, J. Van Dalen, C. De Blois, and L. Kroon, "Estimation of dynamic traffic densities for official statistics," *Transportation Research Record*, vol. 2256, pp. 104–111, 2011.

[21] A. Bhaskar, T. Tsubota, L. M. Kieu, and E. Chung, "Urban traffic state estimation: fusing point and zone based data," *Transportation Research Part C: Emerging Technologies*, vol. 48, pp. 120–142, 2014.

[22] J. Zhang, S. He, W. Wang, and F. Zhan, "Accuracy analysis of freeway traffic speed estimation based on the integration of cellular probe system and loop detectors," *Journal of Intelligent Transportation Systems: Technology, Planning, and Operations*, vol. 19, no. 4, pp. 411–426, 2015.

[23] T. Seo, T. Kusakabe, and Y. Asakura, "Estimation of flow and density using probe vehicles with spacing measurement equipment," *Transportation Research Part C: Emerging Technologies*, vol. 53, pp. 134–150, 2015.

[24] L. Montero, M. Pacheco, J. Barceló, S. Homoceanu, and J. Casanovas, "Case study on cooperative car data for estimating traffic states in an urban network," *Transportation Research Record*, vol. 2594, pp. 127–137, 2016.

[25] T. Seo and T. Kusakabe, "Probe vehicle-based traffic state estimation method with spacing information and conservation law," *Transportation Research Part C: Emerging Technologies*, vol. 59, pp. 391–403, 2015.

[26] D. Krajzewicz, J. Erdmann, M. Behrisch, and L. Bieker, "Recent development and applications of SUMO—Simulation of Urban MObility," *International Journal on Advances in Systems and Measurements*, vol. 5, no. 3 & 4, pp. 128–138, 2012.

[27] DLR and contributors, "Sumo homepage," http://sumo.dlr.de/wiki/Main_Page 2016.

[28] S. Krauß, *Microscopic Modeling of Traffic Flow: Investigation of Collision Free Vehicle Dynamics [Ph.D. thesis]*, Hauptabteilung Mobilitt und Systemtechnik des DLR Köln, 1998.

[29] P. G. Gipps, "A behavioural car-following model for computer simulation," *Transportation Research Part B: Methodological*, vol. 15, no. 2, pp. 105–111, 1981.

[30] D. Krajzewicz, "Traffic simulation with sumo–simulation of urban mobility," in *Fundamentals of Traffic Simulation*, vol. 145, pp. 269–293, Springer, New York, NY, USA, 2010.

[31] G. van Rossum, "Python tutorial, technical report cs-r9526," Tech. Rep., Centrum voor Wiskunde en Informatica (CWI), Amsterdam, Netherlands, May 1995.

[32] G. Evensen, "The ensemble Kalman filter: theoretical formulation and practical implementation," *Ocean Dynamics*, vol. 53, no. 4, pp. 343–367, 2003.

[33] C. Antoniou, M. Ben-Akiva, and H. N. Koutsopoulos, "Nonlinear Kalman filtering algorithms for on-line calibration of dynamic traffic assignment models," *IEEE Transactions on Intelligent Transportation Systems*, vol. 8, no. 4, pp. 661–670, 2007.

5

A Reliability-Based Network Equilibrium Model with Adaptive Risk-Averse Travelers

Lei Zhao,[1,2] Hongzhi Guan,[1,2] Junze Zhu,[1,2] and Yunfeng Wei[1,2]

[1] *College of Architecture and Civil Engineering, Beijing University of Technology, Beijing 100124, China*
[2] *Beijing Key Laboratory of Transportation Engineering, Beijing 100124, China*

Correspondence should be addressed to Lei Zhao; leizhao1205@163.com

Academic Editor: Eneko Osaba

In this paper, route free-flow travel time is taken as the lower bound of route travel time to examine its impacts on budget time and reliability for degradable transportation networks. A truncated probability density distribution with respect to route travel time is proposed and the corresponding travel time budget (TTB) model is derived. The budget time and reliability are compared between TTB models with and without truncated travel time distribution. Under truncated travel time distribution, the risk-averse levels of travelers are adaptive, which are affected by the characteristics of the used routes besides the confidence level of travelers. Then, a TTB-based stochastic user equilibrium (SUE) is developed to model travelers' route choice behavior. Moreover, its equivalent variational inequality (VI) problem is formulated and a route-based algorithm is used to solve the proposed model. Numerical results indicate that route travel time boundary produces a great influence on decision cost and route choice behavior of travelers.

1. Introduction

Most studies indicate that travel time reliability (TTR) is an essential travel decision cost of travelers [1–3]. In particular, Abdel-Aty et al. (1977) pointed out that TTR was one of the major elements influencing commuters' route choice considerations [4]. However, the classical user equilibrium (UE) principle assumes that travelers are risk-neutral and their decision behavior only depends on the mean travel time in deterministic transportation networks. Obviously, it leaves the travel time uncertainty out of consideration and neglects the risk preferences of travelers (i.e., risk aversion and risk seeking). Meanwhile, empirical studies demonstrated that majority of travelers are risk-neutral and they would like to take an additional payment to avoid the congestion and travel risk [5, 6].

In transportation systems, uncertainty is unavoidable, which mainly derives from roadway capacity variation and travel demand fluctuation. Both supply and demand variations bring about travel time uncertainty. In order to model travelers' route choice behavior under uncertainty, a variety of traffic equilibrium models incorporating TTR are developed [7–11]. Typically, Lo et al. (2006) proposed the within budget time reliability (WBTR) model [12]. The notion of travel time budget (TTB) in WBTR model is given to represent the decision cost of travelers. The TTB refers to the sum of the mean travel time and a safety margin to arrive on time. The standard deviation of route travel time is used to measure the risk of route choice. However, the WBTR model disregards the effects of unreliable side of route travel time on route choice. By incorporating simultaneously the reliable and unreliable sides, the mean-excess traffic equilibrium (METE) model was developed by Chen and Zhou (2010) [13]. In the METE model, travelers are assumed to use such a route to make TTR ensured and unreliable impacts can be minimized. Subsequently, Chen et al. (2011) further introduced the perception errors of travelers to the METE model and proposed the stochastic mean-excess traffic equilibrium (SMETE) model [14]. In this case, the perceived travel time distribution is derived to model travelers' route choice behavior rather than the actual one. Different from the previous studies, Watling (2006) considered travelers' schedule delay and proposed a late arrive penalized user equilibrium (LAPUE) model [15]. Recently, the multiobjective equilibrium problems have been intensively investigated [16–18]. In this regard, Wang et al. (2014) provided a biobjective user equilibrium (BUE) condition, which is the state that no traveler can improve either his/her expected travel time or risk or both without worsening the other by unilaterally changing routes [17].

It can be found that most of aforementioned traffic equilibrium models with the TTR are based on the degradable transportation networks. As mentioned above, travel demand fluctuation would also lead to travel time uncertainty. Shao et al. (2006) developed the UE model with demand uncertainty [19]. Then, the model was further extended by considering travelers' perception errors and heterogeneity, and the reliability stochastic user equilibrium (RSUE) model was developed [20]. Lam et al. (2008) assumed the uncertainties mainly due to adverse weather condition and proposed a traffic equilibrium model with doubly uncertainties in supply and demand sides [21]. Zhou and Chen (2008) compared and analyzed the equilibrium results of three UE models with stochastic travel demand, using a simple network to illustrate the differences, and examined how these models address the travel risk [22]. Siu and Lo (2008) formulated a multiclass equilibrium model, in which the demand composes of infrequent travelers and commuters [23]. Sun and Gao (2012) assumed the probability distributions of travel demand and link capacities are unknown and developed the robust traffic equilibrium model [24]. In addition, considered uncertainties in demand and supply are also extended to the network design problems (NDPs) to develop the robust optimization models [25, 26].

In the above models, route travel time are commonly assumed to follow continuous normal distribution or lognormal distribution. However, these distributions imply that the lower boundary of route travel time approaches negative infinity or zero. To the best of our knowledge, route travel time is bounded and there is at least a lower boundary, namely, free-flow travel time. Yan et al. (2015) studied the effects of link speed limits on travelers' route choice and the network performance [27]. This study indicated that the mean and variance of total travel time can be reduced by imposing speed limits, but the total TTB of a network increases. Xu et al. (2017) showed clearly that the minimum travel time equals the link length divided by the speed limit value after imposing a speed limit scheme [28]. They indicated that a speed limit scheme on uncertain road networks would affect the network flow reallocation. In the study, the minimum travel time is derived from speed limits. Nevertheless, the route travel time without a speed limit should also have a lower bound, and the travel time uncertain profile would also change.

This paper aims to examine the impacts of route travel time boundary on travel cost and route choice of travelers. To this end, we firstly propose the probability density function (PDF) with the route free-flow travel time as a truncation of original distribution. Subsequently, the TTB model with truncated travel time distribution is derived and compared with the original TTB model in terms of budget time and TTR. Then, the TTB-based SUE models are developed to model the impacts of route travel time boundary on travelers' route choice.

The remainder of this paper is organized as follows. Section 2 presents the effect of route travel time boundary on PDF. Section 3 develops the TTB model with truncated travel time distribution under uncertainty. Section 4 gives the TTB-based SUE model. Then, we use numerical examples to illustrate the effect of route travel time boundary on network equilibrium in Section 5. Finally, Section 6 summarizes some concluding remarks.

2. Effect of Route Travel Time Boundary on Uncertainty

In this section, the route free-flow time is considered as the lower bound of route travel time, and the effects of route travel time boundary on uncertainty are examined by proposing a truncated PDF of route travel time. The relationships and differences between TTB models with and without route travel time boundary are derived.

Let us consider a stochastic network $G(N, A)$, which is composed of N nodes and A directed links with capacity variations. Let t_a^0 be the free-flow time, v_a the traffic flow, and C_a the capacity of link a, respectively. Then, link travel time $t_a(v_a)$ can be determined by the bureau of public roads (BPR) link performance function, i.e.,

$$t_a(v_a) = t_a^0\left(1 + \beta\left(\frac{v_a}{C_a}\right)^n\right) \tag{1}$$

where β and n are, respectively, the deterministic parameters of the BPR link performance function. Assume that link capacity follows an uniform distribution, $C_a \sim U(\phi_a \bar{c}_a, \bar{c}_a)$, where $\bar{c}_a$ is the design capacity associated with link a and ϕ_a denotes the capacity degradable coefficient, $\phi_a \in [0, 1)$. Further, we suppose that link capacity distributions are independent, and thus route travel time T_r is normally distributed $T_r \sim N(E(T_r), \sigma(T_r))$ based on the central limit theorem. The mean $E(T_r)$ and standard deviation $\sigma(T_r)$ of route travel time are, respectively,

$$E(T_r) = \sum_{a\in A} \delta_{ar}^w E(t_a), \quad \forall r \in R_w,\ w \in W \tag{2}$$

$$\sigma(T_r) = \sum_{a\in A} \delta_{ar}^w \sigma(t_a), \quad \forall r \in R_w,\ w \in W \tag{3}$$

where $E(t_a)$ and $\sigma(t_a)$ are, respectively, the mean and standard deviation of link travel time. δ_{ar}^w is a binary variable, which equals 1 when link a is on route r, and 0 otherwise. R_w is a set of all routes between OD pair w, and W is a set of all OD pairs in the transportation network.

Combining the derivation of the mean and standard deviation of link travel time in [12], (2) and (3) can be rewritten as follows.

$$E\left(T_r\right) = \sum_{a\in A}\left\{\delta_{ar}^{w}\cdot\left[t_a^0+\beta t_a^0 v_\alpha^n \frac{1-\varphi_a^{1-n}}{\overline{c}_a^n\left(1-\varphi_a\right)\left(1-n\right)}\right]\right\} \tag{4}$$

$$\sigma\left(T_r\right) = \sqrt{\sum_{a\in A}\left[\delta_{ar}^{w}\cdot\beta^2\left(t_a^0\right)^2 v_a^{2n}\left\{\frac{1-\varphi_a^{1-2n}}{\overline{c}_a^{2n}\left(1-\varphi_a\right)\left(1-2n\right)}-\left[\frac{1-\varphi_a^{1-n}}{\overline{c}_a^n\left(1-\varphi_a\right)\left(1-n\right)}\right]^2\right\}\right]} \tag{5}$$

2.1. Truncated Route Travel Time Distribution. This paper assumes that the route free-flow time is a determinate constant associated with specific route in the transportation network. Travel time uncertainty is only caused by the link capacity degradation. As derived above, route travel time is normally distributed under such case. A normal distribution implies that the lower bound or minimum route travel time is close to negative infinity. However, the minimum route travel time must be a positive constant, namely, the route free-flow travel time. In other words, the free-flow travel time is the lower bound of route travel time, and route travel time cannot be less than the free-flow travel time. More specifically, the probability of route travel time being less than the free-flow travel time should be zero. Hence, we develop the PDF with the free-flow travel time as a truncation of original distribution without the lower bound, i.e.,

$$f\left(T\mid\mu,\sigma,\overline{t}\right) = \begin{cases}\dfrac{f(T)}{1-F\left(\overline{t}\right)}, & T>\overline{t}\\ 0, & others\end{cases} \tag{6}$$

where $f(\cdot)$ and $F(\cdot)$ are the PDF and CDF of the original normal distribution associated with route travel time T. $\overline{t}$ is the route free-flow travel time and $F(\overline{t}) = \Pr(T\le\overline{t})$. μ and σ are the mean and standard deviation of route travel time, respectively.

For the ease of elaboration, a single-route network with the mean of 20 and the standard deviation of 5 is used. Assume the free-flow travel time is 15. The PDFs of route travel time with and without lower bound are shown in Figure 1. The x-axis shows the change in route travel time. The blue solid line is the PDF curve without the lower bound. The green dotted line represents the PDF curve with the free-flow travel time as a truncation of the original distribution. From this figure, one can see that the original distribution is truncated and becomes taller due to the appearance of fixed denominator. Moreover, the probability equals zero when route travel time with the lower bound is less than free-flow travel time. Obviously, the new PDF still fulfills the conservation property of a valid PDF according to (6). In addition, rather than truncated continuous distributions, other modeling ways can also be used to model the impacts of the lower bound of route travel time. Xu et al. (2018) elaborated the reality and feasibility of truncated travel time distribution and its mathematical tractability [28].

2.2. Mean and Standard Deviation Derivations. Below, we derive the moments of route travel time T with the free-flow travel time as the lower bound. The truncated PDF in (6) can be rewritten as

$$f\left(T\mid\mu,\sigma,\overline{t}\right) = \frac{\left(1/\sqrt{2\pi}\sigma\right)\exp\left(-\left(T-\mu\right)^2/2\sigma^2\right)}{1-\Phi\left(\left(\overline{t}-\mu\right)/\sigma\right)}, \quad \forall T>\overline{t} \tag{7}$$

where $\Phi(\cdot)$ is the CDF of the standard normal distribution. The mean and standard deviation of truncated travel time distribution are derived using the method of moment estimation as follows.

The mean of truncated travel time distribution $E(T\mid T>\overline{t})$ can be derived, i.e.,

$$E\left(T\mid T>\overline{t}\right) = \frac{\int_{\overline{t}}^{+\infty} T\left(1/\sqrt{2\pi}\sigma\right)\exp\left(-\left(T-\mu\right)^2/2\sigma^2\right)dT}{\Phi\left(-\left(\overline{t}-u\right)/\sigma\right)} \tag{8}$$

Let $x=(T-\mu)/\sigma$, we have $T=\mu+x\sigma$ and

$$\begin{aligned}&\int_{\overline{t}}^{+\infty} T\frac{1}{\sqrt{2\pi}\sigma}\exp\left(-\frac{\left(T-\mu\right)^2}{2\sigma^2}\right)dT\\ &= \frac{1}{\sqrt{2\pi}}\int_{(\overline{t}-\mu)/\sigma}^{+\infty}\left(\mu+x\sigma\right)\exp\left(-\frac{x^2}{2}\right)dT\\ &= \mu\Phi\left(\frac{\mu-\overline{t}}{\sigma}\right)+\frac{\sigma}{\sqrt{2\pi}}\exp\left(-\frac{\left(\overline{t}-\mu\right)^2}{2\sigma^2}\right)\end{aligned} \tag{9}$$

Then, substituting (4) into (3), we have the following.

$$E\left(T\mid T>\overline{t}\right) = \mu+\frac{\left(\sigma/\sqrt{2\pi}\right)\exp\left(-\left(\overline{t}-\mu\right)^2/2\sigma^2\right)}{\Phi\left(\left(\mu-\overline{t}\right)/\sigma\right)} \tag{10}$$

Obviously, it can be observed that $E(T\mid T>\overline{t})>\mu$ from (5). It implies that the mean of truncated route travel time moves right and becomes larger compared to the original one.

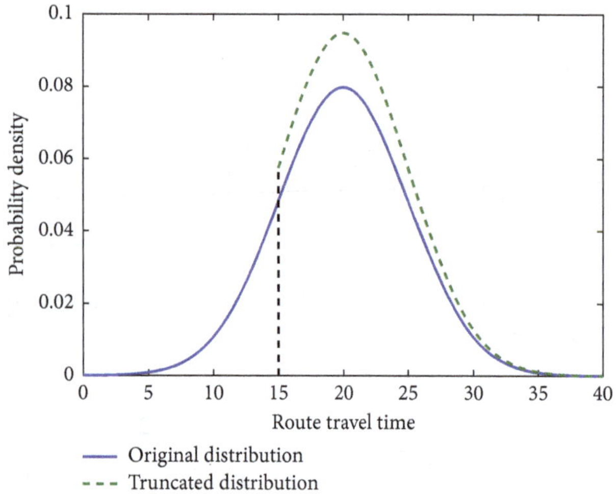

FIGURE 1: PDFs of route travel time with and without lower bound.

Similarly, the secondary moment of truncated travel time distribution can be derived, i.e.,

$$\begin{aligned}
&E\left(T^2 T > \bar{t}\right)\\
&= \frac{\int_{\bar{t}}^{+\infty} T^2 \left(1/\sqrt{2\pi}\sigma\right) \exp\left(-(T-\mu)^2/2\sigma^2\right) dT}{\Phi\left(-(\bar{t}-u)/\sigma\right)}\\
&= \Phi^{-1}\left(-\frac{\bar{t}-u}{\sigma}\right) \int_{(\bar{t}-\mu)/\sigma}^{+\infty} \frac{(\mu + x\sigma)^2}{\sqrt{2\pi}} \exp\left(-\frac{x^2}{2}\right) dx\\
&= \Phi^{-1}\left(-\frac{\bar{t}-u}{\sigma}\right) \left(\int_{(\bar{t}-\mu)/\sigma}^{+\infty} \frac{\mu^2}{\sqrt{2\pi}} \exp\left(-\frac{x^2}{2}\right) dx \right.\\
&\quad + \int_{(\bar{t}-\mu)/\sigma}^{+\infty} \frac{2x\mu\sigma}{\sqrt{2\pi}} \exp\left(-\frac{x^2}{2}\right) dx\\
&\quad \left. + \int_{(\bar{t}-\mu)/\sigma}^{+\infty} \frac{x^2\sigma^2}{\sqrt{2\pi}} \exp\left(-\frac{x^2}{2}\right) dx \right) = \mu^2 + \sigma^2\\
&\quad + \frac{\sigma(\bar{t}+\mu) \exp\left(-(\bar{t}-\mu)^2/2\sigma^2\right)}{\sqrt{2\pi}\Phi\left((\mu-\bar{t})/\sigma\right)}.
\end{aligned} \tag{11}$$

Then, the variance $D(T \mid T > \bar{t})$ and standard deviation $d(T \mid T > \bar{t})$ of truncated travel time distribution are, respectively,

$$D(T \mid T > \bar{t}) = \mathrm{E}\left(T^2 \mid T > \bar{t}\right) - [\mathrm{E}(T \mid T > \bar{t})]^2 \tag{12}$$

$$d(T \mid T > \bar{t}) = \sqrt{D(T \mid T > \bar{t})}. \tag{13}$$

3. TTB-Based Route Choice Model with Truncated Travel Time Distribution

In order to model travelers' route choice behavior under risk, different decision criterions have been proposed, such as TTB, METE, and PMETE. Xu et al. (2014) provided a detailed review on existing traffic equilibrium models under uncertainty [29]. This section restates the concepts of TTB and TTR and illustrates the effects of the route travel time boundary on them.

3.1. TTB without Route Travel Time Boundary

Definition 1. The TTB b_r^w on route r between OD pair w associated with a predefined confidence level ρ^w is defined as the sum of the expected travel time of travelers μ_r^w and a buffer time, the latter of which is the product of the standard deviation σ_r^w and travelers' risk-aversion level λ^w, i.e.,

$$b_r^w = \mu_r^w + \lambda^w \sigma_r^w, \quad r \in R_w, \ w \in W. \tag{14}$$

Definition 2. The TTR refers to the probability that travelers can arrive on time within the TTB b_r^w, i.e.,

$$P\left\{T_r^w \le b_r^w = \mu_r^w + \lambda^w \sigma_r^w\right\} = \rho^w, \quad r \in R_w, \ w \in W. \tag{15}$$

Thus, (15) can also be alternatively written as follows.

$$P\left\{\frac{T_r^w - \mu_r^w}{\sigma_r^w} \le \lambda^w\right\} = \rho^w \tag{16}$$

Mathematically, the risk-aversion level λ^w can be expressed as

$$\lambda^w = \Phi^{-1}\left(\rho^w\right), \quad \forall w \in W \tag{17}$$

where $\Phi^{-1}(\cdot)$ is the inverse function of the CDF of standard normal distribution. Obviously, travelers' risk-aversion level λ^w is only related to their expected on-time arrival probability ρ^w. In other words, the scale of λ^w for users with the same confidence level within one OD pair is identical.

3.2. TTB with Route Travel Time Boundary. The route free-flow time is viewed as a truncation of the original normal distribution, the resultant PDF is significantly different. In this case, the TTB and TTR with travel time boundary need to be derived and discussed.

For ease of the mathematical derivation, we assume that the expected travel time and risk have no changes under the truncated travel time distribution, and the TTB with route travel time boundary is as

$$\widehat{b}_r^w = \mu_r^w + \widehat{\lambda}_r^w \sigma_r^w, \quad \forall r \in R^w, \ w \in W \tag{18}$$

where $\widehat{b}_r^w$ and $\widehat{\lambda}_r^w$ are the TTB and risk-aversion level under truncated travel time distribution, respectively. $\widehat{\lambda}_r^w$ is no longer the constant which is only related to travelers' confidence level and it is also affected by the characteristics used routes (i.e., route travel time distribution and route free-flow travel time). More specifically, travelers' risk-aversion level $\widehat{\lambda}_r^w$ is context-based and adaptive, which would be proved and derived in the following.

According to Definition 2, the TTR under truncated distribution can be expressed as

$$P\left\{\widehat{T}_r^w \le \widehat{b}_r^w\right\} = \rho^w, \quad \forall r \in R^w, \ w \in W \tag{19}$$

Obviously, (19) can also be written as the conditional probability, i.e.,

$$P\{T_r^w \le \hat{b}_r^w \mid T_r^w > T_r^{w,0}\} = \rho^w, \quad \forall r \in R^w,\ w \in W \quad (20)$$

where $T_r^{w,0}$ is the free-flow travel time on route r between OD pair w. Mathematically, (20) can be alternatively written as

$$\frac{P\{T_r^{w,0} < T_r^w \le \hat{b}_r^w\}}{P\{T_r^w > T_r^{w,0}\}} = \rho^w, \quad \forall r \in R^w,\ w \in W \quad (21)$$

According to (21), we have the following.

$$P\{T_r^w \le \hat{b}_r^w\} = \rho^w\left(1 - F\left(T_r^{w,0}\right)\right) + F\left(T_r^{w,0}\right) \quad (22)$$

Then, we can derive

$$F\left(\hat{b}_r^w\right) = \rho^w\left(1 - F\left(T_r^{w,0}\right)\right) + F\left(T_r^{w,0}\right) \quad (23)$$

Let $\rho^w(1 - F(T_r^{w,0})) + F(T_r^{w,0}) = \xi_r^w$, we have

$$\hat{b}_r^w = F^{-1}\left(\xi_r^w\right) \quad (24)$$

where $F^{-1}(\cdot)$ denotes the inverse function of the CDF of normal distribution.

3.3. Effect of Truncated Travel Time Distribution on TTB and TTR. In this subsection, some theoretical analysis on the TTB and TTR changes is given when considering route travel time boundary, and we illustrate the effects of truncated travel time distribution on TTB and TTR.

Proposition 3. *For route r between OD pair w, the truncated travel time distribution makes the corresponding TTR reduce under the same budget time b compared to the original distribution.*

Proof. Assume TTR_{with} and $TTR_{without}$ denote the within budget time reliability with and without travel time boundary, respectively, and the relationship between TTR_{with} and $TTR_{without}$ can be expressed as

$$\begin{aligned} &TTR_{with}(b) - TTR_{without}(b) \\ &= \int_{T_r^{w,0}}^{b} \frac{f(T)}{1 - F\left(T_r^{w,0}\right)} dT - \int_{-\infty}^{b} f(T)\, dT \\ &= \frac{F(b) - F\left(T_r^{w,0}\right)}{1 - F\left(T_r^{w,0}\right)} - F(b) = \frac{F\left(T_r^{w,0}\right)(F(b) - 1)}{1 - F\left(T_r^{w,0}\right)} \\ &< 0 \end{aligned} \quad (25)$$

where $F(b) < 1$ is used in the above inequality. From (25), TTR_{with} is smaller than $TTR_{without}$, and the proof of Proposition 3 is completed. In addition, Proposition 3 also implies that ignoring route travel time boundary would result in overestimating the TTR. □

Proposition 4. *For route r between OD pair w, the truncated travel time distribution provides a longer TTB under the same confidence level ρ^w compared to the original distribution.*

Proof. Assume b_{with} and $b_{without}$ are the TTBs with and without travel time boundary, the WBTR under truncated travel time distribution can be expressed as follows.

$$\begin{aligned} &\int_{T_r^{w,0}}^{b_{with}} \frac{\left(1/\sqrt{2\pi}\sigma_r^w\right)\exp\left(-\left(T - \mu_r^w\right)^2 / 2\left(\sigma_r^w\right)^2\right)}{1 - \Phi\left(\left(T - \mu_r^w\right)/\sigma_r^w\right)} dT \\ &= \rho^w \end{aligned} \quad (26)$$

Then, we transform (26) and have the following.

$$\begin{aligned} F\left(b_{with}\right) &= \rho^w \Phi\left(\frac{\mu_r^w - T_r^{w,0}}{\sigma_r^w}\right) + F\left(T_r^{w,0}\right) \\ &= \rho^w\left(1 - F\left(T_r^{w,0}\right)\right) + F\left(T_r^{w,0}\right) \end{aligned} \quad (27)$$

Similarly, the TTR under nontruncated travel time distribution can be derived, i.e.,

$$\int_{-\infty}^{b_{without}} \frac{1}{\sqrt{2\pi}\sigma_r^w} \exp\left(-\frac{\left(T - \mu_r^w\right)^2}{2\left(\sigma_r^w\right)^2}\right) dT = \rho^w \quad (28)$$

Meanwhile, (28) can be written as

$$F\left(b_{without}\right) = \rho^w \quad (29)$$

According to (27) and (29), we can obtain the difference of budget time between two types of distribution, i.e.,

$$\begin{aligned} &F\left(b_{with}\right) - F\left(b_{without}\right) \\ &= \rho^w\left(1 - F\left(T_r^{w,0}\right)\right) + F\left(T_r^{w,0}\right) - \rho^w \\ &= F\left(T_r^{w,0}\right)\left(1 - \rho^w\right) > 0. \end{aligned} \quad (30)$$

Because the CDF of normal distribution is monotonically increasing, $b_{with} > b_{without}$. Proposition 4 is proved. □

Proposition 5. *Travelers adopt a self-adaptive risk-aversion level $\hat{\lambda}_r^w$ under truncated travel time distribution, which is not only related to travelers' confidence level, but also affected by the used route.*

Proof. Combining (18) and (22), the TTR under truncated travel time distribution can be derieved, i.e.,

$$P\{T_r^w \le \mu_r^w + \hat{\lambda}_r^w \sigma_r^w\} = \rho^w\left(1 - F\left(T_r^{w,0}\right)\right) + F\left(T_r^{w,0}\right) \quad (31)$$

By rearranging terms of LHS in (31), and

$$P\left\{\frac{T_r^w - \mu_r^w}{\sigma_r^w} \le \hat{\lambda}_r^w\right\} = \rho^w\left(1 - F\left(T_r^{w,0}\right)\right) + F\left(T_r^{w,0}\right) \quad (32)$$

Then, $\hat{\lambda}_r^w$ can be derived as follows.

$$\hat{\lambda}_r^w = \Phi^{-1}\left(\xi_r^w\right) \quad (33)$$

Because ξ_r^w is associated with $T_r^{w,0}$ and ρ^w, the risk-aversion level $\hat{\lambda}_r^w$ is not only related to travelers' confidence level, but also affected by the used route. In addition, the difference of risk-aversion levels between different distributions can be expressed as follows.

$$\hat{\lambda}_r^w - \lambda^w = \Phi^{-1}(\xi_r^w) - \Phi^{-1}(\rho^w) \quad (34)$$

Obviously, $\xi_r^w > \rho^w$, and $\Phi^{-1}(\cdot)$ is monotonically increasing function. So we have $\hat{\lambda}_r^w > \lambda^w$. It implies that route travel time boundary has a great effect on travelers' risk attitudes. □

3.4. An Illustrative Example. An example network with two parallel routes connecting one origin and one destination is used to present the difference between the TTB model with and without truncated travel time distribution. In this example, the route travel time is assumed to follow a normal distribution $N(\mu, \sigma)$. The means and standard deviations of two route travel time are assumed to be $\mu_1 = 20$, $\mu_2 = 15$, $\sigma_1 = 5$, $\sigma_2 = 10$. The free-flow travel times are 15 min and 10 min, respectively. All users are assumed to be homogeneous with the same confidence $\rho = 0.90$.

The TTBs of route 1 with and without truncated travel time distribution are 26.89 and 26.41. And those of route 2 are 29.83 and 27.82, respectively. These imply that ignoring the effects of route travel time boundary would result in an exiguity of budget time. Travelers' risk-aversion level is only dependent on their confidence level when route travel time follows a general normal distribution. In this case, travelers' risk-aversion level can be calculated as 1.28. However, travelers' risk-aversion level not only depends on the expected on-time arrival condition, but also is affected by the characteristics of the used routes (i.e., travel time distribution and free-flow time) when considering route travel time boundary, which is a context-based and adaptive constant. The risk-aversion level of travelers using route 1 is 1.38, and that of those using route 2 is 1.48. Then, to reveal the difference of TTR incurred by different travel time distributions, we assume travelers' TTB is a fixed constant and its quantity is 28. The TTRs of route 1 with and without truncated travel time distribution are, respectively, 0.93 and 0.95. The results indicate that neglecting the travel time boundary would overestimate the travel time reliability.

Table 1 shows the effects of scale variations of route free-flow travel time on TTB and travelers' risk-aversion level. From this table, we can see that travelers adopt a longer budget time to ensure on-time arrival when they are faced with higher risk. As the free-flow travel time increases, or as longer minimum travel time is imposed on routes, the values of TTB and risk-aversion level increase. Intuitively, it is clear that the growth of the minimum travel time results in a longer budget time. Thus, travelers become more aversive to risk when a longer budget time is suggested. It can be also found that travelers' risk-aversion levels are different for using different routes. These results indicate that route travel time boundary has a great effect on travelers' TTBs and risk attitudes.

TABLE 1: Effect of route travel time boundary on TTB and traveler' risk attitudes.

Route #	Free-flow travel time	TTB	Risk aversion level
1	6	26.42	1.28
	9	26.45	1.29
	12	26.57	1.31
	15	26.89	1.38
	18	27.55	1.51
2	10	29.83	1.48
	11	30.10	1.51
	12	30.40	1.54
	13	30.73	1.57
	14	31.07	1.61

4. TTB-Based SUE Model

This section models the effects of route travel time boundary on the performance of network equilibrium. We also present the differences between both SUE models with and without route travel time boundary. In the SUE state, no traveler can improve his or her perceived TTB by unilaterally changing route. The perceived TTBs of all used routes are the same and less than those of unused routes.

4.1. TTB-SUE without Route Travel Time Boundary. In the TTB-SUE model, travelers are always inclined to choose the routes with the least perceived TTB. Travelers' route choice behavior follows the random utility maximization (RUM) principle; the perceived TTB B_r^w on route $r \in R_w$ can be expressed as

$$B_r^w = b_r^w + \varepsilon_r^w, \quad r \in R_w, \ w \in W \quad (35)$$

where b_r^w is given by (14) and the random terms ε_r^w is assumed to be identically and independently distributed (i.i.d.) Gumbel variate, $r \in R_w$, $w \in W$. According to the RUM principle, the route choice probability p_r^w on route $r \in R_w$ at equilibrium can be written as

$$p_r^w = \frac{\exp(-\theta b_r^w)}{\sum_{k \in R_w} \exp(-\theta b_k^w)}, \quad r \in R_w, \ w \in W \quad (36)$$

where θ is a dispersion parameter measuring the variability in perceived TTB. A small θ value implies a larger perceived error compared to the shortest route. On the contrary, a large θ value indicates a smaller perceived error associated with the shortest route. That is, travelers would be more inclined to the shortest route when θ value is large enough. The equilibrium flow pattern can be obtained as follows

$$f_r^w = d_w p_r^w = d_w \frac{\exp(-\theta b_r^w)}{\sum_{k \in R_w} \exp(-\theta b_k^w)}, \quad r \in R_w, \ w \in W \quad (37)$$

where d_w is travel demand on OD pair w. Apparently, the equilibrium flow pattern determined by the TTB-SUE model depends upon not only the considered route' TTB but also other alternatives' TTBs.

Mathematically, the perceived TTB B_r^w under the general normal distribution can be expressed as follows.

$$B_r^w = \frac{1}{\theta}(\ln f_r^w + 1) + b_r^w, \quad r \in R_w,\ w \in W \tag{38}$$

4.2. TTB-SUE with Route Travel Time Boundary. Compared to the TTB-SUE without route travel time boundary, the TTB-SUE with route travel time boundary uses a truncated travel time distribution to reflect the effects of route free-flow time on travelers' route choice behavior. In this case, travelers aim to minimize the perceived travel time budget with truncate travel time distribution. The new perceived travel time budget $\widehat{B}_r^w$ can be expressed as

$$\widehat{B}_r^w = \widehat{b}_r^w + \xi_r^w, \quad r \in R_w,\ w \in W \tag{39}$$

where $\widehat{B}_r^w$ is the perceived travel time budget under truncated travel time distribution. $\widehat{b}_r^w$ is given by (18) or (24). The random term ξ_r^w is assumed to be i.i.d. Gumbel variate, $r \in R_w$, $w \in W$. The route choice probability p_r^w on route $r \in R_w$ can be written as follows.

$$p_r^w = \frac{\exp\left(-\theta \widehat{b}_r^w\right)}{\sum_{k \in R_w} \exp\left(-\theta \widehat{b}_k^w\right)}, \quad r \in R_w,\ w \in W \tag{40}$$

Accordingly, we can obtain the equilibrium route flow pattern.

$$f_r^w = d_w p_r^w = \frac{\exp\left(-\theta \widehat{b}_r^w\right)}{\sum_{k \in R_w} \exp\left(-\theta \widehat{b}_k^w\right)}, \quad r \in R_w,\ w \in W \tag{41}$$

The perceived travel time budget $\widehat{B}_r^w$ under truncated travel time distribution can be expressed as follows.

$$\widehat{B}_r^w = \frac{1}{\theta}(\ln f_r^w + 1) + \widehat{b}_r^w, \quad r \in R_w,\ w \in W \tag{42}$$

4.3. The Equivalent VI Formulation. Both TTB-SUE conditions can be written as follows.

$$f_r^w (f_r^w - d_w p_r^w) = 0, \quad f_r^w \ge 0 \tag{43}$$

So, logit-based TTB-SUE problem can be formulated equivalently as the following VI problem.

$$\sum_{w \in W} \sum_{r \in R_w} (f_r^{w*} - p_r^w (b(f_r^{w*})) \cdot d_w)(f_r^w - f_r^{w*}) \ge 0, \quad \forall f_r^w \in \Omega \tag{44}$$

Obviously, (44) can be alternatively written as

$$(\mathbf{f}^* - \mathbf{p}(\mathbf{b}(\mathbf{f}^*)) \cdot \mathbf{d})^{\mathrm{T}} (\mathbf{f} - \mathbf{f}^*) \ge 0, \quad \mathbf{f} \in \Omega \tag{45}$$

where Ω is the feasible flow set defined by the following.

$$\Omega = \left\{ f_r^w \ge 0 : \sum_{r \in R_w} f_r^w = d_w,\ w \in W \right\} \tag{46}$$

Theorem 6. *The VI problem (45) is equivalent to the equilibrium conditions of both TTB-SUE models.*

Proof. The proof of the theorem follows the works of Zhou et al. (2012) [30] and Bekhor et al. (2012) [31]. For the proof of the necessity, we suppose $\mathbf{f}^*$ is a solution of any TTB-SUE model; from the SUE condition in (37) or (41), the VI problem is satisfied naturally. Thus, any solution of both TTB-SUE models is a solution of the proposed VI problem. For the proof of sufficiency, let $\mathbf{f}^*$ be a solution of the VI problem. Without loss of generality, we fix a route $h \in R_v$ within OD pair v and construct a feasible route flow $\mathbf{f}$ such that $f_r^w = f_r^{w*}$, $(r, w) \ne (h, v)$, but $f_h^v \ne f_h^{v*}$. Then substituting them into (40), we have the following.

$$(f_h^{v*} - p_h^v (b(f_h^{v*})) \cdot d_v)^{\mathrm{T}} (f_h^v - f_h^{v*}) \ge 0 \tag{47}$$

For each effective route $h \in R_v$, $v \in W$, it can be deduced that $f_h^v > 0$. So one can obtain the following.

$$f_h^{v*} - p_h^v (b(f_h^{v*})) d_v = 0 \tag{48}$$

Thus, the SUE condition in (37) or (41) is satisfied, and the solution of the proposed VI problem is also the solution of TTB-SUE model. □

Theorem 7. *There exists at least one solution for the proposed VI problem in (45).*

Proof. According to the mathematical formulation of $\mathbf{b}(\mathbf{f})$, we know that it is a continuous function associated with $\mathbf{f}$, and $\mathbf{f}^* - \mathbf{p}(\mathbf{b}(\mathbf{f}^*)) \cdot \mathbf{d}$ is also a continuous function of $\mathbf{f}$. In addition, since Ω is a nonempty, convex, and compact set, the proposed VI problem (45) has at least one solution.

However, the uniqueness of the solution of the equivalent VI problem cannot be guaranteed. The reason is that the solution to logit-based TTB-SUE is not unique in general and the strict monotonicity of the mapping $\mathbf{f}^* - \mathbf{p}(\mathbf{b}(\mathbf{f}^*)) \cdot \mathbf{d}$ cannot be ensured. □

5. Solution Algorithm

The MSA algorithm is employed to resolve the proposed equilibrium problem in this paper. And the algorithm steps are expressed as follows.

Step 1 (initialization). For given OD demand d_w of OD pair w, randomly load a feasible route flow vector $\mathbf{f}^{(1)} \in \Omega$. Set the maximum iteration number N and the convergence tolerance $\varepsilon > 0$. Let the iteration counter $n := 1$.

Step 2 (cost calculation). Determine link flow vector $\mathbf{x}^{(n)}$ according to the relationship between link flows and route flows, and then calculate the mean $E^{(n)}(t_a)$ and standard deviation $\sigma^{(n)}(t_a)$ of link travel time. Accordingly, we can obtain the mean $E^{(n)}(T_r^w)$ and standard deviation $\sigma^{(n)}(T_r^w)$ of route travel time by (4) and (5). Thus, the travel time budget with truncated travel time distribution $\widehat{b}_r^{w(n)}$ can be derived by (24).

Table 2: Route characteristics of the three-route transportation network.

Route #	Free-flow travel time (min)	Capacity (veh/h)	Reliability (φ_a)
1	12	1000	0.5
2	30	2000	0.7
3	40	3000	0.9

Table 3: The equilibrium results of TTB-SUE without travel time boundary.

Route #	f_r^w	μ_r^w	σ_r^w	λ	b_r^w	B_r^w
1	1356.63	12.36	23.22		42.12	50.33
2	2115.83	35.31	4.97	1.28	41.68	50.33
3	1527.54	41.92	0.60		42.00	50.33

Table 4: The equilibrium results of TTB-SUE with travel time boundary.

Route #	f_r^w	μ_r^w	σ_r^w	$\hat{\lambda}_r^w$	$\hat{b}_r^w$	$\hat{B}_r^w$
1	1281.98	12.29	18.52	1.64	42.64	50.79
2	2117.24	35.32	4.98	1.36	42.13	50.79
3	1600.78	42.32	0.07	1.28	42.41	50.79

Step 3 (flow assignment). Set the auxiliary route vector $\overline{\mathbf{f}}^{(n)}$, which can be yielded by (40) and (41).

$$\overline{f}_r^w = d_w p_r^w = \frac{\exp\left(-\theta \hat{b}_r^w\right)}{\sum_{k \in R_w} \exp\left(-\theta \hat{b}_k^w\right)}, \quad r \in R_w,\ w \in W \quad (49)$$

Step 4 (update). The route flow vector $\mathbf{f}^{(n)}$ can be updated by the following.

$$\mathbf{f}^{(n+1)} = \mathbf{f}^{(n)} + \frac{1}{n}\left(\overline{\mathbf{f}}^{(n)} - \mathbf{f}^{(n)}\right) \quad (50)$$

Step 5 (checking the convergence). If the iteration counter n exceeds the maximum iteration number N or (51) is satisfied, the algorithm stops and the present $\mathbf{f}^{(n+1)}$ is the equilibrium results.

$$\frac{\sqrt{\sum_{r \in R_w}\left(f_{rw}^{(n+1)} - f_{rw}^{(n)}\right)^2}}{\sum_{r \in R_w} f_{rw}^{(n)}} \le \varepsilon \quad (51)$$

Otherwise, make $n := n + 1$ and return to Step 2.

6. Numerical Example

6.1. A Three-Route Network. This section uses two transportation networks to demonstrate the proposed model and algorithm. The first network consists of three parallel routes shown in Figure 2. The characteristic values of the link parameters are given in Table 2. The parameters of BPR function in (1) are $\beta = 0.15$ and $n = 4$. The total traffic demand is assumed to be fixed and it is 15,000 vehicles per hour. Without loss of generality, travelers are assumed to be homogeneous with the identical expected on-time arrival probability $\rho = 0.90$.

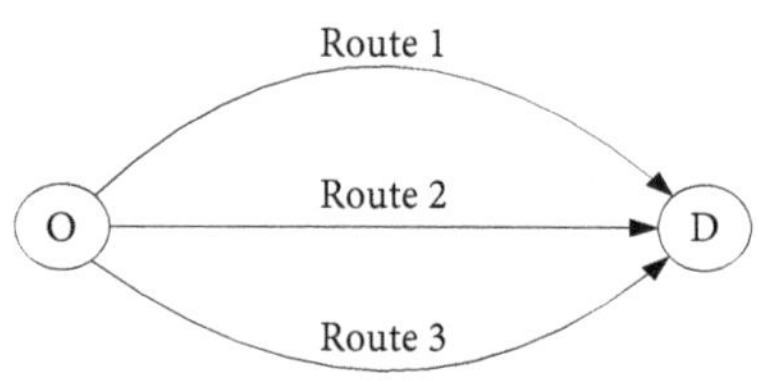

Figure 2: A three-route transportation network.

The dispersion parameter θ is assumed to be 1 in both TTB-SUE models. Tables 3 and 4 provide the equilibrium results of both SUE models. From the tables, one can see that two SUE models give quite different equilibrium results. It implies that route travel time boundary has a great impact on travelers' route choice behavior under uncertainty. In the TTB-SUE model without route travel time boundary, the risk-averse level is 1.28 no matter which route is used, which is only related to travelers' expected on-time arrival probability. In contrast, it is heterogeneous which is also affected by the used routes when considering route travel time boundary. It is plausible that travelers adopt different risk-aversion attitudes for diverse routes because of the physical performance difference between different routes. From the tables, we can see that the risk-aversion level is still 1.28 when using route 3. However, it is distinct when using other two routes, such as 1.64 for route 1 and 1.36 for route 2. As expected, longer travel time budget is suggested by the TTB-SUE model with travel time boundary than that without travel time boundary in terms of each route. In addition, the perceived travel time budgets using different routes are identical at equilibrium, and more perceived travel time budget is caused under truncated travel time distribution.

Figure 3 depicts the effects of the expected on-time arrival probability ρ on travelers' risk-aversion levels using different routes. As ρ increases, travelers' risk-aversion levels increase

Table 5: The equilibrium results of TTB-SUE with truncated travel time distribution.

O-D	Route #	Sequence of nodes	Route flow	Travel time budget	Perceived travel time budget	Expected travel time	Risk	Risk-aversion level
1-3	1	1-5-6-7-11-3	145.37	61.39	67.37	42.23	17.46	1.10
	2	1-5-6-10-11-3	57.90	62.31		49.52	12.39	1.03
	3	1-5-9-10-11-3	44.81	62.57		49.07	12.62	1.07
	4	1-12-6-7-11-3	231.02	60.93		42.91	16.20	1.11
	5	1-12-6-10-11-3	86.63	61.94		42.77	16.80	1.14
	6	1-5-9-13-3	37.27	62.75		50.06	11.72	1.08
4-2	7	4-5-6-7-8-2	260.01	73.37	79.93	47.73	24.05	1.07
	8	4-5-6-7-11-2	130.46	74.06		47.89	23.76	1.10
	9	4-5-6-10-11-2	53.96	74.94		55.18	18.68	1.06
	10	4-9-10-11-2	40.58	75.23		54.73	18.92	1.08
	11	4-5-9-10-11-2	314.99	73.18		55.20	17.88	1.01

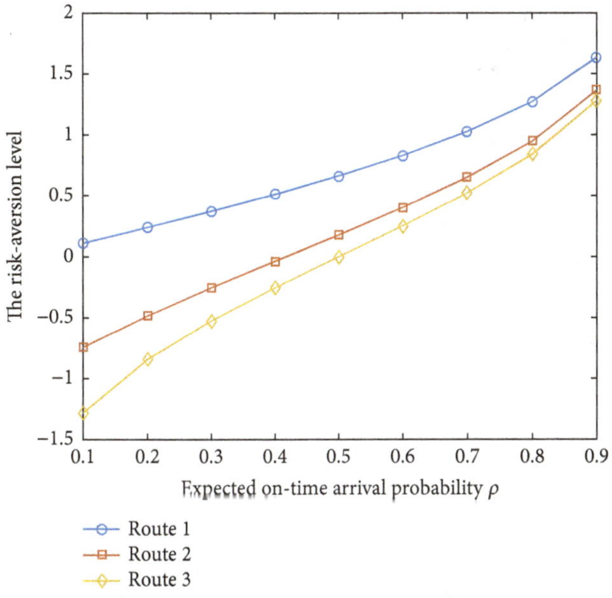

Figure 3: The risk-aversion level against ρ.

for each route. In comparison, travelers using route 1 of the worst reliability adopt largest risk-aversion levels to address the risk, and travelers on route 3 of the best reliability have the smallest risk-aversion attitudes.

Figures 4(a) and 4(b) show the impacts of different boundary scales of route travel time on the expected travel time and risk. Specifically, the expected travel time on three routes is always increasing as the boundary scale becomes larger. Meanwhile, the most reliable route (i.e., route 3) has the longest expected travel time, whereas using the most degradable route (i.e., route 1) spends the shortest expected travel time. As the boundary scale of route travel time increases, travel time risk on three routes is also always increasing. In addition, one can see that travelers using the most reliable route are confronted with the lowest travel time risk, and those using the shortest expected travel time encounter the highest travel time risk. These imply that travelers have to make a tradeoff between expected travel time and risk. These also demonstrate that travelers' risk attitude has to be incorporated into route choice modeling. Figure 4(c) plots the variation of perceived TTB with route travel time boundary in the equilibrium state. From this figure, we can see that the perceived TTB increases constantly when boundary scale of route travel time becomes larger. All of these indicate that route travel time boundary has a great influence on travelers' travel cost (i.e., expected travel time, risk, and perceived travel time budget).

6.2. Nguyen-Dupuis Network. In this section, the Nguyen-Dupuis network is used to demonstrate the solvability of the proposed model with route travel time boundary under uncertainty. This network has 13 nodes, 19 links, and 4 OD pairs. The BPR performance function ($\beta = 0.15$, $n = 4$) in (1) is adopted to compute link travel cost. Link specifications are provided in Figure 5. The potential demands are $d_{12} = 400$, $d_{13} = 800$, $d_{42} = 600$, and $d_{43} = 400$, respectively. The link capacity parameter and discrete parameter are set as $\varphi_a = 0.4$ and $\theta = 1.0$. All travelers in the test network are assumed to be homogeneous with $\rho = 0.8$.

First, the equilibrium results of TTB-SUE with truncated travel time distribution are shown in Table 5. Without loss of generality, we only present the equilibrium results of OD (1-3) and OD (4-2). As expected, the perceived TTBs at the equilibrium on all used routes for each OD pair are equal, which is satisfied with the TTB-based SUE condition. Travelers prefer to choose the routes with shorter TTBs. It can be also observed that the travel time risk decreases with the increase in expected travel time. The risk-aversion level is heterogeneous for different used routes within one OD pair, which is jointly affected by expected on-time arrival probability and route characteristics (i.e., route free-flow travel time and distribution). It is relatively larger for routes with higher risk. However, it does not mean that the larger the travel time risk, the higher the risk-aversion level. As mentioned above, route travel time boundary is a major influence factor of travelers' risk-aversion level.

Second, we further examine the joint impacts of expected on-time arrival probability ρ and discrete parameter θ on perceived TTB in Figures 6 and 7. From these figures, we can

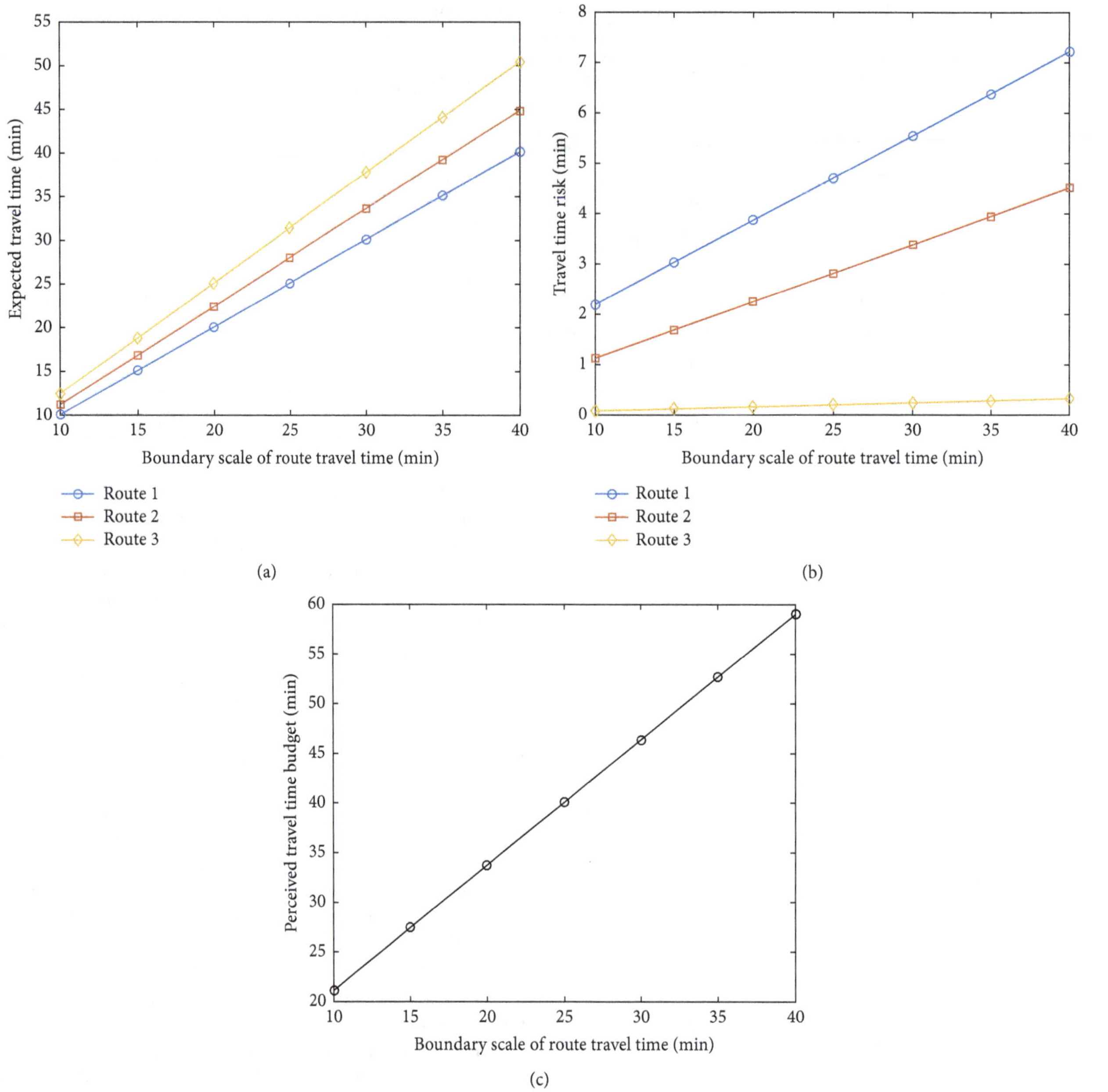

FIGURE 4: Expected travel time, risk, and perceived travel time budget on three routes under different boundary scale of route travel time. (a) Expected travel time, (b) travel time risk, (c) perceived travel time budget.

see that the perceived TTB decreases as θ increases. And the greater the expected on-time arrival probability ρ, the longer the perceived travel time budget. It implies that travelers with high requirement for arriving on time would make a longer budget time to avoid risk.

Then, the self-regulated averaging method (SRAM) proposed by Liu et al. (2009) is adopted to solve the model [32]. In the SRAM, the step size is $1/\tau^{(n)}$, where

$$\tau^{(n)} = \begin{cases} \tau^{(n-1)} + \Gamma, \ \Gamma > 1, & \text{if } \left\| \bar{\mathbf{f}}^{(n)} - \mathbf{f}^{(n)} \right\| \geq \left\| \bar{\mathbf{f}}^{(n-1)} - \mathbf{f}^{(n-1)} \right\| \\ \tau^{(n-1)} + \gamma, \ 0 < \gamma < 1, & \text{if } \left\| \bar{\mathbf{f}}^{(n)} - \mathbf{f}^{(n)} \right\| < \left\| \bar{\mathbf{f}}^{(n-1)} - \mathbf{f}^{(n-1)} \right\|. \end{cases} \quad (52)$$

In the MSA, $\tau^{(n)}$ is enlarged by an increment of one at each iteration. It can result in the next solution keeping farther away from the optimal solution than previous one and extremely slow convergence speed. However, $\tau^{(n)}$ depends on the information between iterations to adjust the choice of step size (i.e., speeding up or slowing down) in the SRAM. Meanwhile, the choice of the step size increment parameters Γ and γ is flexible, e.g., $\Gamma \in [1.5, 2.5]$ and $\gamma \in [0.01, 0.5]$.

Thus, we replace the step size $1/n$ with $1/\tau^{(n)}$ in (45). In addition, we assume the parameters $\theta = 1$, $\rho = 0.9$, $\Gamma = 1.5$, $\gamma = 0.5$, the maximum iteration number $N = 6000$, and the convergence tolerance $\varepsilon = 0.01$. Figure 8 compares the convergence performances of SRAM and MSA. The results

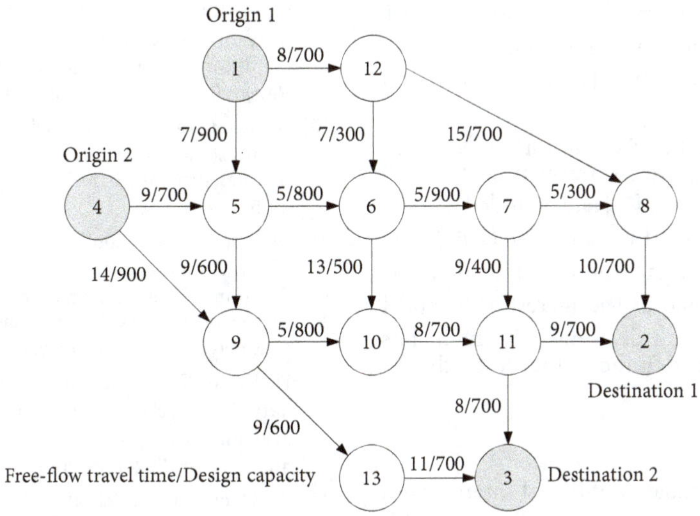

FIGURE 5: The Nguyen-Dupuis network and link specifications.

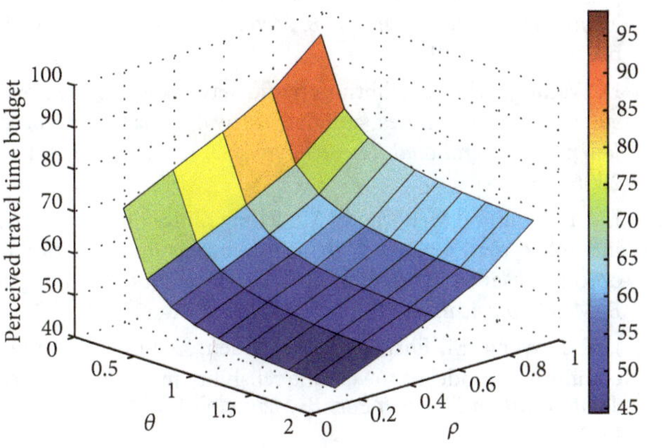

FIGURE 6: Joint impacts of parameters ρ and θ on perceived travel time budget (OD pair 1-3).

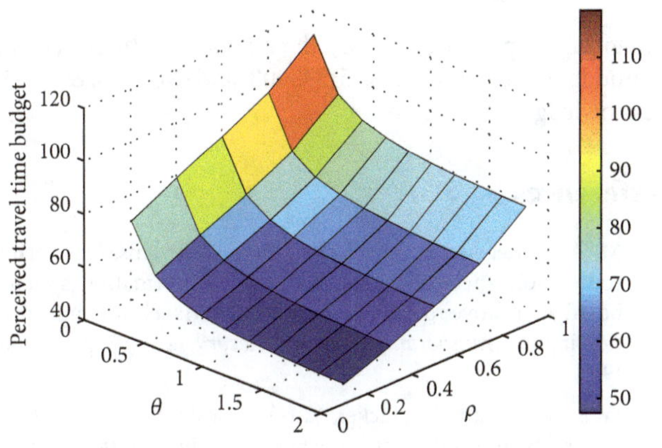

FIGURE 7: Joint impacts of parameters ρ and θ on perceived travel time budget (OD pair 4-2).

show that both SRMA and MSA can solve the model; the model is solvable. It is also clear that SRAM has a faster convergence speed than MSA.

7. Conclusions

This paper considers route free-flow travel time as the lower boundary of route travel time under uncertain networks. The travel time budget model with route travel time boundary is developed and derived. In such case, the budget time required by travelers is no less than that without route travel time boundary. Moreover, it has a lower travel time reliability under the same budget time. Travelers' risk-aversion attitude is no longer only associated with the expected on-time arrival condition, which is self-adaptive and depends on the chosen route.

To model traveler' route choice behavior when considering route travel time boundary, logit-based stochastic user equilibrium model based on travel time budget (TTB-SUE) is developed. A route-based solution algorithm is designed

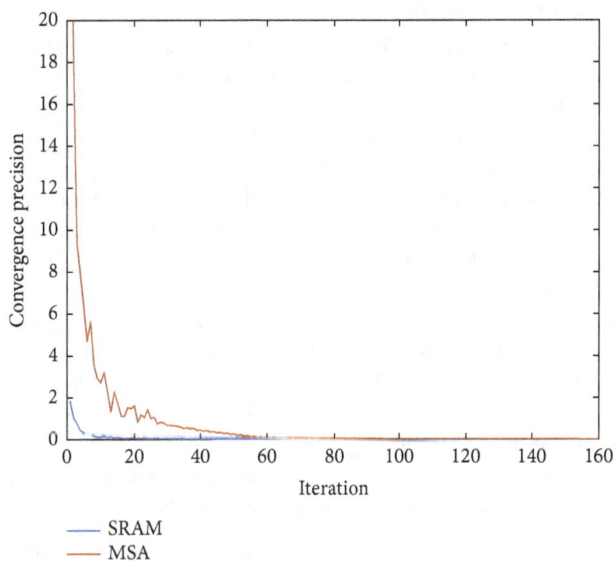

FIGURE 8: Convergence of the algorithms.

to solve the equivalent VI problem of the proposed model. The results indicate that route travel time boundary has a great impact on travelers' expected travel time, travel risk, and perceived travel time budget.

It should be noted that the model can be further extended to model its effects on mean-excess travel time (METT) proposed by Chen and Zhou (2010) [13] and bounded rational confidence level (BRCL) proposed by Sun et al. (2017) [33]. It can be also introduced and incorporated into dynamic travel assignment model. It would also be interesting to model travelers' risk-averse route choice decisions with multiclass users. These issues will be settled in the further research.

Conflicts of Interest

The authors declared no potential conflicts of interest with respect to the research, authorship, and/or publication of this article.

Acknowledgments

Financial support provided by the National Natural Science Foundation of China (No. 51338008 and No. 51378036) is acknowledged.

References

[1] M. A. Abdel-Aty, R. Kitamura, and P. P. Jovanis, "Exploring route choice behavior using geographic information system-based alternative routes and hypothetical travel time information input," *Transportation Research Record*, no. 1493, pp. 74–80, 1995.

[2] W. B. Jackson and J. V. Jucker, "An empirical study of travel time variability and travel choice behavior," *Transportation Science*, vol. 16, no. 4, pp. 460–475, 1982.

[3] T. C. Lam and K. A. Small, "The value of time and reliability: Measurement from a value pricing experiment," *Transportation Research Part E: Logistics and Transportation Review*, vol. 37, no. 2-3, pp. 231–251, 2001.

[4] M. A. Abdel-Aty, R. Kitamura, and P. P. Jovanis, "Using stated preference data for studying the effect of advanced traffic information on drivers' route choice," *Transportation Research Part C: Emerging Technologies*, vol. 5, no. 1, pp. 39–50, 1997.

[5] D. Brownstone, A. Ghosh, T. F. Golob, C. Kazuimi, and D. V. Amelsfort, "Drivers' willingness-to-pray to reduce travel time: Evidence from the San Diego I-15 congestion pricing project," *Transportation Research Part A: Policy and Practice*, vol. 37, no. 4, pp. 373–387, 2003.

[6] H. X. Liu, W. Recker, and A. Chen, "Uncovering the contribution of travel time reliability to dynamic route choice using real-time loop data," *Transportation Research Part A: Policy and Practice*, vol. 38, no. 6, pp. 435–453, 2004.

[7] H. K. Lo and Y. K. Tung, "Network with degradable links: capacity analysis and design," *Transportation Research Part B: Methodological*, vol. 37, no. 4, pp. 345–363, 2003.

[8] R. B. Noland, K. A. Small, P. M. Koskenoja, and X. Chu, "Simulating travel reliability," *Regional Science & Urban Economics*, vol. 28, no. 5, pp. 535–564, 1998.

[9] G. N. Bifulco, G. E. Cantarella, F. Simonelli, and P. Velonà, "Advanced traveller information systems under recurrent traffic conditions: network equilibrium and stability," *Transportation Research Part B: Methodological*, vol. 92, pp. 73–87, 2016.

[10] A. Khani and S. D. Boyles, "An exact algorithm for the mean-standard deviation shortest path problem," *Transportation Research Part B: Methodological*, vol. 81, part 1, pp. 252–266, 2015.

[11] M. Li, N. M. Rouphail, M. Mahmoudi, J. Liu, and X. Zhou, "Multi-scenario optimization approach for assessing the impacts of advanced traffic information under realistic stochastic capacity distributions," *Transportation Research Part C: Emerging Technologies*, vol. 77, pp. 113–133, 2017.

[12] H. K. Lo, X. W. Luo, and B. W. Y. Siu, "Degradable transport network: travel time budget of travelers with heterogeneous risk aversion," *Transportation Research Part B: Methodological*, vol. 40, no. 9, pp. 792–806, 2006.

[13] A. Chen and Z. Zhou, "The α-reliable mean-excess traffic equilibrium model with stochastic travel times," *Transportation Research Part B: Methodological*, vol. 44, no. 4, pp. 493–513, 2010.

[14] A. Chen, Z. Zhou, and W. H. K. Lam, "Modeling stochastic perception error in the mean-excess traffic equilibrium model," *Transportation Research Part B: Methodological*, vol. 45, no. 10, pp. 1619–1640, 2011.

[15] D. Watling, "User equilibrium traffic network assignment with stochastic travel times and late arrival penalty," *European Journal of Operational Research*, vol. 175, no. 3, pp. 1539–1556, 2006.

[16] J. Y. T. Wang and M. Ehrgott, "Modelling route choice behaviour in a tolled road network with a time surplus maximisation bi-objective user equilibrium model," *Transportation Research Part B: Methodological*, vol. 57, pp. 342–360, 2013.

[17] J. Y. T. Wang, M. Ehrgott, and A. Chen, "A bi-objective user equilibrium model of travel time reliability in a road network," *Transportation Research Part B: Methodological*, vol. 66, pp. 4–15, 2014.

[18] C. Sun, L. Cheng, S. Zhu, F. Han, and Z. Chu, "Multi-criteria user equilibrium model considering travel time, travel time reliability and distance," *Transportation Research Part D: Transport and Environment*, 2017.

[19] H. Shao, W. H. K. Lam, Q. Meng, and M. L. Tam, "Demand-driven traffic assignment problem based on travel time reliability," *Transportation Research Record*, no. 1985, pp. 220–230, 2006.

[20] H. Shao, W. H. K. Lam, and M. L. Tam, "A reliability-based stochastic traffic assignment model for network with multiple user classes under uncertainty in demand," *Networks and Spatial Economics*, vol. 6, no. 3-4, pp. 173–204, 2006.

[21] W. H. K. Lam, H. Shao, and A. Sumalee, "Modeling impacts of adverse weather conditions on a road network with uncertainties in demand and supply," *Transportation Research Part B: Methodological*, vol. 42, no. 10, pp. 890–910, 2008.

[22] Z. Zhou and A. Chen, "Comparative analysis of three user equilibrium models under stochastic demand," *Journal of Advanced Transportation*, vol. 42, no. 3, pp. 239–263, 2008.

[23] B. W. Y. Siu and H. K. Lo, "Doubly uncertain transportation network: degradable capacity and stochastic demand," *European Journal of Operational Research*, vol. 191, no. 1, pp. 166–181, 2008.

[24] H. Sun and Z. Gao, "Stochastic Traffic Equilibrium Based on Travel Time Robust Reliability," *Journal of Transportation Systems Engineering and Information Technology*, vol. 12, no. 2, pp. 76–84, 2012.

[25] S. V. Ukkusuri, T. V. Mathew, and S. T. Waller, "Robust transportation network design under demand uncertainty," *Computer-Aided Civil and Infrastructure Engineering*, vol. 22, no. 1, pp. 6–18, 2007.

[26] Y. Yin, S. M. Madanat, and X. Lu, "Robust improvement schemes for road networks under demand uncertainty," *European Journal of Operational Research*, vol. 198, no. 2, pp. 470–479, 2009.

[27] C.-Y. Yan, R. Jiang, Z.-Y. Gao, and H. Shao, "Effect of speed limits in degradable transport networks," *Transportation Research Part C: Emerging Technologies*, vol. 56, pp. 94–119, 2015.

[28] X. Xu, A. Chen, H. K. Lo, and C. Yang, "Modeling the impacts of speed limits on uncertain road networks," *Transportmetrica A: Transport Science*, vol. 14, no. 1-2, pp. 66–88, 2018.

[29] X. Xu, A. Chen, L. Cheng, and H. K. Lo, "Modeling distribution tail in network performance assessment: a mean-excess total travel time risk measure and analytical estimation method," *Transportation Research Part B: Methodological*, vol. 66, pp. 32–49, 2014.

[30] Z. Zhou, A. Chen, and S. Bekhor, "C-logit stochastic user equilibrium model: formulations and solution algorithm," *Transportmetrica*, vol. 8, no. 1, pp. 17–41, 2012.

[31] S. Bekhor, C. Chorus, and T. Toledo, "Stochastic user equilibrium for route choice model based on random regret minimization," *Transportation Research Record*, no. 2284, pp. 100–108, 2012.

[32] H. X. Liu, X. He, and B. He, "Method of successive weighted averages (MSWA) and self-regulated averaging schemes for solving stochastic user equilibrium problem," *Networks and Spatial Economics*, vol. 9, no. 4, pp. 485–503, 2009.

[33] C. Sun, L. Cheng, and J. Ma, "Travel time reliability with boundedly rational travelers," *Transportmetrica A: Transport Science*, vol. 14, no. 3, pp. 210–229, 2018.

6

Reliability-Based Estimation of Traffic Interruption Probability due to Road Waterlogging

Manuel Contreras-Jara,[1] Tomás Echaveguren,[1,2] José Vargas Baecheler,[1] Alondra Chamorro Giné,[2,3] and Hernán de Solminihac Tampier[3,4]

[1] *Laboratory of Transportation Systems Management (GESITRAN), Departamento de Ingeniería Civil, Facultad de Ingeniería, Universidad de Concepción, Edmundo Larenas 219, Concepción, Chile*
[2] *National Research Center for Integrated Natural Disaster Management (CIGIDEN), CONICYT/FONDAP/15110017, Avda. Vicuña Mackenna 4860, Santiago, Chile*
[3] *Department of Construction Engineering and Management, School of Engineering, Pontificia Universidad Católica de Chile, Avda. Vicuña Mackenna 4860, Santiago, Chile*
[4] *Latin American Center of Economic and Social Policies (CLAPES UC), Avda. Libertador Bernardo O'Higgins 440, Piso 13, Santiago, Chile*

Correspondence should be addressed to Tomás Echaveguren; techaveg@udec.cl

Academic Editor: Nirajan Shiwakoti

Floods affect road infrastructure physically and operationally, increase road user costs and road agencies cost, and eventually isolate communities. The research of the effect of floods on vehicular circulation is mainly focused on the stability of vehicles. There are few studies related to the regime of still water in the trafficability. In still water, the speed flow is low and does not compromise the vehicle stability. In this case, the vehicle's wading height becomes relevant. This article proposes a procedure to estimate the traffic interruption probability caused by floods in roads, considering the still water regime. The procedure uses the first-order reliability method to estimate traffic interruption probabilities, based on the difference between the probability density functions (PDF) of still water depth (or waterlogging depth) and vehicle wading height. A specific procedure to estimate the PDF of wading height based in the geometric characteristics of light and heavy vehicles was developed. The PDF for water depth was estimated using water level profile simulation software. The procedure was applied in the south of Chile. The PDF of wading height was obtained from a record of 166,155 vehicles tagged in open road tolls. The PDF of waterlogging depth was obtained from discharge records of 10 fluviometric stations. 42 probability curves were obtained for six vehicle classes and return periods between 4 and 500 years. The still water depth obtained for traffic interruption probability of 1,0 varied between 70 and 90 cm for light vehicles and between 110 and 150 cm for heavy vehicles.

1. Introduction

Floods damage the physical highway assets and affected their operations. Floods are caused by high intensity rains in a short period of time and induce a rapid rise in water level, overflowing the river banks [1]. Floods usually erode the river banks, scour and destabilize the pier and footings of bridges, and weaken the rock fill and rip-rap foundations. The waterlogging is defined as a slow increase of the water level that covers the low lands and flood plains at low water flow speed. If the highways are placed in flood plains, the highways' physical assets experience little damage, but the road network operation can be totally or partially interrupted during and before the rain period, until the water level decreases and the roads can recover the trafficability. The traffic interruption degree depends upon if the water level over the roadway is higher or lower than the vehicle's wading capability and if the water speed is enough or not to destabilize or drag the vehicle. The wading capability depends on the wading height, which is the maximum water level at which a vehicle can ride without affecting the engine, brakes, and the heating, ventilation, and conditioning system (HVAC) and without loss stability because of floating and sliding. Each type of vehicle has a specific wading height. If the water level is lower than the wading height and the water speed is low, vehicles can ride over the flooded road at a reduced speed, increasing the travel time and the operating cost. The consequence is a reduction of the traffic quality in terms of the Level of Service (LOS) of the road, which is measured using the average travel speed and the spend-time-following vehicles in two-lane rural roads and in multilane highways in terms of the traffic density [2]. Also, if the water level and speed are high, vehicles are destabilized and cannot pass through the flooded road. In this case, the loss of LOS is total. Therefore, predicting the possibility that a vehicle can ride over a flooded road depends on the vehicle characteristics and the magnitude of the flood.

There exist several models to predict the water depth over the roadway because of a flood: ad hoc models; hydrological-hydraulic (H-H) 1D, 2D, and 3D models; and GIS-based models [3]. The ad hoc models are empirical and estimate flooded areas (named blue spots) based on historical data. These allow creating flood maps and are an efficient alternative if it is not possible to calibrate more sophisticated models [4–6]. The H-H 1D models used the river and bank's topography, the water flow, and the hydraulic regime to calculate the water level profile and the flooded surface. Examples of this type of models are HEC-RAS and MIKE11 [7, 8]. The H-H 2D models also need data of the flooded area, the ground roughness, the river slope, the rain intensity, and terrain digital models [9]. Examples of these types of models are HEC-RAS 2D, IBER, and MIKE FLOOD [10, 11]. The H-H 3D models estimate the flow speed too but are intensive in data needs because of considering advective and convective transport, the sediment transport, and the wave effect to estimate the water level profile [12]. The GIS-based models integrate some characteristics of the H-H models to the GIS capabilities but also allow estimating physical and economic damage due to floods particularly in urban areas. Examples of these models are HEC-FDA, HAZUS-HM, Damage Scanner, Flemish Model, and Multicolored Manual among others [13–17].

These models previously described only estimate the water depth over the roadway. The water depth depends on the return periods, the river, and floodplains' topography. Therefore, it is plausible to assume that this variable can be characterized with a probability density function. However, to study the probability of traffic interruption, it is also necessary to calculate the wading height, which is a specific property of each vehicle type included in a vehicle fleet. Each type of vehicle has its own engine, brake, and HVAC system, as well as a tire diameter and ground clearance. Therefore, there is a probability function that describes the vehicle's wading height of each brand and model of light cars, pick-up, SUV, trucks, and buses. Because the wading height and the water depth are random, the traffic interruption must be estimated using probabilistic models. In this sense, this paper proposes and applies a procedure to estimate the traffic interruption probability by integrating both variables: still water depth (also named waterlogging depth) and wading height, considering their randomness. In the first part of the paper, the effect of flooding over highway assets is discussed. Next, an analytical procedure based on the first-order reliability method (FORM) is presented. The procedure included the estimation of probability density function (PDF) of water depth and wading height using discharge data and individual vehicle data, respectively. The procedure was applied to a case study in the middle-south of Chile, where the road waterlogging is recurrent year after year.

2. Effects of Floods on Road Infrastructure, Traffic, and Vehicles

The floods affected the road infrastructure that crosses rivers and those parallel to the river banks, such as roads with cut-fill slopes, those placed in floodplains and in flat ground with low infiltration capacity. The damage to the infrastructure is structural and functional. The structural damage is produced by hydraulic erosion. The erosion wake crosses slopes, increasing the scour in bridges and weakening river bank protections. The floods also affect the drainage system, compromising its capacity of moving the water from slopes, embankments, and road subgrades. In the long term this effect reduces the pavement's durability. The functional damage is manifested in a reduction of the LOS of the road network. This reduction can be induced by partial destruction of the pavement surface, reduction of the lane width, loss of lanes available, or roadway flooding. Functional damage does not necessarily interrupt traffic but reduces the average travel speed and eventually only certain type of vehicles can pass through flooded roads. If the traffic is totally interrupted, communities can be isolated if there are no alternative routes.

The level of road traffic interruption caused by floods can be classified in the following scenarios: normal circulation, in which all the vehicle can pass through the flooded road without reducing the speed; restrained circulation, in which all the vehicles can pass through the flooded road but at a low speed; partial interruption, in which only heavy vehicles can pass along the flooded road; and total traffic interruption in which no type of vehicle can pass through the flooded road. The traffic interruption level depends on the magnitude and duration of the flood. It affects the road users, the nonusers, and the highway agencies. Road users experience higher travel times and operating costs, restrained circulation, or road crashes; the nonusers may experience isolation and lack of access to essential or vital services. The highway agency expends additional budget for repairing and/or rebuilding roads, taking a percent of the budget assigned to other maintenance activities, for instance.

If the water covers the roadway, the circulation possibility depends also of the vehicle wading height and weight, the water flow speed, and the water depth over the road. If the water flow speed is low and the wading height is higher (or similar) than the water depth, the vehicle keeps stable, but the HVAC system can be damaged. If the water depth is lower than the wading height the vehicle can ride over the flooded road at a low speed. On the contrary, if the water speed flow is high, there exists a water depth threshold that avoids the vehicle destabilization. If the water depth overcomes this threshold, the vehicle loses stability and is dragged by the water flow and driver loses control of the vehicle and traffic is totally interrupted. In this case, the wading height is not applicable, being more important the water depth threshold.

The research available in the literature about the effect of climate on road traffic can be categorized into two groups: (a) studies that analyze the reduction of traffic speed under adverse weather conditions and (b) studies that analyze the effect of flooding on roads due to rainfall. Examples of the first type of study are found in [18–25], where the authors estimate the percentage of reduction of the average travel speed in relation to the intensity of the rain, during the rain and before the flood. The second group of research is classified in two subgroups: (a) studies of vehicle stability in flooded roads and (b) studies of wading capabilities of vehicles in flooded road with a still water regime. The studies of vehicle stability used hydrodynamic models to estimate the water depth and speed threshold than inducing (or avoiding) sliding, toppling, or floating to vehicles [26–31]. Martinez-Gomariz [32] represented in a speed-depth graph all the results existent in the literature. From the summary of Martinez-Gomariz it was concluded that the water depth-speed thresholds that destabilize light cars ranged between 0,4 and 0,7 m at flow speeds lower than 1 m/s and between 0,2 and 0,5 m at flow speeds between 1 and 6 m/s. In comparison with the vehicle stability studies, the studies of traffic interruption in still waters are scarce. Pyatkova et al. [33] studied the effect on urban traffic on street flooded with water depths between 0,07 and 2,0 m and more. They used the software MIKE FLOOD to simulate flood with different return periods and assumed that streets are closed approximately when water depth reaches between 0,5 and 1,0 m. Yin et al. [34] developed in China a similar study. They elaborate flood maps for return periods between 5 and 100 years, with which water depths between 0,05 m and 0,5 m and more are obtained. They used a water depth threshold between 25 and 35 cm to close streets and study the effect over traffic delays. In Australia, Affum et al. [35] used the concept of limit still water depth to warn drivers about the safe crossing of floodways. It is an indirect measure of wading height. The still water depth thresholds used suggested by Affum et al. for light cars and SUV ranged between 0,3 and 0,5 m. However, they recommended more studies to validate it.

3. Analytical Modelling of Traffic Interruption Probability

3.1. Conceptual Framework. The traffic interruption model proposed considered the uncertainty of the wading height and waterlogging depth through sigmoidal probability curves. The conceptual model is constituted by the three parts described in Figure 1: (a) estimation of the traffic probability interruption in terms of the wading height and waterlogging depth; (b) estimation of the individual probabilities using FORM; and (c) estimation of the PDF for wading height and waterlogging depth. Each point of the probability curve (Box "a" in Figure 1) is obtained by applying FORM (Box "b" in Figure 1) and using a limit state function (LSF) that is the difference between the wading height (H_{wading}) and the waterlogging depth ($H_{waterlogging}$). Both variables are random. The H_{wading} randomness is explained by the variability of the vehicles' geometry, engines, HVAC systems, tire diameter, and ground clearance of each vehicle class (light car, pick-up, SUV, trucks, and buses), brands, and models. The $H_{waterlogging}$ randomness is explained by the water flow magnitude, the return period, the river topography, and the location of the road in relation to the river.

3.2. Estimation of the Probability Density Function of Wading Height. The wading height is defined as the maximum depth of a low-speed water flow at which a vehicle can ride without electric, engine, or HVAC system damage, depending on the vehicle configuration. In a vehicle fleet, there is one wading height for each vehicle brand and model. The vehicle fleet is defined as a set of classes and subclasses existing in a country. Typical vehicle classes and subclasses belonging to a vehicle fleet are shown in Figure 2.

The wading height (H_{wading}) can be obtained using three methods: from manufacturers catalog, by in-field measuring, and by analytical estimates from the geometric properties of vehicles described in the manufacturers' catalog. The analytical procedure is more practical because the manufacturers do not always provide wading height values, and the in-field measurements usually considered small sample sizes that underrepresent the vehicle class, brand, and model present in vehicle fleets. The analytical procedure estimate $H_{wading,sub\text{-}class,brand,model}$ for each vehicle subclass, brand, and model classified according to Figure 2, using (1) and (2), in which $D_{LL,sub\text{-}class,brand,model}$ (in mm) is the external diameter of the tires of each vehicle subclass, brand, and model without considering its deformation because of loads. The values of D_{LL} are obtained from the catalogs of vehicle manufacturers as well as the values of H_{wading} if they are available. The parameter $\alpha_{sub\text{-}class}$ of (1) is estimated by adding all the values of $H_{wading,sub\text{-}class,brand,model}$ and $D_{LL,sub\text{-}class,brand,model}$ available in the catalogs for the "V" vehicle subclasses. Therefore, this parameter is constant for each vehicle subclass. Then, introducing the $\alpha_{sub\text{-}class}$ value in (2), the H_{wading} for each brand and model of vehicle belonging to each subclass can be obtained.

$$\alpha_{sub\text{-}class} = \frac{1}{V}\sum_{i=1}^{i=V}\frac{H_{wading,sub\text{-}class,brand,model}}{D_{LL,sub\text{-}class,brand,model}} \tag{1}$$

$$H_{wading,vehicle\ brand\ and\ model} = \alpha_{sub\text{-}class} D_{LL,sub\text{-}class,brand,model} \tag{2}$$

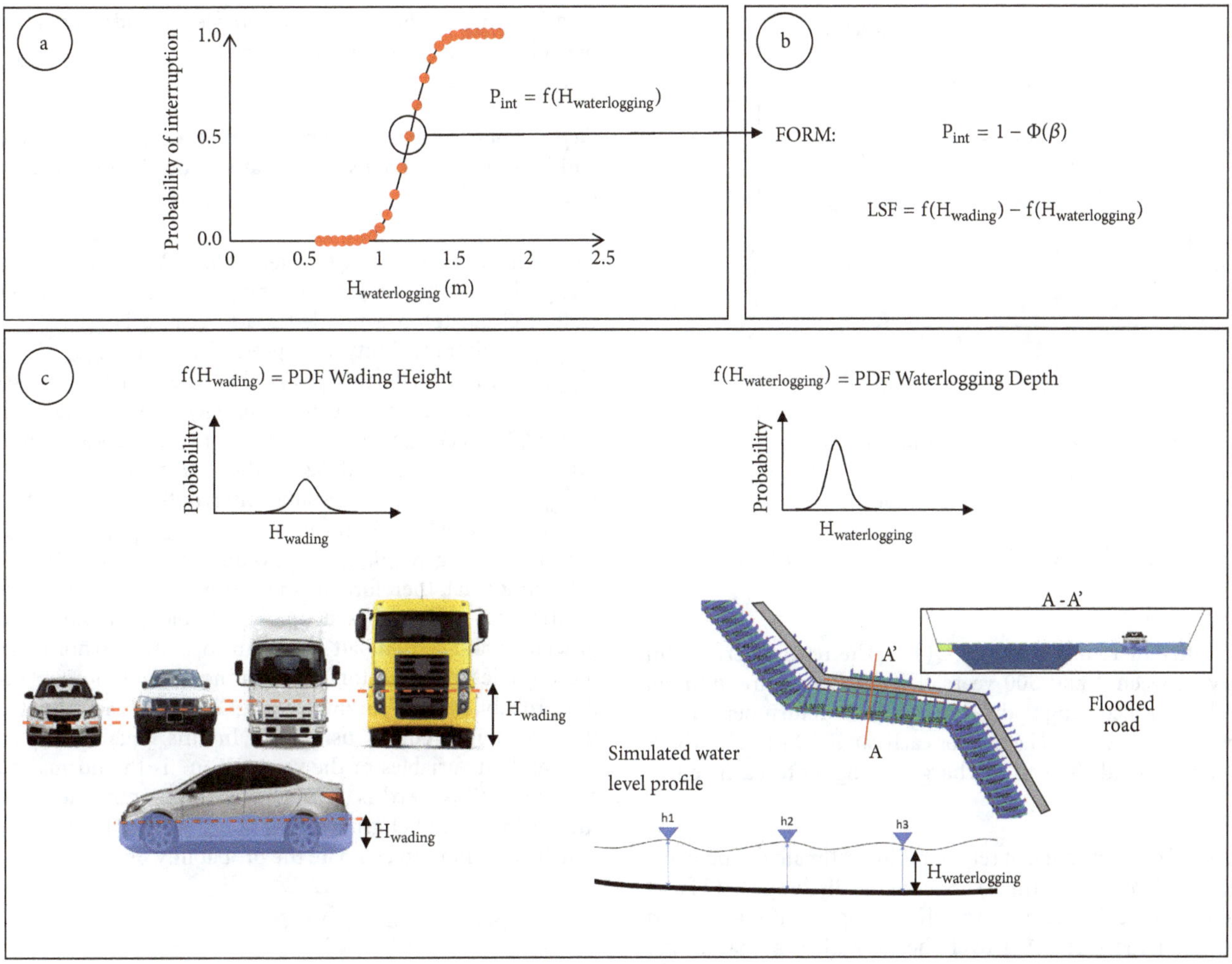

FIGURE 1: Conceptual model for obtaining traffic interruption probability curve caused by waterlogging.

The PDF of H_{wading} is estimated using the standard procedure to fit empirical PDF to continuous PDF: (a) to group the data of H_{wading} of each vehicle class and subclass; (b) to apply goodness of fit (GoF) tests, such as Kolmogorov-Smirnov, Chi square, Anderson-Darling, or Shapiro Wilks to identify the type of probability distribution; (c) to elaborate a ranking of the adjusted PDF according to the GoF test; (d) to select the probability distribution that best fit the empirical PDF of H_{wading}; (d) to estimate the parameters associated with the PDF of fitted H_{wading}.

3.3. Estimation of the Probability Density Function of Water Depth. The waterlogging height ($H_{waterlogging}$) is the difference between the water depth at each river cross-section obtained from the water level profile (EH, in m) and the roadway centerline height of the longitudinal profile (CR, in m). The randomness is explained by the variations of the EH because of the variability of water flow, according to the return period and the river topography. If Ti is the return period (in years) and h_i is the cross-section of the flooded area, Figure 3 indicates that, for the same flooded area, there are different values of $H_{waterlogging}$ depending on the discharge and water level associated with each Ti (Q(Ti)) in each cross-section. This happens because the water level depth and the roadway centerline height vary along the flooded area.

The EH can be obtained by direct measurement or by simulation according to Figure 4. After calculating the EH, the $H_{waterlogging}$ is obtained using (3).

$$H_{waterlogging} = EH - CR \quad (3)$$

To analytically estimate the PDF of $H_{waterlogging}$, a quasi 2D model in which the road cross-section flooded is considered as part of the cross-section of the rivers was used. This way of modelling allows overcoming the limitation of a 1D modelling and at the same time avoid the use of digital elevation models, which not always are available [36]. The procedure considered eight steps:

Step 1. Collect the maximum yearly discharge from available limnimetric or fluviometric stations.

Step 2. Perform a frequency analysis of the time series of maximum discharges and fit the gamma, log normal, and log Pearson II PDFs. The fitting allows estimating the discharge

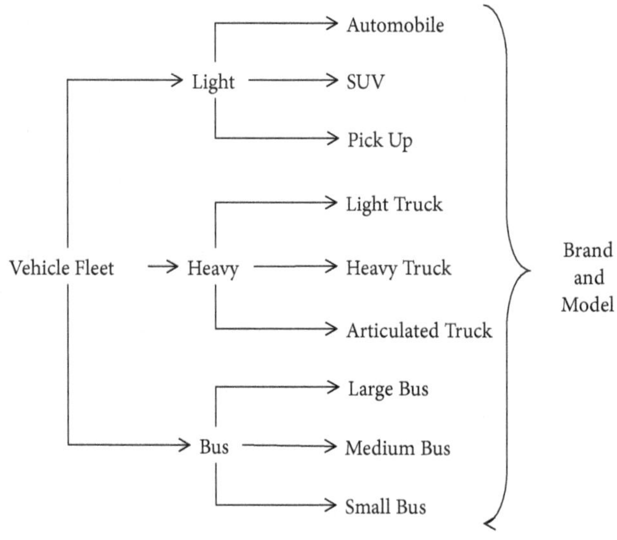

FIGURE 2: Typical classification of a vehicle fleet.

for different return periods (Q(T)). The return period can vary between 4 and 500 years. The discharges are grouped in clusters according to different ranges of return periods for obtaining a set of discharges for each range ($Q_i(T)$), in which "i" represented the i-th discharge belonging to each return period range.

Step 3. For each range of return period estimate the percentage of difference ΔQ_i (in %) between the discharge $Q_i(T)$ and the formative discharge of the river using (4). It represented the discharge increasing over the formative discharge that provoked the overflow. The formative discharge is defined as the most probable flow at a certain recurrence period that shape rivers [37]. It is estimated for a return period of 4 years ($Q_i(T=4)$) according to Niño [38].

$$\Delta Qi = 100\left(\frac{Qi\,(T)}{Qi\,(T=4)} - 1\right) \tag{4}$$

Step 4. Estimate the maximum percentage of discharge increasing (ΔQ_i max(T)) based on the flood's return period T and the percentage of difference (ΔQ_i) using (5).

$$\Delta Q_i \max(T) = \max\left[\Delta Q_i\right] \tag{5}$$

Step 5. Configure the river cross-section and its properties: dimensions, Manning's coefficients, placement of the road, longitudinal slope of the riverbed, and longitudinal slope of the road (see Figure 7).

Step 6. Estimate the discharge Qs_i to be used in the simulation of the water level profile using (6). This equation represented the incremental increasing of discharge from the formative discharge according to the increasing of the return period. Considering the cross-section and the longitudinal bedstep of the river, the Qs_i value allows estimating the water depth (EH of (7)) using hydraulic simulation software.

$$Qs_i = Q_i\,(T=4)\left(1 + \Delta Q_i \max\,(T)\right) \tag{6}$$

Step 7. Simulate the water level profile for different return periods and discharges (Qs_i) and estimating the $H_{waterlogging}$ values using (3).

Step 8. For each simulation run obtain the $H_{waterlogging}$ values and fit a suitable PDF using the same procedure followed to fit H_{wading} values.

3.4. Simulation of the Traffic Interruption Probability Curves. The probability curves estimate the probability that a vehicle cannot circulate on a flooded road because $H_{waterlogging} > H_{wading}$. This probability is expressed as $\Pr(H_{waterlogging} > H_{wading})$. To estimate this probability, the FORM was used [39]. The FORM is based on the LSF: g(X) = G1(X) – G2(X), in which G1 and G2 are nonlinear functions, and X is a vector of random variables. G1 and G2 are the PDF of $H_{waterlogging}$ and H_{wading}, respectively. The traffic interruption probability is defined as $\Pr(g(X) \geq 0)$, meaning that $H_{waterlogging} > H_{wading}$. Given that, in general, g(X) is nonlinear and the PDF of X are nonnormal, therefore, it is necessary to apply a numeric solution for g(X=0) based on the Hasofer and Lind [40] procedure and Rossenblatt's [41] transformation to normalize the variables of the vector X. The numerical solution obtains the probability β_{HL} parameter in g(X)=0 for each point of the probability curves using (7). In this equation, x_i are the random variables of the vector X, u^* is the normalized solution of (7), and α^* is the normalized vector that describes the normalized LSD g(u)=0. The solution of (7) is the β_{HL}, which is an input to estimate the probability using (8).

$$\beta_{HL} = \min_{\{x \in g(X)=0\}} \sqrt{\sum_i x_i^2};$$
$$u^* = -\beta_{HL}\alpha^* = \beta_{HL}\left(\frac{\partial g/\partial x_i}{\sqrt{\sum_{i=1}^{n}\left(\partial g/\partial x_i\right)}}\right) \tag{7}$$

$$P_{int} = 1 - \Phi\left(\beta_{HL}\right) \tag{8}$$

The probability of (8) represents only one data point of the probability curve of Figure 1 (Box a). Therefore, it is necessary to apply multiple runs with different simulation conditions, which is similar to applying a random sampling on the PDF of $H_{waterlogging}$ and H_{wading} but considering the consistency of the physical behavior of the flood phenomenon and its effect over the different vehicle classes.

4. Case Study in the Central Valley of Chile

The procedure was applied in the Central Valley of Chile (see Figure 5) with the objective of illustrating the proposed procedure to obtain probability curves that allow estimating the traffic interruption probability in that area of the country. The most roads placed in this area are located in floodplains and are exposed to low water level increase during floods that can be assimilating to still water. In the study area, there are 10 fluviometric stations located in Maule, Itata, Bio Bio, and Imperial rivers. Also, 44 ORT are located within the road network of the study area.

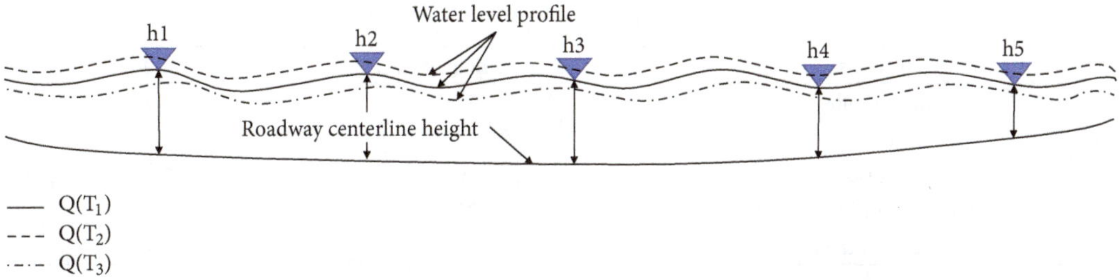

FIGURE 3: Variability of water level depth regarded to different return periods.

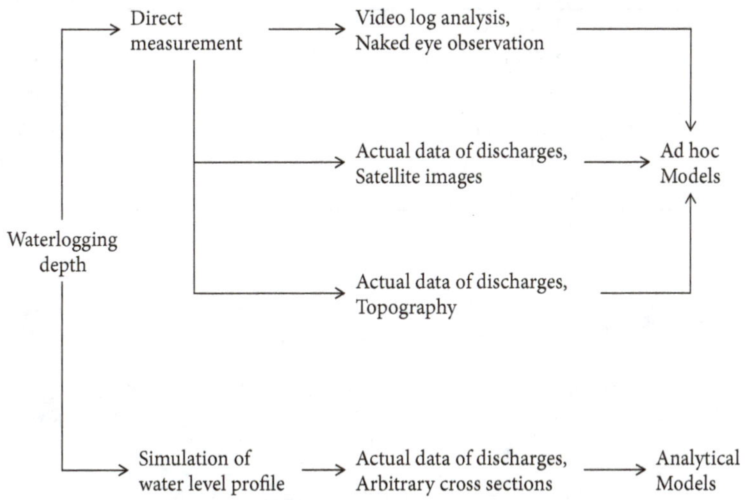

FIGURE 4: Method for obtaining water logging height.

4.1. Wading Height PDF Estimation. A sample of 166,155 vehicles of 3-year records from the ORT provided by the Ministry of Public Works [42] was used to estimate the PDF of H_{wading}. Table 1 summarized the sample categorized by vehicle class and subclass. The frequency diagrams are plotted in Figure 6. The software Easy Fit was used to estimate the PDF of H_{wading} for each vehicle subclass. Table 2 summarizes the results obtained.

4.2. Estimation of the Probability Density Function of the Waterlogging Height. The input data used to estimate the $H_{waterlogging}$ PDF were a generalized river cross-section that included the road cross-section, the maximum yearly discharge data sets, and the parameters used for modelling the water level profile. The geometry of the cross-section is presented in Figure 7. The hydraulic properties used were river Manning's roughness = 0,03; road surface Manning's roughness = 0,016; mean slope of the river bottom = 0,0001 m/m; road longitudinal slope = 0,5%; and section length = 1000 m.

The discharge data from ten fluviometric stations was provided by the Ministry of Public Works of Chile [43]. Figure 8 plots the maximum yearly discharge (in m^3/s) per fluviometric station. The discharge values ranged from 6 to 700 m^3/s and the discharge records between 5 and 60.

The frequency analysis of the discharge data is summarized in Table 3. Using the PDF of Table 3 discharges for return periods (T) of 4, 5, 10, 25, 50, 100, 200, and 500 years were obtained. The discharge estimated for T=4 was 55 m^3/s and was considered as the formative discharge in (4).

Discharges obtained for each return period were the input data used to estimate EH, with quasi-2D modelling using the software HEC-RAS 1D. Figure 9 shows the histograms of the EH obtained, clustered by return period ranges.

The $H_{waterlogging}$ PDF was obtained from the data of Figure 9 by using the Easy Fit software. Table 4 summarized the pdf obtained, classified by return period ranges. All the pdf obtained were normal. The mean and the standard deviation of the PDF are in brackets.

4.3. Traffic Interruption Probability Curves Simulated. To estimate the traffic interruption probability using FORM, a LSF is needed. The LSF was configured using the PDF of $H_{waterlogging}$ and H_{wading}. The probabilities were obtained using the Hasofer-Lind procedure. The simulation considered 6 vehicle classes and 7 return periods. Figure 10 shows the 42 estimated probability curves.

From Figure 10, three patterns were identified: the interruption probability is similar for light automobiles and SUVs; the single unit heavy trucks and the articulated trucks follow

TABLE 1: Vehicle sample used to estimate wading height.

Class	Sub-class	# of vehicle brands	# of models	Sample
Light	Automobile	52	392	87,264
	SUV	42	217	41,440
	Pick up	29	75	31,245
Heavy	Single unit light and heavy trucks	27	136	3,884
	Articulated trucks	29	212	2,322

TABLE 2: PDF of wading height of each vehicle class selected in this study.

Class	Sub-class	Probability density function (PDF)
Light	Automobile	Normal (0,4125; 0,0232)
	Pick up	Lognormal3P (0,606; 0,019; 0,544)
	SUV	Lognormal3P (0,0477; 0,0436; 0,4496)
Heavy	Single unit light truck	Lognormal3P (0,8417; 0,0462; 0,4789)
	Single unit heavy truck	Weibull (1,1323; 0,0568; 0,6261)
	Articulated truck	Normal (1,1986; 0,0692)

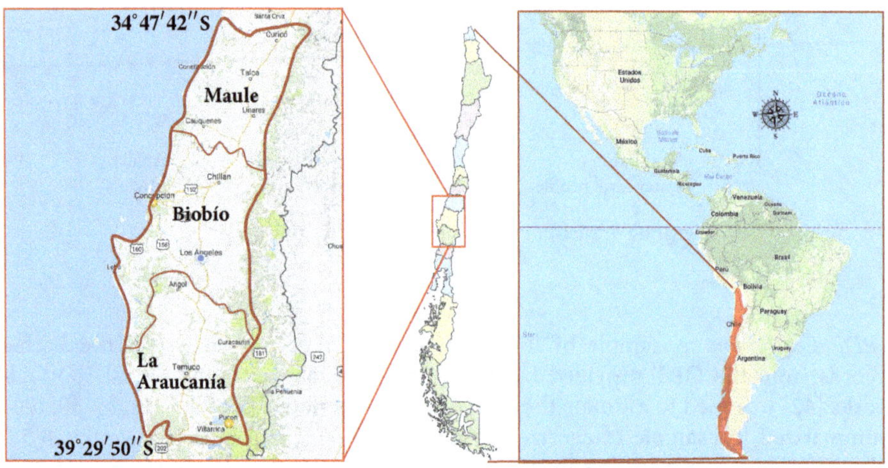

FIGURE 5: Location of the studied area.

TABLE 3: Frequency functions of maximum annual river flows by fluviometric station.

Fluviometric station	Probability density function
Loncomilla in Las Brisas	LogPearson III (6,5066; -0,252; 9,634)
Perquilauquén in Quella	LogPearson III (4,133; -0,379; 8,311)
Perquilauquén in San Manuel	LogNormal (0,105; 7,817; -1898,7)
Diguillin Longitudinal	LogPearson III (16,023; -0,145; 8,667)
Itata in Cholguán	LogNormal (0,041; 8,650; -5116,9)
Duqueco en Villucura	Gamma (5,4995; 109,120)
Mininco in Longitudinal	LogPearson III (7,471; -0,229; 6,630)
Malleco in Collipulli	LogPearson III (9,546; -0,159; 7,345)
Lumaco in Lumaco	LogPearson III (16,266; -0,148; 7,178)
CholChol in Cholchol	LogPearson III (8,155; -0,137; 7,885)

TABLE 4: Summary of PDF of $H_{waterlogging}$.

Return period (years)	PDF of $H_{waterlogging}$
[4 – 5]	Normal (0,0000; 0,06430)
[5 – 10]	Normal (0,1495; 0,10903)
[10 – 25]	Normal (0,3923; 0,09948)
[25 – 50]	Normal (0,6099; 0,09327)
[50 – 100]	Normal (0,7783; 0,07420)
[100 – 200]	Normal (0,9371; 0,08670)
[200 – 500]	Normal (1,0888; 0,07054)

a similar pattern; and the traffic interruption probability of automobiles and SUVs is higher than the same probability for single unit heavy trucks and articulated trucks. This

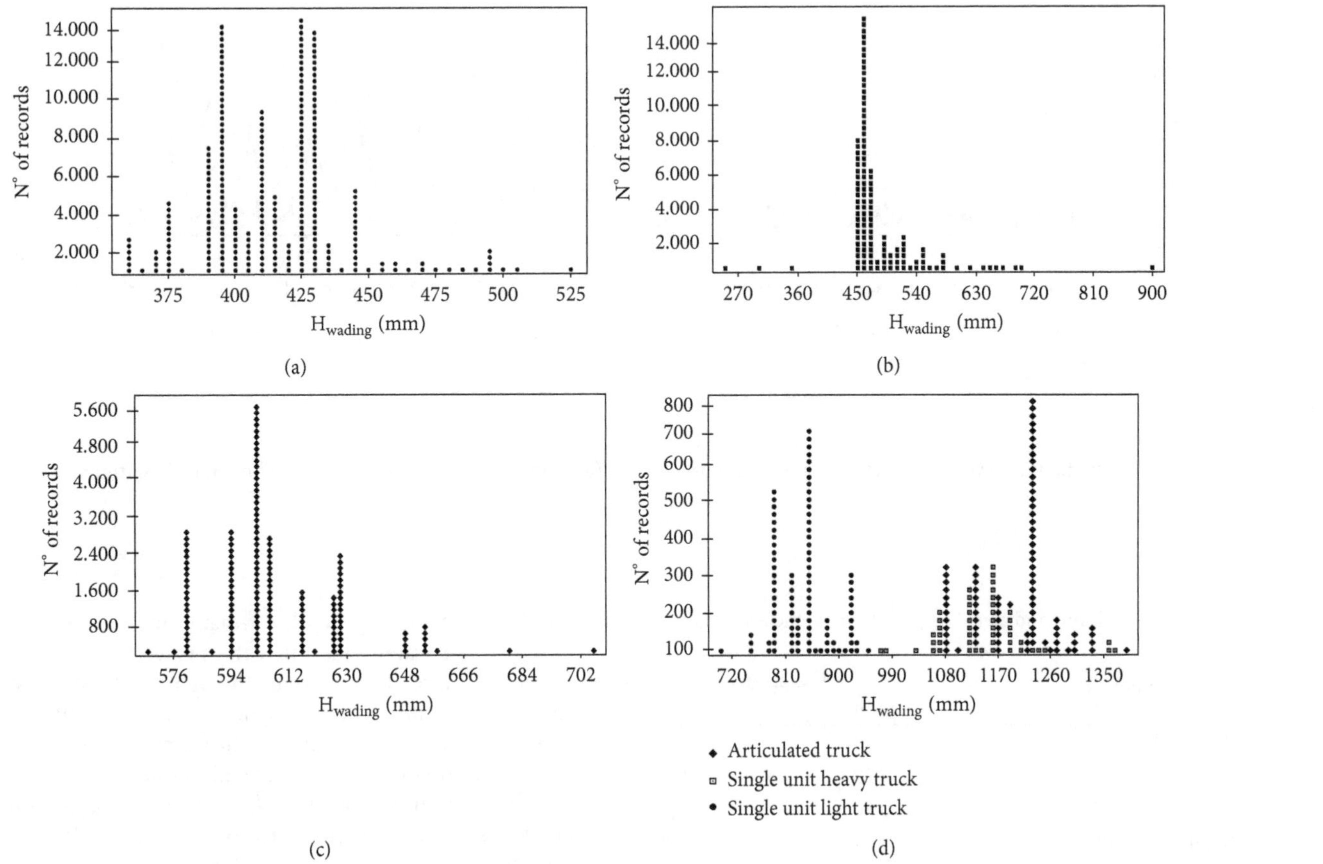

FIGURE 6: Histograms of wading height per vehicle class. (a) Automobile (each dot = 346 data points). (b) SUV (each dot = 386 data points). (c) Pick-up (each dot = 127 data points). (d) Heavy vehicles (each dot =30 data points).

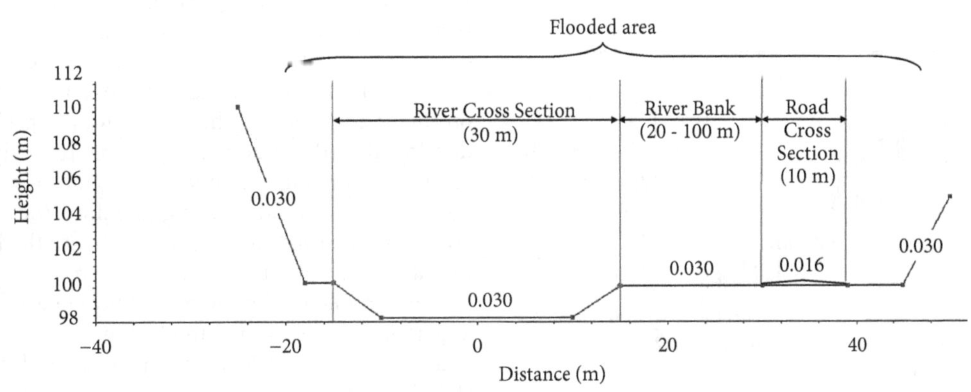

FIGURE 7: Cross-section geometry used for modelling the water level.

result is logical and proves the coherence of the model developed. Using the PDF curves of Figure 10 two values of $H_{waterlogging}$ were obtained: (a) the limit waterlogging depth that is defined as the depth at which the traffic interruption probability is 1,0 and (b) the alert waterlogging depth associated with a traffic interruption probability of 0,5. Both values of $H_{waterlogging}$ can be used to establish an alert system for drivers based on the properties of the local vehicle fleets. Results are summarized in Table 5 per vehicle subclass.

5. Conclusions

The aim of this paper was to propose a procedure based on reliability principles to estimate the traffic interruption probability due to road waterlogging considering very low-speed flow. The proposed procedure was applied in the south of Chile. A data set of 166,155 vehicles and records up to 30 years of discharge data of ten fluviometric stations were used and a total of 42 probability curves were obtained. In this context, the following conclusions were obtained:

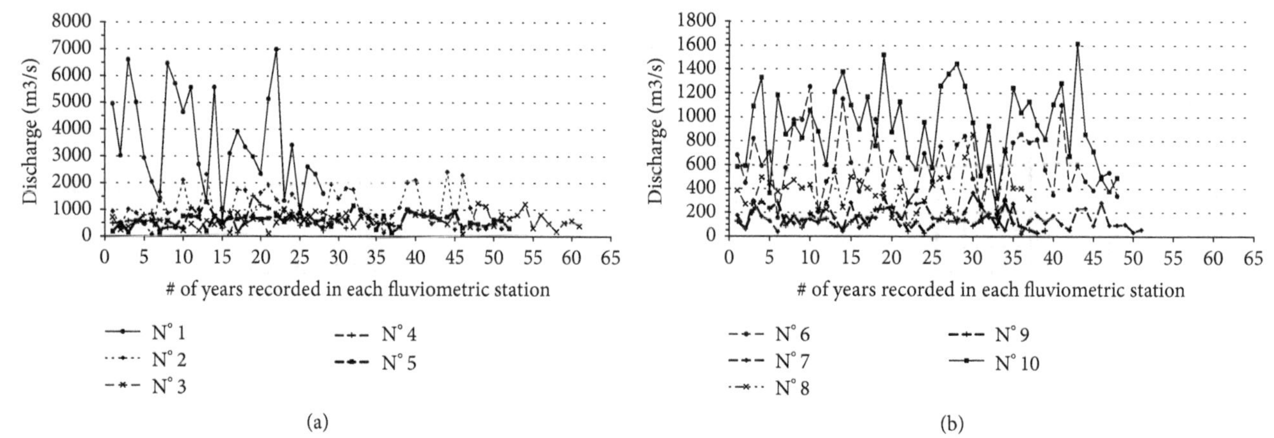

FIGURE 8: Maximum yearly discharge per fluviometric station. (a) Fluviometric stations 1 to 5. (b) Fluviometric stations 6 to 10.

TABLE 5: Limit and alert $H_{waterlogging}$ per vehicle subclass.

Vehicle subclass	$H_{waterlogging}$ (m)	
	Limit	Alert
Automobile	0,73	0,41
SUV	0,77	0,46
Pick-up	0,88	0,57
Single unit light truck	1,14	0,83
Single unit heavy truck	1,44	1,13
Articulated truck	1,53	1,20

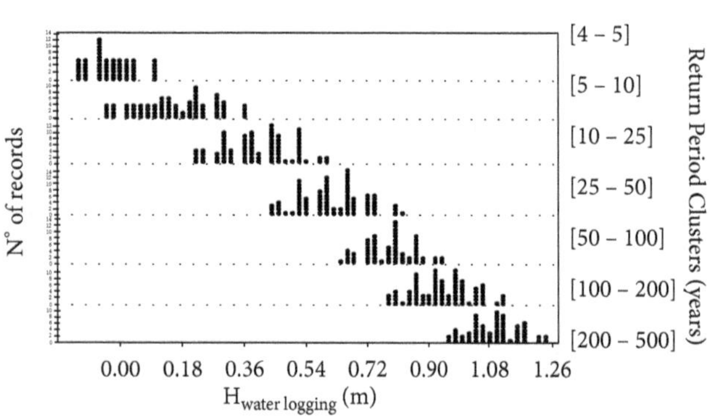

FIGURE 9: Summary of waterlogging height clustered according to return time.

Results obtained show that the waterlogging depth that interrupts traffic with a probability of 1,0 varies between 0,73 and 0,88 m for light vehicles and between 1,14 to 1,53 m for heavy vehicles. Likewise, the waterlogging depth that interrupts traffic with a probability of 0,5 varies between 0,41 and 0,57 m for light vehicles and between 0,83 to 1,20 m for heavy vehicles. These values are higher than the values reported in the literature that varies between 0,3 and 0,5 for light vehicles in still waters.

The waterlogging depth estimated in this study is valid only for still waters, in which the protection of the HVAC system is more relevant than the vehicle stability. At high speed flows, for instance, higher than 1 m/s, the stability turns more relevant and leads to the waterlogging depth being reduced compared to the estimated in this study.

The literature about the effect of waterlogging on road networks used simplified methods to obtain the vehicle's wading height; this is because those literature studies were supported by direct in-field observation, which leads to using small sample sizes. This limitation has been resolved in this research by incorporating the variability of the vehicle's wading height and a large sample of vehicles obtained from records of existent open road tolls in Chilean highways.

Areas of potential flooding can be identified by superimposing flood-prone areas onto road network maps. Based on the composition of the traffic and the probability curves developed in this study, it is possible to analyze the cost of rerouting to justify the implementation of flood protection measures in the road network; therefore, the results obtained are useful for road managers to estimate the budget needed to mitigate the effect of floods on the road network.

For simplicity, the water level profile was obtained using a quasi-two-dimensional modelling, which does not permit incorporating the velocity of the water and its effect on the stability of the vehicles. For that reason, the model is valid only for still waters. To refine the model, a two-dimensional modelling is needed in order to incorporate the effect of water flow speed on the vehicles' stability using hydrodynamic models combined with reliability principles.

The probability curves have enough generality to be used in traffic management policies in flood-prone areas and also demonstrate the need for technical evaluation of need for works of protection against flooding. The critical and limit waterlogging depths calculated are representative of the Central Valley of Chile. To apply this to other areas of the country, only the PDF of the EH would need to be modelled, because the PDF of waterlogging depth is representative of the entire country.

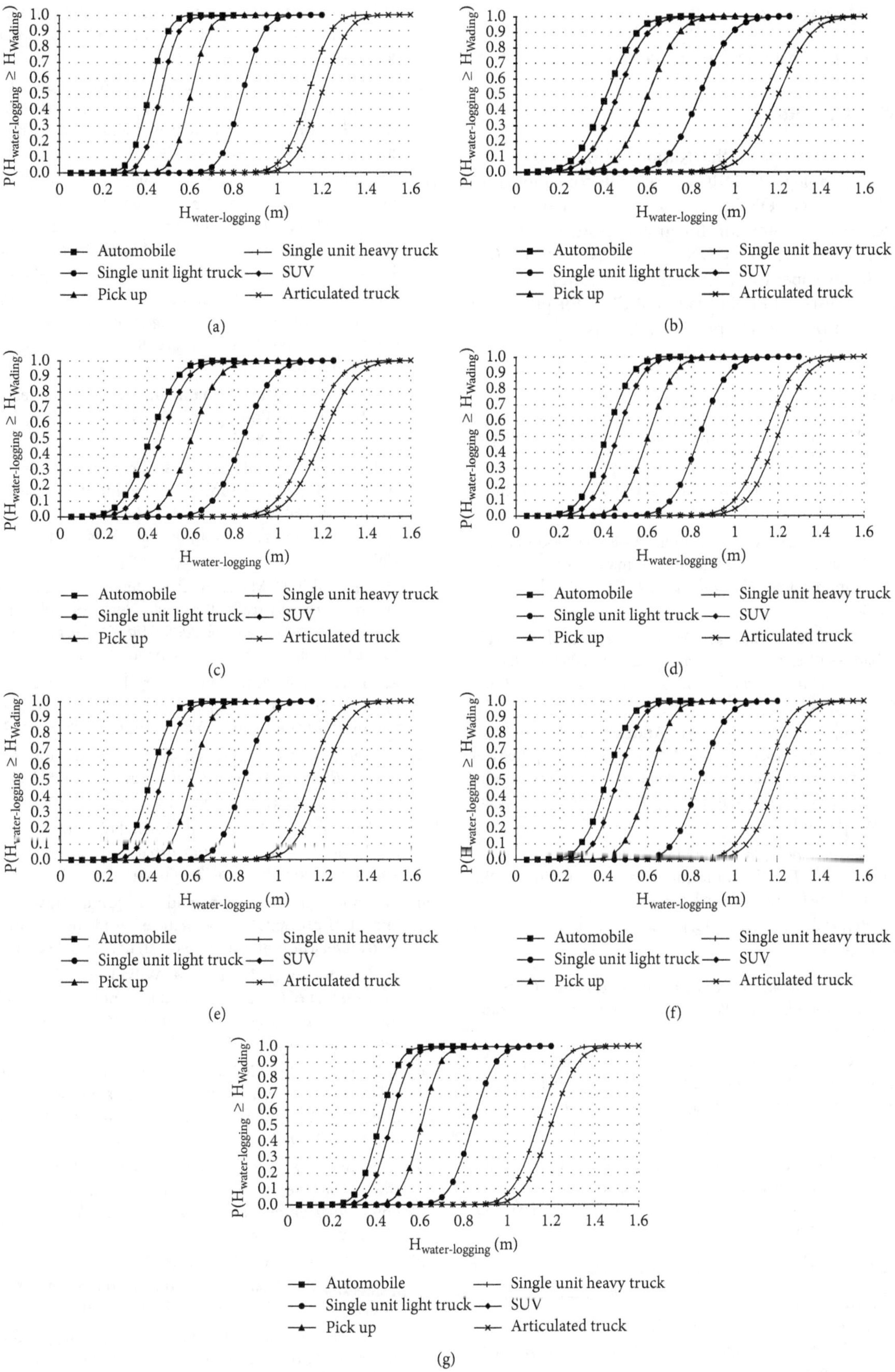

FIGURE 10: Interruption probability curves simulated per return period and vehicle subclass. (a) 4 < T < 5 years. (b) 5 < T < 10 years. (c) 10 < T < 25 years. (d) 25 < T < 50 years. (e) 50 < T < 100 years. (f) 100 < T < 200 years. (g) 200 < T < 500 years.

Conflicts of Interest

The authors declare that they have no conflicts of interest.

Acknowledgments

The authors wish to thank the National Commission for Science and Technology (CONICYT) of Chile for supporting the research projects FONDEF ID14I20309 and thank the National Research Center for Integrated Natural Disaster Management (CIGIDEN), CONICYT/FONDAP/15110017, within which this paper was prepared. The authors also wish to thank the Ministry of Public Works of Chile for providing the data used for the research presented in this paper.

References

[1] W. Kron, "Flood Risk= Hazard∗Values∗Vulnerability," *Water International*, vol. 30, no. 1, pp. 58–68, 2005.

[2] TRB, *Highways Capacity Manual*, Transportation Research Board, 2016.

[3] M. Gall, B. J. Boruff, and S. L. Cutter, "Assessing flood hazard zones in the absence of digital floodplain maps: Comparison of alternative approaches," *Natural Hazards Review*, vol. 8, no. 1, pp. 1–12, 2007.

[4] O. Petrucci and A. A. Pasqua, "Damaging events along roads during bad weather periods: A case study in Calabria (Italy)," *Natural Hazards and Earth System Sciences*, vol. 12, no. 2, pp. 365–378, 2012.

[5] P.-A. Versini, E. Gaume, and H. Andrieu, "Assessment of the susceptibility of roads to flooding based on geographical information - Test in a flash flood prone area (the Gard region, France)," *Natural Hazards and Earth System Sciences*, vol. 10, no. 4, pp. 793–803, 2010.

[6] ICPR, *Atlas of Flood Danger and Potential Damage Due to Extreme Floods of The Rhine*, International Commission for the Protection of the Rhine, Koblenz. Germany.

[7] USACE, "HEC-RAS v. 5.0.3," Tech. Rep., U.S. Army Corps of Engineers, 2016.

[8] DHI, *MIKE11 A Modelling System for Rivers and Channels - Reference Manual*, DHI Water & Environment, Hørsholm, Denmark, 2002.

[9] S. Patro, C. Chatterjee, S. Mohanty, R. Singh, and N. S. Raghuwanshi, "Flood inundation modeling using MIKE FLOOD and remote sensing data," *Journal of the Indian Society of Remote Sensing*, vol. 37, no. 1, pp. 107–118, 2009.

[10] E. Bladé, L. Cea, G. Corestein et al., "Iber: a tool for the numerical simulation of river flow," *Revista Internacional de Métodos Numéricos para Cálculo y Diseño en Ingeniería*, vol. 30, no. 1, pp. 1–10, 2014.

[11] DHI, *MIKE FLOOD 1D-2D MODELLING - User manual*, DHI Water & Environment, Hørsholm, Denmark, 2003.

[12] DHI, *MIKE 3 Estuarine and Coastal Hydraulics and Oceanography-Short Description*, DHI Water & Environment, Hørsholm, Denmark, 2001.

[13] USACE, *HEC-FDA: Flood Damage Reduction Analysis*, U.S. Army Corps of Engineers, 2016.

[14] FEMA, "Multi-hazard loss estimation methodology. Flood Model HAZUS®MH MR4," Technical Manual, Federal Emergency Management Agency, 2003.

[15] F. Klijn, P. J. A. Baan, K. M. De Bruijn, and J. Kwadijk, *Overstromingsrisicos in Nederland in een veranderend klimaat, WL delft hydraulics*, WL delft hydraulics, Delft, Netherlands, 2007.

[16] W. Vanneuville, R. Maddens, C. Collard, P. Bogaert, P. de Maeyer, and M. Antrop, "Impact op mens en economie t.g.v. overstromingen bekeken in het licht van wijzigende hydraulische condities, omgevingsfactoren en klimatologische omstandigheden," Tech. Rep. MIRA/2006/02, Universiteit Gent, Belgium, 2006.

[17] E. C. Penning-Rowsell, C. Johnson, S. Tunstall et al., *The Benefits of Flood And Coastal Risk Management: A Handbook of Assessment Techniques*, Flood Hazard Research Centre. Middlesex University Press, 2005.

[18] M. Kyte, Z. Khatib, P. Shannon, and F. Kitchener, "Effect of weather on free-flow speed," *Transportation Research Record*, no. 1776, pp. 60–68, 2001.

[19] B. L. Smith, K. G. Byrne, R. B. Copperman, S. M. Hennessy, and N. J. Goodall, "An investigation into the impact of rainfall on freeway traffic flow," in *Proceedings of the Proceedings 83th Annual Meeting of the Transportation*, Washington, Wash, USA, 2004.

[20] M. Agarwal, T. H. Maze, and R. Souleyrette, "Impacts of weather on urban freeway traffic flow characteristics and facility capacity," in *Proceedings of the 2005 Mid-Continent Transportation Research Symposium*, pp. 18-19, Ames, Iowa, USA, 2005.

[21] R. Hranac, E. Sterzin, D. Krechmer, H. Rakha, and M. Farzaneh, *Empirical Studies on Traffic Flow in Inclement Weather*, Federal Highway Administration, Washington, Wash, USA, 2006.

[22] M. Sabir, J. van Ommeren, M. J. Koetse, and P. Rietveld, *Welfare Effects of Adverse Weather through Speed Changes on Car Commuting Trips*, Vu University, Amsterdam, Netherlands.

[23] I. Tsapakis, T. Cheng, and A. Bolbol, "Impact of weather conditions on macroscopic urban travel times," *Journal of Transport Geography*, vol. 28, pp. 204–211, 2013.

[24] E. Hooper, L. Chapman, and A. Quinn, "Investigating the impact of precipitation on vehicle speeds on UK motorways," *Meteorological Applications*, vol. 21, no. 2, pp. 194–201, 2014.

[25] M. Pregnolato, A. Ford, S. M. Wilkinson, and R. J. Dawson, "The impact of flooding on road transport: A depth-disruption function," *Transportation Research Part D: Transport and Environment*, vol. 55, pp. 67–81, 2017.

[26] T. D. Shand, R. J. Cox, M. J. Blacka, and G. P. Smith, "Appropriate safety criteria for vehicles. Revision Project 10. Stage 2. Literature Review," Tech. Rep., Australian Rainfall and Runoff, Australia, 2010.

[27] F. Y. Teo, J. Xia, R. A. Falconer, and B. Lin, "Experimental studies on the interaction between vehicles and floodplain flows," *International Journal of River Basin Management*, vol. 10, no. 2, pp. 149–160, 2012.

[28] J. Xia, R. A. Falconer, X. Xiao, and Y. Wang, "Criterion of vehicle stability in floodwaters based on theoretical and experimental studies," *Natural Hazards*, vol. 70, no. 2, pp. 1619–1630, 2014.

[29] J. Xia, F. Teo, R. Falconer, Q. Chen, and S. Deng, "Hydrodynamic experiments on the impacts of vehicle blockages at bridges," *Journal of Flood Risk Management*, vol. 11, pp. S395–S402, 2018.

[30] E. Martínez-Gomariz, M. Gómez, B. Russo, and S. Djordjević, "A new experiments-based methodology to define the stability

threshold for any vehicle exposed to flooding," *Urban Water Journal*, vol. 14, no. 9, pp. 930–939, 2017.

[31] M. Kramer, K. Terheiden, and S. Wieprecht, "Safety criteria for the trafficability of inundated roads in urban floodings," *International Journal of Disaster Risk Reduction*, vol. 17, pp. 77–84, 2016.

[32] E. Martínez-Gomariz, M. Gómez, B. Russo, and S. Djordjević, "Stability criteria for flooded vehicles: a state-of-the-art review," *Journal of Flood Risk Management*, vol. 11, pp. S817–S826, 2018.

[33] K. Pyatkova, A. S. Chen, S. Djordjevic et al., "Flood impacts on road transportation using microscopic traffic modelling technique," in *Proceedings of the SUMO User Conference*, Berlin, Germany, 2015.

[34] J. Yin, D. Yu, Z. Yin, M. Liu, and Q. He, "Evaluating the impact and risk of pluvial flash flood on intra-urban road network: A case study in the city center of Shanghai, China," *Journal of Hydrology*, vol. 537, pp. 138–145, 2016.

[35] J. Affum, G. Giummarra, and H. Cheung, "Safety provisions for floodways over roads," Austroads Report AP-R481-15, 2015.

[36] L. M. Timbe and P. Willems, "Desempeño de modelos hidráulicos 1D y 2D para la simulación de inundaciones," *Maskana*, vol. 2, no. 1, pp. 91–98, 2015.

[37] M. Bolla Pittaluga, R. Luchi, and G. Seminara, "On the equilibrium profile of river beds," *Journal of Geophysical Research: Earth Surface*, vol. 119, no. 2, pp. 317–332, 2014.

[38] Y. Niño, *River Hydraulic and Sediment Transport*, Universidad de Chile, Chile, 2004.

[39] A. Haldar and S. Mahadevan, *Probability, Reliability and Statistical Methods in Engineering Design*, Wiley, New York, NY, USA, 1st edition, 2000.

[40] A. M. Hasofer and N. C. Lind, "An exact and invariant first order reliability format," *Journal of Engineering Mechanical*, vol. 100, pp. 111–121, 1974.

[41] M. Rosenblatt, "Remarks on a multivariate transformation," *Annals of Mathematical Statistics*, vol. 23, pp. 470–472, 1952.

[42] MOP, *Basic Study on Update HDM-4 Parameters*, Ministry of Public Works, Chile, 2017.

[43] MOP, *Información Oficial Hidrometeorológica y de Calidad de Aguas en Línea*, MOP, Chile, 2016, http://snia.dga.cl/BNAConsultas/reportes.

7

Improving Traffic State Prediction Model for Variable Speed Limit Control by Introducing Stochastic Supply and Demand

Yuwei Bie,[1] Mudasser Seraj,[1] Can Zhang,[1] and Tony Z. Qiu[1,2]

[1]*Department of Civil and Environmental Engineering, University of Alberta, Edmonton, Canada*
[2]*Intelligent Transportation Systems Research Center, Wuhan University of Technology, Wuhan, China*

Correspondence should be addressed to Tony Z. Qiu; zhijunqiu@ualberta.ca

Academic Editor: David Z. W. Wang

Variable speed limit (VSL) is becoming recognized as an effective way to improve traffic throughput and road safety. In particular, methods based on traffic state prediction exhibit promising potential to prevent future traffic congestion and collisions. However, field observations indicate that the traffic state prediction model results in nonnegligible error that impacts the next step decision making of VSL. Thus, this paper investigates how to eliminate this prediction error within a VSL environment. In this study, the traffic state prediction model is a second-order traffic flow model named METANET, while the VSL control is model predictive control (MPC) based, and the VSL decision is discrete optimized choice. A simplified version of the switching mode stochastic cell transmission model (SCTM) is integrated with the METANET model to eliminate the prediction error. The performance of the proposed method is assessed using field data from a VSL pilot test in Edmonton, Canada, and is compared with the prediction results of the baseline METANET model during the road test. The results show that during the most congested period the proposed SCTM-METANET model significantly improves the prediction accuracy of regular METANET model.

1. Introduction

The VSL control method was designed to keep the credibility of speed limits even under adverse conditions such as congestion and bottleneck road segments, so that the speed limits still work to maintain traffic safety and the highest possible traffic throughput. This method adjusts the posted speed limit based on real-time road, traffic, and weather conditions [1]. The VSL control method performs well in terms of control scale, control method flexibility, and feasibility. The most commonly used VSL strategies take a reactive approach with simple policies: using this strategy, the VSL decision is triggered by abnormal traffic states detected in real time, such as high traffic flow or low average speed. However, to achieve improved traffic throughput, a more complex control algorithm that includes optimized VSL decision-making is needed. Proactive VSL control goes one step further than reactive VSL control since it is optimized and calculated based on the predicted traffic state, so as to prevent a severe traffic breakdown from happening. Between two general patterns of VSL control, stimulation-reactive pattern and proactive pattern, this paper considers the latter pattern to be more challenging but promising.

In proactive VSL control, an accurate macroscopic level traffic prediction model is the core part. Over the last decades, various traffic prediction models were invented: the author classifies them to statistical and nonstatistical methods [2]. The statistical methods can predict the traffic flow with high accuracy under stable traffic conditions, commonly used methods including the class of Time Series models, such as seasonal Autoregressive Integrated Moving Average model [3] and Kalman Filter State-Space model [4], Neural Network [5], and Nonparametric Regression [6]. However, in the case of VSL control, traffic flow prediction is required under congestion condition at bottleneck road sections where traffic conditions can change abruptly, so that the author chose nonstatistical methods, i.e., the traffic prediction model derived from traffic flow models. Those models have potential to reflect the nature of traffic flow and capture the fluctuation of traffic flow: the traffic flow is treated as fluid that is aggregated by individual vehicles, and three variables are capable of describing the characteristics of traffic stream; they are flow, density, and mean speed. The traffic flow models are classified as first-, second-, or higher-order models, depending on the number of the included differential equations [7]. First-order model is able to describe traffic flow, but not traffic density or speed. Among all first-order models, the most used is Lighthill–Whitham–Richards model [8, 9] using one partial differential equation to describe the vehicle flow conservation law. This was also the first time that traffic flow model is combined with static fundamental diagram. Another representative first-order model is the Cell Transmission Model [10–12], which is a discretized and simplified version of the Lighthill–Whitham–Richards model. The Payne model [13] is the oldest second-order traffic flow model. Besides the flow conservation law equation, the Payne model also includes one partial differential equation that describes mean speed dynamics. This model can replicate traffic phenomenon with higher accuracy. There are also other types of second-order traffic prediction models and combinations, such as variation kinematic waves [14], second-order traffic flow model with Kalman filter [15], and CTM-based second-order traffic flow model with particle filtering [16].

To achieve the highest traffic prediction accuracy, in this paper, a second-order traffic flow model named METANET [17, 18] is chosen. It is a discretized and enhanced version of the Lighthill–Whitham–Richards model combined with the Payne model. As it is a second-order traffic flow model, it is able to predict traffic density, average vehicle speed, and traffic flow by three dynamic functions in the model. METANET model fulfills the simplicity and convenience requirement since it has a space-time discrete, explicit analytical state-space form and allow for convenient discretization intervals [7]. What is more, METANET model is proved by previous studies that it is of high accuracy and relatively easy to calibrate, which makes it one of the most frequently utilized macroscopic traffic flow models in variety of traffic engineering tasks and researchers, and this model is capable of being used for optimal real-time traffic control of freeway traffic.

However, due to the stochastic nature of traffic flow, accurately predicting the traffic state of bottlenecks for VSL control is a complex task, especially during peak hours and for congested road segments. The traffic flow, speed, and density are by nature stochastic and variable, when traffic flow models are used for predicting traffic condition for making VSL control decisions, those stochasticity and variability need to be described and modeled and be added to the original static model to meet the specific high prediction requirement of VSL control. VSL field experiments in this paper have demonstrated that the traffic state prediction model exhibits larger error during the afternoon peak period. The stochastic nature of traffic can be modeled by adding an exogenous disturbance term to each round of calculation. This paper is inspired by the stochastic elements in the stochastic cell transmission model (SCTM) [19] and incorporates them in this study. The most straightforward method to model the stochasticity is to apply a sequential Monte Carlo simulation method to the CTM to mimic the change from free flow traffic conditions to congested conditions [20], but this method has a high computation cost. Another way to incorporate the stochastic element is to set the sending and receiving flow as random variables [21]. However, this method does not function well when traffic is light or congested, for the random terms cannot represent transportation features. In 2011, Sumalee et al. [19] proposed a stochastic CTM (SCTM) to describe the macroscopic dynamics of traffic flow under demand and supply uncertainties. Its switching mode design simplifies the sequential Monte Carlo method, and its stochastic element takes into account the fundamental diagram parameters instead of purely random generated. The SCTM has been implemented in practical research projects, such as dynamic journey time estimation [22] and road network traffic assignment under adverse weather [23]. The cell transmission model (CTM) is a first-order traffic flow model with no exogenous input, and it serves as the density dynamic function in the METANET model. In this paper, the regular density dynamic function in METANET model, also called CTM is replaced with the SCTM. Stochastic elements are added in anticipation of better prediction performance.

Besides traffic prediction model, the VSL control structure also has various choices [24] such as case-based reasoning, knowledge-based systems (in which the policy is based on previous practices), and model-based prediction. To find the optimal control structure that utilizes the traffic state prediction model, a MPC framework [25] is applied in this paper, forming a closed loop. The future traffic state is predicted, and then the optimal VSL decision is made based on prediction results. MPC is an optimal control scheme applied within a rolling horizon framework. Optimal control was successfully applied by Kotsialos et al. [26] to coordinate or integrate traffic control measures. Both optimal control and MPC have the advantage of a controller that generates an optimal control signal or decision according to a user-defined objective function. The objective function in this paper is a weighted summation of total travel time (TTT) and total travel distance (TTD), where the total travel time is minimized and total travel distance is maximized simultaneously.

VSL control has been successfully implemented in European countries and the United States [27]. In Edmonton, Canada, a four-week VSL field test was implemented on a 10-kilometer stretch of freeway, and this field test serves as a case study in this paper. The field data truly reflect the driver behavior and compliance of people in a major western Canadian city. The actual field data and METANET-predicted traffic data are extracted for comparison with the simulated data generated by the proposed model. This paper first integrates the SCTM to the METANET model for VSL control purposes and then tests its feasibility using both field experiment data and computer simulation data. The upgrading of METANET model, the field test result analysis, and the simulation provide references for future local researches.

The following sections of this paper include a literature review of VSL and the original SCTM model; an explanation of the model predictive control- (MPC-) based VSL algorithm; a description of the modified SCTM that is used in this study; presentation of the case study; and, finally, a discussion of the results and conclusion.

2. A Model Predictive Control Based Variable Speed Limit Algorithm

The model description is separated into four parts: the description of traffic state prediction model and the description of fundamental diagram used in this study. Section 2.1 will be notations and assumptions. Section 2.2 will describe in detail the baseline VSL control logic and algorithm. Section 2.3 will present model calibration skills including the calibration of fundamental diagram and the macroscopic traffic prediction model involved in VSL control.

2.1. Notations and Assumptions. For consistency, the freeway is divided into N sections with lengths Δ_i, $i = 1, \ldots, N$, each having at most one on-ramp and off-ramp. All the variables used throughout this paper are defined as follows:

T: data collection interval=20s

λ_i: number of lanes at section i

Δ_i: length of section i

$q_i(k)$: number of vehicles in the freeway section i at time $k \cdot T$ divided by the length Δ_i

$\rho_i(k)$: traffic density at time step k, section i.

$v_i(k)$: average vehicle travel speed at time step k, section i.

$u_i(k)$: the optimized speed limit posted at time step k, section i.

$r_i(k)$: on-ramp flow at time step k, section i

$s_i(k)$: off-ramp flow at time step k, section i

$V(\rho_i(k))$: desire speed in speed dynamics in METANET model

Θ: set of unknown parameters of METANT model

$\rho_{cr,i}$: critical density of FD at section i

$\rho_{jam,i}$: jam density of FD at section i

$v_{f,i}$: free flow speed of FD at section i

w_i: shockwave speed of FD at section i

θ: capacity drop fraction of FD at section i

TTT: total travel time

TTD: total travel distance

α_{TTT}: weight of TTT

α_{TTD}: weight of TTD.

2.2. The MPC-Based VSL Algorithm Using METANET Prediction Model. The MPC-based VSL algorithm is composed of five parts: an original traffic prediction model, control case traffic prediction model, optimizer, objective function, and constraints. As shown in Figure 1, for VSL control road segments, the system uses a modified METANET model for traffic state prediction, and the discrete choice optimizer makes a choice based on predicted traffic states, given the objective function and constraints. After optimization, the optimal VSL choice is shown to operators, and, simultaneously, the VSL value is inputted into the VSL-METANET model again for the final traffic state prediction results to be recorded. Figure 1 shows a block diagram of the VSL control system.

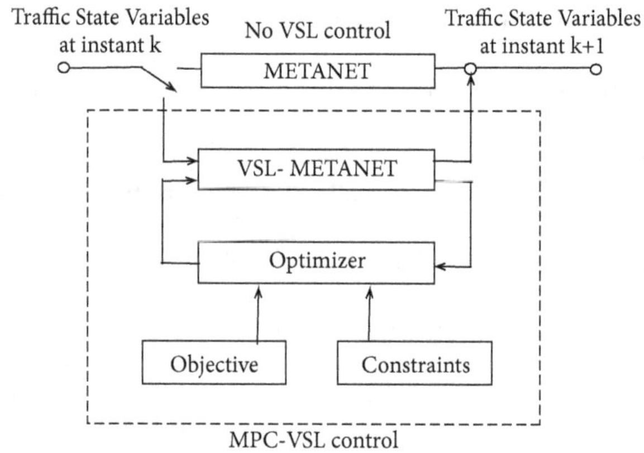

Figure 1: The block diagram of model predictive VSL control.

The METANET model as a second-order traffic flow model introduces speed dynamic functions instead of including only density dynamics to describe flow conservation law. This feature of the METANET model enables it to predict density, speed, and flow variables, respectively, accurately, as well as with a small time interval. The discretized model formulation of the original METANET is written as follows.

Density dynamics (also referred to the regular CTM) are as follows:

$$\rho_i(k+1) = \rho_i(k) + \frac{T}{\Delta_i \lambda_i} \left[\lambda_{i-1} q_{i-1}(k) - \lambda_i q_i(k) + r_i(k) - s_i(k)\right] \tag{1}$$

Speed dynamics are as follows:

$$v_i(k+1) = v_i(k) + \frac{T}{\tau}\left[V(\rho_i(k)) - v_i(k)\right] + \frac{T}{\Delta_i} v_i(k)\left[v_{i-1}(k) - v_i(k)\right] - \frac{\eta \cdot T\left[\rho_{i-1}(k) - \rho_i(k)\right]}{\tau \cdot \Delta_i \cdot \left[\rho_i(k) + \kappa\right]} \tag{2}$$

Flow dynamics are as follows:

$$q_i(k+1) = \rho_i(k+1) \cdot v_i(k+1) \tag{3}$$

where i is the index of links, and $i = 1, 2, \ldots M$, with M representing the number of testbed sections. K is the index of time instants, and T is the calculating time interval, where

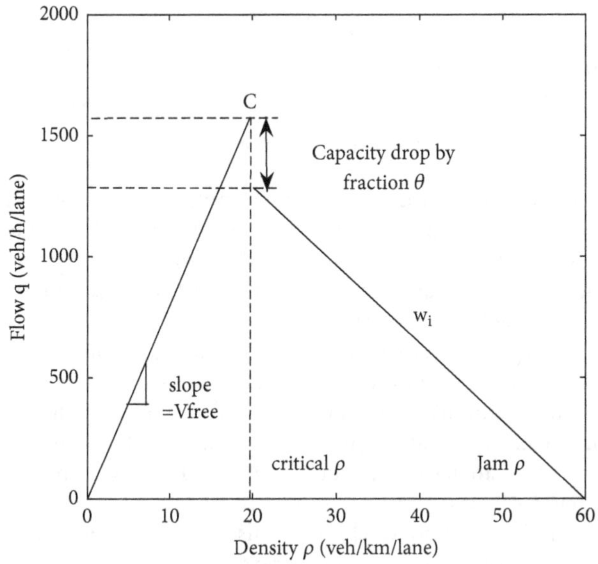

FIGURE 2: The triangular FD assumption.

T=20s. Δ_i represents the segment length of link i, and λ is the lane number of link i. In the model, τ, κ, α, and η (km²/h) are global model parameters calibrated using the historical data. The desired speed $V(\rho_i(k))$ (km/h) in the speed dynamics is represented by the following expression:

$$V(\rho_i(k)) = v_{f,i}\exp\left[-\frac{1}{\alpha}\left(\frac{\rho_i(k)}{\rho_{c,i}}\right)^{\alpha}\right] \quad (4)$$

The triangular fundamental diagram (FD) is assumed in the control algorithm. Figure 2 shows the triangular FD assumption used in this paper. In (4), $v_{f,i}$ is free flow speed and $\rho_{c,i}$ represents critical density, which is the density associated with capacity. In this model, the density dynamics do not involve parameters, and other parameters in the speed dynamics are calibrated globally for goodness of fit. Of the four terms that make up the speed dynamics equation, each term has a physical meaning. In (2), the second term is referred to as the relaxation term, indicating that, with a lag time item τ, the mean speed v of the link "relaxes" to the desired speed, which largely depends on the parameters of the FD. The selection of the desired speed is critical in reflecting driver behavior, and, based on previous practice [28], the format of (4) was chosen. The third term is the convection term used to indicate that vehicles traveling from the upstream link $i-1$ to the current link i adapt their speed gradually rather than instantaneously. The fourth term is the anticipation term, indicating that drivers are always watching the road ahead for changes in traffic. If a driver observes high traffic density in the downstream link $i+1$, the driver then reacts by slowing down. The constant $\kappa > 0$ is added to keep the anticipation term limited when density is low.

The METANET algorithm for the VSL control environment is modified by replacing the desired speed with a modified one. In the control segments, the desired speed is assumed to be the posted advisory speed, rather than the speed determined by the fundamental diagram as in the non-control case. In the VSL-METANET model, the desired speed term then becomes the optimized VSL value, as decided by the optimizer in the MPC system.

Desired speed term when VSL is triggered is as follows:

$$V(\rho_i(k)) = u_i(k) \quad (5)$$

The objective function minimizes a weighted sum of TTT and TTD in a discrete choice fashion. In terms of definition, TTT indicates a weighted sum of density of all time steps and links. Only minimizing TTT produces system bias, wherein the optimizer tends to choose a lower speed limit to decrease the density, so that TTD is maximized simultaneously. In the objective function, weights α_{TTT} and α_{TTD} are applied to TTT and TTD.

$$TTT = T\sum_{j=1}^{pr-1}\sum_{i=1}^{M}\lambda_i L_i \rho_i(k+j) \quad (6)$$

$$TTD = T\sum_{j=1}^{pr-1}\sum_{i=1}^{M}\lambda_i L_i \rho_i(k+j)\, v_i(k+j) \quad (7)$$

$$J = T\sum_{j=1}^{pr-1}\sum_{i=1}^{M}\lambda_i L_i \left[\alpha_{TTT}\rho_i(k+j) - \alpha_{TTD}\rho_i(k+j)\, v_i(k+j)\right] \quad (8)$$

The desired speed under the VSL environment is decided by the optimizer. The optimization is subjected to several constraints of traffic safety, driver acceptance, and traffic flow characteristics [1]. The first constraint is the upper limit of $u_i(k)$, since the regular speed limit is 80 km/h, under congested conditions $u_i(k) \le 80$. The second constraint is to maintain continuous traffic flow even under congestion. From local regulations, the lower bound of the speed limit is 30 km/h, which means $u_i(k) \ge 30$. In Canada, the speed limit is in multiples of 10 km/h; therefore, to ensure safety, the change in speed limit between two time instants is 0 or 10 km/h, represented as $u_i(k) - u_i(k+1) \in \{-10, 0, 10\}$. The optimizer then becomes a discrete choice model.

2.3. Calibrating the FD and METANET Model. The parameter estimation of the FD is based on data collected by conventional loop detectors. For the parameter calibration of the FD, free flow speed $v_{f,i}$, critical density $\rho_{cr,i}$, capacity C_i, jam density $\rho_{jam,i}$, and capacity drop fraction θ must be estimated. For calibration, the data format used is (ρ_i, q_i) data points, where the horizontal coordinate is density and the vertical coordinate is flow, as Figure 2 shows. The procedure for obtaining the necessary parameters from a triangular FD is as follows.

(1) The identification of capacity C_i: in defining the triangular FD, the summit point of the triangle roughly indicates critical density and capacity. All (ρ_i, q_i) points are plotted to find the third largest q_i and then take it as capacity C_i. The largest flow is not chosen because it is normally an extreme outlier; this could be due to detection error so that the third largest flow is empirically chosen as capacity.

(2) The identification of critical density $\rho_{cr,i}$: after determining C_i, the corresponding traffic density ρ_i is then identified as critical density $\rho_{cr,i}$. This identification is based on the assumption of triangular shape FD: the vertical line across C_i intersects the horizontal density axis at critical density $\rho_{cr,i}$. Point $(\rho_{cr,i}, C_i)$ is then the vertex of the triangular FD.

(3) The identification of free flow speed $v_{f,i}$: after defining $\rho_{cr,i}$, the whole dataset can be divided into two parts, the left and right side of the triangle. The left side of the triangle represents uncongested traffic conditions and the right side of the triangle represents congested traffic conditions. Then the slope of each data point distributed on the left side is calculated, and the average is taken as $v_{f,i}$, as shown in the following equation [2]:

$$v_{f,i} = \frac{1}{n}\left(\sum_{ii=1}^{n}\frac{q_{ii}}{\rho_{ii}}\right), \quad \rho_{ii} \in (0, \rho_{cr,i}) \tag{9}$$

(4) The identification of jam density $\rho_{jam,i}$, the capacity drop fraction θ, and shockwave speed w_i: the jam density $\rho_{jam,i}$ represents a theoretical value that indicates when the road section is totally congested and all vehicles have stopped moving. From the dataset, the data point with the largest density value and zero speed is selected as $\rho_{jam,i}$. The slope of the right side, i.e., shockwave speed w_i is determined by least square fit fixing one foot $(\rho_{jam,i}, 0)$. After determining one foot and the slope of the right side, the intersecting point of the regression line and the vertical line across C_i is the congested capacity after a capacity drop fraction of θ.

For the parameter estimation of METANET model, τ, κ, α, and η (km^2/h) are global model parameters to be calibrated. The calibration of parameters is not done in real time, according to the logic that both parameters from FD and METANET model represent aggregated traffic behavior of a period of time, the calibration happens right before the field test and uses traffic data collected in the last 7 days. In doing SCTM-METANET simulation, the same parameter set as in the field test is used. The parameters of METANET model are estimated via the following expression:

$$\Theta^* = \arg\min_{\Theta}\left[\sum_{ii}^{N}\left\{v^{ii}{}_{obs} - f\left(v^{ii}{}_{obs} \mid \Theta\right)\right\}^2 + \sum_{ii}^{N}\left\{\rho^{ii}{}_{obs} - f\left(\rho^{ii}{}_{obs} \mid \Theta\right)\right\}^2\right] \tag{10}$$

3. Integrating Switching-Mode SCTM with METANET Model

The switching-mode SCTM is employed to replace the regular CTM in the METANET model. The traditional CTM, derived from the conservation law of traffic flow, is a single linear system without exogenous input or tuning parameters. When CTM is used for traffic state prediction, after rounds of iteration, the model is farther and farther from reality since during the calculation time the outside traffic condition changes quickly. In this paper, the traffic prediction length in the VSL field test is 5 minutes, containing 15 iterations of calculation, but congestion can form within 3 minutes. To solve this problem, the SCTM introduces exogenous stochastic disturbance terms to each iteration of computing to counteract the prediction error.

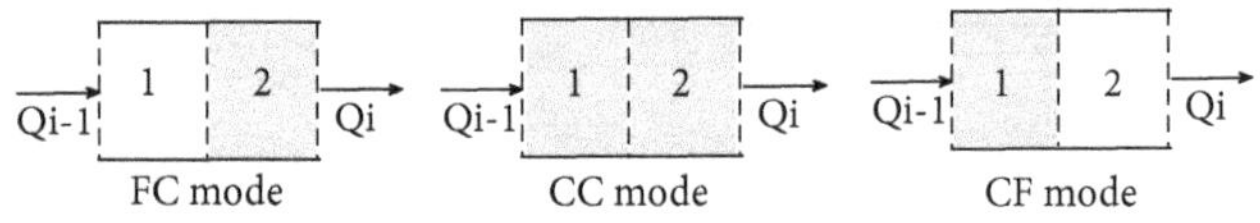

FIGURE 3: Traffic operation modes.

Incorporating stochastic supply and demand into a traffic flow model can be different ways, the FD stochasticity, the boundary condition stochasticity, and stochastic adjusting parameter if necessary. In the case of the SCTM in this paper, the stochastic elements are added in two ways: (1) demand stochasticity, the inflow disturbance term added to the CTM; (2) supply stochasticity also FD stochasticity, each of the disturbance terms containing stochastic FD parameters extracted by statistical technics.

Another feature of the SCTM is that, under different traffic condition modes, the disturbance elements change slightly. The original SCTM has five modes (refer to [19]), and in this paper the number of modes is simplified to three because the CTM is still used for free flow conditions and the wave front of congestion is only from downstream to upstream. The three modes are free flow to congestion (FC) mode, congestion to congestion (CC) mode, and congestion to free flow (CF) mode. Different from the CTM, the SCTM considers the adjacent two cells as a whole and uses vector $P(k) = [\rho_1(k), \rho_2(k)]^T$ in the density dynamics calculation. Please see Figure 3 for the illustration of traffic operation modes where white segment represents for low traffic density free flow condition and grey segment represents for high traffic density congested condition.

When defining the traffic operational modes, a typical SCTM develops complicated probabilistic occurrence models as mode switches [19], and it serves generalized traffic situations. However, when SCTM is implemented in a specific VSL control road segment whose traffic condition is highly repeatable and well understood, the probability expression then decays to a fixed version. The fixed mode switch time points are empirically set based on traffic data, which can be observed in the case study of this paper. Additionally, the switch time considers the suggestion of local authorities. Free flow mode, which is the original density dynamic function in METANET, is in effect 4:00pm to 4:59pm. The FC mode takes effect during 5:00pm to 5:09pm and serves as a transitional stage between free flow mode and congested mode. The CC mode takes place between 5:10pm and 5:49pm. Then the CF mode takes effect between 5:50pm and 5:59pm. From 6:00pm on, the CTM takes over again. In the FC mode, $u(k) = (q_{i-1}(k), r_i(k), s_i(k), q_i(k))^T$, $\omega_2(k) = w_{c,2}(k)$, and $\lambda(k) = (0, \rho_{jam,2}(k))^T$.

Density dynamics (SCTM) under FC mode are

$$P(k+1) = (I + A_2\omega_2(k))\,P(k) + B_2\omega_2(k)\,\lambda(k) + Bu(k) \tag{11}$$

where

$$A_2 = \begin{bmatrix} 0 & \frac{T_s}{l_1} \\ 0 & -\frac{T_s}{l_2} \end{bmatrix},$$

$$B_2 = \begin{bmatrix} 0 & -\frac{T_s}{l_1} \\ 0 & \frac{T_s}{l_2} \end{bmatrix}, \tag{12}$$

$$B = \begin{bmatrix} \frac{T_s}{l_1} & -\frac{T_s}{l_1} & 0 & 0 \\ 0 & 0 & -\frac{T_s}{l_2} & \frac{T_s}{l_2} \end{bmatrix}.$$

After the traffic condition transitions to congestion mode, the CC model applies. In the CC model, $\omega_i(k) = w_i(k)$, and $\lambda(k) = (\rho_{jam,1}(k), \rho_{jam,2}(k))^T$.

Density dynamics (SCTM) under CC mode are

$$P(k+1) = \left(I + \sum_{i=1}^{2} A_i \omega_i(k) \right) P(k) + \sum_{i=1}^{2} B_i \omega_i(k) \lambda(k) + Bu(k) \tag{13}$$

where

$$A_1 = \begin{bmatrix} -\frac{T_s}{l_1} & 0 \\ 0 & 0 \end{bmatrix},$$

$$A_2 = \begin{bmatrix} 0 & \frac{T_s}{l_1} \\ 0 & -\frac{T_s}{l_2} \end{bmatrix}, \tag{14}$$

$$B = \begin{bmatrix} 0 & 0 & 0 & 0 \\ 0 & 0 & -\frac{T_s}{l_2} & -\frac{T_s}{l_2} \end{bmatrix},$$

$$B_i = -A_i, \ i = 1, 2.$$

After the traffic condition transitions out of congestion mode, the CF model applies. In the CF model, $\omega_1(k) = w_1(k)$, $\omega_2(k) = \mathrm{v}_{f,2}(k)$, and $\lambda(k) = (\rho_{jam,1}(k), \mathrm{C}_1(k))^T$.

Density dynamics (SCTM) under CF mode are

$$P(k+1) = \left(I + \sum_{i=1}^{2} A_i \omega_i(k) \right) P(k) + \left(B_0 + \sum_{i=1}^{2} B_i \omega_i(k) \right) \lambda(k) + Bu(k) \tag{15}$$

where

$$A_1 = \begin{bmatrix} -\frac{T_s}{l_1} & 0 \\ 0 & 0 \end{bmatrix},$$

$$A_2 = \begin{bmatrix} 0 & 0 \\ 0 & -\frac{T_s}{l_2} \end{bmatrix}, \tag{16}$$

$$B_0 = \begin{bmatrix} 0 & -\frac{T_s}{l_1} \\ 0 & \frac{T_s}{l_2} \end{bmatrix},$$

$$B_1 = -A_1, \ B = B_2 = 0.$$

4. Case Study

The vehicle detection stations (VDS) on a 10-kilometer corridor of Whitemud Drive in Edmonton, Canada, collect, and store traffic data from dual loop detectors. This section of road plays an important role in people and freight transportation in the city. The VDS system currently has 28 VDS in total, and each station has three or four dual loops. The data recording frequency is 20 seconds. Each dual loop reports the volume q—the number of vehicles crossing the loop detector during a 20-second time interval—and mean speed measurement v, and the occupancy measurement, which cannot be used directly in traffic control and cannot be transformed accurately into density. The accurate density measurement ρ is calculated by $\rho = q/v$. Figure 4 shows the testbed geometry, VDS locations and the location of dynamic message signs (DMS) that will be used to display variable speed limits.

The VSL control decisions are made and sent out by software with the METANET model embedded. When the detected average speed is below the posted speed of 80 km/h, the VSL-METANET prediction is triggered on the corresponding road segment. The prediction step length is 20 seconds and 15 iterations, which means the prediction horizon is 5 minutes. The prediction is calculated every minute. So that all variables for comparison are 1 min aggregated, and when simulating SCTM-METANET the same aggregation level is used. Theoretically speaking, the METANET model can be made to predict farther future such as 10 min and 15 min; however, in real life situation in Edmonton, Canada especially during peak hours, traffic condition changes drastically even within 5 minutes (please see Figures 6(b), 6(c), and 6(d) and Figures 7(b), 7(c), and 7(d)). So that when doing municipal level experiment on freeways, prediction horizon longer than 5 minutes was rarely used. If under non-peak hours when traffic flow is relatively steady, it is possible to recalibrate METANET model and achieve high accuracy prediction for longer horizons.

During real-time VSL control, data from loop detector stations are sent to computers in the traffic management center in the City of Edmonton. The software installed in the computers is responsible for providing speed limit

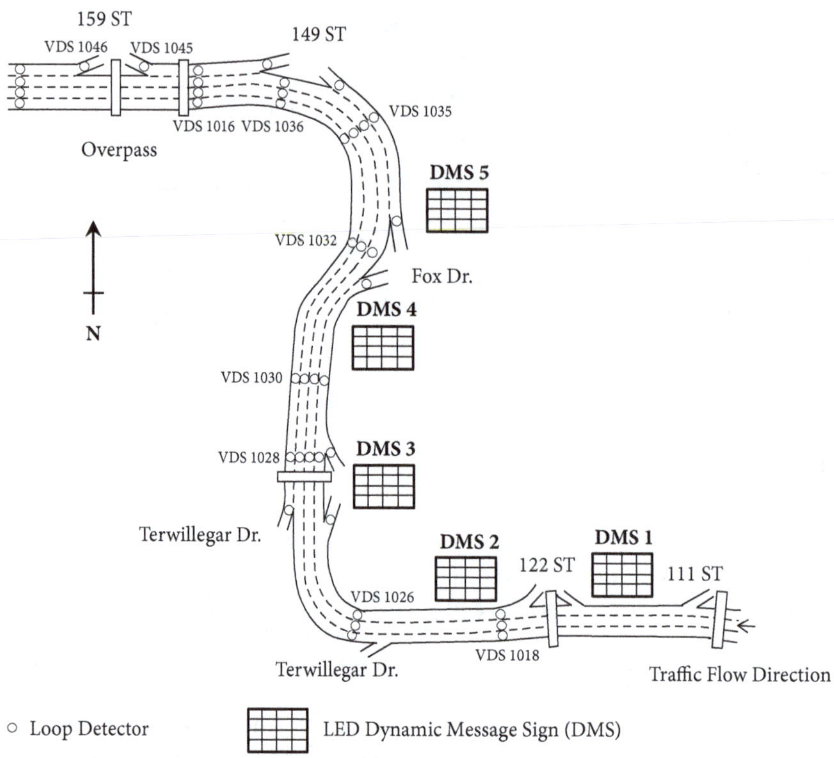

FIGURE 4: Vehicle detection stations and dynamic message signs on Whitemud Drive, Edmonton, Canada.

suggestions. The decisions are then shown on screen with beeping notifications for human operators to check and send to DMS on the road. From August 13 to September 4 of 2015 on weekdays, the City of Edmonton conducted a four-week VSL pilot test on westbound Whitemud Drive. Five DMSs in total were located upstream of predefined congestion-prone points. From the testing period, DMS 1 on August 17 was chosen as a case for this study because a period of normal peak hour congestion occurred and the VSL functioned well. Due to legal reasons, the VSLs were not mandatory to drivers, and they were labeled as "Advisory Driving Speed" on the DMSs.

The performance measurements of VSL in this study are TTT and TTD, which can be taken together as the measure of effectiveness (MOE), as shown in (6) and (7). At each time step, the TTT and TTD are calculated from a weighted sum of the density and flow of all links, in which the weight is the segment length of each link. Figure 5 shows the speed profile of the test day. VSL is observed to help stabilize the speed and the predefined fixed time FC-CC-CF mode switch makes sense in this case.

In the field test, the global parameters of METANET model were calibrated based on historical traffic data of August 2014, and, during field test, those parameters were fixed. The reason for fixing global parameter is to reduce the complexity of online traffic prediction. Table 1 shows the precalibrated global parameters of METANET model.

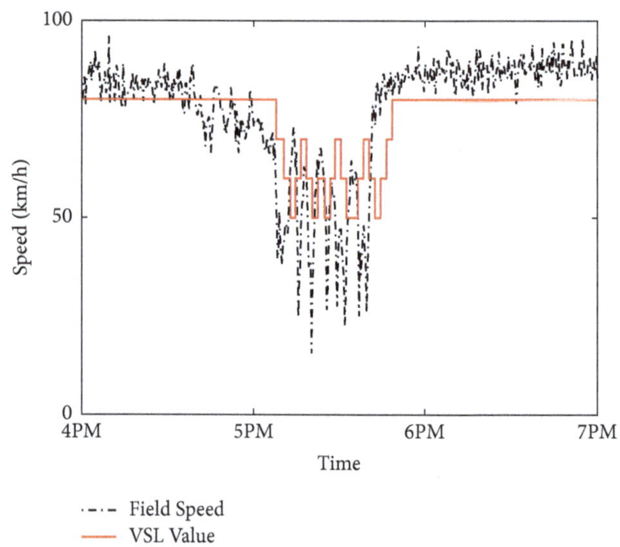

FIGURE 5: Speed profile of case study day Aug.17, 2015.

4.1. The Measurement of Prediction Error and the Performances of SCTM-METANET. The discrepancies between the actual MOE and those predicted can be categorized as structured error and unstructured error. Structured error occurs when fixed parameters, such as segment lengths, are incorrectly measured. Unstructured error is the true error or the error that is found in the prediction model, which usually happens after eliminating structured error.

The structured error (SE) of the predicted MOE is assumed to be proportional to the original predicted MOE, after denying other assumptions such as proportion to

TABLE 1: Parameters used in the field test.

METANET Parameters	η 37.98	κ 10.00	α 2.29	τ 120.00
FD Parameters	w_1, w_2 12	$v_{f,1}, v_{f,2}$ 82	$\rho_{jam,1}, \rho_{jam,2}$ 118	

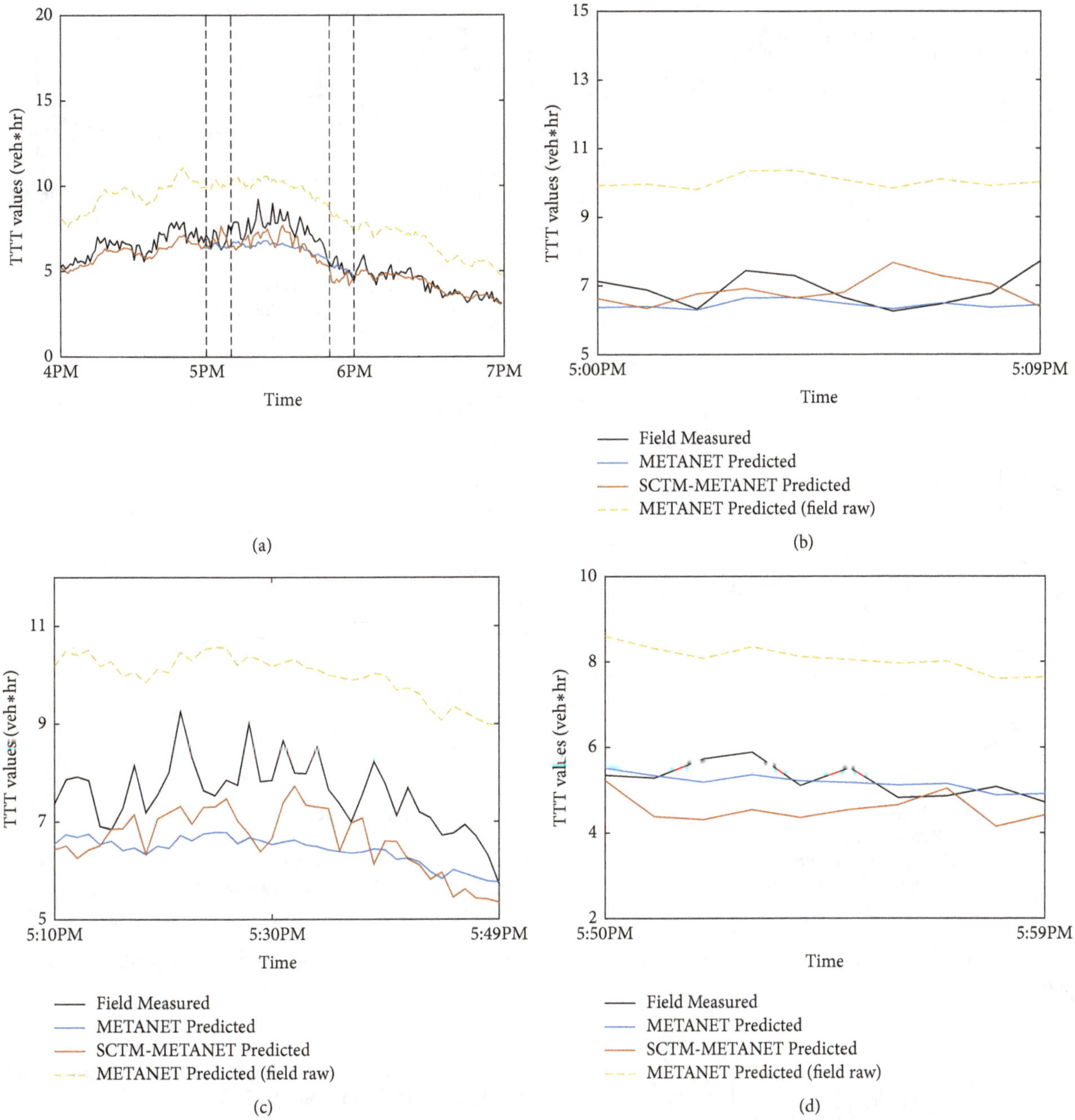

FIGURE 6: (a) Comparing field-measured TTT, original METANET, and SCTM-METANET predicted TTT; (b), (c), (d) detailed view of FC, CC, and CF modes.

measured MOE, fixed value of error, and normal distribution. So that the adjusted predicted MOE is calculated by scaling down original predicted MOE. The reference point is the beginning of the study period 4:00:00PM, meaning the suggestion that at the start point of this time period no structured prediction uncertainty existed. Based on the above understanding, the overall structured uncertainty can be calculated. In Figures 6 and 7, SE refers to the distance between the curve METANET-predicted (field raw) and the curve METANET-predicted.

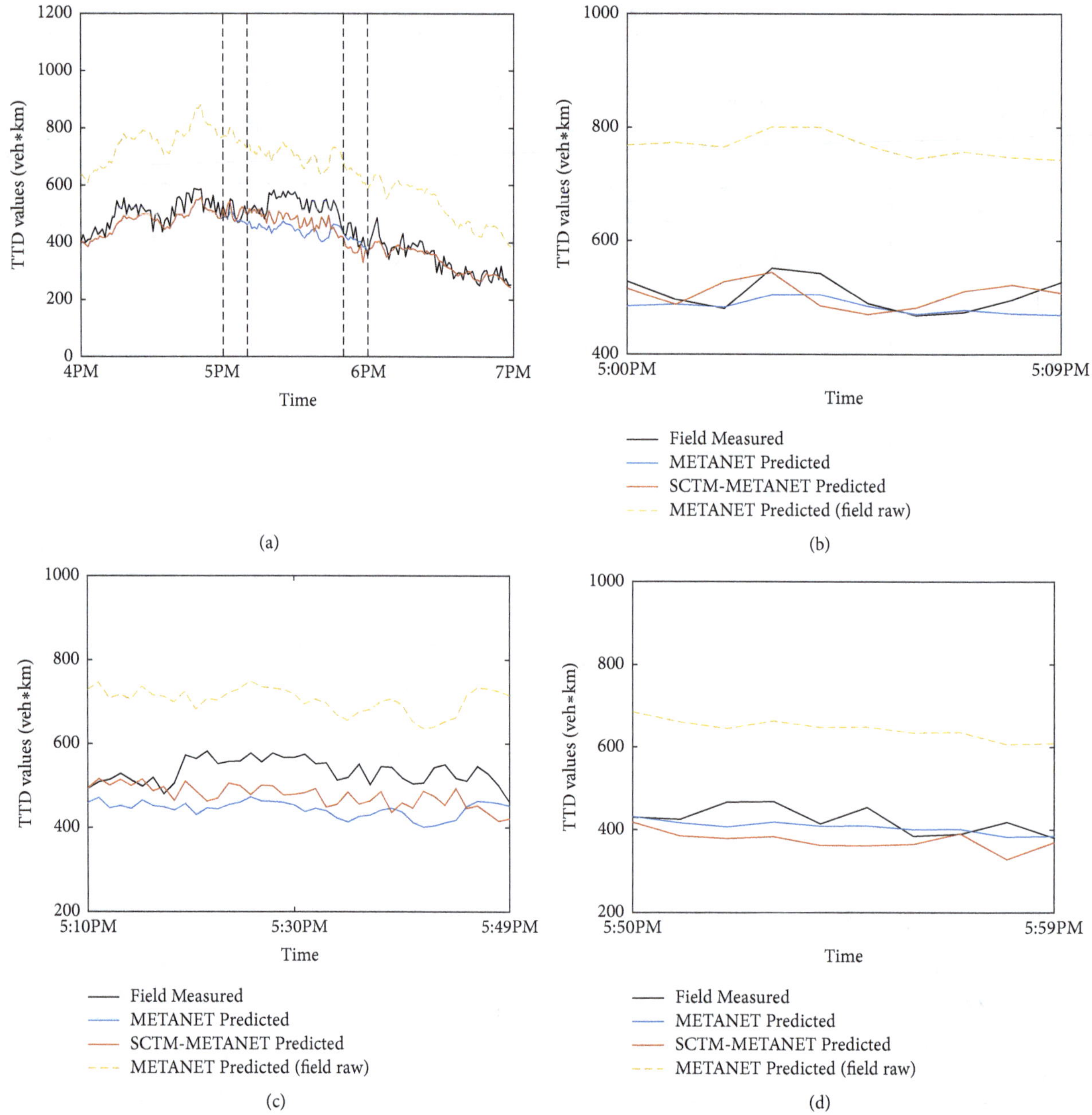

FIGURE 7: (a): Comparing field-measured TTD, original METANET, and SCTM-METANET predicted TTD. (b), (c), (d) detailed view of FC, CC, and CF modes.

$$SE(t) = \left(1 - \left.\frac{Moe_m(t')}{Moe_p(t')}\right|_{t'=4:00:00PM}\right) \cdot Moe_p(t) \quad (17)$$

where $Moe_m(t)$ and $Moe_p(t)$ represent field-measured and METANET-predicted TTT and TTD, respectively, at time step t. After eliminating structured error by scaling down the original predicted MOE, unstructured error (UE) remains, which represents the true performance of the prediction model. In Figures 6 and 7, unstructured error (UE) refers to the difference between the curve METANET-predicted and field-measured. In the following section, the root mean square error (RMSE) of both the original and new models is calculated from the UE.

$$UE(t) = Moe_m(t) - Moe_p(t) \cdot \left.\frac{Moe_m(t')}{Moe_p(t')}\right|_{t'=4:00:00PM} \quad (18)$$

Figures 6 and 7 and Tables 2 and 3 demonstrate the performance of the proposed SCTM-METANET versus the regular METANET. The raw field data predicted by METANET are collected from the field control software, and, after eliminating SE, the data become METANET-predicted data,

TABLE 2: Quantitative comparison of METANET and SCTM-METANET models: TTT.

RMSE of TTT prediction	Free Flow (4:00-4:59pm)	FC Mode (5:00-5:09pm)	CC Mode (5:10-5:49pm)	CF Mode (5:50-5:59pm)	Free Flow (6:00-6:59pm)
Conventional METANET	0.40	0.61	1.27	0.31	0.19
SCTM-METANET	——	0.77	1.09	0.84	——
Improvement (Percentage)	0	-26%	14%	-171%	0
No. of Data Points	60	10	40	10	60

TABLE 3: Quantitative comparison of METANET and SCTM-METANET Models: TTD.

RMSE of TTD prediction	Free Flow (4:00-4:59pm)	FC Period (5:00-5:09pm)	CC Period (5:10-5:49pm)	CF Period (5:50-5:59pm)	Free Flow (6:00-6:59pm)
Conventional METANET	20.0	31.42	95.97	31.53	25.5
SCTM-METANET	——	29.76	67.96	60.69	——
Improvement (Percentage)	0	5%	29%	-92%	0
No. of Data Points	60	10	40	10	60

which are ready to be compared with the computer-simulated SCTM-METANET.

Figure 6 shows that during free flow before 5PM and after 6PM the regular METANET model satisfactorily captures the trend of traffic flow. During 5:10PM and 5:49PM, the most congested period, the original METANET largely underestimates the TTT as well as its fluctuation. After introducing the SCTM into METANET, the predicted TTT not only is closer to reality, but also better matches the oscillating trend of the field-detected TTT. Figures 6(a) and 6(c) show steady improvement of the proposed SCTM-METANET model. Under transitional mode FC and CF, the SCTM-METANET does not show visible improvement of prediction accuracy, which may be due to too few data points. The FC and CF mode last only for 10 minutes, which means each condition has only 30 data points for computation. In summary, the SCTM-METANET outperforms the regular METANET model with noticeable improvements in stability, especially during the most congested period. Those improvements can be observed not only in the figure, but also quantitatively in Table 2.

When focusing on the prediction error during the CC period, the proposed SCTM-METANET model shows a significant 14% improvement calculated from 40 data points. Quantitative comparison also shows that, during the FC mode and CF mode, the proposed model shows unstable performance with negative improvements. The free flow period before and after SCTM-METANET shows stable and best prediction accuracy with the lowest error.

In the prediction of TTD, the SCTM-METANET performs better than for TTT. Figure 7 shows that, during free flow hours before 5PM and after 6PM, the regular METANET model satisfactorily captured the traffic flow trend. Also, during 5:10PM and 5:49PM, the original METANET largely underestimated the TTD and its fluctuation. When using the SCTM-METANET, the predicted TTD was not only closer to reality, but also better matched the oscillating trend of the field-detected TTD. Figures 7(a) and 7(c) show steady improvement for proposed model. Under CF mode, the SCTM-METANET does not show improvement in prediction accuracy, as in the case of predicting TTT. Under FC, the proposed model shows overall improvement. The quantitative measurement of the prediction performance for TTD proves the above observation. During the CC period, the proposed model shows a significant reduction in error (29%), which outperforms the TTT. Under the FC mode, the proposed model shows 5% improvement compared to the baseline model, which is still more satisfying than the case of TTT. The details of quantitative performance comparison can be found in Table 3.

5. Conclusion and Remarks

In this paper, a modified second-order traffic status prediction model, SCTM-METANET, is proposed for improving the performance of the regular METANET model during VSL control. In particular, the proposed model helped to counteract traffic statues prediction error during the most congested period. This study employed a switching mode SCTM model and integrated it with the METANET model so that under each of the FC, CC, and CF modes the SCTM-METANET had slightly different parameters. In addition, unlike the original METANET, the proposed model made predictions in a piecewise manner. The original METANET was already implemented in the VSL field test, and the predicted traffic variables were stored, so the collected field data were considered as ground truth data. The proposed SCTM-METANET was calibrated and simulated by a computer, using field data from the same period. Then the prediction results of the original METANET and SCTM-METANET models could be compared to the ground truth data. Data refining was applied to the field raw data of regular METANET predicted results. The concepts of SE and UE were introduced for clarity, and by eliminating SE the prediction result of regular METANET could then be compared to the SCTM-METANET model. The results clearly show that, during transitional modes FC and CF, the proposed model shows unstable performance, and, during the CC mode, the SCTM-METANET model largely reduces the prediction error of TTT and TTD by 14% and 29%, respectively.

This paper also revealed that that both METANET and SCTM-METANET underestimate the traffic density especially within 5pm~6pm when the traffic mode switches to congestion. The happening of this phenomenon is mainly due to the nature of loop detector system and the basic feature of CTM which is the core part of density prediction in both METANET and SCTM-METANET: the road is segmented as cells and it is assumed that within one cell the speed, density, and flow are all homogeneous, and the homogeneous condition transfers to the next cell. Based on this assumption, in the field the freeway is divided into cells which are approximately 500 m long and the loop detectors installed at the beginning of the cell collects and represents the traffic condition of the whole cell, the data collection interval is 20s. Under noncongested situation, the above hardware settings are abundant for accurately estimating traffic conditions. However, when during peak hour, traffic density, speed, and flow are not homogeneous in 500 m and many vehicles would enter the cell in the 20s data collection interval. It is delightful to find that when upgrading METANET to SCTM-METANET, the underestimation is relaxed although not totally eliminated. That is because of the bringing in of supply and demand stochasticity: the model realizes that there exist more vehicles in the cell than previously estimated.

Future research may build on the results of this study by solving the model's problem of instability during the FC and CF modes. Since the current SCTM-METANET model was calibrated offline and with precalibrated parameters, future work with the SCTM-METANET model will involve obtaining parameters and conducting calibration in real time. In the future research, the utilizing of higher frequency data collection method and more location flexible data collection method such as connected vehicle will shed light upon this situation.

Data Availability

The data used in this research is collected by underpavement loop detectors on a segment of urban freeway called Whitemud Drive in Edmonton, Alberta, Canada. The loop detector data collected in the field is continuously sent to the traffic management center of City of Edmonton. The owner of the data is City of Edmonton, the local government. The data is shared by University of Alberta for research use. In this paper, the loop detector data as well as the posted advisory driving speed data was provided by City of Edmonton. To replicate the research results in this paper, one can ask the author for the raw data set but under the approval of City of Edmonton. And recently, all loop detector data of Edmonton can be visualized and downloaded from a public website, although the highest data frequency raw data cannot be downloaded. Using lower frequency loop detector data will partially reproduce the analysis result in this research. The website mentioned is http://www.its.ualberta.ca/wcpa/homeIndex/index.

Disclosure

The contents of this paper reflect the views of the authors who are responsible for the facts and the accuracy of the data presented herein. The contents do not necessarily reflect the official views or policies of the City of Edmonton and Transport Canada. This paper does not constitute a standard specification or regulation.

Conflicts of Interest

The authors declare that they have no conflicts of interest.

Acknowledgments

The authors would like to thank Wai Cheung, Janis Chow, Michael Vaudan, Adrian Loh, and Rahim Karmali from the traffic operation group in the City of Edmonton for providing the Whitemud Drive loop detector data. This research work was jointly supported by the Natural Sciences and Engineering Research Council (NSERC) of Canada, City of Edmonton, and Transport Canada.

References

[1] M. Hadiuzzaman, T. Z. Qiu, and X.-Y. Lu, "Variable speed limit control design for relieving congestion caused by active bottlenecks," *Journal of Transportation Engineering*, vol. 139, no. 4, pp. 358–370, 2013.

[2] Y. Bie, T. Z. Qiu, C. Zhang, and C. Zhang, "Introducing Weather Factor Modelling into Macro Traffic State Prediction," *Journal of Advanced Transportation*, vol. 2017, Article ID 4879170, 15 pages, 2017.

[3] B. M. Williams and L. A. Hoel, "Modeling and forecasting vehicular traffic flow as a seasonal ARIMA process: theoretical basis and empirical results," *Journal of Transportation Engineering*, vol. 129, no. 6, pp. 664–672, 2003.

[4] C. Antoniou, M. Ben-Akiva, and H. Koutsopoulos, "Online Calibration of Traffic Prediction Models," *Transportation Research Record: Journal of the Transportation Research Board*, pp. 235–245, 2005.

[5] S. Afandizadeh and J. Kianfar, "A Hybrid Neuro-Genetic Approach to Short-Term Traffic Volume Prediction," *International Journal of Civil Engineering*, vol. 7, p. 41, 2009.

[6] B. L. Smith, B. M. Williams, and R. K. Oswald, "Comparison of parametric and nonparametric models for traffic flow forecasting," *Transportation Research Part C: Emerging Technologies*, vol. 10, no. 4, pp. 303–321, 2002.

[7] A. Spiliopoulou, M. Kontorinaki, M. Papageorgiou, and P. Kopelias, "Macroscopic traffic flow model validation at congested freeway off-ramp areas," *Transportation Research Part C: Emerging Technologies*, vol. 41, pp. 18–29, 2014.

[8] M. J. Lighthill and G. B. Whitham, "On kinematic waves. II. A theory of traffic flow on long crowded roads," *Proceedings of the Royal Society A Mathematical, Physical and Engineering Sciences*, vol. 229, pp. 317–345, 1955.

[9] P. I. Richards, "Shock waves on the highway," *Operations Research*, vol. 4, no. 1, pp. 42–51, 1956.

[10] C. F. Daganzo, "The cell transmission model: a dynamic representation of highway traffic consistent with the hydrodynamic

theory," *Transportation Research Part B: Methodological*, vol. 28, no. 4, pp. 269–287, 1994.

[11] C. F. Daganzo, "A behavioral theory of multi-lane traffic flow. Part I: long homogeneous freeway sections," *Transportation Research Part B: Methodological*, vol. 36, no. 2, pp. 131–158, 2002.

[12] C. F. Daganzo, "A behavioral theory of multi-lane traffic flow. Part II: merges and the onset of congestion," *Transportation Research Part B: Methodological*, vol. 36, no. 2, pp. 159–169, 2002.

[13] H. J. Payne, "Models of freeway traffic and control," in *Mathematical Models of Public Systems: Simulation Council Proceedings Series*, pp. 51–61, 1971.

[14] J. A. Laval and C. F. Daganzo, "Lane-changing in traffic streams," *Transportation Research Part B: Methodological*, vol. 40, no. 3, pp. 251–264, 2006.

[15] Y. Wang and M. Papageorgiou, "Real-time freeway traffic state estimation based on extended Kalman filter: a general approach," *Transportation Research Part B: Methodological*, vol. 39, no. 2, pp. 141–167, 2005.

[16] L. Mihaylova, R. Boel, and A. Hegyi, "Freeway traffic estimation within particle filtering framework," *Automatica*, vol. 43, no. 2, pp. 290–300, 2007.

[17] M. Papageorgiou, J.-M. Blosseville, and H. Hadj-Salem, "Macroscopic modelling of traffic flow on the Boulevard Périphérique in Paris," *Transportation Research Part B: Methodological*, vol. 23, no. 1, pp. 29–47, 1989.

[18] M. Papageorgiou, H. Hadj-Salem, J. M. Blosseville, and N. Bhouri, *Control, Computers, Communications in Transportation*, Elsevier, 1990.

[19] A. Sumalee, R. X. Zhong, T. L. Pan, and W. Y. Szeto, "Stochastic cell transmission model (SCTM): a stochastic dynamic traffic model for traffic state surveillance and assignment," *Transportation Research Part B: Methodological*, vol. 45, no. 3, pp. 507–533, 2011.

[20] X. Sun, L. Munoz, and R. Horowitz, "Highway traffic state estimation using improved mixture Kalman filters for effective ramp metering control," in *42nd IEEE International Conference on Decision and Control (IEEE Cat. No. 0 3CH 37475)*, vol. 6, pp. 6333–6338, 2003.

[21] R. Boel and L. Mihaylova, "A compositional stochastic model for real time freeway traffic simulation," *Transportation Research Part B: Methodological*, vol. 40, no. 4, pp. 319–334, 2006.

[22] A. Sumalee, T. Pan, R. Zhong, N. Uno, and N. Indra-Payoong, "Dynamic stochastic journey time estimation and reliability analysis using stochastic cell transmission model: Algorithm and case studies," *Transportation Research Part C: Emerging Technologies*, vol. 35, pp. 263–285, 2013.

[23] A. Sumalee, K. Uchida, and W. H. K. Lam, "Stochastic multimodal transport network under demand uncertainties and adverse weather condition," *Transportation Research Part C: Emerging Technologies*, vol. 19, no. 2, pp. 338–350, 2011.

[24] A. Hegyi, B. De Schutter, and H. Hellendoorn, "Model predictive control for optimal coordination of ramp metering and variable speed limits," *Transportation Research Part C: Emerging Technologies*, vol. 13, no. 3, pp. 185–209, 2005.

[25] Y. Bie, *Implementing Model Predictive Control based Variable Speed Limit on Urban Freeways: Data Imputation, Model Modification and Field Test Analysis*, 2016.

[26] A. Kotsialos, M. Papageorgiou, and F. Middelham, "Optimal coordinated ramp metering with advanced motorway optimal control," *Transportation Research Record*, no. 1748, pp. 55–65, 2001.

[27] L. Elefteriadou, S. S. Washburn, Y. Yin, V. Modi, and C. Letter, *Variable Speed Limit (VSL) - Best Management Practice*, vol. 11, Florida Dep Transp Res Center, 2012.

[28] X. Wang, M. Seraj, Y. Bie, T. Z. Qiu, and L. Niu, "Implementation of variable speed limits: Preliminary test on Whitemud Drive, Edmonton, Canada," *Journal of Transportation Engineering*, vol. 142, no. 12, 2016.

8

Evaluating Highway Traffic Safety: An Integrated Approach

Yanqun Yang,[1] Said M. Easa,[2,3] Zhibin Lin,[4] and Xinyi Zheng[5]

[1] *College of Civil Engineering, Fuzhou University, Fuzhou, Fujian 350108, China*
[2] *Department of Civil Engineering, Ryerson University, Toronto, Ontario, Canada M5B 2K3*
[3] *College of Civil Engineering, Fuzhou University, Fujian 350108, China*
[4] *Durham University Business School, Mill Hill Lane, Durham, DH1 3LB, UK*
[5] *Faculty of Humanities and Social Sciences, Fuzhou University, Fuzhou, Fujian 350108, China*

Correspondence should be addressed to Xinyi Zheng; zhengxinyi@fzu.edu.cn

Academic Editor: Richard S. Tay

This paper presents a novel methodology for determining the overall highway safety level by integrating statistical analysis and analytic network process (ANP) with set pair analysis (SPA) which is applied in the evaluation of the overall highway safety for the first time. The methodology accounts for both quantitative and qualitative factors that contribute to traffic safety. The statistical analysis uses crash, alignment, intersection, and other data to determine the significant indices (variables) that affect safety. These indices are then combined with the planning (qualitative) indices to determine the weights of all indices based on expert opinions using ANP. Finally, the overall safety level of the highway is determined using SPA. The methodology is illustrated using data collected from two highways in China. The results demonstrate that the proposed methodology is sound and reliable. The methodology is applicable to existing or new highways and can help to effectively evaluate the overall safety of a highway and develop long-term strategies for safety improvements.

1. Introduction

With the increased mobility on highways, traffic crashes have substantially increased and safety assessment has become increasingly important [1]. Traffic safety has been among the cutting-edge topics for scientific research [2]. For example, Theofilatos [3] adopts Bayesian and finite mixture logit models to predict the likelihood and severity of road accident in urban arterials. Wang and Huang [2] apply Bayesian hierarchical joint model to evaluate road network safety. Christoforou et al. [4] use multivariate probit models to examine the relationships between a variety of traffic factors and type of crash. Siegrist [5] presents a method for the ex ante estimation of a potential road safety program. Nghiem et al. [6] adopt a state-space time-series model to investigate the determinants of accidents. Kweon [7] uses regression analysis with correction for serial correlations to identify factors affecting the changes in traffic safety. These studies are mainly based on "hard data" or quantitative indices, which do not consider the full spectrum of indices that may affect highway safety. Several advanced analytical methods have been developed in the literature, with consideration of both quantitative and qualitative indices, including analytic network process (ANP) and set pair analysis (SPA). Yet, there is limited attempt that integrates these methods for evaluating overall highway traffic safety, so as to aid local and national government agencies' planning decisions for improving traffic safety.

To address the research gap, the current paper proposes a novel methodology that integrates three analysis methods, i.e., regression analysis, ANP, and SPA, for evaluating highway traffic safety. SPA adopts a system theory using a connection number to process the uncertainty caused by fuzzy, random, and incomplete information as proposed by a Chinese scholar Keqin Zhao in 1989 [8]. SPA is little known outside China, despite the popularity it enjoys in its home country. In contrast, ANP has been extensively applied in a variety of contexts. In the area of safety, AHP has been implemented to evaluate railway traffic safety, ship traffic safety, road traffic safety, and work zone safety [9]. ANP integrates expert judgments into the evaluation process and considers the interdependencies and interactions among evaluation

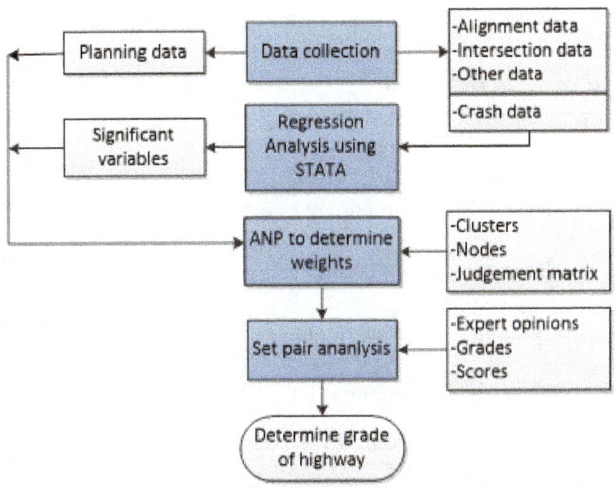

FIGURE 1: Logical steps of proposed integrated methodology.

elements [10, 11]. This important feature has attracted many decision-makers and planners in the fields of urban planning, logistics, supply chain management, and transportation [12].

In this integrated model, the SPA is used to determine the overall safety level of the highway by analyzing the features of a set pair indices from the aspects of identity, discrepancy, and contrary. The method requires two basic inputs: safety indices (or criteria) and their weights. It was decided in this study to develop an objective process for determining these two inputs. The significant safety indices were determined using regression analysis and their weights were determined using the ANP. The three methods are used sequentially, where the output of one method is used as input to the next method. As such, the final decision drawn from the three methods is more objective than a single method. In this integrated model, first, regression analysis is used to predict the statistically significant alignment, intersection, and general indices. Second, the algorithm-based ANP is used to determine the weights of the evaluation indices of highway traffic safety. Finally, SPA is used to determine the grade of the highway based on fuzzy membership functions of the assessed indices and specified criterion grades. Integrating the three analysis methods sequentially provides a more objective process for evaluation of highway safety level.

The following sections describe the methodology and its integrated components and illustrate its application using actual data from a highway in Fujian Province, China. The main conclusions of the study are then presented.

2. Proposed Methodology

The proposed methodology integrates three steps (regression analysis, ANP, and SAP) that are sequentially performed, where the output of one step is used as input to the next step, as shown in Figure 1. Regression analysis uses the quantitative alignment, intersection, and general indices to determine the statistically significant ones.

The ANP uses the significant indices and the qualitative planning indices to determine the indices weights. The SAP uses the weights to determine the overall highway safety. Note that each of the ANP and SAP steps involves the use of expert opinions. The first step is a traditional regression analysis to determine the statistically significant indices that contribute to traffic crashes. As such, the methodology presented in the paper is concerned with determining the overall safety of a highway section, rather than individual elements as compared to the traditional approach. In this respect, it would beneficial for the policy measures in safety highway to decide the safety indices being as aggregate as possible. It was performed using STATA software, described in detail in Gutierrez [13].

The analytic network process integrates expert judgments into the evaluation process and considers the interdependencies and interactions among evaluation elements, as pioneered by Saaty [14] and Saaty and Vargas [11]. This important feature has attracted many decision-makers and planners in the fields of urban planning, logistics, supply chain management, and transportation [12]. In the transportation field, the AHP has been implemented in numerous applications. The applications include the impacts of transit priority on signal coordination [15], project selection and prioritization of pavement preservation [16], prioritizing network level maintenance of pavement segments [17], assessing options to enhance bicycle and transit integration [18], selection and prioritization of intelligent transportation system user service [19], integration multiple criteria decision making to prioritize transportation [20], assessing asphalt pavement construction quality control [21], analytic minimum impedance surface [22], prioritizing traffic calming projects [23], and determining low volume road standards, long-term needs, and environmental risks [24]. In the area of safety, AHP has been implemented to evaluate railway traffic safety, ship traffic safety, road traffic safety, and work zone safety [9, 25].

The set pair analysis adopts a system theory using a connection number to process the uncertainty caused by fuzzy, random, and incomplete information as proposed by a Chinese scholar Keqin Zhao in 1989 [8]. The method has been extensively applied in the field of water resource management and agricultural science. For example, Feng et al. [26] used

Table 1: Fundamental scale for pairwise comparison.

Weight of Importance	Definition	Explanation
1	Equally important	Two activities contribute equally to the objective
3	Moderately more important	An activity is moderately favored over another
5	Strongly more important	An activity is strongly favored over another
7	Very strongly more important	An activity is very strongly favored over another
9	Extremely more important	An activity is extremely more favored than another
2, 4, 6, 8	Intermediate values between two adjacent judgments	Intermediate values are used as needed

SPA to predict changes in water resources trends. Hu and Yang [27] applied SPA for dynamic stochastic multicriteria decision making problems with unknown information on criteria weights. Wang et al. [28] evaluated flood risk based on GIS technology where a risk assessment map of the Beijing River basin was developed using SPA. Ren et al. [29] constructed a comprehensive evaluation model based on improved SPA that better serves reservoir dispatching. Application of SPA in transportation engineering, however, has been rather limited.

The specifics of the two steps involving the implementation of the ANP and SPA are described next.

2.1. Analytic Network Process. The analytic network process is a decision making tool that is suited for nonindependent hierarchy structure as described by [30]. The ANP structure is made up of clusters and nodes. The method involves carrying out pairwise comparisons by an expert who judges how important an index i is when compared to another index j with respect to the overall goal. The verbal judgments of the experts are then transformed into numerical values using a nine-level Likert scale, as shown in Table 1. The odd numbers represent the primary importance weights and the even numbers represent intermediate importance weights. The scale is used to compare input parameters by pairings to determine how much one parameter is more or less important than another. Based on this comparison, an n by n evaluation matrix $A = (a_{ij})$ is established, where n is the number of parameters involved in the decision and a_{ij} is relative value of index i to index j. Mathematically speaking, the evaluation matrix is defined as

$$A = \begin{pmatrix} 1 & a_{12} & \cdots & a_{1n} \\ a_{21} & 1 & \cdots & a_{2n} \\ \cdot & \cdot & \cdots & \cdot \\ \cdot & \cdot & \cdots & \cdot \\ \cdot & \cdot & \cdots & \cdot \\ a_{n1} & a_{n2} & \cdots & 1 \end{pmatrix} \quad (1)$$

$$a_{ij} = \begin{cases} 1, & i = j \\ a_{ij}, a_{j_i}^{-1}, & i \neq j \end{cases} \quad (2)$$

The evaluation matrix is related to the eigenvector (weights) and eigenvalue by

$$Aw = \lambda_{\max} w \quad (3)$$

where A is matrix with elements a_{ij}, w is eigenvector of matrix A with elements w_j, $j = 1, 2, \ldots, n$, which represent the weights of the indices, and $\lambda_{\max}$ is eigenvalue of matrix A. The weights w_j are then normalized to obtain the final weights of the indices as follows:

$$W_j = \frac{w_j}{\sum_{j=1}^{n} w_j} \quad (4)$$

To assess the consistency of the decision-maker in the assignment of the importance weights, a consistency ratio is computed as follows:

$$CR = \frac{CI}{RI} \quad (5)$$

$$CI = \frac{(\lambda_{\max} - n)}{(n-1)} \quad (6)$$

where CR is consistency ratio, CI is consistency index, RI is random consistency index, $\lambda_{\max}$ is maximum eigenvalue, and n is size of the matrix. The value of RI has been defined in the literature based on matrix size, where it ranges from 0 to 1.49 for $n = 2$ to 10, respectively [31]. The consistency ratio helps to identify possible errors in judgment and actual consistency in the judgment itself. It has been suggested that CR should be less than 0.1 which means that the method allows up to 10% error in human judgment during the paired comparison process. If the error is greater than 10%, the experts are asked to revise the pairwise comparisons. To calculate the weights of the parameters, we used Super Decision software developed by Creative Decisions Foundation in this study [32].

2.2. Set Pair Analysis. The set pair analysis is a modified uncertainty theory that considers both certain and uncertain indices as an integrated system and depicts the certainty and uncertainty systematically in terms of three aspects: identity, discrepancy, and contrary. The SPA refers to a pair that consists of two interrelated sets and uses connection degrees to handle, in a unified manner, the uncertainty caused by fuzzy, random, mediation, and incomplete information. The main steps of SPA are as follows: (a) structure the sets in view of the problem, (b) analyze the features of the two sets, and (c) set up a connection-degree formula for the two sets including identity, discrepancy, and contrary degrees.

Let N be the total number of features of the two sets, μ the connection degree of the set pair, S the number of identity features, P the number of contrary features, and F the number of the features of the two sets that are neither identity nor contrary, denoted as discrepancy degree (which equals $N - S - P$). Let the ratios S/N, F/N, and P/N represent the identity, discrepancy, and contrary degrees of the two sets, respectively, and j be the coefficient of the contrary degree (specified as -1). The coefficient of the discrepancy degree i is an uncertain value between -1 and 1; that is, $i\in[-1, 1]$. The uncertainty of the discrepancy degree of the two sets is eliminated when i equals -1 or 1 and will increase as i approaches zero. Therefore, the connection degree of set (A, B), $\mu(A, B)$, is defined as

$$\mu(A,B) = \frac{S}{N} + \frac{F}{N}i + \frac{P}{N}j \quad (7)$$

Let $a = S/N$, $b = F/N$, and $c = P/N$; then (7) becomes

$$\mu = a + bi + cj \quad (8)$$

where a, b, and c satisfy the normalization condition $a + b + c = 1$. The SPA method involves generating the evaluation grade domain V = {Excellent, Very Good, Good, Average, Poor} = {I, II, III, IV, V}. The grade ranges are defined as follows: Grade I [80, 100], Grade II [60, 80], Grade III [40, 60], Grade IV [20, 40], and Grade V [0, 20]. Note that the lower limits of Grades I to IV start from values just greater than those shown in the above grade ranges. The corresponding index values are defined as set B = {100, 80, 60, 40, 20}. According to Feng et al. [33], if the values of the indices are within the specified boundaries of the grade, they are considered identity; if they are within separated boundaries of the grade, they are considered contrary; and if they are within the adjacent boundaries of the grade, they are considered discrepancy.

To illustrate this, consider $\mu(A, B_1)$, for example. Assume that there are S indices falling in Grade I whose weights are $u_1, u_2, \ldots, u_3$, F indices falling in Grade II whose weights are $t_1, t_2, \ldots, t_3$, and P indices falling in Grade III, whose weights are $v_1, v_2, \ldots, v_3$. Then, the connection degree of set (A, B_1) is defined as

$$\mu(A,B_1) = \sum_{i=1}^{S} u_i + \sum_{k=1}^{F} t_k i_k + \sum_{l=1}^{P} v_l j \quad (9)$$

where i_k reflects the discrepancy degree of set (A, B_1) about the evaluation index x_k whose weight is t_k (that is, i_k is the fuzzy connection degree between the evaluation index x_k and Grade II standard value $b_1{}^k$). Then, i_k is defined as

$$i_k = \mu\left(x_k, b_1^k\right) = a_k + b_k i + c_k j \quad (10)$$

The connection degrees of the set pair $\{x_k, b_1{}^k\}$ are defined, respectively, as

$$a_k = \frac{S_1^k S_2^k}{\left(S_1^k + S_2^k\right) x_k} \quad \text{(Identity degree)} \quad (11)$$

$$b_k = \frac{\left(S_2^k - x_k\right)\left(x_k - S_1^k\right)}{\left(S_1^k + S_2^k\right) x_k} \quad \text{(Discrepancy degree)} \quad (12)$$

$$c_k = \frac{x_k}{S_1^k + S_2^k} \quad \text{(Contrary degree)} \quad (13)$$

where a_k, b_k, and c_k are the identity, discrepancy, and contrary degrees, respectively. Note that the identity degree is the approximate degree between x_k and $S_1{}^k$, while the contrary degree represents the approximate degree between x_k and $S_2{}^k$. The terms $S_1{}^k$ and $S_2{}^k$ represent the upper and lower limits of Grade II standard values, respectively.

3. Empirical Application

The proposed methodology was applied to assess the traffic safety level on an actual highway in China. Details on data collection, implementation of the three basic steps of the methodology, and the respective results are described in this section.

3.1. Data Collection. The data collection involved three main tasks: (1) selection and definition of evaluation indices, (2) collection of indices data to calibrate the methodology, and (3) selection of the highway section to be evaluated. For the first task, various indices that affect traffic safety were divided into four categories: planning indices, alignment indices, intersection indices, and general indices (see Table 2).

To ensure that the evaluation methodology is effective, the indices selected were rational, simple, and comprehensive. However, only the significant indices are determined and used for evaluating the highway safety level. For the second task, the indices data were collected for Xia Rong expressway from BK111 + 800 to BK107 (longitude = 116.88°, latitude = 25.17°) and G324 Line Fortress arterial road (longitude = 118.66°, latitude = 24.94°), as shown in Figure 2. The crash statistics (2010-2013) were obtained from the traffic police corps of Fujian Province. There were 117 crashes on the expressway and 150 crashes on the arterial road. Then, the crash data were organized to establish the panel data and perform the regression analysis. Four planning indices that affect highway safety were identified: functional grade, highway classification, land use, and service level. The ANP was subsequently used to analyze these indices along with the significant indices obtained from the statistical analysis.

For the third task, a section of Horizontal Five (HF) highway that has similar characteristics to the highways for which the methodology was calibrated was selected for evaluation. The section extends from Taiding Village station at K15 + 849.258 (longitude = 118.84°, latitude = 26.18°) to Baofeng Village at K50 + 354.648 (longitude = 118.59°, latitude = 26.15°) with a length of about 35 km (Figure 3). The HF highway is a second-class highway with a design speed of

Table 2: Definition of the indices that affect traffic safety.

Index Category	Index Name	Designation	Unit
Planning indices	Functional grade	FG^a	-
	Highway classification	HC	-
	Land use	LU	-
	Service level	SRL	-
Alignment indices	Straight length	SL	m
	Grade length	GL	m
	Horizontal curve radius	R	m
	Sight distance	SD	m
	Longitudinal gradient	G	%
	Superelevation	e	%
Intersection indices	Intersection density	D	Intersections / km
	Sight distance	S	m
	Intersection Type	$Type^b$	-
General indices	Road width	RW	m
	Pavement drainage	PD	-
	Climatic condition	C^c	-
	Truck proportion	T	%

[a]Functional grade, a principle grade designed by the Chinese Highway Management Institution, reflects the service level provided by the highway. It is a surrogate measure of traffic volume.
[b]This is a nominal variable that is equal to 1 for nonsignalized intersections and zero otherwise.
[c]This is the climatic condition at the time of the crash.

(a) Xia Rong Freeway (b) G324 Line Fortress Intersection

Figure 2: Photos of experimental highway and example of intersection.

40 km/h, lane width of 3.5 m, and shoulder width of 1.5 m. The section contains tangents, horizontal curves, and vertical curves. The desirable minimum horizontal curve radius is 100 m, while the absolute minimum radius is 60 m. The recommended and ultimate maximum longitudinal slopes are both 7%. The section passes through numerous villages with intersection density up to five intersections per kilometer. This section is considered suitable for a comprehensive evaluation using the proposed methodology.

3.2. Results of Regression Analysis. Regression analysis was performed to establish the significant alignment, intersection, and general indices (independent variables) that affect traffic safety. The crash was used as the dependent variable, where the index takes a value of 1 if it affects traffic safety and takes a value of zero otherwise. The analysis involved 13 independent indices: five alignment indices, three intersection indices, and four general indices. A sample of the panel data of these indices is shown in Table 3. The table shows crash data involving four crashes on the expressway (1, 3, 116, and 117) and four crashes at the intersections (118-120, 265). As previously mentioned, the total number of crashes used in the regression analysis was 117 crashes.

The results of regression analysis are shown in Table 4. The indices that have p values less than or equal to 0.05 are considered statistically significant. As noted, the significant indices are grade length, horizontal curve radius, sight distance, superelevation, intersection density, and intersection sight distance. The p values of the preceding indices are 0.002, 0.018, 0.022, 0.001, 0.040, and 0.003, respectively. The impacts

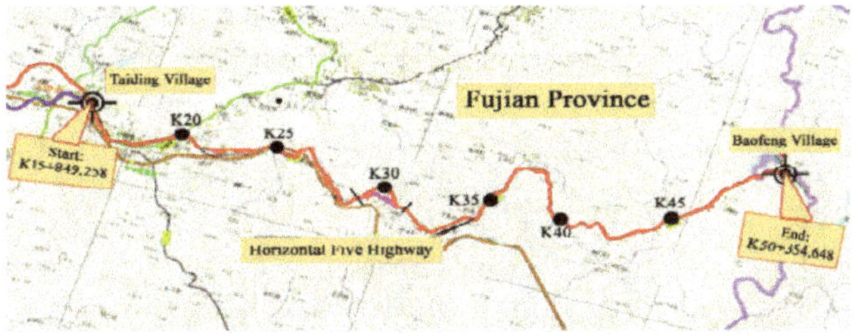

FIGURE 3: Map of Horizontal Five highway used for evaluation. (Source: Fujian Police Department, China)

TABLE 3: Panel data of 13 indices related to selected crashes on the expressway and intersections (total number of crashes is 117)[a].

Crash No.	Alignment Indices						Intersection Indices			General Indices			
	SL	*GL*	*R*	*SD*	*G*	*e*	*D*	*S*	*Type*	*RW*	*PD*	*C*	*T*
2	1	0	1	1	1		1	0	1				
1	0	0	0	0	1	1	-[b]	-	-	1	0	0	0
2	0	1	0	0	0	1	-	-	-	1	0	1	0
...	...	...	...	...	...	...	...	...	...	...	...	...	...
116	1	1	1	0	0	0	-	-	-	0	0	0	0
117	1	1	1	1	0	1	-	-	-	0	0	0	1
118	1	0	1	1	0	0	1	0	1	1	0	0	0
119	1	0	0	1	0	0	1	0	1	1	1	0	0
120	0	0	1	1	0	1	0	1	0	1	1	0	0
...	...	...	...	...	...	...	...	...	...	...	...	...	...
265	0	0	0	1	1	1	0	1	1	0	0	1	1

[a]The first four crashes belong to the expressway and the remaining four crashes belong to intersections.
[b]Not applicable since the corresponding crashes are on the expressway.

of other indices on traffic crashes were not significant at the 95% confidence level. The preceding significant indices are considered further along with the planning indices in the ANP step.

3.3. Results of Analytic Network Process

3.3.1. Building ANP Traffic Safety Model. Based on the relationships among the indices of highway traffic safety, an ANP decision model with inner dependencies was built, in which each cluster was linked to itself through a loop link, as shown in Figure 4. According to this model, traffic safety is affected by alignment indices, intersection indices, and planning indices. The significant alignment indices (*GL*, *R*, *SD*, and *e*) and intersection indices (*D*, *S*), which are quantitative, were previously determined using regression analysis. The planning indices included *FG*, *HC*, *LU*, and *SRL*.

3.3.2. Computing Weights. Since the indices affect highway safety differently, their relative weights were determined using questionnaires completed by 14 experts. The experts were from local and provincial governments, consulting companies, and college professors. The panel composition covered the broad categories of indices, including highway planning, geometric design, and road safety. Each expert was asked to compare all ten indices and rank them based on their experience and knowledge of the HF highway. Most experts were designers who have been involved in the design and construction of numerous highway facilities in the province.

The ANP software was then used to help the respondents evaluate those indices. The software allows the respondents to judge key indices by using a pairwise comparison. In this method, the expert compared two indices at a time, decided which one was more important in promoting highway safety, and set the preference level of the selected index. The scale for pairwise comparison describing the intensity of judgments is previously presented in Table 1. Note that, based on (2), if index *i* has one of the numbers in Table 1 assigned to it when compared with index *j*, then index *j* has the reciprocal value when compared with *i*.

The average of the judgment scores was then used to establish the judgment matrix for comparison of the clusters and the indices within each cluster. The ANP involved two main tasks. First, the judgment matrix is used to establish the respective weights for clusters (Table 5) and indices (Tables 6–12). Note that *SD* and *SL* were not included in the evaluations conducted in Tables 9 and 10, respectively, since they were considered to have little impact on *GL* and *FG*. For each pairwise comparison, the consistency ratio (CR) is shown. As noted, the CR values of all comparisons are less than 0.1, which indicates that all comparisons pass the consistency test.

To illustrate the calculations of the weights in Tables 6–12 (last column), consider Table 6. The judgment matrix **A** is given by

Table 4: Statistical characteristics of indices using regression analysis.

Variable	Coefficient	Std. Error	t-value	p-value
SL	0.7351	0.4117	1.79	0.075
GL[a]	1.1221	0.3577	3.14	0.002
R[a]	-1.2816	0.5374	-2.38	0.018
SD[a]	1.1273	0.4896	2.30	0.022
G	0.7135	0.3892	1.83	0.068
e[a]	1.6778	0.4790	3.50	0.001
RW	-0.2010	0.2792	-0.72	0.472
PD	0.6872	0.5126	1.34	0.181
C	1.0183	0.7619	1.34	0.183
T	-0.0905	0.5944	-0.15	0.879
D[a]	1.2769	0.6175	2.07	0.040
S[a]	-1.4799	0.4912	-3.01	0.003
Type	-0.5654	0.3990	-1.42	0.158
Constant	0.5621	0.4515	1.24	0.214

[a]Statistically significant indices (p value < 0.05).

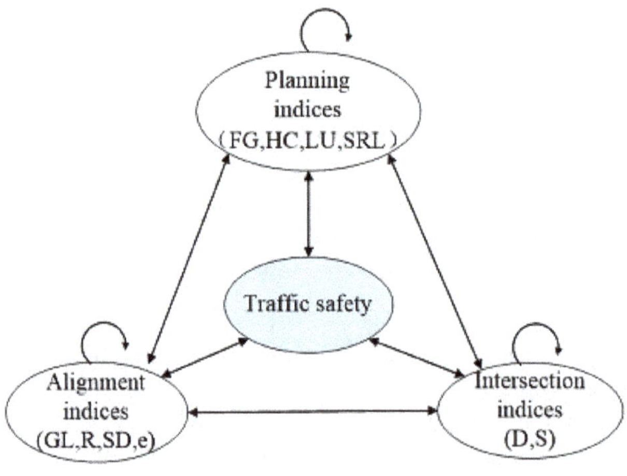

Figure 4: Elements of ANP highway traffic safety model.

Table 5: Cluster comparison for "traffic safety".

Traffic Safety	P	Ali	Inter	W
P	1.00	0.61	1.57	0.2976
Ali	1.64	1.00	3.29	0.5268
Inter	0.63	0.31	1.00	0.1756

Note: $CR = 0.0057 < 0.1$.

Table 6: Comparison for superelevation node.

Ali	*SD*	*R*	*GL*	W
SD	1.00	2.34	2.56	0.5487
R	0.43	1.00	1.29	0.2482
GL	0.39	0.78	1.00	0.20316

Note: $CR = 0.0028 < 0.1$.

$$A = \begin{bmatrix} 1.00 & 2.34 & 2.56 \\ 0.43 & 1.00 & 1.29 \\ 0.39 & 0.78 & 1.00 \end{bmatrix} \qquad (14)$$

Solving (1), the eigenvector $\mathbf{w}$ (or weights w_j, $j = 1, 2, 3$) is given by

$$w = \begin{bmatrix} 0.8633 \\ 0.3905 \\ 0.3196 \end{bmatrix} \qquad (15)$$

The eigenvalue $\lambda_{\max} = 3.0064$. The normalized weights are then given by (2) as

$$W = \begin{bmatrix} 0.5487 \\ 0.2482 \\ 0.2031 \end{bmatrix} \qquad (16)$$

which are the values shown in the last column of Table 6.

Second, the normalized weights in Tables 6–12 make up the unweighted index matrix which should be modified to reflect the weights of the cluster level. Thus, the components

TABLE 7: Comparison for sight distance node.

Ali	*e*	*R*	*GL*	W
e	1.00	3.88	3.75	0.6560
R	0.26	1.00	1.02	0.1722
GL	0.27	0.98	1.00	0.1718

Note: $CR = 0.0003 < 0.1$.

TABLE 8: Comparison for horizontal curve radius node.

Ali	*e*	*SD*	*GL*	W
e	1.00	4.26	3.85	0.6675
SD	0.23	1.00	1.57	0.18801
GL	0.26	0.63	1.00	0.1445

Note: $CR = 0.0321 < 0.1$.

TABLE 9: Comparison for grade length node.

Ali	*e*	*R*	W
e	1.00	3.57	0.7814
R	0.28	1.00	0.2186

Note: $CR = 0.0000 < 0.1$.

TABLE 10: Comparison for functional grade node.

P	*HC*	*LU*	W
HC	1.00	1.96	0.6621
LU	0.51	1.00	0.3379

Note: $CR = 0.0000 < 0.1$.

TABLE 11: Comparison for highway classification node.

P	*FG*	*LU*	*SRL*	W
FG	1.00	3.48	0.95	0.4509
LU	0.29	1.00	0.48	0.1564
SRL	1.05	2.08	1.00	0.3927

Note: $CR = 0.0342 < 0.1$.

TABLE 12: Comparison for land use node.

P	*FG*	*HC*	*SRL*	W
FG	1.00	1.75	0.98	0.3940
HC	0.57	1.00	3.22	0.2630
SRL	1.02	0.90	1.00	0.3430

Note: $CR = 0.0241 < 0.1$.

of the unweighted matrix were multiplied by the corresponding cluster weights to obtain the weighted, normalized matrix. The final weights of the indices are shown in the last column of Table 13.

3.4. Results of Set Pair Analysis. To establish the standard values for assessment grade, the evaluation domain of highway traffic safety was divided into five grades, as previously described. To establish the grade of each index, 30 experts were invited to grade the ten indices, 25 responses were received, and 2 incomplete responses were discarded (the panel was different from that used in the ANP). The assessment results (scores and grades) as well as the weights from the ANP are shown in Table 13. The scores in this table are calculated as follows: $score = \sum (M_i U_i) / 23$, where $i = 1, 2, \ldots, 5$, M_i is number of responses for Grade i and U_i is the upper limit of Grade i. For example, in Table 12 the score of *FG* is calculated as (13 x 100 + 7 x 80 + 3 x 60 + 0 x 40 + 0 x 20) / 23 = 88.7%. Hence, the index *FG* has Grade I since 88.7% lies between 80% and 100%.

To compute the connection degrees, the 10 indices under consideration were denoted as set A to form a set pair with different standard levels set B_i. Using (9)–(13) and the results of Table 13, the connection degrees for each grade were calculated. They were then compared and the grade corresponding to the maximum connection degree represented the grade of highway traffic safety. The results are shown in Tables 14–16. Note that in the equation of fuzzy connection degree (see (10)), i is the coefficient of the discrepancy degree and is specified as 0 and j is the coefficient of the contrary degree and is specified as -1. Based on Tables 14–16, the connection degrees are $\mu(A, B_1) = 0.162$, $\mu(A, B_2) = 0.513$, and $\mu(A, B_3) = -0.395$. To illustrate the calculations, in Table 14, the sum of u_i equals 0.2817, the sum of $(t_k\ i_k)$ equals -0.011, and the sum of $v_l j$ equals -0.1095. Therefore, based on (9), $\mu(A, B_1) = 0.162$. The respective calculations are shown as a footnote for each table. Thus, the order of the connection degrees (from large to small) is $\mu(A, B_2) > \mu(A, B_1) > \mu(A, B_3)$. Since the connection degree between set A and Grade II which is 0.513 is the largest, the safety level of the highway belongs to Grade II (Very Good). This finding is consistent with the overall perception of safety of this highway.

4. Discussion and Conclusions

This paper proposes an advanced methodology for determining the overall highway safety level, integrating both quantitative and qualitative methods: statistical analysis, ANP, and SPA, which is an integrated algorithm method to evaluate the highway geometric design. The logic of study design is firstly to use three analysis (regression, ANP, and SPA) to generate an objective evaluation method and secondly to use the empirical dataset to testify the integrated method reliability as compared to the traditional subjective evaluation. The statistical analysis used crash, alignment, intersection, and other data to determine the significant indices (variables) that affect safety. These indices were then combined with the planning (qualitative) indices to determine the weights of all indices based on expert opinions using ANP theory. Finally, the overall safety level of the highway is determined using SPA. The methodology was illustrated using data collected on two highways in China.

The power of the ANP is that it can account for such factors in the evaluation process. The reason for the ANP success is that it elicits judgments and uses measurement to derive ratio scales. Weights as ratio scales are amenable to performing the basic arithmetic operations of adding within the same scale and meaningfully multiplying different scales. In addition, SPA also involves expert judgments. The

Table 13: Assessment index system and associated grades and weights.

Index	No. of Responses for Grade					Score (%)	Grade	Weight (from ANP)
	I	II	III	IV	V			
FG	13	7	3	0	0	88.7	I	0.0969
HC	9	9	5	0	0	83.5	I	0.0804
LU	12	10	1	0	0	89.6	I	0.0415
SRL	3	6	8	4	2	63.5	II	0.1206
e	8	7	8	0	0	80.0	II	0.2245
R	2	6	7	5	3	59.1	III	0.1095
GL	8	9	6	0	0	81.7	I	0.0629
SD	3	7	7	3	3	63.5	II	0.1086
D	6	6	7	4	0	72.2	II	0.1031
S	4	5	6	5	3	61.7	II	0.0520

Table 14: Results of connection degree $\mu(A, B_1)^A$.

Grade I Index	Weight u_i	Grade II Index	Weight t_k	$i_k = a_k + b_k i + c_k j$					Grade III Index	Weight v_{lj}
				a_{k1}	$b_{k1}i$	$c_{k1}j$	i_k	$t_k i_k$		
FG	*0.0969*	*SRL*	*0.1206*	*0.540*	*0.006i*	*0.454j*	*0.086*	*0.010*	*R*	*0.1095j*
HC	*0.0804*	*e*	*0.2245*	*0.429*	*0.000i*	*0.571j*	*-0.142*	*-0.032*		
LU	*0.0415*	*SD*	*0.1086*	*0.540*	*0.006i*	*0.454j*	*0.086*	*0.009*		
GL	*0.0629*	*D*	*0.1031*	*0.475*	*0.009i*	*0.516j*	*-0.041*	*-0.004*		
		S	*0.0520*	*0.556*	*0.004i*	*0.441j*	*0.115*	*0.006*		

[a] $\Sigma u_i = 0.2817$, $\Sigma t_k i_k = -0.011$, and $\Sigma v_l j = -0.1095$. Therefore, $\mu(A, B_1) = 0.162$.

judgment in both methods is formulated in a systematic way. Expert knowledge improves the derived results and makes the evaluation much more accurate. The feasibility of the proposed methodology is illustrated using HF highway in Fujian Province, China. Regression analysis was used to analyze 13 indices related to alignment, intersection, and general features and six indices were found to be statistically significant. These indices were then combined with four planning indices and their weights were determined using the ANP which involved the use of expert judgments. Based on these weights, the overall traffic safety level of the highway was determined using SPA as Grade II (Very Good).

The methodology presented in the paper focuses on the overall safety of a highway section, rather than individual elements, with the aim of devising practical implications for evaluating the current safety status of highways and for designing new ones. In the proposed methodology, the performance indices are as aggregate as possible. For example, instead of considering the individual elements of combined horizontal and vertical alignments, sight distance was used as a surrogate measure of numerous factors including combined alignments. Similarly, intersection density was used instead of individual intersection characteristics, such as signal features, signal phasing, and intersection geometry. In China, highway classification is considered a key element of road safety. Although different classes may be designed according to standards, lower highway classes are normally associated with higher crashes due to the nature of the traffic mix. Unlike freeways, traffic in lower highway classes includes not only vehicle traffic but also pedestrians, motor bicycles, and bicycles. In addition, traffic behavior and the level of adhering to traffic regulations are different. For these reasons, highway classification was considered as one of the planning indices and the results of the ANP showed that it was an important index.

The computational results demonstrate that the proposed methodology is reasonable, reliable, and applicable for highway safety evaluation, and the evaluation results have significant traffic safety policy implications. Although in this study the methodology was presented in the context of highway safety evaluation, it can be applied to other contexts such as railway, water, and air transportation. This study has some limitations which can be addressed in future research. For example, the weights of the indices have a direct and substantial influence on the assessment results where different weights may lead to a different safety level of the highway. The method used for determining the indices weights is subjective, and the appropriate number of experts used to establish the indices weights could be further explored. Nonetheless, it is hoped that the proposed methodology will be of interest to transportation policy makers, researchers, and practitioners involved in developing long-term strategies for safety improvements.

TABLE 15: Results of connection degree $\mu(A, B_2)^A$.

Grade II Index	Weight u_i	Grade I Index	Weight t_k	a_{k1}	$b_{k1}i$	$c_{k1}j$	i_k	$t_k i_k$	Grade III Index	Weight $v_l j$
			$i_k = a_k + b_k i + c_k j$							
SRL	0.1206	*FG*	0.0969	0.501	0.006i	0.493j	0.008	0.001	*R*	0.1095j
e	0.2245	*HC*	0.0804	0.532	0.004i	0.464j	0.068	0.005		
SD	0.1086	*LU*	0.0415	0.496	0.006i	0.498j	-0.002	0.000		
D	0.1031	*GL*	0.0629	0.544	0.002i	0.454j	0.090	0.006		
S	0.0520									

[a] Σu_i = 0.6088, $\Sigma t_k i_k$ = 0.012, and $\Sigma v_l j$= -0.1095. Therefore, $\mu(A, B_2)$ = 0.513.

TABLE 16: Results of connection degree $\mu(A, B_3)^A$.

Grade III Index	Weight u_i	Grade II Index	Weight t_k	a_{k1}	$b_{k1}i$	$c_{k1}j$	i_k	$t_k i_k$	Grade I Index	Weight $v_l j$
			$i_k = a_k + b_k i + c_k j$							
R	0.1095	*SL*	0.1206	0.378	-0.013i	0.635j	-0.257	-0.031	*FG*	0.0969j
		e	0.2245	0.300	-0.100i	0.800j	-0.500	-0.112	*HC*	0.0804j
		SD	0.1086	0.378	-0.013i	0.635j	-0.257	-0.028	*LU*	0.0415j
		D	0.1031	0.332	-0.054i	0.722j	-0.390	-0.040	*GL*	0.0629j
		S	0.0520	0.389	-0.006i	0.617j	-0.228	-0.012		

[a] Σu_i = 0.1095, $\Sigma t_k i_k$= -0. 223, and $\Sigma v_l j$= -0.2817. Therefore, $\mu(A,B_3)$ = -0.395.

Appendix

Notation

A: Evaluation (judgment) matrix
CI: Consistency index
a, b, and c: Coefficients that equal S/N, F/N, and P/N, respectively
a_k, b_k, and c_k: Identity, discrepancy, and contrary degrees, respectively
F: Number of the discrepancy features of two sets
i: Coefficient of the discrepancy degree i $\in$[-1, 1]
i_k: Fuzzy connection degree between x_k and Grade II standard value $b1_k$
j: Coefficient of the contrary degree (-1)
k: Index for evaluation indices
n: Number of significant indices
N: Total number of features of the two sets
P: Number of contrary features of two sets
RI: Random consistency index
S: Number of identity features of two sets
S_1^k, S_2^k: Upper and lower limits of Grade II standard values, respectively
$t_1, t_2, \ldots, t_3$: Weights of the F indices
$u_1, u_2, \ldots, u_3$: Weights of the S indices
$v_1, v_2, \ldots, v_3$: Weights of the P indices
V: Grade domain {Excellent, Very Good, Good, Average, Poor}
w: Weights of indices 1 to n (eigenvector of matrix A)
W: Normalized weights of indices 1 to n
x_k: Evaluation index k
λ_{max}: Maximum eigenvalue of matrix A
μ: Connection degree of two sets.

Acknowledgments

This work was supported by the Department of Transportation of Fujian Province, China. The authors would like to thank Jiang Jingfang, Fujian Traffic Safety Association of China, for providing the collision data used in this study and for his valuable suggestions.

Conflicts of Interest

There are no conflicts of interest.

References

[1] I. O. Lund and T. Rundmo, "Cross-cultural comparisons of traffic safety, risk perception, attitudes and behaviour," *Safety Science*, vol. 47, no. 4, pp. 547–553, 2009.

[2] J. Wang and H. Huang, "Road network safety evaluation using Bayesian hierarchical joint model," *Accident Analysis & Prevention*, vol. 90, pp. 152–158, 2016.

[3] A. Theofilatos, "Incorporating real-time traffic and weather data to explore road accident likelihood and severity in urban arterials," *Journal of Safety Research*, vol. 61, pp. 9–21, 2017.

[4] Z. Christoforou, S. Cohen, and M. G. Karlaftis, "Identifying crash type propensity using real-time traffic data on freeways," *Journal of Safety Research*, vol. 42, no. 1, pp. 43–50, 2011.

[5] S. Siegrist, "Towards a method to forecast the effectiveness of national road safety programmes," *Safety Science*, vol. 48, no. 9, pp. 1106–1110, 2010.

[6] S. Nghiem, J. J. F. Commandeur, and L. B. Connelly, "Determinants of road traffic safety: New evidence from Australia using

state-space analysis," *Accident Analysis & Prevention*, vol. 94, pp. 65–72, 2016.

[7] Y.-J. Kweon, "What affects annual changes in traffic safety? A macroscopic perspective in Virginia," *Journal of Safety Research*, vol. 53, pp. 17–21, 2015.

[8] Y.-L. Jiang, C.-F. Xu, Y. Yao, and K.-Q. Zhao, "Systems information in set pair analysis and its applications," in *Proceedings of the 2004 International Conference on Machine Learning and Cybernetics*, pp. 1717–1722, Shanghai, China.

[9] M. Hirasawa, A. Takemoto, M. Asano, and T. Takada, "Study on improving worker safety at roadway sites in Japan," *Transportation Research Record*, no. 2015, pp. 41–54, 2007.

[10] T. L. Saaty, "Decision making—the analytic hierarchy and network processes (AHP/ANP)," *Journal of Systems Science and Systems Engineering*, vol. 13, no. 1, pp. 1–35, 2004.

[11] T. L. Saaty and L. G. Vargas, *Decision Making with the Analytic Network Process*, Springer, New York, 2013.

[12] E. Mulliner, K. Smallbone, and V. Maliene, "An assessment of sustainable housing affordability using a multiple criteria decision making method," *OMEGA - The International Journal of Management Science*, vol. 41, no. 2, pp. 270–279, 2013.

[13] R. G. Gutierrez, "Stata," *Wiley Interdisciplinary Reviews: Computational Statistics*, vol. 2, no. 6, pp. 728–733, 2010.

[14] T. L. Saaty, "Fundamentals of the analytic network process — multiple networks with benefits, costs, opportunities and risks," *Journal of Systems Science and Systems Engineering*, vol. 13, no. 3, pp. 348–379, 2004.

[15] S. Q. Huang, A. Shalaby, and R. Bissessar, "Impacts of transit priority on signal coordination: Case study of Toronto, Ontario, Canada," *Transportation Research Record*, vol. 2311, pp. 29–43, 2012.

[16] C. F. Gurganus and N. G. Gharaibeh, "Project selection and prioritization of pavement preservation," *Transportation Research Record*, no. 2292, pp. 36–44, 2012.

[17] J. Farhan and T. F. Fwa, "Use of analytic hierarchy process to prioritize network-level maintenance of pavement segments with multiple distresses," *Transportation Research Record*, no. 2225, pp. 11–20, 2011.

[18] K. J. Krizek and E. W. Stonebraker, "Assessing options to enhance bicycle and transit integration," *Transportation Research Record*, no. 2217, pp. 162–167, 2011.

[19] N. Khademi, A. S. Mohaymany, and J. Shahi, "Intelligent Transportation System User Service Selection and Prioritization," *Transportation Research Record*, vol. 2189, no. 1, pp. 45–55, 2018.

[20] J. Shelton and M. Medina, "Integrated multiple-criteria decision-making method to prioritize transportation projects," *Transportation Research Record*, no. 2174, pp. 51–57, 2010.

[21] D. Guo, W. Zhou, A. Sha, and R. Bai, "Application of uncertainty analytic hierarchy process method for asphalt pavement construction quality control in China," *Transportation Research Record*, no. 2098, pp. 43–50, 2009.

[22] T. Grossardt, K. Bailey, and J. Brumm, "Analytic minimum impedance surface: Geographic information system-based corridor planning methodology," *Transportation Research Record*, no. 1768, pp. 224–232, 2001.

[23] D. P. Guegan, P. T. Martin, and W. D. Cottrell, "Prioritizing traffic-calming projects using the analytic hierarchy process," *Transportation Research Record*, no. 1708, pp. 61–67, 2000.

[24] E. M. Dodson, W. Chung, K. Mills, R. Nall, J. Sessions, and G. Zartman, "Planning to determine low-volume road standards, long-term needs, and environmental risks and trade-offs," *Transportation Research Record*, vol. 1, no. 1989, pp. 11–19, 2007.

[25] G. Bureika, G. Bekintis, L. Liudvinavičius, and G. Vaičiunas, "Applying analytic hierarchy process to assess traffic safety risk of railway infrastructure," *Eksploatacja i Niezawodność*, vol. 15, no. 4, pp. 376–383, 2013.

[26] L. Feng, X. Zhang, and J. Gong, "Statistical forecast of change tendency of water resources based on the set pair analysis," *Hydrology*, vol. 24, no. 2, pp. 11–14.

[27] J. Hu and L. Yang, "Dynamic stochastic multi-criteria decision making method based on cumulative prospect theory and set pair analysis," *Systems Engineering Procedia*, vol. 1, pp. 432–439, 2011.

[28] Z. Wang, H. Ma, C. Lai, and H. Song, "Set pair analysis model based on GIS to evaluation for flood damage risk," in *Proceedings of the 2012 International Conference on Modern Hydraulic Engineering, CMHE 2012*, pp. 196–201, chn, March 2012.

[29] B. Ren, Y. Sun, Z. Zhou, Z. Cheng, and X. Hu, "Comprehensive evaluation model of reservoir operation based on improved set pair analysis," *Transactions of Tianjin University*, vol. 19, no. 1, pp. 25–28, 2013.

[30] T. L. Saaty, *Decision Making for Leaders: the Analytic Hierarchy Process for Decisions in a Complex World, RWS publications*, RWS publications, 1990.

[31] D. Ergu, G. Kou, Y. Shi, and Y. Shi, "Analytic network process in risk assessment and decision analysis," *Computers & Operations Research*, vol. 42, pp. 58–74, 2014.

[32] Creative Decisions Foundation, "Super Decision Software, 2017".

[33] L. H. Feng, G. S. Sang, and W. H. Hong, "Statistical Prediction of Changes in Water Resources Trends Based on Set Pair Analysis," *Water Resources Management*, vol. 28, no. 6, pp. 1703–1711, 2014.

9

Density-Based Statistical Clustering: Enabling Sidefire Ultrasonic Traffic Sensing in Smart Cities

Volker Lücken, Nils Voss, Julien Schreier, Thomas Baag, Michael Gehring, Matthias Raschen, Christian Lanius, Rainer Leupers, and Gerd Ascheid

Institute for Communication Technologies and Embedded Systems (ICE), RWTH Aachen University, Aachen, Germany

Correspondence should be addressed to Volker Lücken; luecken@ice.rwth-aachen.de

Academic Editor: Mehmet Yildirimoglu

Traffic routing is a central challenge in the context of urban areas, with a direct impact on personal mobility, traffic congestion, and air pollution. In the last decade, the possibilities for traffic flow control have improved together with the corresponding management systems. However, the lack of real-time traffic flow information with a city-wide coverage is a major limiting factor for an optimum operation. Smart City concepts seek to tackle these challenges in the future by combining sensing, communications, distributed information, and actuation. This paper presents an integrated approach that combines smart street lamps with traffic sensing technology. More specifically, infrastructure-based ultrasonic sensors, which are deployed together with a street light system, are used for multilane traffic participant detection and classification. Application of these sensors in time-varying reflective environments posed an unresolved problem for many ultrasonic sensing solutions in the past and therefore widely limited the dissemination of this technology. We present a solution using an algorithmic approach that combines statistical standardization with clustering techniques from the field of unsupervised learning. By using a multilevel communication concept, centralized and decentralized traffic information fusion is possible. The evaluation is based on results from automotive test track measurements and several European real-world installations.

1. Introduction

In the context of the Internet of Things (IoT), the integral transformation process of urban infrastructure into intelligent and connected devices plays an important role for future mobility. Cities experience an increasingly high level of road congestion due to the focused traffic coming along with progressing urbanization. This congestion induces a high social, economical, and environmental cost. From a European Union (EU) perspective, it is amounted to about one percent of the GDP [1] and predicted to increase by about 50 percent by the year 2050 [2]. From an environmental point of view, carbon dioxide emissions and air pollution, especially locally in cities, are a significant social and health aspect [3]. With this new unprecedented level of traffic density, offline and online traffic planning including live routing techniques becomes a requisite for larger cities, being useful both for end users and for city planners. Especially in the case of real-time traffic monitoring and the subsequent information provision to drivers and public routing systems, proper conditioning of this traffic information in terms of accuracy, area coverage, and density is paramount. Traffic congestion can only be effectively reduced if this information base, which is used for the routing decisions, is delivered with a sufficient level of quality in these dimensions. Then, with a city-wide traffic status, large-scale end-to-end routing optimizations can be performed. The results can be deployed to the vehicle driver with provision methods such as in-car information systems or centralized traffic guidance for triggering the individual's routing reaction. If this provision is not performed properly either due to an insufficient traffic information quality or improper consideration of group effects in the optimization, even detrimental effects can occur [4]. In this article, these requirements on traffic information quality and density are in focus, building a base for future routing decisions. Many currently used traffic sensing techniques for creating this live data basis will only have a negligible positive impact on the routing decisions if they are distributed sparsely and do not achieve sufficient coverage.

In general, two main types of techniques achieving more comprehensive traffic information have been established in the last years. A first category is the fixed or mobile infrastructural distribution of sensors, like inductive loops and traffic cameras. The second type is represented by systems relying on end-user based distributed information, for example, movement data from smartphone users or information gathered using V2X communication approaches. Details and characteristics of several techniques are discussed in this paper, considering not only metrics like traffic information quality and performance in different scenarios but also possible privacy issues which may arise with certain techniques.

The novel ultrasound-based traffic sensing technique, which is the focus of this paper, belongs to the category of infrastructural sensing. A significant difference to existing ultrasonic sensor systems is the placement of the sensors in a so-called sidefire setup. This means that the sensors are mounted in a height of at least three meters and face the street sideways with downtilted sensors, while being positioned on the side of the street. Previously existing systems were mainly single-lane ultrasonic sensors measuring the height profile top-down, for example, from a special sensor gantry above the street, with a simple first-reflection processing. The monitoring of multiple lanes is possible with our setup, while, at the same time, the requirements in terms of algorithmic evaluation and processing are increased in comparison to these previously existing systems. However, with these improvements, our approach allows operation in a highly reflective acoustic environment, which is present in urban areas and permits simply mounting the sensor on the side of the street. This is enabled by a new approach with a combination of statistical signal processing, clustering, and inference algorithms for traffic participant object detection.

This sidefire ultrasonic traffic sensing technique is part of a development which focuses on intelligent infrastructure solutions, more specifically on intelligent street lights. An exemplary integrated setup is shown in Figure 1. These light modules can be used to replace inner components of street lamps, for example, by retrofitting the widespread type of high-pressure sodium-vapor lamps. Then, they can act as a flexible platform for many applications by employing the processing, communication, and light control actuation capabilities. With the transition to LED-based lighting, which is triggered by high energy costs and several EU directives [5], a large proportion of the existing street lamps will have to be replaced in the next decade, offering potential for an integration of Smart City solutions. The seamless integration into infrastructure allows an almost complete coverage of relevant urban traffic to provide the required coverage density of traffic information. Furthermore, the major problem of power supply for sensing devices is alleviated, as the lamps are connected to the mains power, which eliminates the need for energy-constrained and expensive battery-powered devices.

The rest of this article is organized as follows. Section 2 provides the related work of urban traffic sensing techniques in general and ultrasonic sensing in specific. Section 3 describes the system architecture together with the specific scenario and typical sensor setup. Section 4 provides the algorithmic approach to the sensor signal processing system and object detection inference. In Section 5, the evaluation methodology and framework for reference data are presented. Section 6 gives results of real-world measurements and provides an evaluation from the perspective of detection and classification together with a state-of-the-art technology comparison. Section 7 then concludes the article.

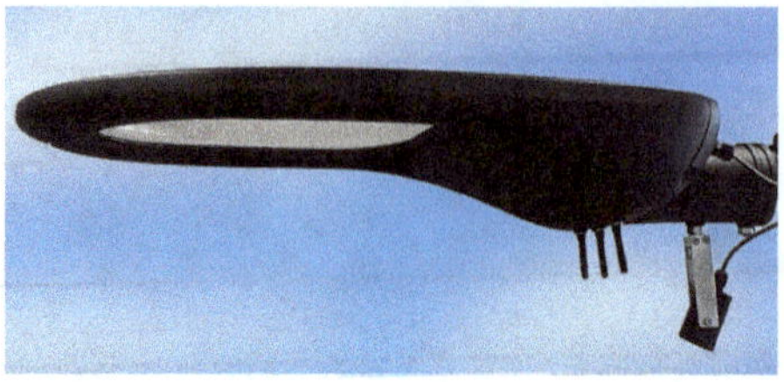

FIGURE 1: Integrated street lamp head with a single ultrasonic sensor head.

2. Related Work

Ultrasonic traffic participant detection falls into the field of traffic sensing with a large number of techniques available in research and in practical application. In this section, we therefore provide a state-of-the-art overview from two perspectives. First, we give a comprehensive overview of traffic monitoring techniques in general with a discussion of their performance, strengths, and weaknesses in order to show the technological gaps and potential. In the second part, specific technical approaches in ultrasonic traffic sensing, which are rare both in concepts and in practical implementations, are given. This application-centric perspective shows the drawbacks and unresolved problems in current state of the art.

2.1. Traffic Monitoring Techniques from a Broad View. In the introductory section, two main categories of traffic information acquisition systems were introduced. These are on the one hand infrastructure-based sensing systems, working in a temporarily or permanently fixed location. On the other hand, the type of end-user assisted sensing systems incorporates distributed information acquired from the drivers, their technical devices, and the vehicle's internal sensors itself. In the following, a state-of-the-art overview together with a technological comparison is given for nowadays most relevant traffic monitoring techniques of both categories. A condensed overview of the properties is also given in Table 1.

The most widely used approach to traffic sensing is the use of infrastructure-based solutions, which represents a mainly centralized approach. In general, authority of the data is kept with the operator and, thus, several of the problems coming with user-centric techniques do not arise. However, the sensing systems have to be deployed over the whole city or at least at critical locations in terms of traffic. Achieving the required density for an extensive urban traffic monitoring is therefore paramount, which brings along high effort and cost as a centralized solution.

TABLE 1: Overview of state-of-the-art traffic monitoring techniques.

Sensing technology	Characteristics (⊕/⊖)	References
Infrastructural sensing techniques		
Infrared sensors (passive and active)	⊕ Vehicle class and speed information acquisition is possible ⊕ Multilane operation possible (active sensors) ⊖ Susceptible to bad weather conditions ⊖ Regular cleaning required (active sensors)	[10, 11]
Camera-based systems	⊕ High detection and classification accuracy ⊖ Privacy and authorization issues ⊖ Susceptible to bad weather and light conditions	[6, 7]
In-pavement sensors (inductive, piezoelectric, and strain)	⊕ High detection accuracy with axle counting ⊖ Invasive and costly installation due to in-pavement integration	[8, 9]
Directional high-end radar sensors	⊕ Multilane multiobject traffic participant detection ⊕ Directional and velocity information ⊖ Cost and complexity of solutions	[10, 12, 13]
Top-down ultrasonic and radar sensors	⊕ Low cost and simple mode of operation ⊖ Only single-lane operation ⊖ Mounting above lane required, for example, on sensor gantry	[10, 14–18]
Proposed sidefire ultrasonic sensing	⊕ Multilane operation possible ⊕ Sidefire mounting allows seamless integration into environment ⊖ No directional information available in sidefire configuration	
User-based distributed traffic monitoring techniques		
Crowdsourcing systems (smartphones and navigation devices)	⊕ High number of devices ⊖ Privacy concerns ⊖ Susceptibility to attacks and data manipulation	[21–23]
V2X-based cooperative systems	⊕ Combination with in-vehicle information and decisions ⊖ Low support and prevalence	[24]

Camera-Based Systems. Frequently, camera-based systems are used for traffic monitoring. Here, we focus on systems with an automated analysis that is based on Computer Vision algorithms. With camera-based systems, a high detection accuracy plus a vehicle classification can be achieved. During nighttime, further solutions are required, for example, artificial illumination of the road and infrared or thermal imaging. Only in bad visibility conditions as fog, snow, or smog, the performance is limited [6]. The main concern with these systems, however, is the issue of privacy arising with video cameras in the public domain. On the one hand, the personal privacy of the citizens is affected; on the other hand, governmental regulations might even forbid the use of cameras in public areas, especially in this high-coverage case. Also, even with only a sparse coverage of video cameras in the public space, acceptance levels amongst citizens are low [7].

In-Pavement Sensors. Sensors in the urban area embedded into the pavement asphalt are often used in city intersection areas for traffic light control. The main in-pavement sensor technologies are inductive loops, piezoelectric sensors, and strain gauges [8]. They can also enable high-accuracy traffic measurements together with a classification by counting the axles of vehicles [9]. As a downside, the installation of pavement-integrated sensors is highly invasive and costly and requires per-lane road work with a possible temporary disturbance of city traffic.

Infrared Sensors. Passive and active infrared sensors are nonintrusive sensor types which can be mounted on the side or directly above the street. They offer detection with additional speed and vehicle class information in some realizations. With active sensors, multilane operation is possible. However, infrared sensors are susceptible to bad weather conditions like rain, snow, or fog. Furthermore, active infrared sensors require regular cleaning which might lead to lane closure during maintenance [10, 11].

Radar Systems. From the basic object detection principle, radar sensors represent the technology most comparable to ultrasonic sensors in the field of traffic sensing. High pulse repetition rates and ranges together with the possibility of obtaining directional information together with velocity data allow the monitoring of multiple traffic lanes and objects independent of the weather conditions [10, 12, 13]. As a major downside, the significant cost of the radar system, including front-end and signal processing, can compromise a city-wide system deployment for a sufficient coverage with these sensors.

Top-Down Ultrasonic and Radar. In the field of radar and ultrasonic sensors, several top-down sensor solutions are available, which require the placement above a specific street lane for single-lane car profile measurements [10, 14–18]. These techniques allow obtaining the complete height profile of vehicles for detection and classification but require a gantry or bridge over the street for mounting the sensors, which is infeasible in many situations. Moreover, only a single lane can be monitored with the existing solutions [10]. For the ultrasonic sensors, the processing is strongly simplified, as with the available systems, only the distance to the first relevant reflection and not the complete impulse response is evaluated. The significantly lower propagation speed and resulting decay time of the impulse responses together with the requirements for a sufficient pulse repetition rate (typically ≥10 Hz) of ultrasonic sensors can lead to self-interference and also to issues with fast-moving traffic participants [18]. For ultrasonic sensors, previous studies have shown that the impact of weather effects such as wind, snow, and rain is only marginal [19, 20].

Crowdsourcing Systems. For the second category of end-user based distributed information systems for traffic monitoring, one of the most present techniques is the crowdsourcing of traffic monitoring to end-user smartphones and navigation devices in order to obtain real-time and high-density traffic flow information [21, 22]. This approach is used in several systems and solutions, for example, by "Google Maps" with the product "Google Traffic" or by "TomTom HD Traffic." As a central advantage, no investments in additional sensors are required and the very high prevalence of these applications on end-user devices can be exploited by actively tracking the devices. However, the live tracking of the users is highly critical in terms of privacy aspects. Furthermore, these systems are susceptible to manipulative techniques and fake information due to trust issues with the user base, which has been demonstrated before [23].

V2X Communications. With the emerging field of V2X (vehicle-to-everything) communications, which is currently focused on on V2V (vehicle-to-vehicle) and V2I (vehicle-to-infrastructure) techniques, new possibilities for traffic information acquisition and exchange arise. Together with in-vehicle information for a direct delivery of inferred routing decisions to the driver and in a fusion with available travel plans on the car navigation device, an optimized information exchange and routing is enabled. Nevertheless, these types of systems are still to be realized in mass market together with suitable application scenarios and it will take time until they are sufficiently established amongst everyday vehicles. Moreover, reservations with regard to privacy and trust are also an important aspect for these systems. In literature, it is shown that, with a sufficient prevalence of cooperative V2V techniques in cars together with a high relaying communication range, dense traffic information can be obtained [24]. Therefore, vehicular communication techniques are a promising approach for the future. Certainly, as long as the density and proportion of vehicles supporting V2X communications is too low, a diverse traffic monitoring approach, which in parallel comprises infrastructural sensing techniques, has to be pursued for several decades.

2.2. Specific Advances in Ultrasonic Traffic Detection. In the previous introductory Section 2.1, the basic top-down ultrasonic sensing technology has already been introduced. However, the practical use is limited, as additional mounting efforts are required, if no infrastructure like bridges above the street or traffic light poles at intersections are available. Moreover, only single-lane detection is possible in this regard. Therefore, these solutions are limited in their practical applicability.

There also exist initial concepts that use ultrasonic sensors mounted on a lateral street position with a so-called sidefire orientation, facing the opposite street side either horizontally or in a downtilted position [19, 25, 26]. In the case of horizontal tilt, even a first simple multilane detection has been realized. However, all these concepts perform only an analysis of the first-reflection distance instead of the whole impulse response, which highly limits the practical use in a variety of scenarios. The urban scenario is a highly reflective time-varying environment, where first reflections of other objects such as moving trees, parking cars, and the scattering of the road surface are always present in the distance range with the relevant traffic participants for detection, the distance region-of-interest for measured signal reflections of objects. In addition, the horizontally facing type of sidefire sensors could be obstructed by other objects and traffic participants in urban scenarios due to their low mounting height.

3. System Concept and Architecture

Sensor systems have specific requirements regarding the setup, scenario, and working environment they are being operated in. In this chapter, the system setup and application concept, together with the generic requirements for our proposed ultrasonic traffic sensing system, are given. The scenario stays the same as initially proposed in Section 1: a Smart City street lamp system platform. Nevertheless, a wide variety of installation and application possibilities are given also without any ties to this integration concept.

3.1. Sensor System Setup. An exemplary setup of the smart street lamp platform was already shown with Figure 1. It features a single mounted ultrasonic sensor attached to the lamp head or the pole and an embedded system integrated into a customized LED-based lamp head. The digital core system contains modules for realizing processing and communication capabilities, enabling distributed processing in a local-level group of lamps in a street by building a dynamic mesh network and also allowing communication on the global and cloud level by employing mobile network machine-to-machine type communication (M2M) functionalities. The central component for the signal processing capabilities consists of a Texas Instruments Sitara system-on-chip (SoC) including an ARM Cortex-A8 core. Typical examples of

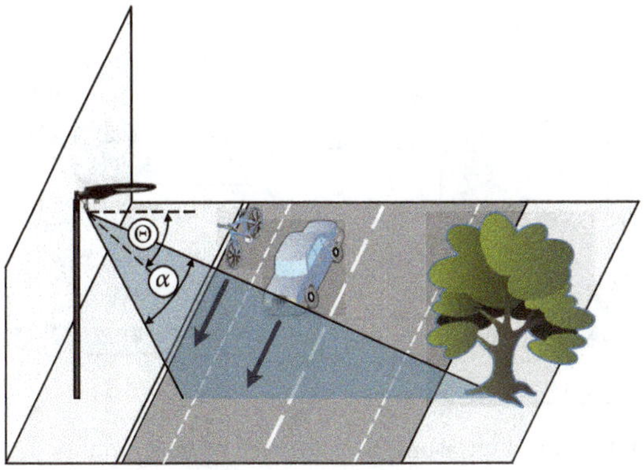

Θ: Downtilt angle
α: Beam width

FIGURE 2: Downtilted sidefire sensor configuration in typical urban environment with two-lane street (with downtilt angle Θ and beam width α).

possible applications with this system are location-based content delivery, attachment of modules for environmental or traffic sensing applications, and energy efficiency optimizations of street lighting. Due to the combination of infrastructure with an extensible platform, the systems can be widely distributed amongst the city, as almost all central areas are covered by street lamps. Furthermore, many of the constraints on processing and communications, which are typically imposed for Smart City traffic monitoring systems in IoT context, can be relaxed, for example, data and refresh rate restrictions due to battery power [5].

The typical seamless integration of the ultrasonic sensing concept into an urban street area is shown in Figure 2. In the so-called downtilted sidefire setup, the sensor is mounted in a height between 3 and 8 m, which is above the typical height of vehicles in urban areas. The sensor head is then oriented diagonally downwards the street, with a downtilt angle Θ between 30° and 80°. The ultrasonic sensor transducer together with the casing has a cone-shaped beam characteristic, with an opening angle α typically between 20° and 80° depending on the scenario and lane coverage requirements. In the exemplary scenario shown in Figure 2, the multitude of reflective elements in the urban environment already becomes apparent. For example, the first reflections and scattering of the street surface can arrive earlier than reflections from relevant objects on distant lanes to the sensor. Additionally, reflective objects can exhibit short- or long-term time variation, like parked cars or trees moving in the wind. In these cases, simple distance-measurement based sensor techniques would already fail.

3.2. Ultrasonic Sensor Platform Module. In order to provide the capabilities for ultrasonic sensing, the core system is extended with a custom sensor platform module, which is shown in Figures 3 and 4. It provides four ultrasonic sensor channels for real-time arbitrary waveform generation and recording in half-duplex mode. The reception quality of the ultrasonic impulse response is with an input stage SNR of above 105 dB in the ultrasonic band. The analog driving circuit for the transducer is integrated into the ultrasonic transducer head itself and is adapted to the capabilities and characteristics of the transducer.

Control of the attached components, signal generation, and preprocessing capabilities is realized with an integrated Xilinx Artix-7 series FPGA. It provides communication capabilities to the core processing system over an Ethernet link, which is also used for the power supply of the sensor system extension. Together with the core system, GPS including the high-precision PPS signal is used for a timing synchronization and exact coordination of multiple sensors in the vicinity of a street. Interference can therefore be coordinated by a global optimization of the sensor timing patterns if the sensor heads are transmitting in the same frequency band.

3.3. General Signal Structure. For simplicity, the signal structure in the single-sensor case is discussed. The transmit signal $s_T(t)$ consists of repeated sequences of the base pulse $h_i(t)$ with an overall duration T_{rep}, yielding a pulse repetition rate of $f_{\text{rep}} = T_{\text{rep}}^{-1}$. This is also shown in Figure 5. T_{rep} is typically chosen between 50 and 150 ms to ensure low self-interference levels between the different impulse response blocks. The repetition time should in general be larger than the RT60 reflections decay time of the acoustic channel [27, p. 98]. The RT60 time is a common measure in indoor acoustics but can also be applied in outdoor scenarios, where it typically yields much shorter values. The general pulse sequence is then defined as follows:

$$s_T(t) = \sum_{n=0}^{\infty} h_n\left(t - n \cdot T_{\text{rep}}\right). \tag{1}$$

The base pulses $h_i(t)$ itself consist of two specific duplexer phases. At the beginning, the transmit duplex mode is used and a transmit pulse part of length T_{wnd} is emitted. Then, the duplex mode is switched to receive mode and records the channel impulse response for the rest of the period $T_{\text{rep}} - T_{\text{wnd}}$ without further transmission. For the investigations in this article, the transmit pulse is chosen to be a windowed signal with a carrier frequency of $f_c = 40$ kHz and an envelope window $h_{\text{wnd},i}(t)$ with length $T_{\text{wnd}} = 2$ ms as a tradeoff between time and frequency resolution for the narrowband signal. The i-th base pulse is then given by the following:

$$h_i(t) = \begin{cases} h_{\text{wnd},i}(t) \cdot \operatorname{Re}\left\{e^{j2\pi f_c t}\right\} & \text{if } 0 \le t \le T_{\text{wnd}}, \\ 0 & \text{otherwise.} \end{cases} \tag{2}$$

With the use of identical repeated transmit pulses in our case, $h_i(t) = h(t)\ \forall i \in \mathbb{N}$. A Blackman window [28] function is chosen for $h_{\text{wnd},i}(t)$ because it does not exhibit discontinuities at the beginning and end of the time domain window, while exhibiting a high stop-band attenuation [29].

4. Processing Concept and Algorithms

The ultrasonic traffic sensing technique presented in this paper is based on a combination of statistical analysis

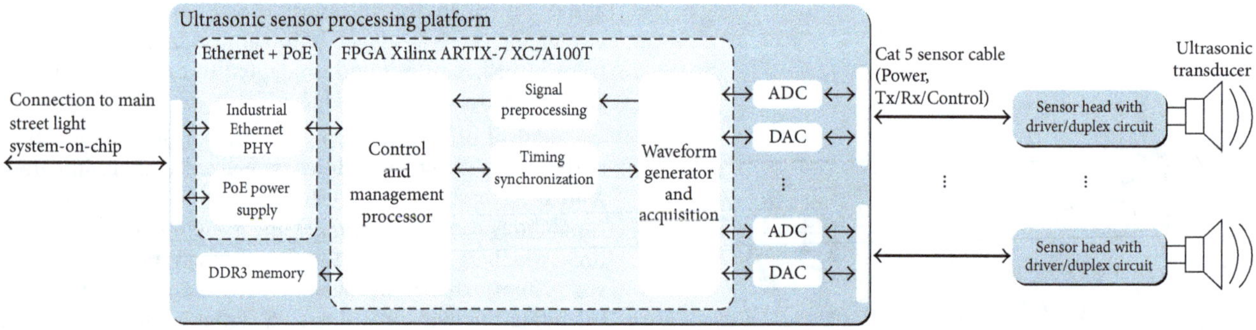

Figure 3: General structure of sensor platform including the FPGA and attached ultrasonic sensors with driving circuits.

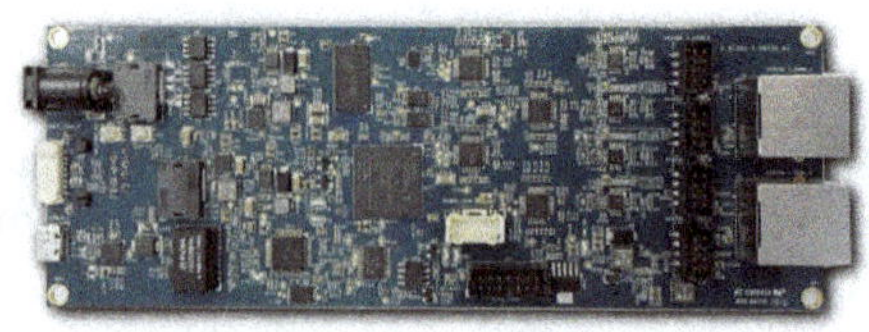

Figure 4: Custom sensor platform which is integrated as an extension module into the intelligent street light.

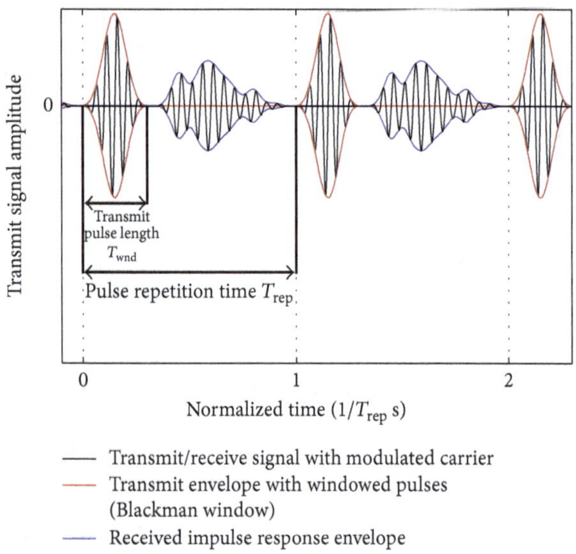

Figure 5: Generalized signal structure with transmit/receive duplexing mode, corresponding envelopes, and modulated signals.

and clustering techniques for object candidate detection. Together with a special signal chain structure, objects representing traffic participants can be detected, together with an extraction of further characteristics and object properties. In this section, the model for the received signal is given first. Then, the characteristics of the reflective environment are modeled and empirically analyzed. Based on these findings, statistical hypothesis testing is performed during operation and the resulting statistical information is combined with a modified clustering algorithm for object boundary detection. Then, the system concept for the extraction of further characteristics is given and the reduced-complexity embedded implementation for real-time operation is discussed.

4.1. Received Signal Model and Preprocessing. With the general signal structure described in Section 3.3, repeated measurements of the acoustic channel are performed with the ultrasonic sensor system. Based on the impulse responses as the one-dimensional received signals, which consist of a large number of reflections in the acoustic environment, object detection and further analyses can be performed together with the time-of-flight distance information.

The received signal $s(t)$ from the sensor head is recorded continuously and then sliced into blocks of length T_{rep}, yielding a series of impulse responses $s_i(t_D)$, with i being the ith measurement interval of length T_{rep}. The parameter t_D is the time position of the specific impulse response, which is mapped to an equivalent time-of-flight distance $d = t_D \cdot c$ given the speed of sound $c = 343\,\text{m/s}$ in the air medium at normal conditions. Due to the round-trip, the real distance to the reflective object is half the equivalent time-of-flight distance. By acquiring information on the current air temperature via sensor measurements or weather information, c can be compensated for the current conditions. During the initial period of $s_i(t)$ with $0 \le t_D \le T_{wnd}$, which is the transmission duplex mode, no useful signal is acquired, while the region $T_{wnd} \le t_D \le T_{rep}$ represents the main signal of interest for further analysis. The single impulse response slices are then given by the following:

$$s_i(t_D) = s\left(i \cdot T_{rep} + t_D\right). \tag{3}$$

The equivalent time-discrete representation is

$$x_i(n_D) = s\left(i \cdot N_{rep} T + n_D \cdot T\right). \tag{4}$$

with the sampling time T and the discrete-time repetition interval length $N_{rep} = T_{rep}/T$.

As a first step of preprocessing of the sliced receive signal on the FPGA, each impulse response is convolved with a bandpass filter $h_{BP}(t)$ centered at the transmit carrier frequency f_C with a bandwidth of 6 kHz for allowing the analysis of Doppler-shifted signals with typical urban vehicle velocities. Then, a time-discrete Hilbert transform $\mathcal{H}$ [30, 31]

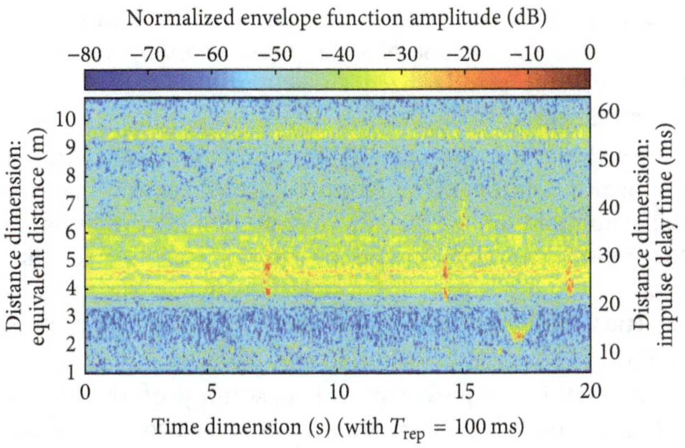

FIGURE 6: Two-dimensional envelope function mapping of received signal $e_{R,i}(n_D)$ (amplitude normalized and in logarithmic scale).

is performed in order to acquire the complex-valued analytic signal:

$$x_i(n_D) = (s_i * h_{\text{BP}})(n_D) + j \cdot \mathscr{H}\{(s_i * h_{\text{BP}})(n_D)\} \quad (5)$$

together with the instantaneous amplitude or real-valued envelope signal

$$e_i''(n_D) = |x_i(n_D)|. \quad (6)$$

This envelope signal $e_i''(n_D)$ is further processed to facilitate object detection in the following. Only in case of Doppler-based motion and velocity estimation, time-frequency information such as the instantaneous phase and frequency has to be incorporated. Both its two discrete dimensions, i being the index of the impulse response (from now on called time dimension, with an equivalent sampling rate of f_{rep}) and n_D being the distance-equivalent impulse response time (from now on called distance dimension, with an equivalent sampling rate $f_S = 1/T$), are derived from a single original time dimension in the received signal by slicing. Therefore, the distance dimension is fixed and limited to $0 \leq n_D \leq N_{\text{rep}} - 1$, while the causal time dimension in real-world application is extending with f_{rep}. With this two-dimensional mapping, object detection algorithms become applicable which consider a vicinity in both dimensions. This means that effects of an object creating a peak at several neighboring distance and time points are now properly represented as neighbors in the 2D space, in contrast to the original one-dimensional recorded sensor signal.

An example for the envelope signal $e_{R,i}''(n_D)$ is given in Figure 6, with a normalized logarithmic color mapping of the amplitude for improved visualization. The equivalent distance and impulse response time mappings are shown on the ordinate for reference purposes. In this plot, the scattering and reflections from the street surface can already be seen in a distance range of 4 to 6 m in a comb-like pattern. Also, around the time positions at 7 s, 14 s, 15 s, 17 s, and 19 s, strong reflections from objects passing the sensor are visible.

4.2. Signal Statistics and Standardization. Given the high complexity of the reflective patterns in the sensor data, the statistics of the signal need to be described properly in order to perform an outlier or anomaly detection, which would indicate the presence of objects in the sensor field. The goal is to incorporate all noise components (measurement noise and other acoustic emissions in the ultrasonic band) and also the typical static and time-varying reflections (e.g., moving leaves of a tree caused by wind and reflection patterns of the street) for every specific distance point of the signal into the statistics. This allows the classification of segments of newly acquired impulse responses in terms of the data following the a priori distribution, called base distribution, which was estimated in the past, or whether an outlier is present.

In contrast to typical binary decision problems, here, only this base distribution without any presence of an object can be estimated, representing the null hypothesis. The alternative hypothesis of object presence is however different for every object. This problem can be solved using parametric or nonparametric hypothesis testing techniques for properly representing the underlying base distributions of the signal points and testing the outlier probabilities. However, for the real-time implementation on the embedded system of the sensor setup, we present a simpler and less computationally complex approach that is based on the distance-wise standardization of the signal based on the calculated statistics in a predefined time window in the past. These standardization techniques are often used in the field of data science as a preprocessing step for feature scaling. The distance-wise standardized envelope signal $e_i''(n_D)$ is calculated based on the information from the statistics memory with a window length of $N_{T,w}$ time dimension steps in the past (equivalent to a time range of $N_{T,w} \cdot T_{\text{rep}}$ in continuous time), yielding $e_i'(n_D)$:

$$e_i'(n_D) = \frac{e_i''(n_D) - \mu_{n_D}(i)}{\sigma_{n_D}(i)} \quad (7)$$

$$= \frac{e_i''(n_D) - \overbrace{N_{T,w}^{-1} \sum_{k=i-N_{T,w}}^{i} e_k''(n_D)}^{\text{Windowed mean } \mu_{n_D}(i) \text{ for distance } n_D}}{\underbrace{\sqrt{(N_{T,w}-1)^{-1} \sum_{m=i-N_{T,w}}^{i} \left[e_m''(n_D) - N_{T,w}^{-1} \sum_{k=i-N_{T,w}}^{i} e_k''(n_D)\right]^2}}_{\text{Windowed standard deviation } \sigma_{n_D}(i) \text{ for distance } n_D}}, \quad (8)$$

$$0 \leq n_D \leq N_{\text{rep}} - 1, \; i \geq N_{T,w}.$$

Under the simplified assumption that the underlying process of $e_i''(n_D)$ (with no objects present) for a given distance n_D is an i.i.d. Gaussian process $X_{n_D} \sim \mathcal{N}(\mu_{n_D}, \sigma_{n_D}^2)$, the specific distance ranges of $e_i'(n_D)$ will follow the standard normal distribution $Y_{n_D} \sim \mathcal{N}(0,1)$. The assumption is especially found to be valid for direct reflections in the closer range, shown in studies from other acoustic environments like underwater [32]. As a general approximation, it will later be used for approximatively calculating the initial detection thresholds for a given false alarm rate.

For the object detection, only the right tail of the distribution is of interest for outlier detection, as shadowing effects are not considered. Therefore, the complete standardized distribution data $e_i'(n_D)$ is clipped for values below zero, resulting in the final envelope $e_i(n_D)$ with statistical standardization:

$$e_i(n_D) = \max\left(e_i'(n_D), 0\right). \quad (9)$$

This yields a variable with a standard normal distribution left-censored at 0. The resulting probability distribution function (PDF) $f(x)$ can be written as follows:

$$f(x) = \begin{cases} \frac{1}{\sqrt{2\pi}} e^{-x^2/2} + \frac{1}{2}\delta(x) & x \geq 0, \\ 0 & \text{otherwise;} \end{cases} \quad (10)$$

and the cumulative distribution function (CDF) $F(x)$ is

$$F(x) = \begin{cases} \frac{1}{2}\operatorname{erf}\left(\frac{x}{\sqrt{2}}\right) + \frac{1}{2} & x \geq 0, \\ 0 & \text{otherwise,} \end{cases} \quad (11)$$

where $\operatorname{erf}(x)$ is the Gauss error function. The first terms of $f(x)$ and $F(x)$ are similar to a scaled half-normal distribution [33], with the addition of the clipped values being statistically censored and not truncated.

To summarize, $e_i(n_D)$ is now used for all further object detection and processing together with the clustering techniques in the following. Simply put, it yields the positive outlier level of a newly acquired data point by the distance to the calculated mean in a measure of the multiple of standard deviations. This is based on the statistical history in a specific time interval and calculated separately for each discrete distance point. By having a limited time window for the statistics, the sensor system can adapt to changing environments without needs for recalibration. Optionally, for achieving a better robustness and generalization, these distance-wise statistics can be blurred with neighboring distance points.

4.3. Object Boundary Detection with Density-Based Statistical Clustering (DBStaC). As a next step, the standardized data given in the previous section is used for object detection. The two-dimensional data $e_i(n_D)$ is passed through a modified density-based clustering algorithm based on the original DBSCAN (Density-Based Spatial Clustering of Applications with Noise [34]), which is able to perform clustering on noisy data by aggregating core points of data peaks. In our case, these data peaks can be the result of statistical outliers due to physical objects passing by the sensor, in comparison with the base distribution of the signal when no objects are present. Our proposed modified algorithm in combination with the input preprocessing steps from (7) and (9), called *density-based statistical clustering* (DBStaC), can be used on both statistical hypothesis test results, and in our case of a simplified algorithm, data standardized based on the a priori statistics.

4.3.1. Choice of Clustering Algorithm. The choice of a suitable clustering algorithm from the field of unsupervised learning techniques for our application was subject to several requirements. While many clustering algorithms can aggregate nearby points in a space with arbitrary dimensionality and sparsely distributed data points, it is required here to incorporate the weight of data points, coming from our the nonsparse standardized and clipped data field $e_i(n_D)$ described before in Section 4.2.

Furthermore, the algorithm should be able to detect important data peaks based on a local density, while, for the whole cluster, no penalty for an infinite extension in the time dimension should be given. This is necessary because the dwell time of an object in the sensor range is arbitrary, especially due to different movement speeds of traffic participants or even in a traffic jam situation.

As a third requirement, an anisotropy of the cluster properties and shape is desired, as we have two dimensions that are time (index of different impulse responses) and distance (specific point in the impulse response itself), which have highly different characteristics. A typical scenario where anisotropy would not be required is 2D image processing, where both image dimensions have similar properties.

4.3.2. Original Definition of DBSCAN. First, we want to provide a basic understanding of the original DBSCAN algorithm in general and the specific terms coming with the algorithmic definition. Then, we can introduce the modifications for our specific application. Readers interested in the complete formal description are referred to [34, 35], on which the following original definitions are based.

Core Point. The general principle of DBSCAN is the analysis of a set D of points in a k-dimensional space. Then, for any given point $\mathbf{p} \in D$, the ε-neighborhood of this point is evaluated, meaning that all points $\mathbf{q} \in D$ with a distance $\operatorname{dist}(\mathbf{p}, \mathbf{q}) \leq \varepsilon$ belong to the neighborhood of $\mathbf{p}$, denoted $\overline{N}$. The distance metric $\operatorname{dist}(\mathbf{p}, \mathbf{q})$ is typically chosen to be the L_1 or L_2 norm, and ε is preset. This results in the score $\overline{n}_\varepsilon$ of a point $\mathbf{p}$ being the number of other points $\mathbf{q}$ falling into the ε-neighborhood. Then, the point $\mathbf{p}$ is called core point if $\overline{n}_\varepsilon \geq \text{MinPts}$, with MinPts being a preset threshold for core point detection. All points in the ε-neighborhood of a core point are part of a cluster set C, and they are called border points if they are not core points themselves.

Direct Density-Reachability. A point $\mathbf{p}$ is directly density-reachable from a point $\mathbf{q}$ if $\mathbf{q}$ is a core point and $\mathbf{p}$ is in the ε-neighborhood of $\mathbf{q}$. This property is symmetric if $\mathbf{p}$ and $\mathbf{q}$ are a pair of core points [34].

Density-Reachability. A point $\mathbf{p}$ is density-reachable from a point $\mathbf{q}$ if a chain of n points $\mathbf{p}_1, \dots, \mathbf{p}_n$ with $\mathbf{p}_1 = \mathbf{q}$, $\mathbf{p}_n = \mathbf{p}$ exists where every point $\mathbf{p}_{i+1}$ is directly density-reachable from $\mathbf{p}_i$, $\forall i \in [1 \cdots n-1]$ [34].

Density-Connectivity. A point $\mathbf{p}$ is density-connected to another point $\mathbf{q}$ if there is a point $\mathbf{o}$ such that both $\mathbf{p}$ and $\mathbf{q}$ are density-reachable from $\mathbf{o}$ [34].

Cluster Definition. Given the set of all points D, a cluster C is a nonempty subset of D which satisfies the following conditions [34]:

(1) $\forall \mathbf{p}, \mathbf{q}$: if $\mathbf{p} \in C$ and $\mathbf{q}$ is density-reachable from $\mathbf{p}$, then $\mathbf{q} \in C$ (maximality).

(2) $\forall \mathbf{p}, \mathbf{q} \in C$: $\mathbf{p}$ is density-connected to $\mathbf{q}$ (connectivity).

Any cluster C is then uniquely determined by its core points.

4.3.3. Modified Core Point Definition for DBStaC. In our special case, the algorithm is modified to suit the needs for clustering our two-dimensional space; therefore $k = 2$ in order to represent the time and distance dimension of our acquired data. Furthermore, our data points are weighted by their standardized value. Therefore, for a point $\mathbf{p} = (i, n_D)$, the weight $e_i(n_D)$ is assigned, which is not an option in the classic DBSCAN algorithm. Instead of using the dist metric for determining the ε-neighborhood, anisotropy is introduced by choosing the ε-neighborhood as a rectangular area with dimension $(2\varepsilon_T + 1) \times (2\varepsilon_D + 1)$, symmetrically placed around the position (i, n_D) of the core point, with ε_T and ε_D given. Then, all points $\mathbf{q}(i', n_D')$ with $|i' - i| \leq \varepsilon_T \wedge |n_D' - n_D| \leq \varepsilon_D$ belong to the ε-neighborhood N of $\mathbf{p} = (i, n_D)$ and $\mathbf{p}$ is a core point if

$$\sum_{t=i-\varepsilon_T}^{i+\varepsilon_T} \sum_{d=n_D-\varepsilon_D}^{n_D+\varepsilon_D} e_t(d) \geq \text{MinSum}. \qquad (12)$$

With this definition, all data points of our two-dimensional data set are now taken into account, not just by their count, but by the sum of their assigned weights in the specific ε-neighborhood. The implementation of the final clustering is omitted at this point and widely described in literature [36]. With the original algorithm implemented, adaptions only have to be made in the core point definition and neighborhood metrics.

4.3.4. False Alarm Rate. In the clustering process, every element of $e_i(n_D)$ is tested for the core point property, requiring both compensation for multiple comparisons and an approximation of the false alarm rate for a single point itself. For the latter, we will provide measures for determining the false alarm rate under the approximative distribution assumptions from Section 4.2. With the threshold decision from (12) for a core point, given MinSum and the ε-neighborhood with

$$A = (2\varepsilon_T + 1)(2\varepsilon_D + 1) \qquad (13)$$

points in the rectangular region, we want to calculate the false alarm probability:

$$P_f = P\left(S(i, n_D) \geq \text{MinSum} \mid \overline{O}\right), \qquad (14)$$

where $\overline{O}$ is the event that no relevant object reflection yielding a core point is present, meaning that all summands follow the base distribution without outliers. $S(i, n_D)$ is the sum of weights in the ε-neighborhood:

$$S(i, n_D) = \sum_{t=i-\varepsilon_T}^{i+\varepsilon_T} \sum_{d=n_D-\varepsilon_D}^{n_D+\varepsilon_D} e_t(d). \qquad (15)$$

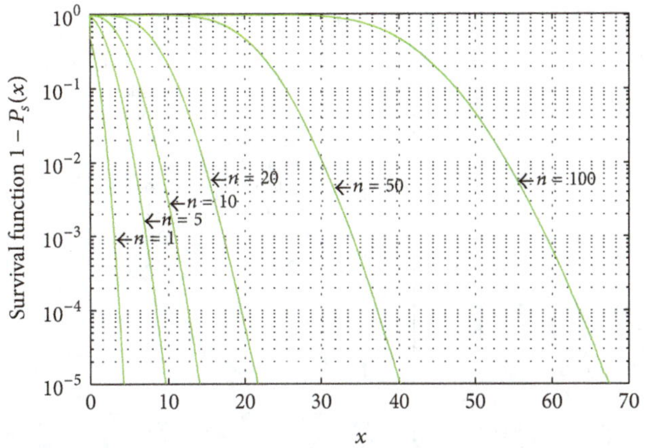

FIGURE 7: Survival function $1 - P_s(x)$ for distribution $p_s(x)$ with n-fold convolution of $f(x)$.

In Section 4.2, we showed that under the assumption that the original envelope signal $e_i''(n_D)$ is coming from a process with normal distribution, $e_i'(n_D)$ is standard-normally distributed after standardization, and $e_i(n_D)$ follows a left-censored standard normal distribution. If the data points $e_i(n_D)$ $\forall i, n_D$ and the resulting $S(i, n_D)$ are also hypothesized to be statistically independent and exhibit the identical PDF $f(x)$ from (10), we can write for the resulting PDF $p_S(x)$ of the summation of $n = A$ points in (15):

$$p_S(x) = \underbrace{(f * \cdots * f)}_{n \text{ times}}(x) = f^{*n}(x). \qquad (16)$$

Please note that the statistical independence of all $S(i, n_D)$ is not given in the first place because the same data points in the two-dimensional field are affected by the sliding window in several i and n_D positions. Then, adjacent $S(i, n_D)$ are correlated and a single peak in $e_i(n_D)$ could yield multiple peaks in $S(i, n_D)$ resulting in core points. However, as we seek to calculate the false alarm rate in general for this specific algorithm, adjacent core points would be clustered into a single cluster, raising only false alarm for a single object if $P_f \ll 1$.

Finally, neither for the left-censored standard normal distribution $f(x)$ nor for the related half-normal distribution in general, the result of k-fold convolution is known or given in literature in closed form for $k > 2$. Therefore, the final calculation is based on Monte Carlo simulations of these random processes. For the false alarm probability analysis, we are interested in the survival function (SF) $1 - P_s(x)$, where $P_s(x)$ is the CDF of $p_s(x)$. With $x =$ MinSum, the survival function yields the false alarm probability per data point. The results from the Monte Carlo simulations for the survival function are shown in Figure 7 and were also verified with an additional false alarm Monte Carlo simulation.

4.3.5. Choice of Clustering Neighborhood. The neighborhood choice in the time direction, ε_T, affects the detection robustness and separability of vehicles with length l_{Veh} following each other in a gap distance l_{Gap}, with speed v and passing the sensor lobe which approximately covers a section of length

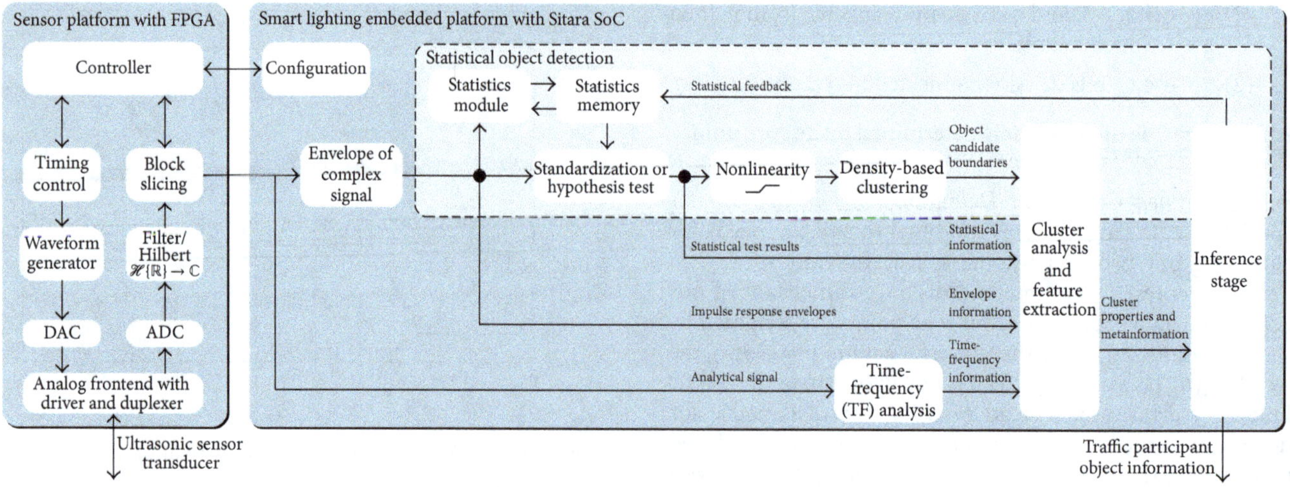

FIGURE 8: Signal chain with acquisition, analysis, and inference stages on both platforms.

l_{Lobe} on the specific lane. Given the pulse repetition rate T_{rep}, no vehicle is covered by the sensors during the gap for n_{Gap} pulse times:

$$n_{\text{Gap}} = \frac{\left(l_{\text{Gap}} - l_{\text{Lobe}}\right)}{\left(v \cdot T_{\text{rep}}\right)}. \tag{17}$$

Also, considering the detection of a single vehicle, the vehicle is sampled in the lobe for n_{Veh} pulses:

$$n_{\text{Veh}} = \frac{\left(l_{\text{Veh}} + l_{\text{Lobe}}\right)}{\left(v \cdot T_{\text{rep}}\right)}. \tag{18}$$

Given ε_T, the full neighborhood extension in time direction is $2\varepsilon_T + 1$ pulses. Therefore, $n_{\text{Gap}} \geq 2\varepsilon_T + 1$ is recommended for a good separation of the vehicle clusters in time direction in order to have no overlapping core points. Typical values of the time neighborhood are $0 \leq \varepsilon_T \leq 2$. With an urban area example of $\varepsilon_T = 1$, $l_{\text{Veh}} = 4.5$ m, $v = 40$ km/h, and $l_{\text{Lobe}} = 1.5$ m, the recommended gap length becomes $l_{\text{Gap}} \geq 4.83$ m and a vehicle is covered by $n_{\text{Veh}} = 5.4$ pulse times. The second neighborhood parameter ε_D in distance direction is mostly relevant when it comes to multilane separation of multiple vehicles. In general, our approach in this paper is to learn the typical vehicle speed and distance profile without a direct calculation, but by learning the scenario for a short training period (e.g., one hour) and then setting the neighborhood parameters in a parameter optimization, which is described in the following section.

4.4. Signal Processing Concept. In Figure 8, the previously described statistical object detection is integrated into the complete object detection concept with feature extraction and inference. It allows detecting candidates for detected objects representing traffic participants, which are then used to gather further information from several signal processing paths. The preprocessing and main processing stages are also divided in hardware into the sensor platform FPGA and the Sitara SoC, respectively.

4.4.1. Preprocessing. In contrast to the sensor platform structure described previously in Section 3, here, the signal flow and processing are in focus. As can be seen in the left part of Figure 8, the sensor platform is configured by the main system on the Sitara SoC and performs the whole signal transmission and reception. The preprocessing steps of filtering, Hilbert transformation, and possible sample rate adjustments are performed using a dedicated architecture on the FPGA, before slicing the blocks from the different sensors into single impulse responses, which are then transmitted to the main processing system with the Sitara SoC.

4.4.2. Object Detection and Extraction. On the right side of Figure 8, the complete object detection and feature extraction concept is shown. It is based on four main information paths going into the cluster analysis and feature extraction block at the end, whose output is then used for inference of detected traffic participants. By performing a fusion of these four paths, all necessary information for the analysis can be provided. The different components of the processing system are described in the following.

Statistical Object Detection with Resulting Object Candidate Boundaries. The basic principle of object clustering on the standardized data for boundary detection has already been described before in Sections 4.2 and 4.3. The only addition is an optional nonlinearity after standardization or the optional hypothesis testing (which is not considered here). The nonlinearity can be modified to perform further transformations on the standardized data. With the DBStaC algorithm, the envelope signal at the beginning of the statistical object detection block is analyzed by the statistics module together with its statistics memory. Only a predetermined amount of past data can be stored in this fixed-size memory for calculating the base statistics. The boundaries of resulting clusters of the algorithm are taken as boundaries for further analyses in the time-distance-field of the different data paths. Inside each cluster's area which was determined by the object boundary information path, the cluster analysis stage performs an analysis of the statistical, envelope, and time-frequency-information path.

Statistical Feedback. The statistical feedback path coming from the inference stage is important to keep the base distribution in the statistics memory accurate to be mostly restricted to data without objects present. This is an optional feature and especially helpful with high fluctuations of traffic density in front of the sensor, as the detected objects are removed from the statistics memory and the outlier analysis of the DBStaC-based object detection is more accurate and robust against drift.

Statistical Information and Envelope Information. The standardized data and the original envelope signal are directly fed into the cluster analysis stage. For feature extraction and accurate information on the object reflection characteristics, properties like the accurate position of the object's first reflection and the mathematical two-dimensional center of mass of the signal or the geometry are analyzed. Especially with a priori geometric information on the system setup in a specific scenario, effects caused by the beam width of the transducer can be exploited. Even if a car moves perfectly sideways through the cone-shaped beam of the sensor, it forms a typical arch-shaped pattern, which could already be seen in the initial example signal, Figure 6. This arc-shaped reflection pattern is also known from marine sonar systems as "fish arc" and caused by the fact that at the point in time when a vehicle first enters the beam, the reflection path is diagonal and is not fully straight until the car is directly in front of the sensor. This can be used to support the information on vehicle size and type.

Time-Frequency Information. The time-frequency information is mostly irrelevant as long as the sensor is oriented perpendicular to the vehicle movement on the street. However, in case that the sensor is oriented partially sideways or a second sensor with such orientation is available, velocity information can be gathered by analyzing the instantaneous frequency of the signal for Doppler shift estimation.

Inference Stage. The inference stage performs the final decision whether the cluster detected in the DBStaC stage originated from a traffic participant in the sensor range and allows the estimation of further object parameters and a simple classification of the vehicle type using a supervised learning stage with a Support Vector Machine. As our analyses in this article are focused on the DBStaC object candidate detection algorithm itself, no further rejection of objects is performed by the inference stage in this case. Only objects detected outside of the specific lanes, for example, on the sidewalk, are rejected.

4.5. Implementation Details. The main parts of the processing chain shown on the right side of Figure 8 were implemented in C++, including optimized versions for embedded system operations, while parts of the postprocessing and learning techniques as part of the cluster analysis and inference stages were implemented in Python. The core processing module can be used for both algorithmic and performance analysis purposes on general Linux/BSD systems and the embedded Linux system running on the ARM Cortex A8 as part of the Sitara SoC. For algorithmic evaluation and parameter optimization in the next section, capabilities of the core module will also be used as part of an evaluation system. The numerical precision requirements for embedded system operation were relaxed from 64-bit floating point to 32-bit floating point, largely without compromising performance, in order to exploit the NEON floating point accelerator of the Sitara SoC. For the processing chain with the DBStaC algorithm, real-time operation is then possible for two attached ultrasonic sensors. This is the case for typical parameter sets, as the algorithmic runtime of DBSCAN is not fixed and is largely dependent on the choice of parameters (size of ε-neighborhood and decision thresholds) and traffic situation yielding different cluster sizes. Edge cases like extremely large clusters, for example, due to a traffic jam situation with very long dwell times of an object in the sensor range, have to be taken care of in the practical implementations.

For the sensor platform with the FPGA shown on the left side of Figure 8, dedicated accelerators for filtering and processing are used together with a MicroBlaze soft core for the coordination of waveform transmission and data acquisition. The configuration is controlled by the main processing system side with the Sitara SoC. A timing reference signal is acquired using GPS together with the high-precision PPS reference, which allows global synchronization of multiple sensors attached to multiple platforms, with an optimization of the transmission and interference patterns. Exchange of resulting information, for example, for sensor fusion and distributed processing for trajectory calculation of objects passing by multiple systems, is possible on both the level of a local wireless meshing between the systems and a global communication using cellular networks.

5. Evaluation Scenarios and Methodology

Evaluation of the complete system concept for traffic participant detection requires analyses both in real-world scenarios and in isolated conditions with synthetic scenarios, such as on automotive test tracks. A variety of European-wide test measurements have been performed with the system, described in Section 5.2, which will be analyzed and discussed in Section 6. For the reference data acquisition, a video camera is running in parallel which is afterwards used to label, annotate, and classify traffic participants on the time line. These labels can then be utilized to evaluate the performance in terms of several metrics for quantifying detection performance, parameter estimation, and classification. In general, it is desirable to have specific correctness information for every object instead of just taking the overall numbers of detected objects for the analysis, which is supported by an exact pairing of detected objects and reference data in the following. Then, parameter sets for the system are optimized with regard to different performance metrics using a central (hyper)parameter optimizer as part of an evaluation framework, which is described in the following. This process can also be deployed to a high-performance cluster (HPC).

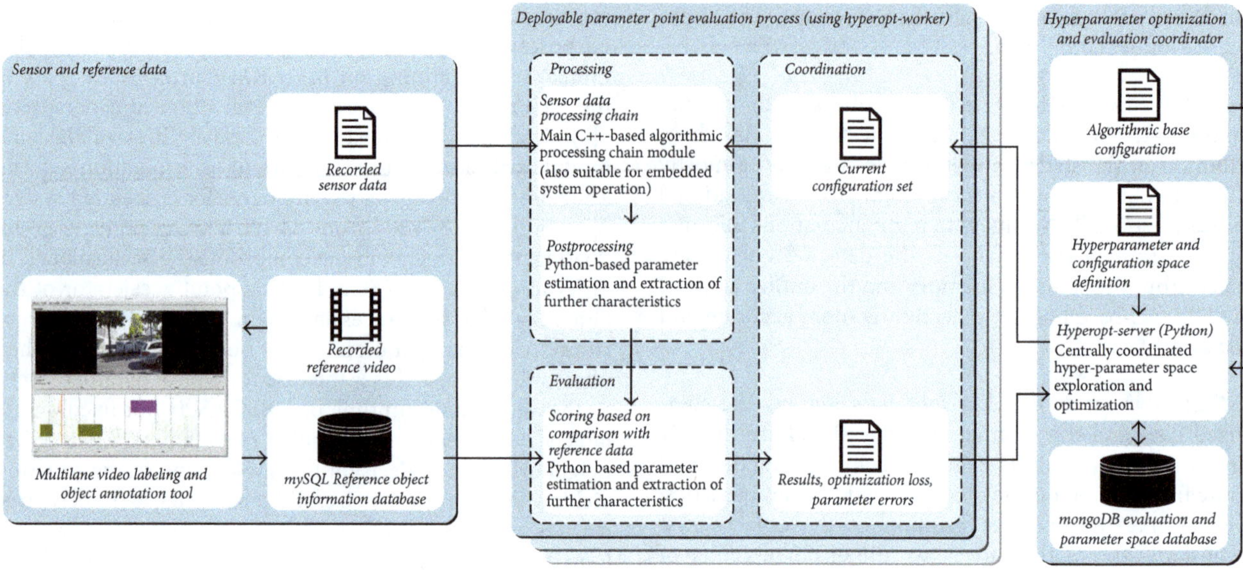

FIGURE 9: Evaluation framework and methodology for parameter space exploration.

5.1. Parameter Space Exploration and Optimization Methodology. In the following, the framework for parameter space exploration of the whole system with its corresponding algorithms is specified. The basic structure is shown in Figure 9.

Parameter Set Evaluation Concept. As a main principle, (hyper)parameter sets are given to a worker shown in the central block in the diagram. Each worker then evaluates a single parameter set (deployed with the *Coordination* subsystem) for real-world recorded sensor data using the C++-based main processing chain and Python-based postprocessing as part of the *Processing* subsystem. The processing principles were described in the implementation description of this article in Section 4.5. Then, in the *Evaluation* subsystem, the objects resulting from the processing stage are taken together with data from an object reference database. This allows the analysis of several performance metrics, such as the detection F_1 score and the correct lane assignment, which yields a performance score or optimization loss as a quality indicator of the parameter set. This is then reported back to the *Coordination* subsystem of the worker.

Reference Database. On the left side of the diagram, the sensor and reference data are shown. Reference object data is stored in a mySQL database. A reference video is recorded in parallel to a sensor data recording. This video is then manually processed using a labeling tool which is also shown in the diagram. The tool allows labeling traffic participant objects on multiple lanes in a timeline view and adding annotations to these objects such as the vehicle type, vehicle size, movement direction, and velocity, which is stored in the database. By analyzing the raw video material itself, the tool also provides the activity level in the video material which supports finding the next vehicle passing by in less frequented time ranges.

Cluster-Label-Pairing for Scoring. Both the resulting clusters from the processing of the raw sensor data and the objects in the reference database are events extended in the time dimension, with a specific start and end time. In the detection performance analysis of the system, each cluster (in case of perfect detection) would be paired with the corresponding reference label object. As the exact cluster-label-pair assignment is not the case in a real scenario, there can be ambiguities or false-positive/false-negative cases. Therefore, a bipartite cluster-label-graph is built up from the cluster and reference data, with edges between the nodes of specific cluster-label candidate pairs. Then, a maximum cardinality bipartite matching is performed. With the resulting cluster-label-pairs and the known sets of all reference and cluster objects, all scorings such as precision, recall, accuracy, and F_1-score can be calculated and reported back for the specific parameter set.

Coordination of (Hyper)parameter Sets. Given the possibility of using a worker process to evaluate parameter sets with the described procedure, the exploration of the (hyper)parameter space needs to be coordinated and optimized. This is achieved by using the hyperopt Python package [37], which allows the definition of several (hyper)parameters with their according distributions and properties and then to automatically explore this space using simulated annealing, random search, and tree of Parzen estimator techniques for optimization. The parameter sets and their according loss results are stored in a MongoDB database. New parameter sets are integrated into a base configuration file which is then assigned to the worker process. In order to speed up the optimization and training procedure, a large number of workers are deployed on an HPC. Typically, good convergence is achieved after 10,000 optimization iterations. If a faster convergence is desired, starting points for the parameters can be calculated with the investigations done in Section 4.3. Typical

TABLE 2: Overview of test scenarios and their characteristics.

Scenario name	Number of lanes	Sensor distance range[1]	Sensor mounting height	Typical speed range	Pulse interval T_{rep}	Description
Urban1	2	7.5 m	3.5 m	20–50 km/h	100 ms	Urban alley street with trees on both sides and parked cars on the adjacent side, mixed use by cars, buses and bicyclists, sensor placement in a horizontal distance of 1 m to curb
Urban2	2	8 m	3.5 m	20–40 km/h	100 ms	Sensor placement behind sidewalk, distance to curb 2 m, mixed use by cars, buses and bicyclists, reflective trash containers on adjacent side
Urban3	2	7 m	5 m	20–50 km/h	100 ms	Typical "urban canyon" with large buildings on both side close to the street and narrow sidewalks, distance to curb 20 cm
TestTrack	-	>15 m	3.5 m	5–100 km/h	50–100 ms	Automotive test track without reflective objects, only asphalted road surface in sensor range, low rate of vehicles passing by with highly different speeds

[1]The distance range is the distance from the ground projection point of the sensor position to the curb side or delimiter of the adjacent street side. For the calculation of the time-of-flight equivalent distance, both this sensor distance range and the mounting height have to be incorporated.

(hyper)parameters, which are part of the optimization process, are as follows:

(i) Dimensions of ε-environment $(\varepsilon_T, \varepsilon_D)$ for the DBStaC algorithm and the core point threshold level MinSum (see Section 4.3.3)

(ii) Properties of the statistics memory, such as the statistics memory size (past information, see Section 4.4.2) and a storage subsampling factor

(iii) Optional use of statistical feedback (see Section 4.4.2) in the scenario for base distribution stabilization

(iv) Optional clipping threshold for calculated statistical norms in standardization to stabilize against outliers

(v) Optional use of matched filtering for the received envelope signal

5.2. Test Scenarios. In Table 2, the different test scenarios are given with their characteristics and description, covering a variety of urban environments and synthetic measurements on an automotive test track. The scenarios are identified by a specific name and will be referenced in the following, when performance results are given.

6. Results and Discussion

Results of the field tests and evaluations are given in this chapter. First, the according performance metrics are introduced and specified, followed by the results for the different scenarios and finally a discussion of the results.

6.1. Evaluation Metrics. In this paper, the main focus lies on the detection of traffic participants as objects and their correct assignment to the lanes of the street. For describing performance in our object detection scenario, no calculation of the true-negative count is possible, as we can have an arbitrary number of objects in our data timeline. Therefore, the widely used F_1 score is chosen as the main performance metric for detection, being the harmonic mean of precision and recall. The precision P is defined as the number of true positives (T_P) divided by all positive outcomes:

$$P = \frac{T_P}{T_P + F_P}. \quad (19)$$

In contrast, the recall R is defined as the number of true positives divided by the sum of true positive and false-negative (F_N) outcomes:

$$P = \frac{T_P}{T_P + F_N}. \quad (20)$$

Then, the F_1 score is defined as follows:

$$F_1 = 2\frac{P \cdot R}{P + R} = \frac{2 \cdot T_P}{2 \cdot T_P + F_N + F_P}. \quad (21)$$

A further advantage is that the combination of precision and recall into the F_1 score allows being independent of requirements biases, which can be a focus on either precision or recall performance, for example, when more false positives (F_P) or negatives are acceptable in a specific scenario. F_1 score calculation only requires the information of true positives, false positives, and false negatives. These numbers are calculated based on the bipartite cluster-label-graph assignment introduced in Section 5.1.

The second important metric is the assignment of objects to the correct lanes for determining the direction of the traffic flow. With multiple lanes as a multiclass problem, the F_1 score cannot be used directly. As an alternative, the weighted F_1 score is used, which first calculates the F_1 score for every lane (correct or incorrect assignment to specific lane) and then calculates a weighted mean of all lane scores, with the number of true reference object instances as the weight.

Table 3: Field test evaluation results for different scenarios.

Scenario name	Object count	F_1 score detection	Weighted F_1 score lane assignment
Urban1	224	0.98	0.91
Urban2	133	0.92	0.95
Urban3	305	0.97	0.98
TestTrack	17	0.94	–

6.2. Discussion of Test Results. The performance evaluation results for the different scenario with optimized parameters are shown in Table 3. Both detection and lane assignment scores are always larger than 0.9, being a very good performance level in real scenarios and satisfying the initially stated high requirements on traffic information quality. The single-sensor ultrasonic traffic participant detection is therefore possible with high performance without exploiting any directional information of the signal. This is facilitated by the algorithms proposed in this paper, which solved several prevailing problems in ultrasonic sensing techniques.

The multilane assignment performance is also very high and of special importance. As no directional or velocity information is available with these sensors, the known information of a fixed lane's driving direction yields the final traffic flow information with direction. A possible problem is showing up in the scenario *Urban1* with the lowest weighted F_1 score for assignment of 0.91: people were driving in the middle of the two-lane street several times, which compromised lane assignment performance.

Further future long-term analyses are required to analyze the long-term stability in scenarios with a highly fluctuating traffic flow and day-and-night cycle. However, with the system structure which relies on the standardized data together with the statistics memory, the sensitivity to long-term sensor drift and also production variations is limited. Further investigation is also required on the impact of the statistical feedback for stabilization of the statistical base information. As the parameter optimization is based on only a few parameters, the susceptibility to overfitting effects is limited in the shown results. Still, for future investigations, also the test performance with regard to future measurements in the same scenario is of high interest.

7. Conclusion and Future Work

In this paper, it was shown that sidefire ultrasonic sensing is a viable option for multilane traffic participant sensing, giving very good performance results in real-world urban scenarios with a single sensor. The functionality is enabled by a novel combination of standardization techniques based on windowed statistics and a modified density-based clustering algorithm, together called DBStaC. Several evaluations were performed, both in real urban environments and for special cases on an automotive test track. The proposed system is integrated into an evaluation and parameter optimization methodology, whose reference system is flexible to be extended with further object characteristics in future.

The ultrasonic sensor system deployed together with the street lamp platform was presented to be a Smart City technology suitable for high-coverage deployment in cities, enabling traffic monitoring and further applications. With this platform, distributed processing and sensor fusion can expand the possibilities of using available traffic participant information even more, for example, for future applications like trajectory prediction. Further future work can be the use of ultrasonic sensor arrays for acquiring directional and velocity information. In the field of algorithmic improvements, the investigation of more advanced hypothesis testing techniques and channel statistics analyses could yield further performance improvements. Also, the comparison of the presented algorithmic concept with state-of-the art supervised machine learning algorithms such as recurrent neural networks is planned.

Conflicts of Interest

The authors declare that they have no conflicts of interest.

References

[1] H. van Essen and A. van Grinsven, "Final report appendix 9: Interaction of GHG policy for transport with congestion and accessibility policies," Project Report, EU Transport GHG 2050, 2012.

[2] "Roadmap to a single european transport area - towards a competitive and resource efficient transport system," White Paper, European Commission, 2011.

[3] H. Mayer, "Air pollution in cities," *Atmospheric Environment*, vol. 33, no. 24-25, pp. 4029–4037, 1999.

[4] R. Arnott, A. de Palma, and R. Lindsey, "Does providing information to drivers reduce traffic congestion?" *Transportation Research Part A: General*, vol. 25, no. 5, pp. 309–318, 1991.

[5] A. Zanella, N. Bui, A. P. Castellani, L. Vangelista, and M. Zorzi, "Internet of things for smart cities," *IEEE Internet of Things Journal*, vol. 1, no. 1, pp. 22–32, 2014.

[6] N. Buch, S. A. Velastin, and J. Orwell, "A review of computer vision techniques for the analysis of urban traffic," *IEEE Transactions on Intelligent Transportation Systems*, vol. 12, no. 3, pp. 920–939, 2011.

[7] S. Himmel, M. Ziefle, and K. Arning, *From Living Space to Urban Quarter: Acceptance of ICT Monitoring Solutions in an Ageing Society*, Springer, Berlin, Germany, 2013.

[8] W. Xue, L. Wang, and D. Wang, "A prototype integrated monitoring system for pavement and traffic based on an embedded sensing network," *IEEE Transactions on Intelligent Transportation Systems*, vol. 16, no. 3, pp. 1380–1390, 2015.

[9] J. Gajda, R. Sroka, M. Stencel, A. Wajda, and T. Zeglen, "A vehicle classification based on inductive loop detectors," in *Proceedings of the 18th IEEE Instrumentation and Measurement Technology Conference. Rediscovering Measurement in the Age of Informatics*, pp. 460–464, May 2001.

[10] L. Klein, D. Gibson, and M. Mills, *Traffic Detector Handbook*, vol. 1, no. FHWA-HRT-06-108, Federal Highway Administration, Turner-Fairbank Highway Research Center, 3rd edition, 2006.

[11] M. Hallenbeck and H. Weinblatt, *Equipment for Collecting Traffic Load Data*, Transportation Research Board, NCHRP Report 509, 2004.

[12] N. Garber and L. Hoel, *Traffic and Highway Engineering, SI Edition*, Cengage Learning, 2014.

[13] R. Möbus and U. Kolbe, "Multi-target multi-object tracking, sensor fusion of radar and infrared," in *Proceedings of the IEEE Intelligent Vehicles Symposium*, pp. 732–737, IEEE, June 2004.

[14] I. Urazghildiiev, R. Ragnarsson, P. Ridderström et al., "Vehicle classification based on the radar measurement of height profiles," *IEEE Transactions on Intelligent Transportation Systems*, vol. 8, no. 2, pp. 245–253, 2007.

[15] I. Urazghildiiev, R. Ragnarsson, K. Wallin, A. Rydberg, P. Ridderstrom, and E. Ojefors, "A vehicle classification system based on microwave radar measurement of height profiles," in *Proceedings of the 2002 International Radar Conference*, pp. 409–413, Edinburgh, UK, October 2002.

[16] J. Tao, S. Chen, L. Yang, and Y. Hu, "Ultrasonic technique based on neural networks in vehicle modulation recognition," in *Proceedings of the The 7th International IEEE Conference on Intelligent Transportation Systems*, pp. 201–204, October 2004.

[17] E. U. Warriach and C. Claudel, "Poster abstract: A machine learning approach for vehicle classification using passive infrared and ultrasonic sensors," in *Proceedings of the 12th International Conference on Information Processing in Sensor Networks, IPSN 2013 - Part of CPSWeek 2013*, pp. 333-334, April 2013.

[18] T. Matsuo, Y. Kaneko, and M. Matano, "Introduction of intelligent vehicle detection sensors," in *Proceedings of the 1999 Intelligent Transportation Systems Conference*, pp. 709–713, Tokyo, Japan.

[19] H. Kim, J.-H. Lee, S.-W. Kim, J.-I. Ko, and D. Cho, "Ultrasonic Vehicle Detector for Side-Fire Implementation and Extensive Results Including Harsh Conditions," *IEEE Transactions on Intelligent Transportation Systems*, vol. 2, no. 3, pp. 127–134, 2001.

[20] V. Agarwal, N. V. Murali, and C. Chandramouli, "A cost-effective ultrasonic sensor-based driver-assistance system for congested traffic conditions," *IEEE Transactions on Intelligent Transportation Systems*, vol. 10, no. 3, pp. 486–498, 2009.

[21] C. Heipke, "Crowdsourcing geospatial data," *ISPRS Journal of Photogrammetry and Remote Sensing*, vol. 65, no. 6, pp. 550–557, 2010.

[22] S. Hind and A. Gekker, "Outsmarting traffic, together: Driving as social navigation," *Exchanges: the Warwick Research Journal*, vol. 1, no. 2, pp. 165–180, 2014.

[23] M. B. Sinai, N. Partush, S. Yadid, and E. Yahav, "Exploiting social navigation," https://arxiv.org/abs/1410.0151v1.

[24] R. Bauza, J. Gozalvez, and J. Sanchez-Soriano, "Road traffic congestion detection through cooperative Vehicle-to-Vehicle communications," in *Proceedings of the 35th Annual IEEE Conference on Local Computer Networks, LCN 2010*, pp. 606–612, October 2010.

[25] S. Jeon, E. Kwon, and I. Jung, "Traffic measurement on multiple drive lanes with wireless ultrasonic sensors," *Sensors*, vol. 14, no. 12, pp. 22891–22906, 2014.

[26] Y. Jo and I. Jung, "Analysis of vehicle detection with WSN-based ultrasonic sensors," *Sensors*, vol. 14, no. 8, pp. 14050–14069, 2014.

[27] H. Kuttruff, *Room Acoustics*, Spon Press, New York, NY, USA, 5th edition, 2009.

[28] A. V. Oppenheim, R. W. Schafer, and J. R. Buck, *Discrete-Time Signal Processing*, Prentice-Hall, Inc., Upper Saddle River, NJ, USA, 2nd edition, 1999.

[29] F. J. Harris, "On the use of windows for harmonic analysis with the discrete Fourier transform," *Proceedings of the IEEE*, vol. 66, no. 1, pp. 51–83, 1978.

[30] B. Gold, A. V. Oppenheim, and C. M. Rader, "Theory and implementation of the discrete Hilbert transform," *Symposium on Computer Processing in Communications*, vol. 19, 1970.

[31] M. Meissner, "Accuracy issues of discrete hubert transform in identification of instantaneous parameters of vibration signals," *Acta Physica Polonica A*, vol. 121, no. 1A, pp. A164–A167, 2012.

[32] L. Xu and T. Xu, *Digital Underwater Acoustic Communications*, Elsevier Science, 2016.

[33] S. C. Kumbhakar and C. A. K. Lovell, *Stochastic Frontier Analysis*, Cambridge University Press, Cambridge, UK, 2003.

[34] M. Ester, H.-P. Kriegel, J. Sander, and X. Xu, "A density-based algorithm for discovering clusters a density-based algorithm for discovering clusters in large spatial databases with noise," in *Proceedings of the Second International Conference on Knowledge Discovery and Data Mining, ser. KDD'96*, pp. 226–231, AAAI Press, 1996.

[35] J. Gan and Y. Tao, "DBSCAN revisited: Mis-claim, un-fixability, and approximation," in *Proceedings of the 2015 ACM SIGMOD International Conference on Management of Data, ser. SIGMOD'15*, pp. 519–530, ACM, New York, NY, USA, 2015.

[36] E. Schubert, J. Sander, M. Ester, H. P. Kriegel, and X. Xu, "DBSCAN Revisited, Revisited," *ACM Transactions on Database Systems (TODS)*, vol. 42, no. 3, pp. 1–21, 2017.

[37] J. Bergstra, D. Yamins, and D. D. Cox, "Hyperopt: A Python library for optimizing the hyperparameters of machine learning algorithms," in *Proceedings of the 12th Python in Science Conference*, S. van der Walt, J. Millman, and K. Huff, Eds., 2013.

10

An Improved Deep Learning Model for Traffic Crash Prediction

Chunjiao Dong, [1,2,3] **Chunfu Shao**, [1,2] **Juan Li**, [1] **and Zhihua Xiong** [1]

[1] *MOE Key Laboratory for Urban Transportation Complex Systems Theory and Technology, Beijing Jiaotong University, Beijing 100044, China*
[2] *Key Laboratory of Industry of Big Data Application Technology for Comprehensive Transport, Ministry of Transport, Beijing Jiaotong University, Beijing 100044, China*
[3] *Center for Transportation Research, Tickle College of Engineering, University of Tennessee, 600 Henley Street, Knoxville, 37996, Tennessee, USA*

Correspondence should be addressed to Chunjiao Dong; cjdong@bjtu.edu.cn

Guest Editor: Hamzeh Khazaei

Machine-learning technology powers many aspects of modern society. Compared to the conventional machine learning techniques that were limited in processing natural data in the raw form, deep learning allows computational models to learn representations of data with multiple levels of abstraction. In this study, an improved deep learning model is proposed to explore the complex interactions among roadways, traffic, environmental elements, and traffic crashes. The proposed model includes two modules, an unsupervised feature learning module to identify functional network between the explanatory variables and the feature representations and a supervised fine tuning module to perform traffic crash prediction. To address the unobserved heterogeneity issues in the traffic crash prediction, a multivariate negative binomial (MVNB) model is embedding into the supervised fine tuning module as a regression layer. The proposed model was applied to the dataset that was collected from Knox County in Tennessee to validate the performances. The results indicate that the feature learning module identifies relational information between the explanatory variables and the feature representations, which reduces the dimensionality of the input and preserves the original information. The proposed model that includes the MVNB regression layer in the supervised fine tuning module can better account for differential distribution patterns in traffic crashes across injury severities and provides superior traffic crash predictions. The findings suggest that the proposed model is a superior alternative for traffic crash predictions and the average accuracy of the prediction that was measured by RMSD can be improved by 84.58% and 158.27% compared to the deep learning model without the regression layer and the SVM model, respectively.

1. Introduction

Road traffic injuries are a leading cause of preventable death, especially, for the young people. In the United States, traffic crashes were the number one cause of death among people from 16 to 24 years old for each year from 2012 to 2014 [1]. In 2015, the nation lost 35,092 people in traffic crashes, a 7.2-percent increase from 32,744 in 2014, which is the largest percentage increase in nearly 50 years [2]. This is an average of approximately 96 people being killed on the nation's roadways every day of the year, and an average of more than four people per hour. In other words, one person dies on roadways every 15 minutes.

To understand the relationship between the influence factors and traffic crash outcomes, with the extracted data from police reports and state highway-asset-management databases, the analyses of traffic safety estimate and predicate the likelihood of a traffic crash. The number of crashes occurring on a defined spatial entity over a specific time period (for example, the number of crashes per year occurring at a roadway intersection, over a specified roadway segment, or in a region) would be considered as the dependent variables and some of the many factors affecting the likelihood of a traffic crash are analyzed and examined (see [3–5] for a comprehensive review). Though more and more factors that are relevant to the traffic crashes have been incorporated and the proposed models became more and more sophisticated, there are still some factors that are not available to the researchers and the models result in bias estimations and erroneous predictions. In this study, we proposed an innovative approach for traffic crash prediction, which incorporates a multivariate regression layer into a dynamic deep learning model that contains an unsupervised feature learning module and a supervised fine tuning module governing the state dynamics to improve the performances of prediction.

2. Literature Review

The statistical methodologies, such as the Poisson, negative binomial (NB), and their variants in univariate and multivariate regression frameworks, have been successfully applied in crash count analyses [4, 6–8], which attempt to deal with the data and methodological issues associated with traffic crash estimations and predictions, and enhance our understanding on the relationship between the influence factors and traffic crash outcomes. However, current research in traffic safety indicates that the applied statistical modeling fails when dealing with complex and highly nonlinear data [9], which could suggest that the relationship between the influence factors and traffic crash outcomes is more complicated than can be captured by a single statistical approach. In addition, most of the statistical methods are based on some strong assumptions, such as specifying a priori and the error distribution. Moreover, a problematic issue is multicollinearity, i.e., the high degree of correlation between two or more independent variables. Furthermore, statistical models have difficulty when dealing with outliers, missing or noisy data [10].

To deal with the limitations of statistical methodologies, the machine learning methods, including Artificial Neural Network (ANN), Support Vector Machine (SVM) models, and deep learning models, have been applied to various traffic safety problems and used as data analytic methods because of their ability to work with massive amounts of multidimensional data. In addition, because of the modeling flexibility, learning and generalization ability, and good predictive ability, the machine learning has been considered as generic, accurate, and convenient mathematical models in the field of traffic safety.

Because the commonly used Poisson or NB regression models assume the predefined underlying relationship between dependent and independent variables and the violation of the assumption would lead to erroneous estimation, ANN and Bayesian neural network (BNN) models have been employed to analyze the traffic safety problems for many years. Although both ANN and BNN models have similar multilevel network structures, they are different in predicting the outcome variables. For ANN, the weights are assumed to fix. However, the weights of BNN follow a probability distribution and the prediction needs to be integrated over all the probability weights. Basically, the ANN can be characterized by three features: network architecture, model of a neuron, and learning algorithms. Chang [11] compared the performances of NB regression model and ANN in crash frequency analyses. The results showed that ANN is a consistent alternative method for analyzing crash frequency. Abdelwahab and Abdel-Aty [12] employed two well-known ANN paradigms [the multilayer perceptron and radial basis functions (RBF) neural networks] to analyze the traffic safety of toll plazas and evaluate the impacts of electronic toll collection (ETC) systems on highway safety. The performance of ANN was compared with calibrated logit models. Modeling results showed that the RBF neural network was the best model for analyzing driver injury severity. Xie, Lord, and Zhang [13] evaluated the application of BNN models for predicting traffic crashes by using data collected on rural frontage roads in Texas. The results showed that back-propagation neural network (BPNN) and BNN models perform better than the NB regression model in terms of traffic crash prediction. The results also showed that BNNs could be used to address other issues in highway safety, such as the development of crash modification factors, and enhance the prediction capabilities for evaluating different highway design alternatives. Kunt, Aghayan, and Noii [14] employed a genetic algorithm (GA), pattern search, and ANN models to predict the severity of freeway traffic crashes. The results showed that the ANN provided the best predictions. Jadaan, Al-Fayyad, and Gammoh [15] developed a traffic crash prediction model using the ANN simulation with the purpose of identifying its suitability for predicting traffic crashes under Jordanian conditions. The results demonstrated that the estimated traffic crashes are close to actual traffic crashes. Akin and Akbas [16] proposed an ANN model to predict intersection crashes in Macomb County of the State of Michigan. The predictive capability of the ANN model was determined by classifying the crashes into these types: fatal, injury, and property damage only (PDO) crashes. The results were very promising and showed that ANN model is capable of providing an accurate prediction (90.9%) of the crash types. In summary, though ANN and BNN models show better linear/nonlinear approximation properties than traditional statistical approaches, these models often cannot be generalized to other data sets [3].

The SVM models have recently been introduced for traffic safety analyses [17, 18], which are a new class of models that are based on statistical learning theory and structural risk minimization [19]. These models are supposed to approximate any multivariate function to any desired degree of accuracy with a set of related supervised learning methods. Li et al. [17] evaluated the application of SVM models for predicting motor vehicle crashes. The results showed that SVM models predict crash data more effectively and accurately than traditional NB models. In addition, the findings indicated that the SVM models provide better (or at least comparable) performance than BPNN models and do not over-fit the data. To identify the relationship between severe crashes and the explanatory variables and enhance model goodness-of-fit, Yu and Abdel-Aty [20] developed three models to analyze crash injury severity, which include a fixed parameter logit model, a SVM model, and a random parameter logit model. The results showed that the SVM models and the random parameter models provide superior model fits compared to the fixed parameter logit model. Findings also demonstrate that it is important to consider possible nonlinearity and individual heterogeneity when analyzing traffic crashes. Chen et al. [21] employed the SVM models to investigate driver injury severity patterns in rollover crashes using two-year crash data collected in New Mexico. The results showed that the SVM models produce reasonable predictions and the polynomial kernel outperforms the Gaussian RBF kernel. Dong, Huang, and Zheng [22] proposed a SVM model to handle multidimensional spatial data in crash prediction. The results showed that the SVM models outperform the nonspatial models in terms of model fitting and predictive performance. In addition, the SVM models provide better goodness-of-fit compared with Bayesian spatial model with conditional autoregressive prior when utilizing the whole dataset as the samples. Ren and Zhou [23] proposed a novel approach that combines particle swarm optimization (PSO) and SVM for traffic safety prediction. The results showed that the predictions of PSO-SVM are better than that from BP neural network. Yu and Abdel-Aty [24] proposed the SVM models with different kernel functions to evaluate real-time crash risk. The results showed that the SVM model with RBF kernel outperformed the SVM model with linear kernel and Bayesian logistic regression model. In addition, the findings showed that smaller sample size could improve the classification accuracy of the SVM models and variable selection procedure is needed prior to the SVM model development. Overall, it has been found that the SVM models showed better or comparable results to the outcomes predicted by ANN/BNN and other statistical models [19]. However, like ANN and BNN, the SVM models often cannot be generalized to other data sets and they all tend to behave as black-boxes, which cannot provide the interpretable parameters as statistical models do.

Other than the ANN/BNN and SVM models, other machine learning methods have been introduced in traffic safety analyses. Abdel-Aty and Haleem [25] introduced a recently developed machine learning technique—multivariate adaptive regression splines (MARS) to predict vehicle angle crashes using extensive data collected on unsignalized intersections in Florida. The results showed that MARS outperformed the NB models. The proposed MARS models showed promising results after screening the covariates using random forest. The findings suggested that MARS is an efficient technique for predicting crashes at unsignalized intersections.

Deep learning is a recently developed branch of machine learning method and has been successfully applied in speech recognition, visual object recognition, object detection, and many other domains such as drug discovery and genomics [26, 27]. Compared to the conventional machine learning techniques that were limited in their ability to process natural data in their raw form, deep learning constructs computational models aiming to extract inherent features in data from the lowest level to the highest level. Though the deep learning methods have shown outstanding performances in many applications [26], the applications of deep learning in the field of transportation are relatively few and only focusing on the topic of traffic flow prediction [28–30]. In this study, we proposed an improved deep learning model for traffic crash prediction. The proposed model includes two modules: an unsupervised feature learning module and a supervised fine tuning module. To discover nonlinear relationship between the investigated variables and identify the impacts of influence factors on traffic crashes for roadway network, a DAE model is proposed in the unsupervised feature learning module to learn the features of explanatory variables. In addition, a multivariate negative binomial (MVNB) regression is embedding into the supervised fine tuning module to address the heterogeneity issues. The proposed model performances are evaluated by comparing to the deep learning model without the MVNB layer and SVM models by using five-year data from Knox County in Tennessee.

3. The Modeling Framework Formulation

A novel model is proposed for the traffic crash prediction and Figure 1 illustrates the modeling framework. The proposed model includes two modules. One is the unsupervised feature learning module and another is the supervised fine tuning module. The obtained encoded feature representations from the unsupervised feature module are used as the input for the supervised fine tuning module.

3.1. Unsupervised Feature Learning Module. Compared to the commonly used deep learning architectures including deep belief network [31], stacked autoencoder [32], and convolutional neural networks [27], the symmetrical neural networks in an unsupervised manner have shown better performances, which can automatically learn an appropriate sparse feature representation from the raw data [33]. The unsupervised feature learning module includes a denoising autoencoder (DAE) model to learn the underlying structure of the dynamic pattern among the characteristics of roadway, traffic, and environment. With the explanatory variables, such as roadway geometric design features, traffic factors, pavement factors, and environmental characteristics as the input, the designed DAE model can identify the nonlinear relationship between the investigated variables in an unsupervised and hierarchical manner and the robust feature representations can be obtained. In addition, the designed DAE model can encode the explanatory variables into an embedding low-dimensional space. The proposed DAE model contains a visible layer, a hidden layer, an output layer, and a reconstruction layer. Unlike the conventional DAE model with K hidden layers [34], the proposed model uses a reconstruction error optimizing the output layer and the noisy input layer to generate higher level representations.

Assume $\Phi = \{[\mathbf{v}_i, \mathbf{u}_i]\}_{i=1}^{n}$ is a training set that contains n roadway entities, where $\mathbf{v}_i \in \mathfrak{R}^D$ is the explanatory variable vector with dimension D and $\mathbf{u}_i$ is the multivariate traffic safety outcomes for roadway entity i. Given the training set, the proposed DAE model is trained to develop a robust feature representation by reconstructing the

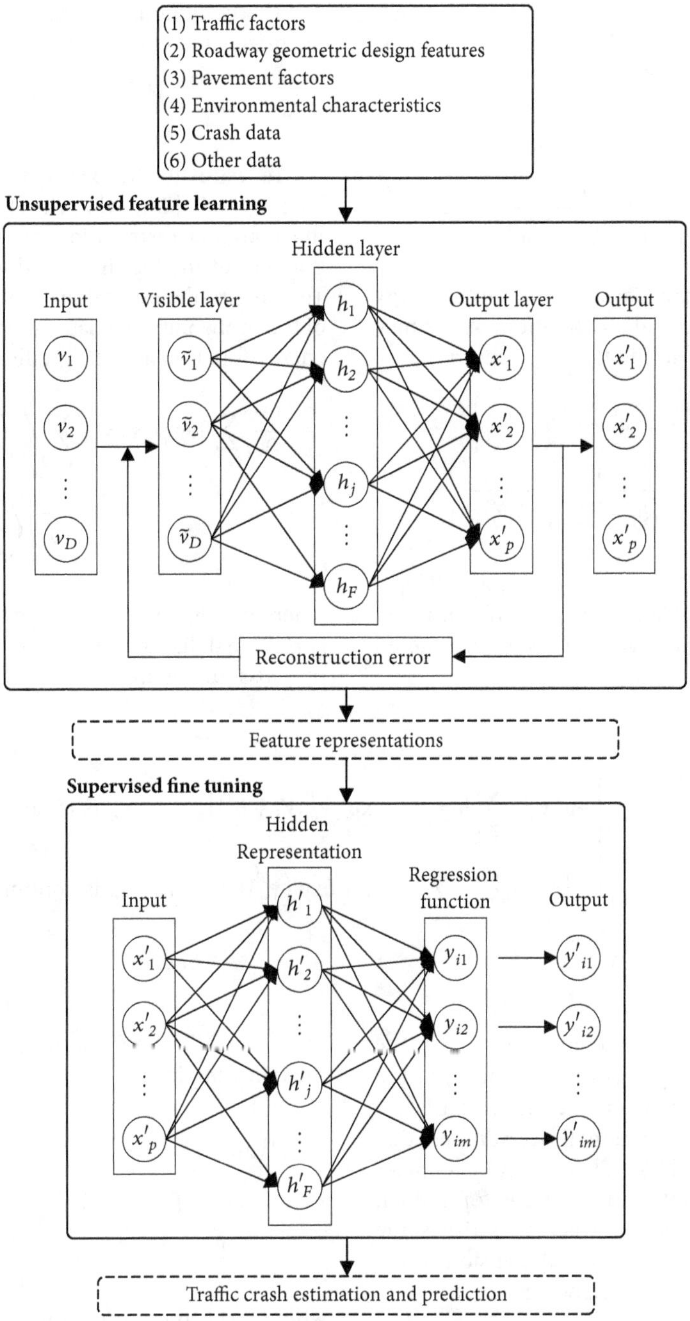

FIGURE 1: Flowchart of the proposed model for traffic crash prediction.

input $\mathbf{v}_i$ from its noisy corrupted version $\tilde{v}_i$, as shown in Figure 1.

There are D units in the visible layer and F units in the hidden layer and the proposed model can be defined by a parameter set $\Theta = \{\mathbf{W}, \mathbf{a}, \mathbf{b}, \mathbf{c}\}$, where $\mathbf{W} = [W_{ij}] \in \mathfrak{R}^{D\times F}$ is the interlayer connection weights, $\mathbf{a} = [a_i] \in \mathfrak{R}^{D}$ is the visible self-interactions or biases, $\mathbf{b} = [b_j] \in \mathfrak{R}^{F}$ is the hidden biases, and $\mathbf{c} = [c_k] \in \mathfrak{R}^{K}$ is the reconstruction error. The joint probability distribution between the noisy input variables and hidden variables is defined as

$$P(\mathbf{v}, \mathbf{h}; \Theta) = \frac{1}{Z(\Theta)} \exp\left[-\varepsilon(\mathbf{v}, \mathbf{h}; \Theta)\right] \quad (1)$$

where $\varepsilon(\mathbf{v}, \mathbf{h}; \Theta)$ is an energy function defined by symmetric interactions between the noisy input variable, hidden variables, and a set of interaction parameters Θ; $Z(\Theta)$ is a normalized factor.

For a binary variable, a Bernoulli-Bernoulli energy function [34, 35] and the conditional distribution of a single stochastic hidden variable are given by

$$\varepsilon(\mathbf{v},\mathbf{h};\Theta) = \mathbf{v}^T\mathbf{W}\mathbf{h} - \mathbf{a}^T\mathbf{v} - \mathbf{b}^T\mathbf{h}$$
$$P\left(h_j = 1 \mid \mathbf{v};\Theta\right) = \frac{1}{1+\exp\left[-\left(b_j + \sum_{i=1}^{D} W_{ij}v_i\right)\right]} \tag{2}$$

For a continuous variable, a Gaussian-Bernoulli energy function [31] and the conditional distribution of a single stochastic hidden variable are given by

$$\varepsilon(\mathbf{v},\mathbf{h};\Theta) = -\mathbf{v}'^T\mathbf{W}\mathbf{h} + \mathbf{a}'^T\mathbf{a}' - \mathbf{b}^T\mathbf{h}$$
$$P\left(h_j = 1 \mid \mathbf{v};\Theta\right) = \frac{1}{1+\exp\left[-\left(b_j + \sum_{i=1}^{D} \left(W_{ij}v_i/\sigma_i\right)\right)\right]} \tag{3}$$

where $\mathbf{v}' = (v_1/\sigma_1, v_2/\sigma_2, \cdots, v_i/\sigma_i, \cdots v_D/\sigma_D)^T$, $\mathbf{a}' = ((v_1 - a_1)/\sigma_1, (v_2-a_2)/\sigma_2, \cdots, (v_i-a_i)/\sigma_i, \cdots, (v_D-a_D)/\sigma_D)$, and σ_i is the standard deviation of the i-th visible variable v_i centered on the bias a_i.

The conditional distribution over hidden units can be factorized and computed by

$$P(\mathbf{h} \mid \mathbf{v};\Theta) = \prod_{j=1}^{F} P\left(h_j \mid \mathbf{v};\Theta\right) \tag{4}$$

To estimate the parameters, the method proposed by Hjelm et al. [35], which maximizes the log-likelihood of the marginal distribution of the hidden units to find the gradient of the log-likelihood, is employed in this research. To simplify the estimation process, the free energy in terms of the probability at a data point $\mathbf{v}_n$ can be used to replace the energy function and the gradient has the following form:

$$\frac{\partial}{\partial\Theta}\sum_{\mathbf{v}\in D}\log p(\mathbf{v}) = -\sum_{\mathbf{v}\in D}\left\langle \frac{\partial}{\partial\Theta}\varphi(\mathbf{v},\mathbf{h};\Theta)\right\rangle_{p(\mathbf{h}|\mathbf{v};\Theta)} + \sum_{\mathbf{v}\in D}\left\langle \frac{\partial}{\partial\Theta}\varphi(\mathbf{v},\mathbf{h};\Theta)\right\rangle_{p(\mathbf{v},\mathbf{h}|\Theta)} \tag{5}$$

where $\varphi(\mathbf{v},\mathbf{h};\Theta) = -\log\sum_h \exp[-\varepsilon(\mathbf{v} = \mathbf{v}_n, \mathbf{h};\Theta)]$, and the conditional distribution over hidden units should be replaced by a loss function.

$$\varphi(\mathbf{h} \mid \mathbf{v}_n) = \begin{cases} -\mathbf{a}^T\mathbf{v}_n - \sum_{j=1}^{F}\log\left[1+\exp\left(\mathbf{v}_n^T W_j + b_j\right)\right] & \mathbf{v}_n \text{ is binary} \\ -\left\|\mathbf{a}-\mathbf{v}_n\right\|^2 - \sum_{j=1}^{F}\log\left[1+\exp\left(\mathbf{v}_n^T W_j + b_j\right)\right] & \mathbf{v}_n \text{ is continuous} \end{cases} \tag{6}$$

Considering the hidden layer as the input layer for the outcome layer and outcome layer as the hidden layer, the joint probability distribution between the hidden variables and outcome variables, energy function, the conditional distributions of the outcome variables and units, and parameter estimation process can be obtained as those for hidden layer. The model will stop training when c_k satisfies the reconstruction error requirements or the dimension of the feature representation achieves the designed goals.

3.2. Supervised Fine Tuning Module. The supervised fine tuning module is a supervised fine tuning procedure that includes a regression layer on the top of the resulting hidden representation layers to estimate the likelihood of the crash occurrences, as shown in Figure 1. The obtained encoded feature representations from the unsupervised feature module are used as the input for the supervised fine tuning module. To jointly estimate the occurrence likelihood for more than one type of crashes simultaneously and address the potential heterogeneity issues in the interdependent crash data, a multivariate negative binomial (MVNB) model is used in the supervised fine tuning module to estimate and predict the traffic crashes across injury severities. Assume $\mathbf{y}_i=(y_{i1}, y_{i2},\ldots, y_{im})'$ is a vector of crash occurrence likelihood for roadway entities i, which includes m types of crashes. The particular NB regression model employed in this study has the following form:

$$p\left(y_{ij}\right) = \frac{\Gamma\left(\lambda_{ij}/\sigma + y_{ij}\right)}{\Gamma\left(\lambda_{ij}/\sigma\right)\Gamma\left(y_{ij}+1\right)}\left(\frac{1}{1+\sigma}\right)^{\lambda_{ij}/\sigma}\left(\frac{\sigma}{1+\sigma}\right)^{y_{ij}} \tag{7}$$

where $\Gamma(\cdot)$ is the gamma function, y_{ij} is the crash number of crash type j for roadway segment i and $E[y_{ij}]=\lambda_{ij}=\exp(\boldsymbol{\beta}_j\mathbf{X}_{ij}+\varepsilon_{ij})$. $\exp(\varepsilon ij)$ is a multivariate gamma-distributed error term with mean 1 and variance α^{-1}.

As described in Shi and Valdez [36] and Anastasopoulos et al. [37], with the expected number of crashes λ_{ij}, the MVNB model has a joint probability function:

$$p\left(y_{1j}, y_{2j}, \cdots, y_{ij}\right) = \sum_{k=0}^{\min(y_{1j},y_{2j},\cdots,y_{ij})} p(k)\prod_{i=1}^{N} p\left(y_{ij}-k\right) \tag{8}$$

The model parameters can be estimated by maximizing the log-likelihood function:

$$L = \sum_{i,j} \left\{ \begin{array}{c} \log\left[\Gamma\left(\sigma^{-1} + y_{ij}\right)\right] - \log\left[\Gamma\left(\sigma^{-1}\right)\right] - \log\left(y_{ij}!\right) \\ -\sigma^{-1}\ln\left[1 + \sigma\exp\left(\boldsymbol{\beta}_j \mathbf{X}_{ij}\right)\right] - y_{ij}\log\left[1 + \sigma^{-1}\exp\left(\boldsymbol{\beta}_j \mathbf{X}_{ij}\right)\right] \end{array} \right\} \tag{9}$$

The MVNB regression layer is added on the top of the resulting hidden representation layers to perform traffic crash prediction. This yields a deep learning model tailored to a task-specific supervised learning. Then we fine tune the module 2 using backpropagation by minimizing the following cost function:

$$f(\boldsymbol{\theta}) = -\frac{1}{n}\sum_{i=1}^{n}\sum_{j=1}^{m} s\left(y_{ij}' = y_{ij}\right)\log\left[\frac{\exp\left(h_\theta\left(x_i\right)\right)}{\sum_{l=1}^{F}\exp\left(h_\theta\left(x_i\right)\right)}\right] + \frac{\alpha}{2p}\left[\left\|W_{regression}\right\|_F^2 + \sum_{l=1}^{F}\left\|W_l\right\|_F^2\right] \tag{10}$$

where s(·) is an indicator function, if $y_{ij}'=y_{ij}$, then s(·)=1; otherwise, s(·)=0, α is a regularization parameter, and $\|\|_F$ is the Frobenius norm. The first term refers to the cross entropy loss for the regression layer, the second term is the weight decay penalty, and $h_\theta(x_i)$ is the output of deep learning for an input x_i'.

The cost function is minimized with a min-batch gradient descent algorithm [27]. The parameters in module 2 $\boldsymbol{\theta} = \{\mathbf{W}_l, \mathbf{b}, \mathbf{W}_{regression}\}$ are estimated by initializing the weights $\mathbf{W}_{regression}$ of the regression layer to small random values and the weights of the F hidden layers are initialized by the encoding weights obtained in the unsupervised feature learning module.

4. Data

The data are obtained from the Tennessee Roadway Information Management System (TRIMS) and the Pavement Management System (PMS), which are maintained by the Tennessee Department of Transportation (TDOT). The dataset includes crash data, traffic factors, geometric design features, pavement factors, and environmental characteristics. The traffic, geometric, pavement, and environmental characteristics are linked to the crash data through the common variable *id_number*. An extensive and comprehensive data screening that includes cleaning, consistency, and accuracy checks is processed and performed to ensure the data are useable, reliable, and valid for the analyses. After the initial data screening, in total 635 roadway segments in Knox County are chosen for the analyses. For the selected roadway segments, each of them has a completed dataset that links to the crash data. In other words, the dataset contains detailed information on traffic factors, geometric design features, pavement characteristics, and environment factors.

In TRIMS, the crash data have been classified into five categories, fatal crashes, incapacitating injury crashes, nonincapacitating injury crashes, possible-injury crashes, and PDO crashes. Because the category of fatal crashes has only a few observations, the categories of fatal crashes and incapacitating injury crashes have been combined and referred to as major injury crashes. The possible-injury crashes and PDO crashes have been combined and referred to as no-injury crashes. The nonincapacitating injury crashes are referred to as minor injury crashes. A few of pervious literature [38–40] has used a similar classification for injury outcomes. For those selected 635 roadway segments, from 2010 to 2014, a total of 5365 traffic crashes were reported by the police officers, which include 135 (2.51%) major injury crashes, 1312 (24.46%) minor injury crashes, and 3917 (73.02%) PDO crashes. Individual roadway segment experienced from 0 to 23 crashes per year with a mean of 1.54 and a standard deviation of 1.89. As expected, a significant amount of zeros is observed. The dependent variables and their descriptive statistics are shown in Table 1. The descriptive statistics of continuous independent variables and categorical independent variables are shown in Tables 1 and 2, respectively.

The considered traffic factors include the logarithm of annual average daily traffic (AADT) per lane, truck traffic percentage, and posted speed limits. Roadway segment AADT per lane from 2010 to 2014 varies from 851 to 32,359 vehicles with a mean of 3,388.44 and a standard deviation of 5495.41. Other than the TRIMS dataset maintained by the DDOT, traffic flow information can be obtained from https://www.tdot.tn.gov/APPLICATIONS/traffichistory, which is an AADT map providing traffic volumes based on a 24-hour, two-directional count at a given location. The website also provides the traffic history of any specific count station. The variable of posted speed limit has a mean of 38.65 and a standard deviation of 6.69 with a minimum value of 30 and a maximum value of 70. The truck traffic percentage varies from 1 to 33 with a mean of 6.71 and a standard deviation of 4.98.

Important measurements of geometric design features considered in this study include segment length, degree of horizontal curvature, median widths, outsider shoulder widths, number of through lanes, lane widths, number of left-turn lanes, median types, and shoulder type. Among them, the segment length, degree of horizontal curvature, median widths, and outsider shoulder widths are considered as the continuous variables and the others are considered as the categorical variables. Other than the traffic factors and geometric design features, the impacts of pavement surface characteristics are considered to better address traffic safety issues for roadway design and maintenance. The considered pavement surface characteristics include international roughness index (IRI) and rut depth (RD). The analyzed IRI varies from 25.45 to 182.58 with a mean of 65.85 and a standard deviation of 27.75, which is calculated

TABLE 1: Summary statistics of analyzed continuous variables.

Variable	Mean	Std. Dev.	Min.	Max.
Independent variable				
The number of major injury crashes per year per roadway segment	0.04	0.32	0	3
The number of minor injury crashes per year per roadway segment	0.41	0.54	0	9
The number of no injury crashes per year per roadway segment	1.23	1.74	0	22
Traffic factors				
The logarithm of AADT per lane	3.53	3.74	2.93	4.51
Truck traffic percentage	6.71	4.98	1	33
Posted speed limits	38.65	6.69	30	70
Geometric design features				
Segment length (miles)	0.81	1.03	0.02	12.31
Degree of horizontal curvature	1.51	3.23	0	14.00
Median widths	1.12	2.02	0	12
Outside shoulder widths	3.06	1.88	3.52	8
Pavement factors				
International roughness index	65.85	27.75	25.45	182.58
Rut depth (in.)	0.13	0.05	0.06	0.41

TABLE 2: Summary statistics of analyzed categorical variables.

Variable	Category	Frequency	Percent
Geometric design features			
Number of through lanes	3	585	18.43
	2	1660	52.28
	1	930	29.29
Lane widths (ft)	12	915	28.82
	11	1650	51.97
	10	610	19.21
Number of left-turn lanes	2	920	28.98
	1	650	20.47
	0	1605	50.55
Median type	1 for non-traversable median	1185	37.32
	0 for traversable median	1990	62.68
Shoulder type	2 for pavement	760	23.94
	1 for gravel	1595	50.24
	0 for dirt	820	25.83
Environmental factors			
Terrain type	1 for mountainous	1206	37.97
	0 for rolling	1969	62.03
Land use type	2 for residential	1622	51.10
	1 for commercial	775	24.41
	0 for rural	778	24.49
Indicator for lighting	1 for lighting exists on the roadway segments	1304	41.06
	0 for others	1871	58.94

using a quarter car vehicle math model and the response is accumulated to yield a roughness index with units of slope (in/mi). Another pavement condition indicator is the RD, which is measured at roadway speeds with a laser/inertial profilograph. The analyzed RD varies from 0.06 to 0.41 with a mean of 0.13 and a standard deviation of 0.05.

The environmental factors, including terrain types, lighting condition, and land use type, are considered. Two terrain types are examined, which include rolling terrace (62.03%) and mountainous terrace (37.97%). Lighting condition was considered as a category variable, which indicated whether lighting devises are provided at the roadway segments.

Three types of land use are considered, including commercial (24.41%), rural (24.49%), and residential (51.10). These variables are considered because they might have potential significant effects on traffic safety.

5. Modelling Results

The MATLAB was employed for model development. Four-year data, from 2010 to 2013, were used as the training set and one-year data, the year of 2014, were used as the testing set. In order to obtain the model with superior performance, module 1 was developed using 9 Gaussian visible units, 13 binary visible units, and a number of hyperbolic tangent hidden units ranging from 32 to 128 in steps of 2. The number of hidden units was setting based on two rules, greater than the input data dimensionality and the powers of two. The parameters for learning rate and weight decay were selected to optimize reduction of reconstruction error over training. Module 1 was trained with a sample size of 2540 (four-year crash data) to allow for full convergence of the parameters. The input data were processed by using module 1 to capture the relationship between traffic factors, geometric design features, pavement conditions, and environmental characteristics. Module 2 was developed using 4 Gaussian visible units and a number of hyperbolic tangent hidden units. The initial number of hyperbolic tangent hidden units ranging from 8 to 32 was tested and examined. The learning procedure will stop when the number of feature representations is achieved to four or the reconstruction error is less than 0.01.

5.1. Results of Unsupervised Feature Learning Module. For the final model, the reconstruction error is less than 0.01 and four hidden layers are included. The weights between the input layer and output layer can be calculated as $\mathbf{W}_{32\times16}\mathbf{W}_{16\times8}\mathbf{W}_{8\times4}=\mathbf{W}_{32\times4}$. The results are shown in Table 3. The negative sign represents a crash-prone condition and a positive sign represents a safe-prone condition. The valued number indicates an evaluation score. Because the feature learning module identifies relational information between input variables and output feature representations, the connection weights between visible units and hidden units can be interpreted as the functional networks [35]. The results show that each of the output units is significant associated with traffic factors, geometric factors, pavement factors, and environmental factors.

Since the proposed feature learning module has symmetric connection between visible and hidden layers and the units in both layers have the probabilistic characteristics, the proposed feature learning module also can be called as an auto-encoder. Thus, the values of output units can be interpreted as a feature representation and the results are shown in Table 4. The number of output units is smaller than the number of visible units, which indicates the feature representation with the values of output units has reduced the dimensionality of the input, but still preserving the original information. Hence the four output units are defined as traffic feature representation, geometric feature representation, pavement feature representation, and environmental feature representation.

The results show that the signs of the means of four feature representations are negative, which indicate that the feature representations are associated with crash-prone condition. In other words, the current traffic, geometric, pavement, and environmental features are the main factors that attribute to the risk of crash occurrences. The findings indicate that the traffic, geometric, pavement, and environmental factors have a direct influence on traffic safety and need to be improved.

The traffic feature representations have a wide range with a minimum value of -4.420 and a maximum value of 6.650, which indicate the traffic factors have a significant impact on traffic safety. The pavement factors have comparative impacts on traffic safety with a minimum value of -12.076 and a maximum value of 6.753. The ranges of geometric and environmental feature representations are from -1.386 to 1.280 and from -1.805 to 0.847, respectively, which indicate that the geometric and environmental feature representations have comparative effects on traffic safety. Compared to those crash modeling techniques that use only geometric design features as the input factors, the current research reveals new insights that would benefit the development of updated guidelines.

5.2. Results of the Supervised Fine Tuning Module. For the supervised fine tuning module, the visible units of the input layer use the feature representations as the input and the crash counts across injury severities are used as the training target. The aggregate weights between the input layer and output layer are shown in Table 5. The results show that the traffic and geometric feature representations have positive effects and pavement and environmental feature representations have negative effects on traffic crashes across injury severities. The findings indicate that decreasing the values of traffic and geometric feature representations will increase the likelihood of crashes and increasing the values of geometric and environmental feature representations will increase the likelihood of crashes. The comparison results show that traffic feature representation and geometric feature representation have significant impact on PDO crashes. The pavement feature representation and environmental feature representation have significant impact on minor injury crashes. Among four feature representations, the geometric feature representation, pavement feature representation, and traffic feature representation have most direct impacts on major injury, minor injury, and PDO crashes, respectively.

Considering the data in the year of 2014 as the input variables, the developed deep learning model is used to predict the crash counts for the year of 2014. To validate the superiority of the proposed models, the predicted results are compared to the observed values. In addition, a deep learning model without the regression layer and a support vector machine (SVM) model are also developed to predict crash counts across injury severities. There are two key issues related to the development of SVR models, kernel selection, and parameters optimization. To address the nonlinear relationship between the outcomes and attributes, the commonly

TABLE 3: The functional network between the input variables and the feature representations.

Variable	Traffic feature representation	Geometric feature representation	Pavement feature representation	Environmental feature representation
Traffic factor				
The logarithm of AADT per lane	-0.529	-0.007	-0.016	-0.034
Truck traffic percentage	-0.659	-0.002	-0.024	-0.010
Posted speed limits	-0.030	-0.010	0.175	-0.047
Geometric factors				
Segment length (miles)	0.007	-0.580	-0.007	-0.003
Degree of horizontal curvature	0.002	-0.918	-0.006	-0.009
Median widths	0.002	0.826	0.068	0.015
Outside shoulder widths	0.005	0.419	0.016	0.064
Number of through lanes				
3 lanes	-0.021	-0.891	-0.006	-0.026
2 lanes	-0.015	-0.864	-0.022	-0.060
Lane widths				
12 ft	0.004	0.776	0.005	0.001
11 ft	0.009	0.592	0.076	0.004
Number of left-turn lanes				
2 left-turn lanes	-0.003	-0.212	-0.019	-0.017
1 left-turn lane	0.002	-0.783	0.028	0.016
Median type				
Non-traversable median	0.013	0.247	0.006	0.007
Shoulder type				
Pavement	-0.035	-0.144	-0.027	-0.016
Gravel	0.005	-0.743	0.011	0.060
Pavement factor				
International roughness index (in./mile)	0.101	0.007	-0.183	-0.001
Rut depth (in.)	-0.004	-0.004	-0.341	-0.038
Environmental factor				
Terrain type				
Mountainous	-0.014	-0.011	-0.010	-0.954
Land use type				
Residential	0.042	0.015	0.011	0.847
Commercial	0.058	0.001	0.006	-0.915
Indicator for lighting				
Lighting exists on roadway	0.018	0.008	0.027	0.781

TABLE 4: Summary statistics of feature representations.

Statistics	Mean	Std. Dev.	Min.	Max.
Traffic feature representation	-0.697	1.464	-4.420	6.650
Geometric feature representation	-1.620	0.521	-1.386	1.280
Pavement feature representation	-5.388	2.441	-12.076	6.753
Environmental feature representation	-2.167	0.442	-1.805	0.847

TABLE 5: The functional network between the feature representation and different crash types.

Variable	Major injury crashes	Minor injury crashes	PDO crashes
Traffic feature representation	-0.109	-0.421	-1.535
Geometric feature representation	0.129	0.261	0.539
Pavement feature representation	-0.033	-0.872	-0.315
Environmental feature representation	0.113	0.298	0.146

used RBF kernel is chosen to develop the SVM model. To precisely reflect the performance on regressing unknown data and prevent the overfitting problem, the k-fold cross-validation approach is employed for optimizing the two parameters in RBF kernels—C and ε [41, 42]. The optimized parameters are (42.30, 37.09) and the optimized SVM model yields to a training mean absolute error (MAE) less than 0.01 and a training root-mean-squared deviation (RMSD) less than 5%.

To evaluate the accuracy and reliability of the prediction results, the predicted means for each injury severity level by the proposed model, the deep learning model without the regression layer, and the SVM model are compared to the observed means and two evaluation measures are used, which include MAE and RMSD. The comparison and validation results are shown in Table 6. Results in Table 6 indicate that the predicted means from the proposed model (0.110, 0.573, 1.335, and 2.019 for major injury, minor injury, no-injury, and all crashes, respectively) are very close to the observed means (0.096, 0.556, 1.306, and 1.986). We believe that a very importation feature of the proposed model is that the included regression layer can provide a good estimate of the chance that the roadway segment is in the crash-free state or some crash-prone propensity states.

For all the observed samples, the proposed model results in a MAE of 0.030, 0.080, 0.071, and 0.150 and a RMSD of 17.298%, 29.961%, 27.206%, and 43.652% for major injury, minor injury, PDO, and all crashes, respectively. The deep learning mode without the regression layer results in a MAE of 0.043, 0.202, 0.405, and 0.520 and a RMSD of 20.620%, 51.741%, 65.086%, and 82.862% for major injury, minor injury, PDO, and all crashes, respectively. The SVM model results in a MAE of 0.055, 0.257, 0.471, and 0.660 and a RMSD of 26.022%, 61.350%, 72.416%, and 96.636% for major injury, minor injury, PDO, and all crashes, respectively. The results suggest that the proposed model predicts better than the deep learning model without the regression layer and the SVM model. In summary, compared to the deep learning model without the regression layer and the SVM model, the proposed model results in the smallest prediction MAE and RMSD, no matter for the crash types across injury severities. We hypothesize that the proposed model better addresses the issue of heterogeneity and allows for excess zero counts in correlated data.

The findings indicate that the predictions from the proposed model have significant improvements over all comparison models, in both accuracy and robustness. The best-performing result of the proposed model for major injury crashes has a MAE of 0.030, which indicates an 42.105% and 84.211% improvement from the deep learning model without the regression layer and the SVM model, respectively. The proposed models perform worse for the minor injury crashes, with a 0.080 RMSD for all observed samples. However, it is still better than the evaluation measurements from the deep learning model without the regression layer and the SVM model, which are 0.202 and 0.257, respectively. This represents a MAE improvement of 150.980% and 219.608% for minor injury crash prediction. For the PDO crashes, the MAE improvements of the proposed models over the comparison models range from 471.111% to 564.444%. For all the observed samples, the MAE improvements range from 247.368% to 341.053%. Clearly, the improvement is significant for the traffic crash predictions. Therefore, the proposed model seems to be a better alternative for crash count predictions.

The proposed model has better performances in terms of small error variances than the comparison models, since the regression model is imbedding into the proposed model. The overall performances of the proposed model for all crashes show an 89.824% RMSD improvement over the deep learning mode without regression layer and an 121.378% RMSD improvement over the SVM model. It is clear that the predictions obtained from the proposed models are superior to those obtained from the comparison models. The greatest difference is demonstrated for the PDO crashes where the proposed model yields a RMSD of 27.206% compared to a 65.086% RMSD value from the deep learning model without the regression layer and a 72.416% RMSD value from the

Table 6: Results of roadway segment crash count prediction.

	Major injury	Minor injury	No injury	Total
Observation mean	0.096	0.556	1.306	1.986
Proposed model				
Estimated mean	0.110	0.573	1.335	2.019
MAE	0.030	0.080	0.071	0.150
RMSD (%)	17.298	29.961	27.206	43.652
Deep learning model without the regression layer				
Estimated mean	0.101	0.556	1.332	1.989
MAE	0.043	0.202	0.405	0.520
RMSD (%)	20.620	51.741	65.086	82.862
SVM model				
Estimated mean	0.101	0.513	1.329	1.943
MAE	0.055	0.257	0.471	0.660
RMSD (%)	26.022	61.350	72.416	96.636

SVM model. The differences in the proposed model and the SVM model for the minor injury crashes are also significant (29.961% versus 61.350%).

6. Conclusions

Because traffic crashes are a big concern of the public, agencies, and policy makers and result in countless fatalities and injuries, there is a need to perform a comprehensive analysis that aims to understand the relationship between the influence factors and traffic crash outcomes. In this study, we presented an innovative approach for traffic crash prediction. Methodologically, we demonstrated a novel deep learning technique embedded within a multivariate regression model can be used to identify relationship between the examined variables and the traffic crashes. Future applications of this approach have the potential to provide insights into basic questions regarding roadway spatial and temporal dynamic function and practical questions regarding countermeasures. The investigation results provide sufficient evidence for the following conclusions:

(1) The results show that the feature learning module identifies relational information between input variables and output feature representations. The findings indicate that the feature representations have reduced the dimensionality of the input, but still preserving the original information.

(2) The proposed model that includes a MVNB regression layer in the supervised fine tuning module can better account for different patterns in crash data across injury severities and provide superior traffic crash predictions. In addition, the proposed model can perform multivariate estimations simultaneously with a superior fit.

(3) The proposed model has superior performances in terms of prediction power compared to the deep learning model without a regression layer and the SVM model. The overall performances of the proposed model for all crashes show an 89.824% RMSD improvement over the deep learning model without a regression layer and an 121.378% RMSD improvement over the SVM model.

(4) The findings suggest that the proposed model is a superior alternative for traffic crash predictions. The proposed model can better account for heterogeneity issues in traffic crash prediction.

The proposed models can perform traffic crash prediction for a given facility. The proposed methodology could be applied to other roadway networks if appropriate attribute variables are available. Traffic and transportation engineering agencies can employ the proposed models with relative cases and develop them to their needs to obtain traffic crash predictions for various time periods. Further investigation of the proposed models includes the predictions of spatial-temporal dynamic pattern in crash data.

Conflicts of Interest

The authors declare that they have no conflicts of interest.

Acknowledgments

Special thanks are due to TDOT for providing the TRIMS data. This research is supported by funding provided by the Southeastern Transportation Center—a Regional UTC funded by the USDOT—Research and Innovative Technology Administration. Additional funding was provided by the National Natural Science Foundation of China (Grant Nos. 51678044, 51338008, 71621001, 71210001, and 71501011).

References

[1] NHTSA, *Motor Vehicle Traffic Crashes as a Leading Cause of Death in the United States, 2012-2014*, Department of Transportation, National Highway Traffic Safety Administration, Traffic Safety Facts, DOT HS 812 297, 2016.

[2] NHTSA, *2015 Motor Vehicle Crashes: Overview*, U.S. Department of Transportation, National Highway Traffic Safety Administration, Traffic Safety Facts, DOT HS 812 318, 2016.

[3] D. Lord and F. Mannering, "The statistical analysis of crash-frequency data: a review and assessment of methodological alternatives," *Transportation Research Part A: Policy and Practice*, vol. 44, no. 5, pp. 291–305, 2010.

[4] F. L. Mannering and C. R. Bhat, "Analytic methods in accident research: methodological frontier and future directions," *Analytic Methods in Accident Research*, vol. 1, pp. 1–22, 2014.

[5] F. L. Mannering, V. Shankar, and C. R. Bhat, "Unobserved heterogeneity and the statistical analysis of highway accident data," *Analytic Methods in Accident Research*, vol. 11, pp. 1–16, 2016.

[6] J. C. Milton, V. N. Shankar, and F. L. Mannering, "Highway accident severities and the mixed logit model: an exploratory empirical analysis," *Accident Analysis & Prevention*, vol. 40, no. 1, pp. 260–266, 2008.

[7] C. Dong, D. B. Clarke, X. Yan, A. Khattak, and B. Huang, "Multivariate random-parameters zero-inflated negative binomial regression model: an application to estimate crash frequencies at intersections," *Accident Analysis & Prevention*, vol. 70, pp. 320–329, 2014a.

[8] C. Dong, D. B. Clarke, S. S. Nambisan, and B. Huang, "Analyzing injury crashes using random-parameter bivariate regression models," *Transportmetrica A: Transport Science*, vol. 12, no. 9, pp. 794–810, 2016.

[9] M. G. Karlaftis and E. I. Vlahogianni, "Statistical methods versus neural networks in transportation research: differences, similarities and some insights," *Transportation Research Part C: Emerging Technologies*, vol. 19, no. 3, pp. 387–399, 2011.

[10] J. C. Principe, N. R. Euliano, and W. C. Lefebvre, *Neural and Adaptive Systems: Fundamentals through Simulations*, vol. 672, Wiley, New York, NY, USA, 2000.

[11] L.-Y. Chang, "Analysis of freeway accident frequencies: Negative binomial regression versus artificial neural network," *Safety Science*, vol. 43, no. 8, pp. 541–557, 2005.

[12] H. T. Abdelwahab and M. A. Abdel-Aty, "Artificial neural networks and logit models for traffic safety analysis of toll plazas," *Transportation Research Record*, no. 1784, pp. 115–125, 2002.

[13] Y. Xie, D. Lord, and Y. Zhang, "Predicting motor vehicle collisions using Bayesian neural network models: an empirical analysis," *Accident Analysis & Prevention*, vol. 39, no. 5, pp. 922–933, 2007.

[14] M. M. Kunt, I. Aghayan, and N. Noii, "Prediction for traffic accident severity: Comparing the artificial neural network, genetic algorithm, combined genetic algorithm and pattern search methods," *Transport*, vol. 26, no. 4, pp. 353–366, 2011.

[15] K. S. Jadaan, M. Al-Fayyad, and H. F. Gammoh, "Prediction of Road Traffic Accidents in Jordan using Artificial Neural Network (ANN)," *Journal of Traffic and Logistics Engineering*, vol. 2, no. 2, pp. 92–94, 2014.

[16] D. Akin and B. Akbaş, "A neural network (NN) model to predict intersection crashes based upon driver, vehicle and roadway surface characteristics," *Scientific Research and Essays*, vol. 5, no. 19, pp. 2837–2847, 2010.

[17] X. Li, D. Lord, Y. Zhang, and Y. Xie, "Predicting motor vehicle crashes using Support Vector Machine models," *Accident Analysis & Prevention*, vol. 40, no. 4, pp. 1611–1618, 2008.

[18] Y. Zhang and Y. Xie, "Forecasting of short-term freeway volume with v-support vector machines," *Transportation Research Record*, vol. 2024, no. 1, pp. 92–99, 2007.

[19] V. Kecman, "Support Vector Machines – An Introduction," in *Support Vector Machines: Theory and Applications*, vol. 177 of *Studies in Fuzziness and Soft Computing*, pp. 1–47, Springer Berlin Heidelberg, Berlin, Heidelberg, 2005.

[20] R. Yu and M. Abdel-Aty, "Analyzing crash injury severity for a mountainous freeway incorporating real-time traffic and weather data," *Safety Science*, vol. 63, pp. 50–56, 2014.

[21] C. Chen, G. Zhang, Z. Qian, R. A. Tarefder, and Z. Tian, "Investigating driver injury severity patterns in rollover crashes using support vector machine models," *Accident Analysis & Prevention*, vol. 90, pp. 128–139, 2016.

[22] N. Dong, H. Huang, and L. Zheng, "Support vector machine in crash prediction at the level of traffic analysis zones: assessing the spatial proximity effects," *Accident Analysis & Prevention*, vol. 82, pp. 192–198, 2015.

[23] G. Ren and Z. Zhou, "Traffic safety forecasting method by particle swarm optimization and support vector machine," *Expert Systems with Applications*, vol. 38, no. 8, pp. 10420–10424, 2011.

[24] R. Yu and M. Abdel-Aty, "Utilizing support vector machine in real-time crash risk evaluation," *Accident Analysis & Prevention*, vol. 51, pp. 252–259, 2013.

[25] M. Abdel-Aty and K. Haleem, "Analyzing angle crashes at unsignalized intersections using machine learning techniques," *Accident Analysis & Prevention*, vol. 43, no. 1, pp. 461–470, 2011.

[26] Y. LeCun, Y. Bengio, and G. Hinton, "Deep learning," *Nature*, vol. 521, no. 7553, pp. 436–444, 2015.

[27] J. Schmidhuber, "Deep learning in neural networks: an overview," *Neural Networks*, vol. 61, pp. 85–117, 2015.

[28] W. Huang, G. Song, H. Hong, and K. Xie, "Deep architecture for traffic flow prediction: deep belief networks with multitask learning," *IEEE Transactions on Intelligent Transportation Systems*, vol. 15, no. 5, pp. 2191–2201, 2014.

[29] Y. Lv, Y. Duan, W. Kang, Z. Li, and F.-Y. Wang, "Traffic Flow Prediction with Big Data: A Deep Learning Approach," *IEEE Transactions on Intelligent Transportation Systems*, vol. 16, no. 2, pp. 865–873, 2015.

[30] N. G. Polson and V. O. Sokolov, "Deep learning for short-term traffic flow prediction," *Transportation Research Part C: Emerging Technologies*, vol. 79, pp. 1–17, 2017.

[31] G. E. Hinton, S. Osindero, and Y. Teh, "A fast learning algorithm for deep belief nets," *Neural Computation*, vol. 18, no. 7, pp. 1527–1554, 2006.

[32] P. Swietojanski, A. Ghoshal, and S. Renals, "Convolutional neural networks for distant speech recognition," *IEEE Signal Processing Letters*, vol. 21, no. 9, pp. 1120–1124, 2014.

[33] M. M. A. Rahhal, Y. Bazi, H. Alhichri, N. Alajlan, F. Melgani, and R. R. Yager, "Deep learning approach for active classification of electrocardiogram signals," *Information Sciences*, vol. 345, pp. 340–354, 2016.

[34] H.-I. Suk, C.-Y. Wee, S.-W. Lee, and D. Shen, "State-space model with deep learning for functional dynamics estimation in resting-state fMRI," *NeuroImage*, vol. 129, pp. 292–307, 2016.

[35] R. D. Hjelm, V. D. Calhoun, R. Salakhutdinov, E. A. Allen, T. Adali, and S. M. Plis, "Restricted Boltzmann machines for neuroimaging: An application in identifying intrinsic networks," *NeuroImage*, vol. 96, pp. 245–260, 2014.

[36] P. Shi and E. A. Valdez, "Multivariate negative binomial models for insurance claim counts," *Insurance: Mathematics & Economics*, vol. 55, pp. 18–29, 2014.

[37] P. C. Anastasopoulos, V. N. Shankar, J. E. Haddock, and F. L. Mannering, "A multivariate tobit analysis of highway accident-injury-severity rates," *Accident Analysis & Prevention*, vol. 45, pp. 110–119, 2012.

[38] L. Chang and J. Chien, "Analysis of driver injury severity in truck-involved accidents using a non-parametric classification tree model," *Safety Science*, vol. 51, no. 1, pp. 17–22, 2013.

[39] J. Pahukula, S. Hernandez, and A. Unnikrishnan, "A time of day analysis of crashes involving large trucks in urban areas," *Accident Analysis & Prevention*, vol. 75, pp. 155–163, 2015.

[40] Q. Wu, F. Chen, and G. H. Zhang, "Mixed logit model-based driver injury severity investigations in single- and multi-vehicle crashes on rural two-lane highways," *Accident Analysis & Prevention*, vol. 72, pp. 105–115, 2014.

[41] N. Cristianini and J. Shawe-Taylor, *An Introduction to Support Vector Machines*, Cambridge University Press, Cambridge,UK, 2000.

[42] C. Campbell, "Kernel methods: A survey of current techniques," *Neurocomputing*, vol. 48, pp. 63–84, 2002.

11

A Review of the Self-Adaptive Traffic Signal Control System based on Future Traffic Environment

Yizhe Wang, Xiaoguang Yang, Hailun Liang, and Yangdong Liu

Key Laboratory of Road and Traffic Engineering of the Ministry of Education, Tongji University, 4800 Cao'an Road, Shanghai 201804, China

Correspondence should be addressed to Xiaoguang Yang; yangxg@tongji.edu.cn

Academic Editor: Ludovic Leclercq

The self-adaptive traffic signal control system serves as an effective measure for relieving urban traffic congestion. The system is capable of adjusting the signal timing parameters in real time according to the seasonal changes and short-term fluctuation of traffic demand, resulting in improvement of the efficiency of traffic operation on urban road networks. The development of information technologies on computing science, autonomous driving, vehicle-to-vehicle, and mobile Internet has created a sufficient abundance of acquisition means for traffic data. Great improvements for data acquisition include the increase of available amount of holographic data, available data types, and accuracy. The article investigates the development of commonly used self-adaptive signal control systems in the world, their technical characteristics, the current research status of self-adaptive control methods, and the signal control methods for heterogeneous traffic flow composed of connected vehicles and autonomous vehicles. Finally, the article concluded that signal control based on multiagent reinforcement learning is a kind of closed-loop feedback adaptive control method, which outperforms many counterparts in terms of real-time characteristic, accuracy, and self-learning and therefore will be an important research focus of control method in future due to the property of "model-free" and "self-learning" that well accommodates the abundance of traffic information data. Besides, it will also provide an entry point and technical support for the development of Vehicle-to-X systems, Internet of vehicles, and autonomous driving industries. Therefore, the related achievements of the adaptive control system for the future traffic environment have extremely broad application prospects.

1. Introduction

The amount of motor vehicles and correspondent travel demand are continuously increasing with economic and social development. The frequent occurrence of traffic congestion in urban road network has negative impacts on economy and environment. Due to the limited land resources of large cities and restrictions to transportation infrastructure construction from socioeconomic factors, to apply traffic management and control measures in a reasonable and effective way, improve the efficiency of existing transportation facilities, and accommodate the growing traffic demand in big cities have become significant research contents for counteracting urban traffic congestion.

Traffic control is one of the most important technical means to regulate traffic flow, improve the congestion, and even reduce emissions. Its progress and development has always been accompanied by the development of information technology, computer technology, and system science. The self-adaptive control system can adjust the signal timing parameters in real time according to the control target of the manager (such as the minimum delay of the intersection) and the arrival characteristics of the traffic flow at the intersection. Compared with timing control and actuated control, the self-adaptive control system can make better use of the overall traffic capacity of the road network and effectively improve the efficiency of road network traffic.

The traffic data collected by the current traffic control system using induction loop detector and other existing sensors is limited. With the advancement of the wireless communication technologies and the development of the vehicle-to-vehicle (V2V) and vehicle to infrastructure (V2I) systems, called Connected Vehicle or V2X, there is an opportunity to optimize the operation of urban traffic network by cooperation between traffic signal control and driving behaviors. This dissertation proposed a series of cooperative

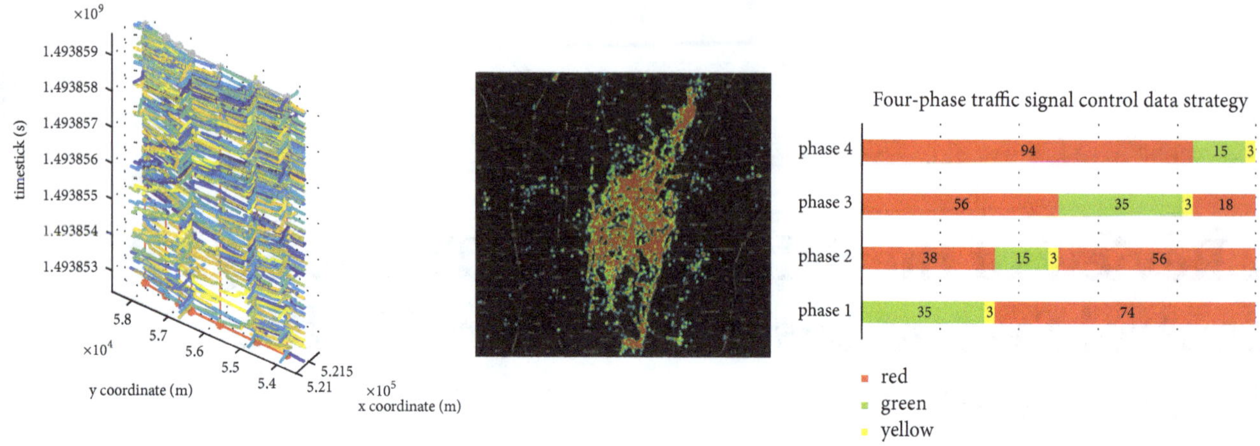

FIGURE 1: New traffic data environment.

optimization methods for urban streets traffic control and driving assistant under the V2X concept. In addition to the existing induction loop detector technology, the video, infrared, radar, floating cars, and other acquisition technologies and equipment provide urban traffic control system with a network of dynamic acquisition traffic flow status data and controller state data, which greatly enriched the information environment and provides more possibilities for the informationalized and intelligent application research. Urban traffic control is entering the data-rich period of multisource holographic network traffic data from the period with only data of cross-section traffic flow.

Recent advances in traffic control methods have led to flexible control strategies for use in an adaptive traffic control system [1]. Metropolitan road traffic digitized and informationalized infrastructure and related system construction has been developed rapidly in the past decade. At the same time, the emergence of intelligent connected vehicles and automated vehicle jointly build a future traffic travel environment, whose abilities of individual information access and perception as well as the performance of response time and interactive behavior are significantly different from conventional artificial driving vehicles. However, the current self-adaptive traffic signal control system cannot effectively utilize these abundant real-time traffic data, and its theory, methods, and techniques have clearly lagged far behind the progress of its key basic technologies [2]. Therefore, the research of data-driven feedback self-adaptive coordination control in data-rich environment is proposed and actively explored by researchers [3].

Moreover, with the continuous improvement of the theory and technology of intelligent control and nonmodel control, the concept of traffic control is changing under the new traffic data environment shown as Figure 1. The researchers hope that the model of the control system is based on the data model identification rather than the existing mechanism model [4]. Besides, they hope that the system is based on real-time monitoring data rather than the traffic forecast data [5] and the control system can automatically adjust the control strategy instead of the manual intervention [6].

2. The Development History and Deficiency of the Existing Traffic Self-Adaptive Control System

2.1. The Development History of the Existing Traffic Self-Adaptive Control System. According to NCHRP, more than 20 self-adaptive traffic control systems have been developed by transportation research institutes and enterprises worldwide, but less than half systems have been put into use [7]. According to the system's ability to adapt to the environment and the level of intelligent decision-making, Gartner et al. proposed the evolution of urban transport control system development level in 1995 [8], as shown in Figure 2.

The first-generation self-adaptive control system adopts the multi-time timing control of fine division of period, or completely isolated self-adaptive control, to realize the simple regulation of traffic flow. Take the multi-period timing control system as an example, which divides the traffic flow arriving within a day into multiple periods (such as peak, nonpeak), taking into account changes in daily traffic demand to optimize the signal timing scheme in different periods of time each day, using the comprehensive performance index method or green wave band timing method to optimize and generate a signal timing scheme library [10]. According to the number of weeks and control period, traffic controller can directly select the appropriate offline scheme from the scheme library.

The second-generation traffic signal control system dynamically adjusts the parameters of the signal timing scheme (signal period, green signal ratio, and phase difference). Compared with the timing and induction coordination control system, the second-generation system greatly improved the flexibility and adaptive adjustment ability of the control system. Typical second-generation control systems include SCATS [11] and SCOOT [12].

The third-generation control system uses the similar idea as the second generation to dynamically adjust the signal timing parameter in response to the fluctuation of the time-varying traffic flow at the intersection. Typical third-generation control systems include OPAC [13] and RHODES

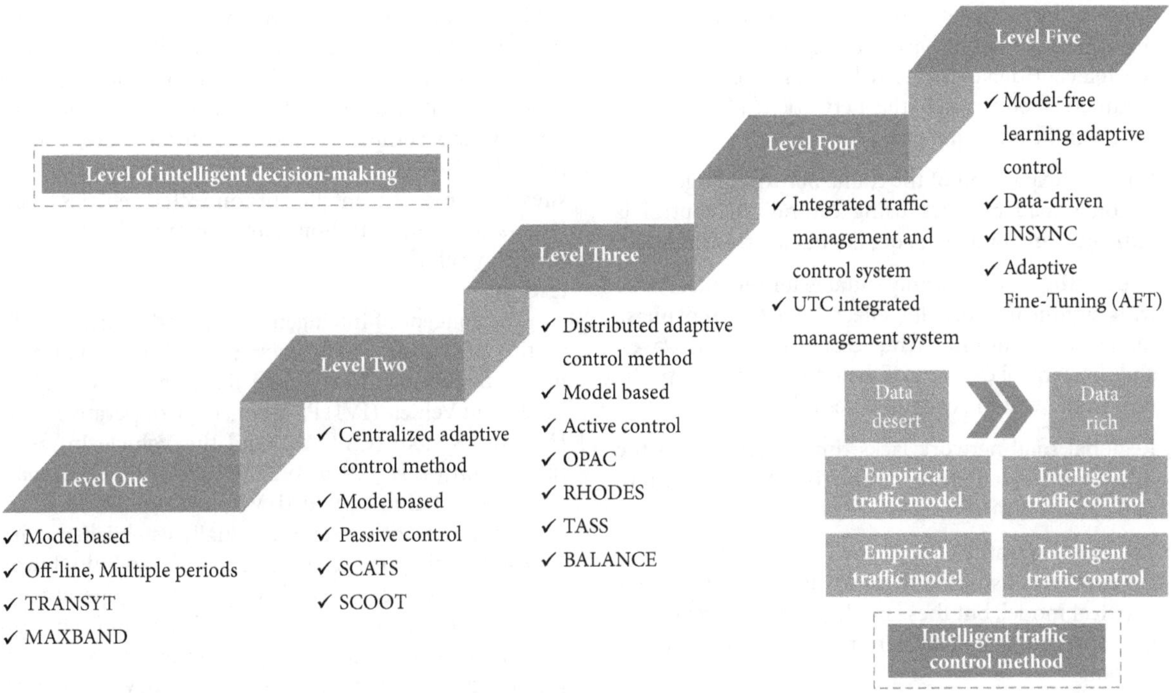

FIGURE 2: Urban traffic self-adaptive control system development process [9].

[14]. Kosmatopoulos et al. chose three traffic networks with quite different traffic and control infrastructure characteristics: Chania, Greece (23 junctions); Southampton, UK (53 junctions); and Munich, Germany (25 junctions), where it has been compared to the respective resident real-time signal control strategies TASS, SCOOT, and BALANCE. The main conclusion drawn from this high-effort inter-European undertaking is that traffic-responsive urban control is an easy-to-implement, interoperable, low-cost real-time signal control strategy whose performance, after very limited fine-tuning, proved to be better or, at least, similar to the ones achieved by long-standing strategies that were in most cases very well fine-tuned over the years in the specific networks [15].

The fourth-generation self-adaptive traffic signal control system is an integrated traffic management and control system, which can realize the integrated management of network traffic and maximize the technical and performance advantages of multiple subsystems [16]. It integrates self-adaptive traffic signal control system and other ITS traffic management systems with system hardware and software integration technology, like dynamic process models of combined traffic assignment and control with different signal updating strategies [17]. It is committed to building an efficient urban traffic control integrated management system to achieve the integration of mobile network management so that it can provide better decision support for local government decision-making [18].

The fifth-generation self-adaptive traffic signal control system is based on the abilities of self-learning and high efficiency calculation in automated vehicles and regular vehicles environment [19]. Based on the empirical information and real-time traffic condition, the fifth-generation adaptive traffic signal control system learns the traffic control knowledge independently and reduces the computational burden of decision optimization intelligently. As of June 2014, InSync system has been applied in 1350 intersections in more than 100 cities across the United States and has become the fastest growing self-adaptive traffic control in the United States, which is also recommended by the FWHA currently [20]. Manolis et al. have developed and evaluated, both by means of theoretical analysis and extensive simulation experiments, a new methodology which fully automatically takes over the manual tuning and calibration procedure. Most importantly, this new methodology, called adaptive fine-tuning (AFT), achieved to improve the performance of the system and compensate the effect of the continuous changes of its behavior that may be due to either internal or external factors. The results from AFT real-life application demonstrated that it was capable of significantly improving the performance of the system in a safe and robust manner. Moreover, the real-life results exhibited the capability of AFT to efficiently adapt and compensated in cases of changes in the system behavior, even if these changes were significant [21].

2.2. The Deficiency and Expectation of the Existing Traffic Self-Adaptive Control System. Each generation system not only inherited the excellent characteristics of the previous generation traffic system, but also moves forward continuously to promote the evolution of traffic control technology under the support of the key basic technology and the guide of the new traffic control strategy. However, there are some shortcomings of the existing self-adaptive traffic control theory, method, and technology with fixed period, as follows:

(1) The existing model of static traffic prediction and timing scheme does not have learning ability. Therefore, the relevant departments will recalibrate the model parameters only when the network traffic patterns have significantly changed.

(2) With the expansion of the traffic network, large-scale regional road network using centralized control is difficult to guarantee the quality of data transmission.

(3) The existing system is only suitable for regional traffic with significant corridor effect (due to fixed phase sequence, it can only achieve one-way green wave), and the control capacity of the typical network traffic flow in the vast majority of cities is limited.

(4) Regional road network lacks timely response of the actual traffic fluctuation so that it is difficult to achieve real-time control.

(5) The existing traffic control methods mostly simplify the control constraints to establish the precise mathematical model, but these methods are different from the actual traffic flow conditions and the control effect is poor.

(6) The system requires a lot of human intervention, and the professional and technical personnel are needed to optimize and maintain the system due to the problem of localized migration process.

Many of the existing self-adaptive traffic control systems use the traffic model to predict the evolution of the network traffic flow under the condition of limited traffic flow data and then use the comprehensive index method to optimize the signal timing parameters. Therefore, volume prediction is an essential part. Associated with the prediction are two aspects: resolution and accuracy. It is imperative to study the relationship and tradeoff between the control strategy, prediction resolution, and its associated error, which are crucial to the development of self-adaptive traffic control systems. In a word, it is the inevitable option to study the theory and method of urban road traffic adaptive control in the future traffic data-rich environment.

3. Research on Traffic Signal Control System Based on Future Traffic Environment

3.1. The Overview of Future Traffic Environment Composition and Development. The composition of the regular vehicle (RV) traffic flow is also changed by the emergence and mixing of the Connected Vehicle (CV) and the autonomous vehicle (AV). It is foreseeable that the car traffic flow will consist of conventional vehicle, CV, and AV in the next few decades. In the "China-made 2025" national strategic plan, it is clearly put forward that China should master the overall driving technology and the key technologies to basically complete the transformation and upgrading of the automobile industry in 2025. The National Highway Traffic Safety Administration (NHTSA) in 2014 enforces that new US vehicles must have networking capabilities [22]. Nowadays, connected vehicles with highly self-driving functions (such as Google's driverless vehicles, Tesla autopilot) and networked communications functions (such as the generic Cadillac CTS 2016) have completed several different driving conditions experiment or have been put into the market, and a variety of domestic and foreign auto companies and institutions have also entered the field of research. The United States establishes seven test sites to promote the intelligent connected vehicles testing and large-scale demonstration. Now, Nevada, Michigan, and so forth have allowed driverless vehicles to enter public road for testing.

The concept of intelligent connected vehicle was formed in the 1990s, known as cooperative infrastructure vehicle in the beginning. The United States began organizing the Intelligent Vehicle (IVI) Program, the Cooperative Automatic Highway System (CVHAS), and the Vehicle Infrastructure Integration (VII) [23] in 1998. In 2007, the US Department of Transportation renamed VII to IntelliDrive. Michigan, California, and other states gradually established connected vehicle test platform from 2012. In 2004 to 2010, Europe has developed PreVENT, SAFESPOT, CVIS (Cooperative Vehicle Infrastructure Systems) [24], COOPERS, and other projects to develop key technologies of connected vehicle system. Japan began to build VICS in 1991 and developed Smartway [25] project from 2004 so far. These systems and projects have entered the stage of large-scale system applications and related technology policy development. In recent years, under the support of the National Natural Science Foundation of China (NSFC) and the National High Technology Development Program (863 Project), Tongji University, Tsinghua University, Beihang University, and National University of Defense Technology and other academic and industrial institutions have developed several connected vehicle prototypes and test systems based on short-range communication [26]. With the continuous development of related fields such as mobile Internet and Internet of Things, cooperative infrastructure vehicle system and its application have become the new trend of the intelligent transportation system. Besides, the necessity of carrying out research on relevant theories, technologies, standards, policies, and regulations has become a broad consensus. Vehicle-road/vehicle-vehicle communication and traffic safety technology based on cooperative infrastructure vehicle system has become a research focus at this stage.

The earliest research on autonomous vehicles began in the 1980s, represented by the Navlab Self-Driving Vehicle [27] of Carnegie Mellon University and the ALV (Autonomous Land Vehicle) project of the US Defense Advanced Research Projects Agency (DAPRA) [28]. In 1995, Carnegie Mellon University developed the autonomous vehicle Navlab-5, completing a self-driving experiment of nearly 5,000 kilometers across the US, of which 98.2% was completed by the automatic driving system [29]. In 2008, Stanford University developed the driverless vehicle Junior, which can independently plan the path and realize its precise positioning, perceive other social vehicles and interact, and can achieve driving behaviors such as lane changing, U-turn, and parking [30]. In 2010, the ARGO autonomous vehicle, which was developed by Professor Alberto Broggi of the University of Parma, Italy, equipped with laser radar, camera, global positioning equipment, and so on, was exhibited at the Shanghai World Expo after more than 80 days' travel from the Italian Palma to Shanghai [31]. The red flag HQ3 autonomous vehicle, which is developed by National Defense University of Science and Technology, completed the 286 km self-driving experiment from Changsha to Wuhan, with less than 1% manual intervention mileage of the total mileage [32]. Besides, Google, Nissan, Tesla, GM, Ford, and other companies are also involved in the study of autonomous vehicle, but the technical details of the study are usually not disclosed [33]. In the theoretical study, Levinson et al. [34] optimized the existing automatic driving system, which enabled the vehicle adapted to a variety of lighting, weather, and traffic conditions, to a certain extent, overcoming the challenges of narrow roads, crosswalks, and signal intersections.

3.2. Research Status of Traffic Signal Control System Based on Future Traffic Environment. Following the methods of earliest fixed signal timing and offline delay calculation proposed by Webster [35], the traffic signal control system has evolved from offline to online control, from point to network control, from fixed-time to self-adaptive control. With the development of intelligent transportation system, the research of a new generation of traffic control technology based on multisource heterogeneous data has been gradually started [36]. In recent years, the research on signal control based on cooperative infrastructure vehicle system has become the frontier field of domestic and foreign traffic control theory and application [37]. Professor Yang's team at Tongji University launched the project "Research on the Next Generation of Traffic Control Technology at Intersection Based on CVIS" in 2010 to 2012 under the support of the National Natural Science Foundation, which analyzed multitarget control mechanism, vehicle-road/vehicle-vehicle communication method, and the prototype of signal control integrated platform with a single intersection as the object. The research of intelligent connected vehicle was mainly focused on the optimization methods of traffic safety, such as collision warning [38] and lane changing assistance [39]. With the concept of active safety and traffic signal control problems being put forward, driving optimization strategy for efficiency and emission reduction, such as the speed guidance strategy considering the signal light state [40], eco driving strategy [41], and so on, has been widely studied. Besides, to meet the special needs like emergency rescue vehicles and bus priority, the multimode signal priority control system considering the real-time status of special vehicles has also been put forward and achieved initial implementation [42]. Automatic driving research mainly focuses on the data collection and forecasting problem of mixed traffic flow [43] and the local optimization method based on the rolling optimization strategy [44]. Most of the optimization targets adopt efficiency-related indicators such as the least delay, the least number of stops, or the shortest across time [45]. About the control effect evaluation, most of the research output optimization control effect based on the secondary developed traditional simulation software [46]. Researches show that the traffic control which considers the mixed traffic flow of the connected vehicle and autonomous vehicle can effectively improve the traffic efficiency of the intersection, compared to the conventional traffic flow control [43].

4. Development Status of Urban Traffic Signal Self-Adaptive Control Method

Traffic congestion in urban road and freeway networks leads to a strong degradation of the network infrastructure and accordingly reduced throughput, which can be countered via suitable control measures and strategies. The traffic signal control method is evolved along with the combination of modern control theory, artificial intelligence theory, traffic information technology, and traffic engineering technology. Because the modern control theory is based on the basic assumption that the mathematical or nominal model of the controlled object is precisely known, the method is collectively referred to as Model Based Control (MBC) theory and method [47]. In the last 20 years, Artificial Intelligence (AI) theory and methods, which were represented by agents, neural networks, fuzzy logic, and group intelligence, were gradually mature. Diakaki et al. presented the design approach, the objectives, the development, the advantages, and some application results of the traffic-responsive urban control (TUC) strategy. Based on a store-and-forward modeling of the urban network traffic and using the linear-quadratic regulator theory, the design of TUC led to a multivariable regulator for traffic-responsive coordinated network-wide signal control that is particularly suitable also for saturated traffic conditions [48].

4.1. Traffic Self-Adaptive Control Method Based on Mathematical Model. According to the status and function of the traffic forecasting module in the traffic control model, the typical traffic signal MBC control method includes Travel-Time Responsive (CTR) traffic signal control algorithm [49], Predictive Model Control [50], Arrival–Discharge Process, and Storage-Forward Response Control. Arrival–Discharge Process signal control algorithm is based on dynamic programming and the optimization of signal policy is performed using a certain performance measure involving delays, queue lengths, and queue storage ratios [51]. Storage-Forward Response Control using a real-time monitoring data of arrival and leaving traffic flow to simulate the movement of the vehicle platoon and realize the predictive control [52].

According to the different control targets, the traffic signal MBC control method can be divided into a coordinated control method based on comprehensive performance index and a coordinated control method based on green wave band. The comprehensive index method, represented by TRANSYT [53], considers the delay, the number of stops, and the length of queues to obtain the best overall efficiency of the network. The green wave band method is designed to maximize the number of nonstop platoons of the main line, and the typical arterial coordinated control method based on green wave band includes the maximum green wave band method MAXBAND and the multi-green wave band with variable band width method MULTIBAND [54].

4.2. Traffic Self-Adaptive Control Method Based on Intelligent Computing. Some researchers think that optimization of traffic lights in a congested network is formulated as a linear programming problem [55]. However, considering the complexity of the internal structure of the urban regional traffic system and the external operation environment, it is impossible to establish the precise mathematical model. There are many challenges in effectively integrating signal timing tools with dynamic traffic assignment software systems, such as data availability, exchange format, and system coupling [56]. However, little effort has been put in developing control frameworks that are aimed not only at improving the average performance of the system, but also at improving the system robustness and reliability. In the past 10 years, artificial intelligence computing technology simulates human reasoning and learning process and controls the optimal control strategy in the process of interaction between traffic controller and road environment. Fuzzy logic, group intelligence algorithms, and neural network control dominate the many traffic control methods based on intelligent computing.

4.2.1. Fuzzy Logic. The fuzzy control of urban traffic signal is one of the effective connotative solutions to solve the urban traffic problem. Pham et al. proposed a fuzzy logic control for the integrated signal operation of a diamond interchange and its ramp meter, to improve traffic flows on surface streets and motorway. This fuzzy logic diamond interchange (FLDI) comprises three modules: fuzzy phase timing (FPT) module that controls the green time extension of the current phase, phase logic selection (PLS) module that decides the next phase based on the predefined phase sequence or phase logic, and fuzzy ramp-metering (FRM) module that determines the cycle time of the ramp meter based on current traffic volumes and conditions of the surface streets and the motorways [57]. To improve the level of fuzzy controller's skill of problem-solving, the multilevel fuzzy and other structural models have been proposed and developed from the single-point control to regional traffic control [58, 59].

4.2.2. Group Intelligence. Genetic algorithm (Genetic Algorithms, GA) and Ant colony optimization (ACO) Particle Swarm Optimization (PSO) are the most widely used strategies for simulating the social behavior of biology. Zhao et al. developed a mathematical model for a traffic equilibrium network, in which optimization of lane reorganization and traffic control strategies were integrated in a unified framework. A genetic algorithm (GA) based heuristic is used to yield meta-optimal solutions to the model. Results from extensive numerical analyses reveal the promising property of the proposed model in enhancing network capacity and reducing congestion [16]. Li and Schonfeld presented a hybrid algorithm based on simulated annealing (SA) and a genetic algorithm (GA) for arterial signal timing optimization. A decoding scheme was proposed that exploits our prior expectations about efficient solutions, namely, that the optimal green time distribution should reflect the proportion of the critical lane volumes of each phase. The numerical results indicated that the SA-GA algorithm outperforms both SA and GA in terms of solution quality and convergence rate [60].

4.2.3. Neural Network. The neural network controller creates the optimal timing based on the real-time detection of traffic system information and weather conditions. Spall and Chin proposed a system-level self-adaptive signal control (S-TRAC) method based on Artificial Neural Network (ANN) [61]. Hoogendoorn et al. proposed a new control framework based on the notion of controlled Markov processes, which explicitly take into account the uncertainty in predicted traffic conditions and system performance [62]. In contrast to traditional optimal control approaches, the objective function can include general statistic of the random system performance, such as the mean, standard deviation, or 95-percentile. Srinivasan et al. developed a multiagent unsupervised flow response signal control model based on a hybrid neural network approach, which used a multilevel online learning process to update and adjust its knowledge base and decision-making mechanism. The results showed that the new model significantly improves the traffic situation when the complexity of the scene increases, and the average delay was reduced by 78% and the average stopping time was reduced by 85% compared with the existing credit control algorithm [63]. Kosmatopoulos et al. studied the approximation and learning properties of one class of recurrent networks, known as high-order neural networks, and applied these architectures to the identification of dynamical systems. It was clear that if enough high-order connections were allowed, then this network was capable of approximating arbitrary dynamical systems [64].

4.3. The Review of Current Traffic Self-Adaptive Control Methods. Since the revolutionary development of traffic information technology provides the urban traffic control system with rich data which implied information such as system state changes and process control, the researchers began to think about data-driven traffic control methods, such as how to effectively use these data in the condition of lacking accurate model of a controlled system to achieve the optimal control of the system and the production process and thus overcome the defects of traditional MBC method. The group intelligence algorithm needs to design the efficient heuristic rule according to the background of the practical application problem, and the group intelligence optimization has the deficiencies of long computing time and local optimum [65]. The most critical problem is that the intelligent computing method has no autonomous learning ability, which needs the neural network and the fuzzy logic to increase the phase optimization module so that it can greatly increase the computational complexity of the agent.

Under the condition that the information of the traffic system cannot be completely obtained, the internal mechanism of the system cannot be fully understood, and the precise dynamic model of the controlled object cannot be established, the traffic control method based on the data-driven has the characteristics of modeless, self-learning, simple structure, small computation, and so on. It can solve the deficiencies of the MBC method, such as the need of the precise mathematical model and lack of self-learning ability. It is an inevitable choice for the development and application of self-adaptive feedback control theory of urban regional traffic feedback in rich data environment [9].

5. The Future Development Trend of Traffic Self-Adaptive Signal Control System

Regional traffic control is an important unit in urban traffic control. At present, regional traffic control problem not only has a long-time congestion phenomenon at peak time, but also has obvious ability of grooming in peak time. The traditional traffic signal control method not only lacks the adaptability for the variable traffic flow, but also relies heavily on the traffic model. The existing adaptive/neural/fuzzy control methodologies cannot be used towards the development of a systematic, automated fine-tuning procedure for general large-scale nonlinear control systems due to the strict assumptions they impose on the controlled system dynamics [66].

5.1. Reinforcement Learning Control Based on Data-Driven Method. The large-scale application of emerging technologies such as video information, probe vehicle, connected vehicle, and autonomous vehicle in the transportation industry has broken through the existing traffic data collecting mode. Reinforcement Learning, as a typical "model-free, self-learning" iterative data-driven method, is applicable of regional traffic control based on multiagent reinforcement learning [67]. The controller adopting this method has the ability to perceive the environmental state and select the optimal action according to the target. Different from other machine learning methods, in the control process, reinforcement learning only judges and evaluates the advantages and disadvantages of the changes of environmental states after selecting the action through reinforcement signals obtained by perceiving the environment, rather than intervening with the specific generation procedure. Reinforcement learning algorithm can acquire knowledge in the process of decision-making and evaluation and balance the knowledge exploration and utilization to achieve optimal strategy. Reinforcement learning algorithm has been used in many applications and successfully applied to the traffic signal control of single-point, artery, and regional transportation [68].

At present, Q-learning algorithm is one of the most frequently used methods in the fields of reinforcement learning, proposed by Watkins in 1989 [69]. Q-learning algorithm is widely used in the fields of control, depending on the update mode of its special value function.

In Q-learning, the solution formula of the mainstream value function is as follows:

$$Q(s_t, a_t) \longleftarrow Q(s_t, a_t) + \alpha\left[\left(r_{t+1} + \gamma \max_{a\in A} Q(s_{t+1}, a) - Q(s_t, a_t)\right)\right]. \quad (1)$$

According to the formula, at the moment of t, the state of Q-learning is s_t. If the taken action is a_t, the corresponding value function will be $Q(s_t, a_t)$. The update of the value function is determined by three factors. The first is the current value of the action state value function, $Q(s_t, a_t)$, that needs to be updated. The second is to control the corresponding maximum value of all Q-values of actions in the postexecution state of $st+1$, and the third is the immediate return, $rt + 1$, after the action. Besides, there are also two model parameters, learning rate $\alpha \in [0,1]$ and discount factor $\gamma \in (0,1]$. The former is used to balance the relationship between the learning and utilization of the algorithm. When $a \rightarrow 1$, the controller tends to explore new knowledge; otherwise it will use the existing knowledge. The latter is used to coordinate the present relationship with the future. When $\gamma \rightarrow 1$, the controller tends to consider the future return, and when $\gamma \rightarrow 0$, the controller mainly considers immediate return.

El-Tantawy et al. proposed a multiagent coordinated RL-TSC control system called MARLIN-ATSC. This is a model-free system based on the Q-Learning algorithm [70]. Kosmatopoulos and Kouvelas introduced and analyzed, both by means of mathematical arguments and simulation experiments, a new learning/adaptive algorithm that can provide convergent, efficient, and safe fine-tuning of general large-scale nonlinear control systems [66]. Abdulhai et al. studied the single-point two-phase intersection control model based on Q-Learning algorithm and a signal control method which performs better than fixed timing scheme when flow rate changes significantly was found [71].

5.2. Research on Traffic Control Based on Adaptive Performance Optimization. The objectives, approach, advantages, and some application results of recent extensions of the traffic-responsive urban control (TUC) strategy were presented by Diakaki et al. Based on well-known methods of the automatic control theory, TUC allows for traffic-responsive coordinated signal control of large-scale urban networks that is particularly efficient under saturated traffic conditions [72]. Adaptive optimization (AO) schemes based on stochastic approximation principles such as the Random Directions Kiefer–Wolfowitz (RDKW), the Simultaneous Perturbation Stochastic Approximation (SPSA), and the adaptive fine-tuning (AFT) algorithms possess the serious disadvantage of not guaranteeing satisfactory transient behavior due to their requirement for using random or random-like perturbations of the parameter vector. Kosmatopoulos introduced and analyzed a new algorithm for alleviating this problem. Application of the proposed scheme to the adaptive optimization of a large-scale, complex control system demonstrates the efficiency of the proposed scheme [73].

Aboudolas et al. presented a methodology that the traffic flow process was modeled by use of the store-and-forward modeling paradigm, and the problem of network-wide signal control (including all constraints) was formulated as a quadratic-programming problem that aims at minimizing and balancing the link queues so as to minimize the risk of queue spillback [74]. Kouvelas et al. investigated the adaptive fine-tuning algorithm for determining the set of design parameters of two distinct mutually interacting modules of the traffic-responsive urban control (TUC) strategy, i.e.,

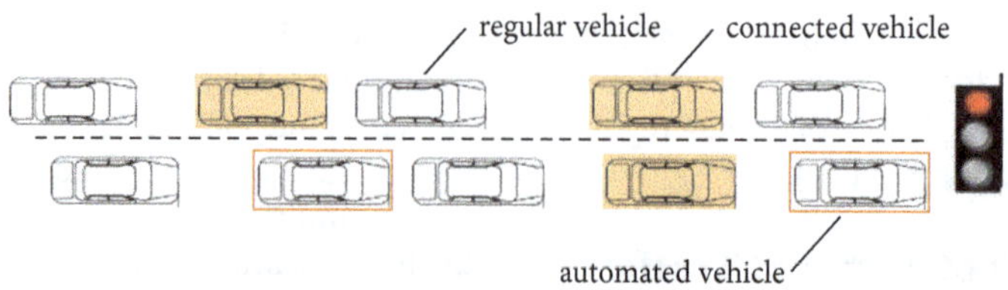

FIGURE 3: Multimodal traffic streams consisting of RVs, CVs, and AVs.

split and cycle, for the large-scale urban road network of the city of Chania, Greece. Simulation results are presented, demonstrating that the network performance in terms of the daily mean speed, which is attained by the proposed adaptive optimization methodology, is significantly better than the original TUC system in the case in which the aforementioned design parameters are manually fine-tuned to virtual perfection by the system operators [75].

5.3. Research on Traffic Control Based on Connected Vehicles and Automated Vehicles Environment. Multimodal vehicles consisting of RVs (regular vehicles), CVs (connected vehicles), and AVs (automated vehicles) develop a persistent multimodal traffic streams shown in Figure 3, which changes the traffic composition of urban road network. Information network of connected vehicles and automated vehicles affected the driving behavior and the demand of traffic stream controlling; it is urgent to do research on the control theory of multimodal traffic streams. The adaptive traffic control strategy aims to respond to real-time traffic demand through current and predicted future traffic flow data modeling. Compared with the traffic flow and occupancy information provided by the fixed coil detector in the traditional traffic environment, the adaptive traffic control system in the V2X environment can collect more detailed data such as vehicle position, speed, queuing length, and stopping time so that it received the attention of many scholars. In the adaptive control strategy under V2X environment, the concept of "rolling horizon" is widely used. Goodall et al. proposed a prediction algorithm based on micro-simulation to implement a distributed adaptive control strategy. This strategy is based on real-time vehicle data and determines 15 seconds as the optimal solution within the rolling optimization time window through simulation. The test results showed that, compared with the induction coordination control, this strategy has obvious advantages in the case of traffic accidents and sudden changes in traffic demand [76].

In the real world, due to the large-scale installation of vehicle-road cooperative communication equipment in existing vehicles, the applicability of various traffic signal control optimization strategies will be directly affected by the penetration rate of vehicle-mounted communication equipment. Therefore, many studies have considered the penetration rate of vehicle-mounted communication equipment into the optimization model. Feng et al. proposed an Estimation of Location and Speed (ELVS) algorithm based on the prevalence rate of different vehicle-mounted communication devices and used it for real-time optimization of traffic signal. Simulation test results showed that the application of this algorithm can reduce the total vehicle delay of 16.33% in the highest penetration rate of vehicle-mounted communication equipment [77]. Guler et al. proposed a traffic control optimization algorithm for unidirectional traffic intersections where some vehicles have communication conditions [78]. Lee et al. integrated vehicle-based real-time communication data and traditional traffic detection data to propose a cumulative travel-time responsive (CTR) real-time traffic control method [79]. Ma et al. evaluated the safety effectiveness of adaptive traffic signal control using the empirical Bayes method. This analysis examined 47 urban or suburban intersections where adaptive traffic signal control was deployed in Virginia using 235 site-years of before data and 66 site-years of after data. It was concluded that adaptive traffic signal control installation can potentially reduce total crashes at highway intersections and that public agencies should consider safety and mobility benefits when justifying adaptive traffic signal control projects [80]. Mandava et al. proposed a vehicle speed planning method based on signal control status of intersections. By proposing speed recommendations to the driver, the probability of not stopping passing the intersection can be increased [81].

5.4. A Review of the Development of Traffic Signal Self-Adaptive Control System Based on Oriented Future Traffic Environment. Above all, many scholars and research institutes have done a lot of research on implementation technology, traffic status analysis, and experimental methods of V2X and automated vehicles. A preliminary study of the traffic control methods for isolated intersections and other special needs under the condition of V2X has been conducted. But the research mainly focused on the construction of physical systems, adaptive analysis, preliminary theoretical research, and single or hybrid (CVs mixed with RVs or AVs mixed with RVs) vehicle traffic flow study. The theory and experimental methods of multimode vehicle traffic flow cooperative control have not been formed yet. Therefore, the study on the traffic control and coordination problems of multimode intersections under urban traffic conditions is still in its infancy. The acquisition technology and means of network traffic control data are developing in the direction of "full spatio-temporal, high-precision, diversified, and high-quality," and it is possible to acquire a large amount of data generated during the interaction between the road traffic control system and the environment. Therefore, the vehicle terminals of connected vehicles and autonomous vehicles can respond to signal control schemes, surrounding vehicles, and road network conditions, resulting in significant changes in the operational mechanism of multimode vehicle traffic flow.

The traffic control system learns independently throughout the whole control process, which is a kind of closed-loop feedback self-adaptive control based on the control effect. As for traffic control scope, the multiagent reinforcement learning algorithm can realize the precise reasoning of the optimal joint action of the intersection with the benefit of the networked dynamic traffic data acquisition and interaction technology. The reinforcement learning algorithm itself only needs the input and output data rather than the forms of data collection. The system has good compatibility with existing and emerging traffic control systems and technologies. At the same time, under the condition of multimode vehicle traffic flow, it can more abundantly and accurately collect information such as the position and speed of the vehicle and can directly guide the connected vehicles and control the autonomous vehicles. Therefore, four control modes, including traffic signal control, connected vehicles guidance, automated vehicles control, and indirect control of conventional vehicles through connected vehicles and automated vehicles, can be formed. The collaborative control mechanism is in urgent need of research. The related research can not only creatively establish a multimode vehicle traffic flow theory, but also provide a key support for the development of a new generation of traffic control system for solving multimode vehicle traffic flow control problems. It is of great strategic significance and practical value for occupying a commanding height in this field.

6. Conclusion

The multimode traffic flow consisting of conventional vehicles, intelligent connected vehicles, and automated vehicles is gradually becoming the norm throughout the world. Therefore, it is imperative to build a new generation of traffic control systems to meet its development and application needs. Most of the existing traffic control systems adopt "prior" feedforward control method or delay-based limited information control method. The control effect depends on the accuracy of the model describing the actual traffic environment, and it cannot learn and adjust the control knowledge online based on the feedback of the control effect. Obviously, the large-scale development and application of new technologies such as floating vehicles, vehicle-to-vehicle communication (V2V), Internet of vehicles (V2I), V2X, and automatic driving will greatly promote the development of the technical route of urban traffic control systems from the data-poor era to the data-rich era. The real-time detection of the spatiotemporal data based on urban road network traffic status can provide rich and high-quality basic data and fine-grained assessment of control effects for traffic control. In the face of the main defects encountered in the existing self-adaptive traffic control system, relying on abundant traffic control data and using the data-driven approach to delve a closed-loop feedback self-adaptive control system with better uncertainty response capability and higher intelligent decision-making level are inevitable result of the objective needs of the development and application of traffic control and advanced infrastructure technologies. Also, it can provide support for the interaction between the traffic control system and the multimode traffic flow.

Therefore, facing the limitations and major shortcomings of existing traffic signal control systems, relying on a wealth of traffic control interaction conditions and data, and developing a collaborative control system with a high degree of refinement, precision, and better responsiveness and intelligence are the objective need and development direction of traffic control technology. Although the outcome of this paper is multi-intersection coordination control theory under the oriented future traffic environment, it can provide scientific support for the development of future road network traffic control systems and can be widely used in new generation traffic control systems. Also, it can improve road network efficiency to a greater extent, reduce traffic operation costs, prevent and mitigate traffic congestion at intersections, and reduce energy consumption and emissions. Traffic signal control based on reinforcement learning is a true sense of closed-loop feedback self-adaptive control and the instantaneity, accuracy, and self-learning can be guaranteed, which will be one of the future research trends. Besides, it will also provide an entry point and technical support for the development of V2X systems, Internet of vehicles, and autonomous driving industries. Therefore, the related achievements of the adaptive control system for the future traffic environment have extremely broad application prospects.

Conflicts of Interest

The authors declare that there are no conflicts of interest regarding the publication of this paper.

Acknowledgments

The research is supported by Project of National Natural Science Foundation of China (Project no. 61773293) and Key Project of National Natural Science Foundation of China (Project no. 51238008).

References

[1] H. K. Lo and H. F. Chow, "Adaptive traffic control system: Control strategy, prediction, resolution, and accuracy," *Journal of Advanced Transportation*, vol. 36, no. 3, pp. 323–347, 2002.

[2] A. Hamilton, B. Waterson, T. Cherrett, A. Robinson, and I. Snell, "The evolution of urban traffic control: changing policy and technology," *Transportation Planning and Technology*, vol. 36, no. 1, pp. 24–43, 2013.

[3] J. Zhang, F.-Y. Wang, K. Wang, W.-H. Lin, X. Xu, and C. Chen, "Data-driven intelligent transportation systems: A survey," *IEEE Transactions on Intelligent Transportation Systems*, vol. 12, no. 4, pp. 1624–1639, 2011.

[4] S. Lin, Q.-J. Kong, and Q. Huang, "A model-based demand-balancing control for dynamically divided multiple urban subnetworks," *Journal of Advanced Transportation*, vol. 50, no. 6, pp. 1046–1060, 2016.

[5] D. Gettman, S. G. Shelby, L. Head, D. M. Bullock, and N. Soyke, "Data-driven algorithms for real-time adaptive tuning of offsets in coordinated traffic signal systems," *Transportation Research Record*, no. 2035, pp. 1–9, 2007.

[6] A. Csikõs, T. Tettamanti, and I. Varga, "Nonlinear gating control for urban road traffic network using the network fundamental diagram," *Journal of Advanced Transportation*, vol. 49, no. 5, pp. 597–615, 2015.

[7] A. Stevanovic, "Adaptive traffic control systems: Domestic and foreign state of practice," NCHRP Synthesis 403, Transportation Research Board, Washington, D.C., 2010.

[8] N. H. Gartner, C. Stamatiadis, and P. J. Tarnoff, "Development of advanced traffic signal control strategies for intelligent transportation systems: multilevel design," *Transportation Research Record*, no. 1494, pp. 98–105, 1995.

[9] W. Yang, *Study on Self-adaptive Coordinated Control Based on Close-loop Feedback for Urban Traffic Signals*, Tongji university, Shanghai, 2015.

[10] J. D. C. Little, M. D. Kelson, and N. H. Gartner, "MAXBAND: a program for setting signal on arteries and triangular network," *Transportation Research Record*, no. 795, pp. 40–46, 1981.

[11] G. Sims A, *The Sydney coordinated adaptive traffic system. Engineering Foundation Conference on Research Directions in Computer Control of Urban Traffic Systems*, Calif, USA, 1979.

[12] Y.-T. Wu and C.-H. Ho, "The development of Taiwan arterial traffic-adaptive signal control system and its field test: A Taiwan experience," *Journal of Advanced Transportation*, vol. 43, no. 4, pp. 455–480, 2009.

[13] N. H. Gartner, "OPAC: A demand-responsive strategy for traffic signal control," *Transportation Research Record*, vol. 1983, no. 906, pp. 75–81, 1983.

[14] P. Mirchandani and L. Head, "A real-time traffic signal control system: architecture, algorithms, and analysis," *Transportation Research Part C: Emerging Technologies*, vol. 9, no. 6, pp. 415–432, 2001.

[15] E. Kosmatopoulos, M. Papageorgiou, C. Bielefeldt et al., "International comparative field evaluation of a traffic-responsive signal control strategy in three cities," *Transportation Research Part A: Policy and Practice*, vol. 40, no. 5, pp. 399–413, 2006.

[16] J. Zhao, Y. Liu, and P. Li, "A network enhancement model with integrated lane reorganization and traffic control strategies," *Journal of Advanced Transportation*, vol. 50, no. 6, pp. 1090–1110, 2016.

[17] C. Meneguzzer, "Dynamic process models of combined traffic assignment and control with different signal updating strategies," *Journal of Advanced Transportation*, vol. 46, no. 4, pp. 351–365, 2012.

[18] M. Papageorgiou, M. Ben-Akiva, J. Bottom et al., "Chapter 11 ITS and Traffic Management," in *Transportation*, vol. 14 of *Handbooks in Operations Research and Management Science*, pp. 715–774, Elsevier, 2007.

[19] N. Jiang, "Optimal signal design for mixed equilibrium networks with autonomous and regular vehicles," *Journal of Advanced Transportation*, vol. 2017, pp. 1–13, 2017.

[20] J. Hu, M. D. Fontaine, B. B. Park, and J. Ma, "Field evaluations of an adaptive traffic signal—Using private-sector probe data," *Journal of Transportation Engineering*, vol. 142, no. 1, Article ID 04015033, 2016.

[21] D. Manolis, I. Papamichail, E. B. Kosmatopoulos, and M. Papageorgiou, "Automated tuning of ITS management and control systems: Results from real-life experiments," *Transportation Research Part C: Emerging Technologies*, vol. 66, pp. 119–135, 2016.

[22] J. Harding, G. Powell R, R. Yoon et al., "Vehicle-to-vehicle communications: Readiness of V2V technology for application," Report No. DOT HS 812 014, National Highway Traffic Safety Administration, Washington, DC, 2014.

[23] S. Peirce and R. Mauri, *Vehicle-infrastructure Integration(VII) Initiative Benefit-cost Analysis: Pre-testing Estimates*, US Department of Transportation, Washington DC, 2007.

[24] P. Kompfner, Z. Jeftic, and M. Andrade, "Cooperative vehicle-infrastructure systems," *Proposal for Information Society Technologies 4th Call*, 2005.

[25] A. Fujimoto, K. Sakai, M. Ogawa et al., "Toward realization of smartway in japan," in *Proceedings of the 15th World Congress on Intelligent Transport Systems and ITS America Annual Meeting 2008*, pp. 6361–6369, USA, November 2008.

[26] X. Yang, L. Huang, and Y. Wang, "A prototype of a cooperative vehicle infrastructure system: Proof of concept-case study in Tongji University," in *Proceedings of the Transportation Research Board 91st Annual Meeting*, 2012.

[27] C. Thorpe, M. H. Hebert, T. Kanade, and S. A. Shafer, "Vision and navigation for the carnegie-mellon navlab," *IEEE Transactions on Pattern Analysis and Machine Intelligence*, vol. 10, no. 3, pp. 362–373, 1988.

[28] "Defense Advanced Research Projects Agency, Strategic Computing Program Second Annual Report, Arlington VA, 1986".

[29] R. Sukthankar, J. Hancock, and S. Baluja, "Adaptive intelligent vehicle modules for tactical driving," in *AAAI Workshop on Intelligent Adaptive Agents*, pp. 13–22, 1996.

[30] M. Montemerlo, J. Becker, and S. Shat, "Junior: the Stanford entry in the urban challenge," *Journal of Field Robotics*, vol. 25, no. 9, pp. 569–597, 2008.

[31] M. Bertozzi, A. Broggi, E. Cardarelli, R. I. Fedriga, L. Mazzei, and P. P. Porta, "VIAC expedition toward autonomous mobility," *IEEE Robotics and Automation Magazine*, vol. 18, no. 3, pp. 120–124, 2011.

[32] P. Wang, *The research of Unmanned Intelligent Car Navigation System*, Xi'an Technological University, Xi'an, China, 2012.

[33] H. Martínez-Barberá and D. Herrero-Pérez, "Multilayer distributed intelligent control of an autonomous car," *Transportation Research Part C: Emerging Technologies*, vol. 39, pp. 94–112, 2014.

[34] J. Levinson, J. Askelavd, J. Becker, and etal., "Towards fully autonomous driving:Systems and algorithms," *Intelligent Vehicles Symposium (IV)*, pp. 163–168.

[35] F. V. Webster, *Traffic Signal Settings*, 1958.

[36] S. Chatterjee and B. Timande, "RFID based Traffic Control System using ARM processor perspective Kolkata and other densely populated cities," *International Journal of Advanced Research in Computer Science and Electronics Engineering*, vol. 1, no. 10, pp. 32–35, 2012.

[37] Z. Sun and J. Ban, "Vehicle trajectory reconstruction for signalized intersections using mobile traffic sensors," *Transportation Research Part C: Emerging Technologies*, vol. 36, pp. 268–283, 2013.

[38] M. Maile and L. Delgrossi, "Cooperative intersection collision avoidance system for violations (CICAS-V) for avoidance of violation-based intersection crashes. Enhanced Safety of Vehicles, 2009".

[39] X. Yang G, L. Huang Y, Y. Wang S et al., "A novel lane change and overtaking assist system design and implementation based on vehicle-to-vehicle communication," *Journal of Highway and Transportation Research and Development*, vol. 29, no. 11, pp. 120–124, 2012.

[40] P. Li K, X. Yang G, and W. Wu, "Modeling vehicle speed guidance at signalized intersections under intellidriver," *Traffic Information and Security*, vol. 30, no. 3, pp. 136–140, 2012.

[41] H. Rakha and R. K. Kamalanathsharma, "Eco-driving at signalized intersections using V2I communication," in *Proceedings of the 14th IEEE International Intelligent Transportation Systems Conference (ITSC '11)*, pp. 341–346, IEEE, Washington, DC, USA, October 2011.

[42] Y. Wang, Z. Wu, X. Yang, and L. Huang, "Design and implementation of an emergency vehicle signal preemption system based on cooperative vehicle-infrastructure technology," *Advances in Mechanical Engineering*, vol. 2013, Article ID 834976, pp. 1–10, 2013.

[43] J. Kolodko and L. Vlacic, "Cooperative autonomous driving at the intelligent control systems laboratory," *IEEE Intelligent Systems*, vol. 18, no. 4, pp. 8–11, 2003.

[44] X. Packiaraj, *A Decentralized PID-Based Approach for The Collaborative Driving of Autonomous Vehicles*, Department of Mechanical Engineering, Dalhousie University, Halifax, UK, 2009.

[45] J. Sun and S. Chen, "Dynamic speed guidance for active highway signal coordination: Roadside against in-Car strategies," *IET Intelligent Transport Systems*, vol. 7, no. 4, pp. 432–444, 2013.

[46] R. Sawant N, *Longitudinal Vehicle Speed Controller for Autonomous Driving in Urban Stop-and-Go Traffic Situations*, The Ohio State University, 2010.

[47] M. Papageorgiou, C. Diakaki, V. Dinopoulou, A. Kotsialos, and Y. Wang, "Review of road traffic control strategies," *Proceedings of the IEEE*, vol. 91, no. 12, pp. 2043–2065, 2003.

[48] C. Diakaki, M. Papageorgiou, and K. Aboudolas, "A multivariable regulator approach to traffic-responsive network- Wide signal control," *Control Engineering Practice*, vol. 10, no. 2, pp. 183–195, 2002.

[49] S. Choi, B. B. Park, J. Lee, H. Lee, and S. H. Son, "Field implementation feasibility study of cumulative travel-time responsive (CTR) traffic signal control algorithm," *Journal of Advanced Transportation*, vol. 50, no. 8, pp. 2226–2238, 2016.

[50] W. Huang, F. Viti, and C. M. J. Tampère, "A dual control approach for repeated anticipatory traffic control with estimation of network flow sensitivity," *Journal of Advanced Transportation*, vol. 50, no. 7, pp. 1386–1412, 2016.

[51] F. C. Fang and L. Elefteriadou, "Modeling and simulation of vehicle projection arrival-discharge process in adaptive traffic signal controls," *Journal of Advanced Transportation*, vol. 44, no. 3, pp. 176–192, 2010.

[52] K. Aboudolas, M. Papageorgiou, and E. Kosmatopoulos, "Store-and-forward based methods for the signal control problem in large-scale congested urban road networks," *Transportation Research Part C: Emerging Technologies*, vol. 17, no. 2, pp. 163–174, 2009.

[53] I. Robertson D, "TRANSYT: A Traffic Network Study Tool," Road Research Laboratory Report, 1969.

[54] N. H. Gartner and C. Stamatiadis, "Arterial-based control of traffic flow in urban grid networks," *Mathematical and Computer Modelling*, vol. 35, no. 5-6, pp. 657–671, 2002.

[55] J. Eddelbüttel and M. Cremer, "A new algorithm for optimal signal control in congested networks," *Journal of Advanced Transportation*, vol. 28, no. 3, pp. 275–297, 1994.

[56] M. Zlatkovic and X. Zhou, "Integration of signal timing estimation model and dynamic traffic assignment in feedback loops: System design and case study," *Journal of Advanced Transportation*, vol. 49, no. 6, pp. 683–699, 2015.

[57] V. C. Pham, F. Alam, J. Potgieter, F. C. Fang, and W. L. Xu, "Integrated fuzzy signal and ramp-metering at a diamond interchange," *Journal of Advanced Transportation*, vol. 47, no. 4, pp. 413–434, 2013.

[58] W. Yang, L. Zhang, and Z. He, "Rolling-horizon optimization-based three-stage fuzzy logic controller for urban traffic signals," *Journal of Tongji University Natural Science*, vol. 42, no. 12, pp. 1846–1853, 2014.

[59] I. Kosonen, "Multi-agent fuzzy signal control based on real-time simulation," *Transportation Research Part C: Emerging Technologies*, vol. 11, no. 5, pp. 389–403, 2003.

[60] Z. Li and P. Schonfeld, "Hybrid simulated annealing and genetic algorithm for optimizing arterial signal timings under oversaturated traffic conditions," *Journal of Advanced Transportation*, vol. 49, no. 1, pp. 153–170, 2015.

[61] J. C. Spall and D. C. Chin, "Traffic-responsive signal timing for system-wide traffic control," *Transportation Research Part C: Emerging Technologies*, vol. 5, no. 3-4, pp. 153–163, 1997.

[62] S. P. Hoogendoorn, V. L. Knoop, and H. J. Van Zuylen, "Robust control of traffic networks under uncertain conditions," *Journal of Advanced Transportation*, vol. 42, no. 3, pp. 357–377, 2008.

[63] D. Srinivasan, M. C. Choy, and R. L. Cheu, "Neural networks for real-time traffic signal control," *IEEE Transactions on Intelligent Transportation Systems*, vol. 7, no. 3, pp. 261–272, 2006.

[64] E. B. Kosmatopoulos, M. M. Polycarpou, M. A. Christodoulou, and P. A. Ioannou, "High-order neural network structures for identification of dynamical systems," *IEEE Transactions on Neural Networks and Learning Systems*, vol. 6, no. 2, pp. 422–431, 1995.

[65] Y.-C. Chiou and Y.-F. Huang, "Stepwise genetic fuzzy logic signal control under mixed traffic conditions," *Journal of Advanced Transportation*, vol. 47, no. 1, pp. 43–60, 2013.

[66] E. B. Kosmatopoulos and A. Kouvelas, "Large scale nonlinear control system fine-tuning through learning," *IEEE Transactions on Neural Networks and Learning Systems*, vol. 20, no. 6, pp. 1009–1023, 2009.

[67] B. Abdulhai and L. Kattan, "Reinforcement learning: Introduction to theory and potential for transport applications," *Canadian Journal of Civil Engineering*, vol. 30, no. 6, pp. 981–991, 2003.

[68] S. El-Tantawy and B. Abdulhai, "Towards multi-agent reinforcement learning for integrated network of optimal traffic controllers (MARLIN-OTC)," *Transportation Letters*, vol. 2, no. 2, pp. 89–110, 2010.

[69] Y. Bie, D. Wang, D. Ma, and Z. Zhu, "A distributed algorithm for signal coordination of multiple agents with embedded platoon dispersion model," *Journal of Southeast University*, vol. 27, no. 3, pp. 311–315, 2011.

[70] S. El-Tantawy, B. Abdulhai, and H. Abdelgawad, "Multiagent reinforcement learning for integrated network of adaptive traffic signal controllers (MARLIN-ATSC): methodology and large-scale application on downtown toronto," *IEEE Transactions on Intelligent Transportation Systems*, vol. 14, no. 3, pp. 1140–1150, 2013.

[71] B. Abdulhai, R. Pringle, and G. J. Karakoulas, "Reinforcement learning for true adaptive traffic signal control," *Journal of Transportation Engineering*, vol. 129, no. 3, pp. 278–285, 2003.

[72] C. Diakaki, V. Dinopoulou, K. Aboudolas et al., "Extensions and new applications of the traffic-responsive urban control strategy: Coordinated signal control for urban networks," *Transportation Research Record Journal of the Transportation Research Board*, vol. 1856, pp. 202–211, 2003.

[73] E. B. Kosmatopoulos, "An adaptive optimization scheme with satisfactory transient performance," *Automatica*, vol. 45, no. 3, pp. 716–723, 2009.

[74] K. Aboudolas, M. Papageorgiou, A. Kouvelas, and E. Kosmatopoulos, "A rolling-horizon quadratic-programming approach to the signal control problem in large-scale congested urban road networks," *Transportation Research Part C: Emerging Technologies*, vol. 18, no. 5, pp. 680–694, 2010.

[75] A. Kouvelas, K. Aboudolas, E. B. Kosmatopoulos, and M. Papageorgiou, "Adaptive performance optimization for large-scale traffic control systems," *IEEE Transactions on Intelligent Transportation Systems*, vol. 12, no. 4, pp. 1434–1445, 2011.

[76] N. Goodall, B. Smith, and B. Park, "Traffic signal control with connected vehicles," *Transportation Research Record Journal of the Transportation Research Board*, vol. 2381, pp. 65–72, 2013.

[77] Y. Feng, K. L. Head, S. Khoshmagham, and M. Zamanipour, "A real-time adaptive signal control in a connected vehicle environment," *Transportation Research Part C: Emerging Technologies*, vol. 55, pp. 460–473, 2015.

[78] S. I. Guler, M. Menendez, and L. Meier, "Using connected vehicle technology to improve the efficiency of intersections," *Transportation Research Part C: Emerging Technologies*, vol. 46, pp. 121–131, 2014.

[79] J. Lee, B. Park, and I. Yun, "Cumulative travel-time responsive real-time intersection control algorithm in the connected vehicle environment," *Journal of Transportation Engineering*, vol. 139, no. 10, pp. 1020–1029, 2013.

[80] J. Ma, M. D. Fontaine, F. Zhou, J. Hu, D. K. Hale, and M. O. Clements, "Estimation of crash modification factors for an adaptive traffic-signal control system," *Journal of Transportation Engineering*, vol. 142, no. 12, Article ID 04016061, 2016.

[81] S. Mandava, K. Boriboonsomsin, and M. Barth, "Arterial velocity planning based on traffic signal information under light traffic conditions," in *Proceedings of the 12th International IEEE Conference on Intelligent Transportation Systems*, pp. 160–165, October 2009.

12

Effects of Users' Bounded Rationality on a Traffic Network Performance

S. F. A. Batista, [1] **Chuan-Lin Zhao**, [2] **and Ludovic Leclercq** [1]

[1] *Univ. Lyon, ENTPE, IFSTTAR, LICIT, F-69518, Lyon, France*
[2] *School of Civil and Transportation Engineering, Beijing University of Civil Engineering and Architecture, Beijing 100044, China*

Correspondence should be addressed to Ludovic Leclercq; ludovic.leclercq@entpe.fr

Academic Editor: Zhi-Chun Li

In this paper, we revisit the principle of bounded rationality applied to dynamic traffic assignment to evaluate its influences on network performance. We investigate the influence of different types of bounded rational user behavior on (i) route flows at equilibrium and (ii) network performance in terms of its internal, inflow, and outflow capacities. We consider the implementation of a bounded rational framework based on Monte Carlo simulation. A Lighthill-Whitham-Richards (LWR) mesoscopic traffic simulator is considered to calculate time-dependent route costs that account for congestion, spillback, and shock-wave effects. Network equilibrium is calculated using the Method of Successive Averages. As a benchmark, the results are compared against both Deterministic and Stochastic User Equilibrium. To model different types of bounded rational user behavior we consider two definitions of user search order (indifferent and strict preferences) and two settings of the indifference band. We also test the framework on a toy Braess network to gain insight into changes in the route flows at equilibrium for both search orders and increasing values of aspiration levels.

1. Introduction

The first notions of traffic assignment were introduced by Wardrop [1]. According to the first Wardrop principle, users aim to minimize their personal route travel times. This leads to a network equilibrium called the Deterministic User Equilibrium (DUE) and it is that most commonly used in dynamic traffic assignment (DTA) problems. Under DUE conditions, no user can decrease his/her own travel time by unilaterally switching routes. However, the first Wardrop principle assumes that users are perfectly rational and perceive all routes and network traffic states perfectly although information on route travel times (i.e., traffic states) is not necessarily perfect. To overcome this problem, Daganzo and Sheffi [2] and Daganzo [3] introduced the Stochastic User Equilibrium (SUE), to take into account the uncertainty of route travel times. The Multinomial Logit and C-Logit are the Random Utility models (RUM) most commonly used in DTA problems. Nonetheless, both these models present several limitations when dealing with correlations between routes. In this study we focus in particular on the Probit model solved using Monte Carlo simulations [4].

Revealed [5] and stated [6] preference surveys show that users tend to choose suboptimal routes instead of optimal ones [7]. We emphasize that a suboptimal route is understood as a route with a longer travel time than the minimum one for the origin-destination (od) pair. In the literature on static traffic assignment, there are other alternative model frameworks that take into account different types of user behavior. One example is the Prospect Theory [8, 9] which considers the users risk-seeking and risk-aversion behavior. It was adapted to the context of route choice by Avineri [10]. In the Prospect Theory, users evaluate the different routes in terms of time prospect and choose the route with the maximum prospect. Users are risk-averse when confronted with prospects of gains and risk-seekers when confronted with prospects of losses and are more sensitive to losses than gains (*loss effect*). Another example is the Regret Theory [11, 12]. The users aim to minimize their regret with respect to the nonselected routes [13, 14]. If the users choose the route with the minimum travel time, they will feel joy or feel regret otherwise. Another example is the notion of bounded rationality introduced by the seminal works of Simon [15–18]. He stated that users choices are driven by aspiration levels (AL), which represent a set of goal or target variables that should be achieved or exceeded for the users satisfaction. In his original idea, the user searches until a satisfactory alternative is found. This term used to describe this process was coined by Simon as *satisficing*, which stands for the combination of satisfy and suffice. In this study, we focus on the application of the notion of bounded rationality in a dynamic context, by considering distributions of route travel times and a traffic simulator. The goal of this study is to investigate the influence of bounded rational user behavior on individual route flows and network performance. This type of study is very important for decision-making in transportation planning.

Mahmassani and Chang [22] discussed the first notion of bounded rationality applied to traffic assignment, but no mathematical formulation was given. To define users AL, Mahmassani and Chang [22] introduced the concept of *indifference band* (IB), where a route is *satisficing* if the difference between its travel cost and that of the best available route is lower than a given threshold (or IB). The implementation of bounded rationality in traffic assignment is challenging as (i) the calibration of the AL is context dependent [23] and (ii) the BR-UE solutions are not unique [24–26]. Thus, to analyze the BR-UE solutions, some authors have focused on the analysis of the best and worst BR-UE flows of the network [24, 25, 27]. Moreover, the AL can change from user to user. A thorough review of bounded rationality in traffic assignment was provided in Di and Liu [28]. There are two main ingredients that dictate bounded rational network equilibrium: (i) the definition of the AL that dictates whether a route is *satisficing* or not and (ii) the users search order that defines how users are guided in their choice of a *satisficing* route.

For a route to be considered as *satisficing*, its route utility must satisfy

$$U_k \leq AL^{od}, \quad \forall k \in \Omega^{od} \wedge \forall (o,d) \in \Xi \tag{1}$$

where U_k is the perceived route utility; AL^{od} is the aspiration level we consider in this paper, to be defined at the od level; Ω^{od} is the route choice set for the od pair; and Ξ is the set of all od pairs of the network.

The AL^{od} can be calibrated exogenously by route choice surveys or calibrated endogenously by explicit formulations. The most commonly used definition is based on the concept of indifference band [22, 29, 30]:

$$AL^{od} = \min\left(\overrightarrow{V}\right) + \Delta^{od}, \quad \forall (o,d) \in \Xi \tag{2}$$

where $\overrightarrow{V}$ is a vector containing all deterministic route utilities V_k, $\forall k \in \Omega^{od} \wedge \forall (o,d) \in \Xi$, and Ω^{od} contains N routes. Δ^{od} is the tolerance or IB at the od level.

Ge and Zhou [20] propose a variable definition of the IB (Δ^{od}):

$$AL^{od} = \min\left(\overrightarrow{V}\right) + \max\left(\left|U_p - U_q\right| \delta_p \delta_q\right), \quad \forall p,q \in \Omega^{od} \wedge p \neq q \wedge \forall (o,d) \in \Xi \tag{3}$$

where δ_p and δ_q are dummy variables that equal 1 if routes p and q belong to Ω^{od}, respectively.

Ge et al. [31] analyzed the BR-DUE equilibrium, considering exogenously fixed AL and fixed and endogenously variable AL. In their model framework, the authors showed that the DUE is a special case of the BR-DUE and discussed the existence conditions of the BR-DUE. However, the uncertainty on the travel times was not considered.

Di and Liu [28] highlighted that a bounded rational behavior can be due to the users habits and inertia or their cognitive costs or individual preferences. In this paper, we focus our attention on the users preferences as a bounded rational behavior to define the search order for the *satisficing* alternatives. Zhao and Huang [19] defined a search order based on a strict preference order for all users sharing the same od pair. This strong assumption allowed obtaining unique BR-UE solutions. To the authors knowledge, the framework of Zhao and Huang [19] has never been tested in a dynamic context, i.e., considering a traffic simulator and time-dependent path costs. In addition, its dynamic implementation using a traffic simulator is highly challenging because it requires solving suboptimization problems to calibrate the AL of the sub-most preferred routes. Thus, a framework capable of solving the global optimization problem is required and discussed further on in this paper. On the other hand, users may also have an indifferent preference for any of the *satisficing* routes (i.e., that satisfy (1)). This is adopted from the notions discussed in [32]. In this case, we consider the fact that all users sharing the same od pair have a similar indifference preference. The choice is modeled by uniform random sampling of any of the *satisficing* routes. Users are then assigned to the *satisficing* route sampled.

Szeto and Lo [30] (the BR-UE [22] and Tolerance-Based Dynamic User Optimum Principle [30] have been used interchangeably in the traffic assignment literature; for the sake of simplicity, we refer to both as BR-UE) discussed an analytical BR-UE dynamic traffic assignment model. The authors proposed a route swapping algorithm, but no clear definition of the users search order was discussed. Instead, the authors targeted certain users on the most congested routes and switched them to less congested ones for each od pair. Moreover, the BR-UE solutions were not unique. Han et al. [33] discussed a dynamic simultaneous departure time and route choice bounded rational framework. However, neither of these frameworks included travel time distributions. In this paper, we revisit the notions of bounded rationality by considering the distribution of travel times rather than deterministic values.

The literature includes a large number of applications of a bounded rational framework to static [25] and dynamic traffic assignment [30, 33], transportation planning [34], traffic policy-making [35], congestion pricing [24], and traffic safety [36]. However, to the authors knowledge, there is no study in the literature that investigates the influence of users preferences (indifferent and strict) for a bounded rational behavior on individual route flows and network performance in terms of the internal level of congestion and inflow and outflow capacities. The goal of this paper is to fill this gap. We consider time-dependent path costs that account for congestion, shock-waves, and spillback effects calculated using a mesoscopic Lighthill-Whitham-Richards (LWR) model [21]. A spillback effect is the reduction of a link capacity that spreads over other connected links in the network. To model bounded rationality behavior, we relax the definition of the search order of the DUE and SUE frameworks [4]. In both the DUE and SUE cases, users are assigned to the routes with the minimum travel times based on an all-or-nothing procedure. The search order is relaxed to account for the users indifferent and strict preferences. In the case of the indifferent preference search order, users present indifference behavior when choosing any of the *satisficing* routes, whereas in the case of the strict preference search order [19], users are assigned to the most preferred route if this route is perceived as *satisficing* (see (1)), or to the first sub-most preferred route that satisfies (1). We make use of Monte Carlo simulations [4] to account for travel times distribution and consider the classical Method of Successive Averages to calculate the network equilibrium. First, we test the bounded rationality methodology in a toy Braess network and consider a simple linear static and flow dependent utility function. We then consider the two settings of the search order previously mentioned and the AL^{od} defined exogenously. These initial tests allow acquiring insight into how the route flows at equilibrium change according to the two definitions of the search order and increasing values of AL^{od}. Second, for the dynamic implementation, we also consider the two settings of the users search order (i.e., indifferent and strict preferences) and the concept of the IB ((2) and (3)) to define the AL^{od}. The dynamic tests are performed on a Manhattan network. We investigate the influence of the definition of the search order on the individual route flows and analyze the network performance in terms of the internal, inflow, and outflow capacities, given the two search orders and different values of the AL^{od}. The results are compared against both DUE and SUE as benchmarks.

This paper is organized as follows. In Section 2, we discuss the bounded rational model framework considered in this paper. In Section 3, we discuss a simple static test scenario on the Braess network, considering both the indifferent and strict preferences search order. In Section 4, we discuss the influence of the bounded rationality behavior on the network performance also considering the two search orders. In Section 5, we outline the conclusions of this paper.

2. Bounded Rational Framework

The analysis of the effect of users' behavior on network performance in terms of its internal inflow and outflow capacities is very important for policy-makers, in particular when determining policies aimed at increasing network performance. In this paper, we focus on two types of bounded rational user behavior.

We start by introducing the general formulation of the route utilities. The perceived route utility, U_k, is

$$U_k = V_k + \epsilon_k, \quad \forall k \in \Omega^{od} \wedge \forall (o,d) \in \Xi \qquad (4)$$

where V_k is the deterministic route utility and ϵ_k is the uncertainty or error term as often referred to in the literature.

The DUE assumes that users are utility minimizers and the error terms ϵ_k are set to 0. Users are assigned based on an all-or-nothing procedure to the route with the minimum travel time. In the case of the SUE, users are also utility minimizers, but they perceive travel times with uncertainties, meaning that the error terms ϵ_k are not 0. Theoretically, the Probit model [2] is the most attractive model for solving the SUE. However, it requires the computation of a covariance matrix and integrating the multivariate normal distribution. The complexity of the computation increases with the number of routes per od pair. An alternative to this is to use Monte Carlo to consider the distributions of route travel times [4]. We consider that the error terms are defined at the link level (i.e., $\epsilon_a, \forall a \in \Gamma$) instead at the route level. This allows capturing existing correlations between different routes sharing the same links. Sheffi [4] proposed to consider link travel times that are normal distributed and to truncate the negative travel times. This skews the distributions and a positive defined travel times distribution is to be preferred. In this paper, we consider that the terms ϵ_a are gamma distributed following Nielsen [37]. The principle of the Monte Carlo simulations is to discretize the error terms ϵ_a into M samples or draws and locally solve DUE problems. For each error draw, the deterministic utility for route k is defined as $U_k^m, m = 1, \ldots, M$. In short, DUE is solved one for all considering the utility U_k for each route. The SUE corresponds to the average of M Monte Carlo trials, where the utilities U_k^m are adjusted for each trial considering the link error terms values. It is sufficient to explain how we extend DUE to account for bounded rational (BR-DUE), as the extension of SUE is exactly similar; i.e., we have to solve a BR-DUE problem for each Monte Carlo trial of link error terms and then average the results to get the BR-SUE solutions. Mahmassani and Chang [22] introduced the first notions of bounded rationality applied to route choice. Lou et al. [24] and Di et al. [25] formulated the BR-DUE mathematically, but without defining preference rules among *satisficing* routes. Under BR-DUE, all users are satisfied with their choices and no longer consider switching routes. It should be noted that the DUE is an extreme case of the BR-DUE (i.e., when $\Delta^{od} = 0$). The idea is that users are assigned to *satisficing* routes instead of optimal routes (i.e., routes with the minimum perceived travel times). A *satisficing* route should satisfy (1); i.e., $U_k \leq AL^{od}, \forall k \in \Omega^{od} \wedge \forall (o,d) \in \Xi$. Let us define ω^{od} as the set of *satisficing* routes. The users are then assigned to one route depending on the preference rule. In this paper, we consider that the search order is defined according to the user's preferences (indifferent and strict preferences, see below). The previous description of BR-SUE was given based on Monte Carlo trials as it helps to make the connection with

BR-DUE. We can also provide a formal definition of the BR-SUE independently of the solution method. Let P_k be the probability that route k is chosen for one od pair:

$$\begin{aligned} P_k &= P\big(\text{"}U_k \le AL^{od}\text{"} \\ &\quad \cap \text{choosing route k according to the preference rule}\big) \\ &= P\left(U_k \le AL^{od}\right) \\ &\quad \times P\big(\text{choosing route k according to the preference rule} \setminus \text{"}U_k \\ &\quad \le AL^{od}\text{"}\big) \end{aligned} \quad (5)$$

The second part of the equation is the conditional probability of route k being chosen when k is perceived as *satisficing*. Its value depends on the preference rule. We consider two behavioral rules to represent the user's search order. We consider indifferent preferences, where users are uniformly assigned to *satisficing* routes. In this case, P(choosing route k according to the preference rule$\setminus$"$U_k \le AL^{od}$") $= 1/|\omega^{od}|$, where $|\cdot|$ is the number of *satisficing* routes listed in ω^{od}. This number is at least equal to 1, as route k is *satisficing*. A close form for this probability cannot be provided as the number of other *satisficing* routes than k is a random variable whose distribution is intractable in the general case. Interestingly, Monte Carlo trials for link error terms make the calculation of $|\omega^{od}|$ straightforward, for each Monte Carlo iteration step since all *satisficing* paths are then known. We also consider a strict preference order [19], where users are selecting a *satisficing* route according to a predefined list of preferences Υ^{od}. The users are always choosing the *satisficing* route of highest rank in this list Υ^{od}. Again, the probability P(choosing route k according to the preference rule$\setminus$"$U_k \le AL^{od}$") does not have a closed form, but Monte Carlo trials permit determining the unique assignment at each iteration.

The idea of the strict preference order was introduced by Zhao and Huang [19], to deal with the nonuniqueness of the equilibrium solution. However, we highlight two main differences between our methodology and that discussed in Zhao and Huang [19]. First, we consider the fact that routes are *satisficing* if and only if their perceived utility satisfies $U_k = V_k \le AL^{od}$, $\forall k \in \Omega^{od} \wedge \forall (o,d) \in \Xi$ for the BR-DUE or $U_k^m \le AL^{od}$, $\forall m = 1,\ldots,M \wedge \forall k \in \Omega^{od} \wedge \forall (o,d) \in \Xi$ for the BR-SUE. Zhao and Huang [19] consider that routes are *satisficing* according to the strict preference order; i.e., the users are first assigned to the most preferred route and then consecutively to the sub-preferred routes, until all the users are assigned. Second, we consider that AL is defined at the od level (i.e., AL^{od}), while Zhao and Huang [19] consider its definition at the route level. We also assume that all users sharing the same od pair have the same AL^{od}. We consider that it is more realistic from the user's perspective to set a global AL^{od} instead of establishing AL for the sub-preferred routes based on the most preferred ones.

In this paper, we consider the two definitions of AL^{od} as defined in (2) and (3).

To reach a solution for the BR-SUE and simulate the probability term P(route k is satisficing) (see (5)), we consider Monte Carlo simulations as discussed in Sheffi [4] and the classical Method of Successive Averages (MSA). The MSA solves a fixed point problem and is commonly used in traffic assignment to solve both the DUE and SUE [4]. The Monte Carlo simulations consist in discretizing the distributions of the link travel times into M samples or draws and solving BR-DUE problems locally. For each discretization, we identify the *satisficing* routes and assign the users based on an all-or-nothing assignment following the search order established. If the search order is considered to be the indifferent preferences, all the users are assigned randomly to any of the *satisficing* routes. On the other hand, if the search order follows a strict user preference order, all users are assigned to the first *satisficing* route found on this strict sequence of preferences. The new temporary route flows, Q_k^*, correspond to averaging all the local BR-DUE solutions. This corresponds to the temporary route flows Q_k^*, which will be used to update the new route flows Q_k^{j+1} at iteration $j+1$, as

$$Q_k^{j+1} = Q_k^j + \alpha_j \left\{Q_k^* - Q_k^j\right\}, \quad \forall k \in \Omega^{od} \wedge \forall (o,d) \in \Xi \quad (6)$$

where Q_k^j represent the route flows at iteration j of the MSA and α_j is the descent step. This process is repeated at every descent step of the MSA algorithm.

The sequence of descent steps α_j guarantees the convergence of the MSA. For the theoretical convergence of the algorithm, the following two conditions must be satisfied [4]:

$$\sum_{j=1}^{\infty} \alpha_j = \infty \quad (7)$$

$$\lim_{j\longrightarrow\infty} \alpha_j \longrightarrow 0 \quad (8)$$

One definition of α_j that satisfies both of the previous conditions is $\alpha_j = 1/j$. We consider this definition of α_j in this paper. Other definitions of the descent step size are discussed in the literature [38–41].

A commonly used convergence or stopping criterion is based on the comparison between the current and the previous descent step of the MSA that must be lower to a predefined threshold. Instead we consider the number of violations $N(\lambda)$ and the relative Gap [42]. $N(\lambda)$ represents the number of cases where $|Q_k^{j+1} - Q_k^j|$ is higher than a predefined path convergence threshold Φ. Note that Φ represents an upper bound. The convergence of the algorithm is achieved if $N(\lambda) \le \Phi$. The relative Gap for the DUE is as follows [42]:

$$Gap = \frac{\sum_o \sum_d \sum_{k\in\Omega^{od}} Q_k^{od}\left(V_k^{od} - \min\left(V^{od}\right)\right)}{\sum_o \sum_d Q^{od} \min\left(V^{od}\right)} \quad (9)$$

where Q^{od} is the total demand for the od pair; and V_k^{od} is the average travel time of route k; and $\min(V^{od})$ is the minimum route travel time for the od pair.

The Gap function (see (9)) represents the difference between the travel costs and the equilibrium travel costs. Thus, under perfect DUE conditions, $Gap = 0$. This means that all users choose the routes with the minimum travel

Input the AL^{od} (if they are set exogenously).
Input the network, demand scenario and simulation duration T.
Calculate the route choice set Ω^{od} for each od pair.
Perform an initial network loading.
Set $N(\lambda) > \Phi$ and $Gap^{BRUE} > tol$. Initialize $j = 1, \eta, \zeta, \alpha_j = 1$ and $Q_k^{j=1} = 0$.
Set the MSA stopping criterion tol.
while $Gap \geq tol$ **or** $N(\lambda) \geq \Phi$ **or** $j \leq N_{max}$ **do**
  Set $Q_k^j = Q_k^{j+1}$.
  If set endogenously, update the AL^{od} based on Eq. (2) or Eq. (3).
  Perform M error samplings at the link level, based on η and ζ.
  for m=1 to M **do**
    Compute the route utilities.
    Determine the *satisficing* routes and update ω^{od}, $\forall k \in \Omega^{od} \wedge \forall(o,d) \in \Xi$.
    Based on the defined search order, perform an all-or-nothing assignment. If the search order is the indifferent preferences, repeat A times the all-or-nothing assignment, for the BR-SUE case.
    If $\omega^{od} = \emptyset$, all users are assigned to the minimum utility route.
  end
  Update the new route flows Q_k^*, $\forall k \in \Omega^{od} \wedge \forall(o,d) \in \Xi$, based on an averaging of the users choices over all error samples.
  Update the route flows according to Eq. (6).
  Run the LWR mesoscopic simulator [21].
  Based on the link travel times, fit a gamma distribution to update η and ζ.
  Calculate the Gap (Eq. (9)) and the number of violations $N(\lambda)$.
  Update $\alpha_j = 1/j$.
  Set $j = j + 1$.
end
Save the route flows: Q_k^{j+1}, $\forall k \in \Omega^{od} \wedge \forall(o,d) \in \Xi$.

ALGORITHM 1: Dynamic implementation algorithm of the *satisficing* model.

times. Under SUE conditions, the Gap is higher than 0, however small. In this case not all users choose the routes with the minimum travel times. In the case of bounded rationality, the Gap value increases as AL^{od} increases. The Gap function is also a measure of how close users are to the equilibrium route travel times (or T^{UE}). The definition of the Gap as defined in (9) is valid for DUE and SUE and informs on how far we are from the DUE. For both the BR-DUE and BR-SUE convergence, we modify the Gap function as follows:

$$Gap^{BRUE} = \frac{\sum_o \sum_d \sum_{k\in\omega^{od}} Q_k^{od} \max\left(V_k^{od} - AL^{od}, 0\right)}{\sum_o \sum_d Q^{od} AL^{od}} \quad (10)$$

Thus, under BR-DUE or BR-SUE conditions, the Gap is about 0 if $\min(V^{od}) \leq V_k^{od} \leq AL^{od}$, $\forall(o,d) \in \Xi$ and the equilibrium condition is fulfilled. Note that, throughout the paper, we use the definition of the Gap as in (9) as an indicator that measures how far the bounded rational equilibria are from the DUE and (10) as the equilibrium convergence criterion for the MSA.

We present the solution algorithm of this framework in Algorithm 1. Note that the difference between Algorithm 1 and that proposed by Sheffi [4] is that we assign the users to *satisficing* routes instead of routes with the minimal travel times. They are assigned to these *satisficing* routes according to one of the search orders discussed previously (i.e., indifferent or strict preferences) at every descent step of the MSA. The first step before entering the MSA loop consists in calculating the route choice set Ω^{od}, for each od pair. It defines the set of routes for the users choices. We then perform an initial loading on these routes and consider the number of violations, the Gap (see (10)), and the maximum number of iterations for the MSA convergence criteria. The corresponding tol, Φ, and N_{max} are set. It is also necessary to define the input scale (η) and shape (ζ) parameters of the link travel time gamma distributions for the first Monte Carlo simulations. We then enter in the MSA loop and the AL^{od} is first updated based on the average route travel times (see (2) or (3)). The next step consists in performing the link error sampling considering the η and ζ parameters. This is done through Monte Carlo simulations. The algorithm then loops over all the M error samples and locally solves the BR-DUE problems. For each sample, the route utilities are computed to identify the *satisficing* routes based on AL^{od}. This defines the *satisficing* set of routes ω^{od}. Users are assigned based on the predefined search order (indifferent or strict preferences) based on an all-or-nothing procedure to one route in ω^{od}. It should be noted that in the case of solving the BR-SUE and taking the indifferent preferences into account, it is necessary to repeat the all-or-nothing assignment on the *satisficing* routes A times. The users choices for the local BR-DUE correspond to averaging the previous choices. By applying the law of large numbers, when A is large, we converge to the same average values. The new temporary route flows Q_k^*

Table 1: Notations used in this paper.

k	Route.
a	Link.
o	Origin.
d	Destination.
Q^{od}	Total demand for the od pair.
Ξ	Set of all od pairs of the network.
U_k	Perceived route utility.
V_k	Deterministic route utility.
ϵ_k	Uncertainty or error term.
ϵ_a	Link error term.
Ω^{od}	Route choice set for od pair.
N	Number of routes listed in the route choice set per each od pair.
Q_k	Flow of route k.
q_a	Link flow.
δ_{ak}	Dummy variable that equals 1 if link a belongs to route k.
Γ_a	Set of links that define the graph.
AL^{od}	Aspiration level defined at the od level.
$\overrightarrow{V}$	Vector containing all deterministic route utilities.
Δ^{od}	Tolerance or Indifference Band.
δ_p	Dummy variable that equals 1 if route p is listed on the choice set.
δ_q	Dummy variable that equals 1 if route q is listed on the choice set.
ω^{od}	Set of *satisficing* routes.
T^{UE}	User Equilibrium route travel time.
Υ^{od}	Users' strict preference order for the *od* pair.
M	Number of draws for the error term discretization.
$N(\lambda)$	Number of violations.
Φ	Pre-defined path convergence threshold.
η and ζ	Scale and shape parameters of a gamma distribution.
j	Iterative counter of the MSA method.
tol	Tolerance for the stopping criterium of the MSWA-I.
N_{max}	Maximum number of descent step iterations.
T^a_{ff}	Free-flow travel time of link a.
T	Simulation period.
A	Number of assignment repetitions for BR-SUE and indifferent preferences search order.

correspond to the average of all the local BR-DUE choices. The new route flows Q_k^{j+1} are updated according to (6) and the network loading is updated. To determine time-dependent link costs that consider congestion, shock-waves, and spillback effects, we run an LWR mesoscopic traffic simulator [21]. The link travel time distributions are obtained based on the simulated vehicle travel times. To update η and ζ, we fit a gamma distribution to each link travel time distribution. The updated values of η and ζ will be used to perform the error samplings in the next MSA descent step. The Gap^{BRUE} (see (10)) and number of violations $N(\lambda)$ are updated based on the new average route travel times through the individual vehicles travel times. This process is repeated until convergence is achieved. Note that Algorithm 1 also allows solving the BR-DUE by setting $\eta = 0$ and $\zeta = 0$.

In Table 1 we summarize the notations of all symbols and variables used in this paper.

3. First Tests on a Toy Network

We first test the bounded rational model framework discussed in the previous section, on a toy Braess network, and consider a static flow dependent utility function. The goal of these simple initial tests is to assess and acquire insight into how the route flows at equilibrium change according to the two definitions of the search order (i.e., indifferent and strict preference search order) and increasing values of AL^{od}. The AL^{od} are defined exogenously.

3.1. Definition of the Test Network. For the first test, we consider that the perceived travel times (i.e., route utility) depend only on route flows and route free-flow travel times. We resort to the following definition of the perceived route utility, $U_k(Q_k)$:

$$U_k(Q_k) = \sum_{a\in\Gamma_a} (V_a(q_a) + \epsilon_a)\,\delta_{ak}, \quad \forall k \in \Omega^{od} \tag{11}$$

TABLE 2: BR-DUE route flows for different values of the AL^{od}. The Gap values represent average values based on 1000 repetitions of the BR-DUE calculations.

Δ^{od}	AL^{od}	Q_1/Q^{od}	Q_2/Q^{od}	Q_3/Q^{od}	U_1	U_2	U_3	$\overline{Gap}$
0	46.6	0.17 ± 0.00	0.00 ± 0.00	0.83 ± 0.00	46.7	58.3	46.7	0.00
1	47.0	0.20 ± 0.00	0.00 ± 0.00	0.80 ± 0.00	47.0	58.0	46.0	0.00
2	47.3	0.23 ± 0.00	0.00 ± 0.00	0.77 ± 0.00	47.3	57.7	45.4	0.01
3	47.7	0.27 ± 0.00	0.00 ± 0.00	0.73 ± 0.00	47.7	57.3	44.7	0.02
4	48.0	0.30 ± 0.00	0.00 ± 0.00	0.70 ± 0.00	48.0	57.0	44.0	0.03
5	50.0	0.22 ± 0.16	0.00 ± 0.00	0.78 ± 0.16	47.2	57.8	45.6	0.06
15	60.0	0.41 ± 0.19	0.01 ± 0.01	0.58 ± 0.19	49.0	56.0	41.6	0.12
20	65.0	0.34 ± 0.14	0.17 ± 0.07	0.49 ± 0.21	46.7	58.3	39.8	0.19
30	66.3	0.37 ± 0.26	0.25 ± 0.15	0.37 ± 0.26	46.2	58.8	37.5	0.34
40	76.3	0.33 ± 0.27	0.35 ± 0.28	0.32 ± 0.26	44.8	60.2	36.5	0.46
50	86.3	0.34 ± 0.27	0.33 ± 0.26	0.33 ± 0.26	45.1	59.9	36.5	0.44
75	105.0	0.33 ± 0.27	0.33 ± 0.27	0.33 ± 0.27	45.0	60.0	36.6	0.45
100	136.5	0.33 ± 0.26	0.33 ± 0.26	0.34 ± 0.27	45.1	59.9	36.8	0.43

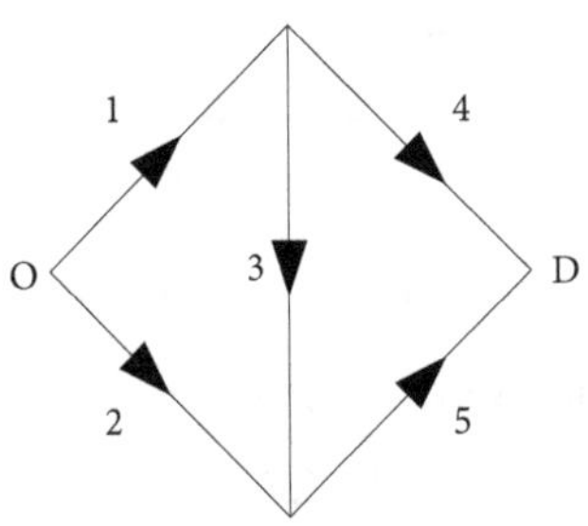

FIGURE 1: Braess network.

where $V_a(q_a) = \theta_1 T_{ff}^a + \theta_2 q_a$ and $q_a = \sum_{l\in a} \delta_{al} Q_l$; T_{ff}^a is the link free-flow travel time of link a; q_a is the flow of link a; δ_{al} is a dummy variable that equals 1 if route l uses link a, or 0 otherwise; Q_l is the total flow of route l; and ϵ_a is the error term associated with link a. θ_1 and θ_2 are parameters set to 1.

For the static tests, we consider the Braess network (Figure 1). The choice set is $\Omega_k = \{1,4;2,5;1,3,5\}$, $\forall k = 1,2,3$. The following sets of link free-flow travel times are considered: $T_{ff}^1 = 5$, $T_{ff}^2 = 45$, $T_{ff}^3 = 10$, $T_{ff}^4 = 30$, and $T_{ff}^5 = 5$.

For the MSA convergence, we set $tol = 10^{-2}$, $\Phi = 0$, and $N_{max} = 10000$. A total demand of $Q^{od} = 10$ is considered. As a reference of user perfect rationality, we consider the DUE. Note that the ϵ_a terms are set to 0 for both the DUE and the BR-DUE calculations. For the BR-SUE calculations, we consider a gamma distribution with a shape parameter set to $\eta = 1$ and a scale parameter set to $\zeta = 4$; and a total of $M = 2000$ error samples. The BR-DUE and BR-SUE are calculated based on Algorithm 1, except that the link travel times are not updated considering the LWR mesoscopic traffic simulator [21]. Instead, we consider the route utility as defined in (11). The link travel times are only sampled at the initiation of the MSA descent procedure.

3.2. Indifferent Preference Search Order and Exogenous AL^{od}. In this section, we analyze both the BR-DUE and BR-SUE results considering the users indifferent preference search order and the AL^{od} defined as in (2). We also analyze the algorithms convergence towards the equilibrium solution through the Gap function. To define the search order, we consider a uniform distribution to simulate the users choices among the set of *satisficing* routes ω^{od}. This procedure must be repeated many times to reach convergence by the law of large numbers (on average) with the same solution of route flows. Then, for each value of AL^{od}, we repeat the assignment procedure 1000 times and calculate the average route flows and corresponding standard deviation. We do so for both the BR-DUE and BR-SUE calculations. First, under DUE conditions, only routes 1 and 3 are used. This means that $T^{UE} = U_1 = U_3 < U_2$ and corresponds to the route flows: $Q_1 = 1.7$, $Q_2 = 0$, and $Q_3 = 8.3$. Note that T^{UE} is the route travel times at the User Equilibrium. Under SUE conditions, only routes 1 and 3 are *satisficing*. But, due to the users perception of travel times, there is a residual flow on route 2. The route flows are $Q_1 = 3.5$, $Q_2 = 0.3$, and $Q_3 = 6.2$.

We first analyze the BR-DUE results, calculated for increasing values of Δ^{od}. These results are listed in Table 2. The first test consists in reproducing the perfect rationality behavior, by setting $\Delta^{od} = 0 \Longrightarrow AL^{od} = T^{UE} = 46.7$. The route flows under BR-DUE are similar to the DUE. Then, to analyze the equilibrium results for increasing values of $\Delta^{od} \in [0, +\infty[$, we must first identify the critical points for the BR-DUE, that is to say the utility values of each route when the total demand Q^{od} is assigned to each of the routes. We first consider $Q_1 = 10$, $Q_2 = 0$, and $Q_3 = 0$, which yields $U_1 = 55$. Similarly for route 2, $U_2(Q_1 = 0, Q_2 = 10, Q_3 = 0) = 60$, and for route 3, $U_3(Q_1 = 0, Q_2 = 0, Q_3 = 10) = 50$. These critical points play an important role in analyzing the equilibrium solutions. The minimum of the critical points indicates the value of AL^{od} from which the objective function is no longer convex. We analyze the BR-DUE route flows for increasing values of AL^{od} in more detail. For $AL^{od} \in [T^{UE}, 50[$, the users switch from route 3 to route 1. Note that the users switch from the *satisficing* routes with higher route flows to the ones with

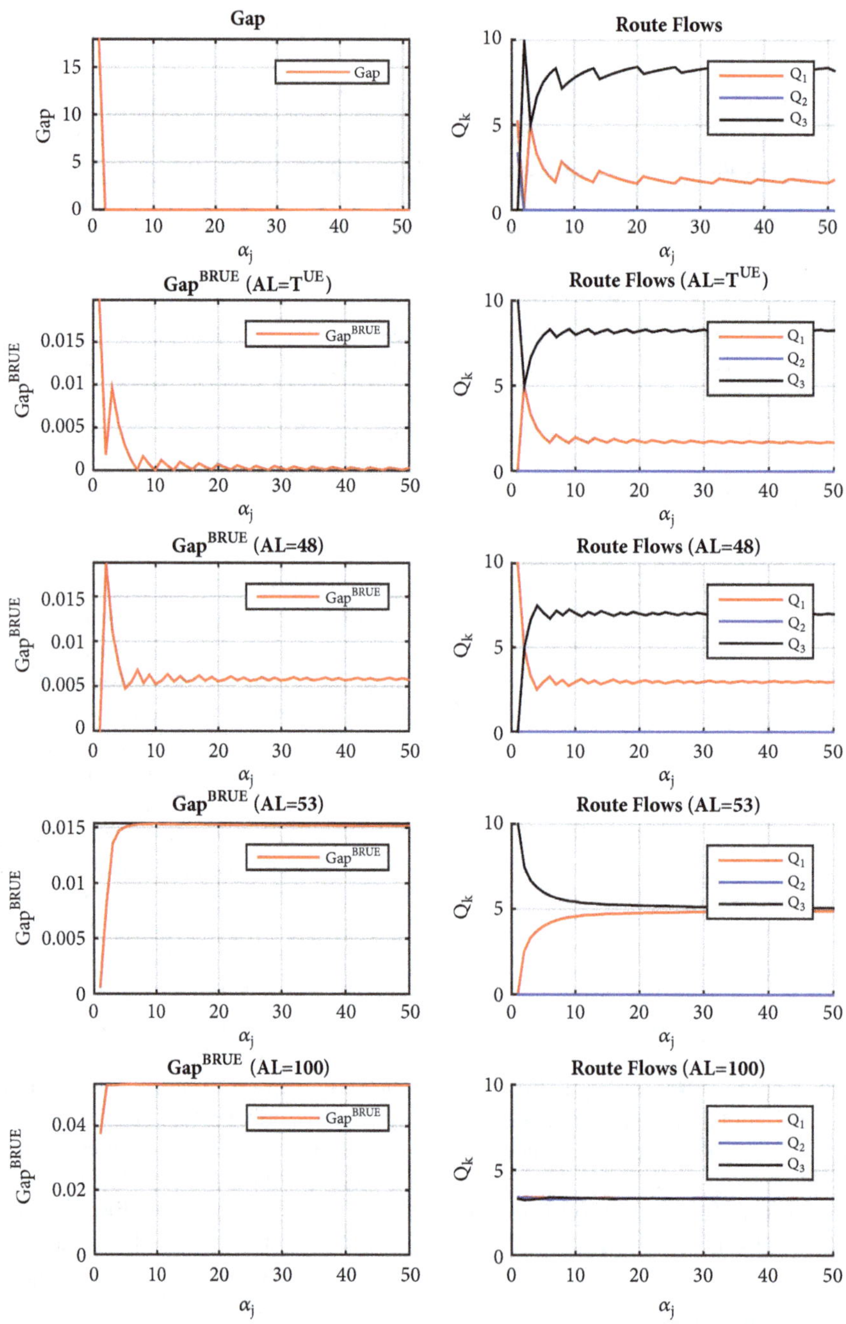

FIGURE 2: Gap (left) and route flows (right) as a function of the increasing number of MSA descent steps j for the DUE and several values of $AL^{od} = T^{UE}, 48, 53, 100$.

lower route flows. For $AL^{od} \in [50, 55[$, the algorithm does not converge to the same solution as evidenced by the standard deviation values listed in Table 2. For example, for $AL^{od} = 50$, two feasible solutions are found: $(Q_1 = 5, Q_2 = 0, Q_3 = 5)$ which yields $(U_1 = 50, U_2 = 55, U_3 = 40)$; and $(Q_1 = 0, Q_2 = 0, Q_3 = 10)$ that yields $(U_1 = 45, U_2 = 60, U_3 = 50)$. The convergence of the algorithm to any of these feasible solutions depends on the initial loading of the network for the MSA algorithm. This explains why we do not converge to the same set of route flows for $AL^{od} \geq 50$. For $AL^{od} \in [55, 60[$, route 2 becomes *satisficing* and users switch from routes 3 and 1 to route 2. For $AL^{od} \geq 60$, the route flows will converge to 1/3 as the value of AL^{od} increases. This represents the users indifference for choosing any of the *satisficing* routes. Under SUE conditions the values in Table 2 are obtained.

We investigate the algorithms convergence for different values of Δ^{od}, which corresponds to different values of AL^{od}, as shown in Figure 2. To do this, we consider a total of 50 descent steps j of the MSA algorithm, despite the convergence criterion of $Gap \leq 10^{-2}$ being verified for a

TABLE 3: BR-SUE route flows for different values of the AL^{od}.

Δ^{od}	AL^{od}	Q_1/Q^{od}	Q_2/Q^{od}	Q_3/Q^{od}	U_1	U_2	U_3	$\overline{Gap}$
0	42.6	0.34 ± 0.00	0.03 ± 0.00	0.63 ± 0.00	48.1	56.9	42.5	0.06
1	43.5	0.34 ± 0.00	0.03 ± 0.00	0.63 ± 0.00	48.1	56.9	42.5	0.06
2	44.6	0.34 ± 0.00	0.03 ± 0.00	0.63 ± 0.00	48.1	56.9	42.5	0.06
3	45.5	0.34 ± 0.00	0.03 ± 0.00	0.63 ± 0.00	48.1	56.9	42.5	0.06
4	46.5	0.34 ± 0.00	0.03 ± 0.00	0.63 ± 0.00	48.1	56.9	42.5	0.06
5	47.6	0.34 ± 0.00	0.03 ± 0.00	0.63 ± 0.00	48.1	56.9	42.5	0.06
15	56.4	0.40 ± 0.00	0.03 ± 0.00	0.57 ± 0.00	48.7	56.3	41.3	0.08
20	59.7	0.42 ± 0.00	0.09 ± 0.00	0.49 ± 0.00	48.3	56.7	39.8	0.13
30	67.7	0.39 ± 0.01	0.23 ± 0.00	0.39 ± 0.00	46.6	58.4	37.7	0.22
40	76.9	0.35 ± 0.00	0.32 ± 0.01	0.35 ± 0.00	45.4	59.6	36.9	0.27
50	86.8	0.33 ± 0.00	0.33 ± 0.00	0.34 ± 0.00	45.1	59.9	36.7	0.29
75	111.9	0.33 ± 0.00	0.33 ± 0.00	0.33 ± 0.00	45.0	60.0	36.7	0.29
100	136.7	0.33 ± 0.00	0.33 ± 0.00	0.33 ± 0.00	45.0	60.0	36.7	0.29

lower number of j. This allows observing that the solution no longer changes after the convergence criterion is satisfied. In Figure 2, we show the evolution of the Gap and route flows for increasing values of j, for the DUE and $AL^{od} = T^{UE}, 48, 53, 100$. For all four cases, the Gap value converges to a constant value for increasing values of j as well as the route flows.

We analyze the BR-SUE results for increasing values of Δ^{od}. These results are listed in Table 3. We first confirm that the route flows under BR-SUE and SUE are similar, by setting $\Delta^{od} = 0$. For $AL^{od} \in [42.6, 55]$, the users change from route 3 to route 1. For $AL^{od} \geq 57$, route 2 becomes *satisficing* and the users also start choosing this route. The comparison results of the BR-DUE and BR-SUE for $AL^{od} \geq 65$ are of particular interest. In both cases, the route flows converge to 1/3 when the value of AL^{od} is sufficiently large. This represents the users indifference for choosing any of the *satisficing* behaviors, since all the routes comply with the condition defined by (1). The effect of the perception of the route travel times explains the small differences verified in the route flows between the BR-DUE and BR-SUE for the same value of AL^{od}. For example, it is interesting to consider the case where $\Delta^{od} = 40$. The BR-DUE route flows are ($Q_1 = 3.3; Q_2 = 3.5; Q_3 = 3.2$) and the BR-SUE route flows are ($Q_1 = 3.5; Q_2 = 3.2; Q_3 = 3.5$). It can also be seen from the BR-SUE results shown in Table 3 that the MSA algorithm converges to the same solution, as evidenced by the standard deviation of the route flows. To solve the BR-SUE, we solve M BR-DUE problems locally, and since the search order is the indifference preferences, the users are assigned A times for each BR-DUE problem. By applying the law of large numbers, we converge on average to the same solution when A is sufficiently large.

We also observe that, for both the BR-DUE and BR-SUE results, the average $\overline{Gap}$ values increase as we increase AL^{od}, as expected. This represents the effect of the *satisficing* behavior, where users choose *satisficing* routes instead of the routes with the shortest travel times.

3.3. Strict Preference Search Order and Exogenous AL^{od}. In this section, we analyze the BR-DUE results calculated considering an exogenous definition of AL^{od} and a strict preference search order [19]. We calculate the BR-DUE results for the Braess network (Figure 1), considering the six possible strict preference search orders (Υ^{od}, $\forall k = 1, 2, 3$). For the calculations, we consider our bounded rational model framework and the model discussed by Zhao and Huang [19]. The mathematical methodology discussed by Zhao and Huang [19] is not suitable for a dynamic implementation considering a traffic simulator. This is because it requires solving suboptimization problems to calibrate the AL of the sub-preferred routes. In this section, we compare the route flows calculated considering the two frameworks and the six preference search orders. The AL^{od} are set exogenously according to each strict preference order. We apply the model discussed by Zhao and Huang [19], considering that the utility of the most preferred route is equal to the value of the aspiration level AL^{od}; and the route flows of the remaining routes correspond to the result obtained by solving the sub-UE problem, as done by Zhao and Huang [19]. The results are listed in Table 4.

It can be seen that when $AL^{od} = T^{UE} = 46.6$, the route flows obtained for both models are equivalent to the DUE result. Thus, both models are able to reproduce the users perfect rationality whatever the preference order.

Consider the first strict preference order $\Upsilon^{od} = 1, 2, 3$. For $47.5 \leq AL^{od} < 50$, in both models the users switch directly from route 3 to the most preferred route 1. The flows on these two routes are equal for both models. But in our model, route 2 is not selected because it is not considered as *satisficing* whereas in the model of Zhao and Huang [19], the assignment problem solved is $U_1 = A_1$ and $U_2 = U_3$. Under sub-User Equilibrium (sub-UE) conditions, $U_2 > U_3$. This means that the remaining users that have not chosen the most preferred route 1 will choose route 3. In our model, the users are assigned to the most preferred route until the *satisficing* condition (see (1)) is satisfied. Users are then assigned to the

TABLE 4: Route flow distribution for the Braess network for different values of the AL and under BR-DUE conditions. A set of strict preferences is considered for the search order.

	Our model						Zhao and Huang [19]					
AL^{od}	Q_1/Q^{od}	Q_2/Q^{od}	Q_3/Q^{od}	U_1	U_2	U_3	Q_1/Q^{od}	Q_2/Q^{od}	Q_3/Q^{od}	U_1	U_2	U_3
Preference order $\Upsilon^{od} = 1\ 2\ 3$												
46.6	0.17	0.00	0.83	46.7	58.3	46.7	0.17	0.00	0.83	46.7	58.3	46.6
47.5	0.25	0.00	0.75	47.5	57.5	45.0	0.25	0.00	0.75	47.5	57.5	45.0
50.0	0.50	0.00	0.50	50.0	55.0	40.0	0.50	0.00	0.50	50.0	55.0	40.0
52.5	0.87	0.13	0.00	52.5	52.5	30.0	0.75	0.00	0.25	52.5	52.5	35.0
55.0	1.00	0.00	0.00	55.0	50.0	30.0	1.00	0.00	0.00	55.0	50.0	30.0
Preference order $\Upsilon^{od} = 1\ 3\ 2$												
46.6	0.17	0.00	0.83	46.7	58.3	46.7	0.17	0.00	0.83	46.7	58.3	46.6
47.5	0.25	0.00	0.75	47.5	57.5	45.0	0.25	0.00	0.75	47.5	57.5	45.0
50.0	0.50	0.00	0.50	50.0	55.0	40.0	0.50	0.00	0.50	50.0	55.0	40.0
52.5	0.75	0.00	0.25	52.5	52.5	35.0	0.75	0.00	0.25	52.5	52.5	35.0
55.0	1.00	0.00	0.00	55.0	50.0	30.0	1.00	0.00	0.00	55.0	50.0	30.0
Preference order $\Upsilon^{od} = 2\ 1\ 3$												
46.6	0.17	0.00	0.83	46.7	58.3	46.7	~	~	~	~	~	~
47.5	0.25	0.00	0.75	47.5	57.5	45.0	~	~	~	~	~	~
50.0	0.50	0.00	0.50	50.0	55.0	40.0	~	~	~	~	~	~
55.0	0.75	0.25	0.00	50.0	55.0	30.0	0.50	0.00	0.50	50.0	55.0	40.0
60.0	0.50	0.50	0.00	45.0	60.0	30.0	0.125	0.125	0.75	45.0	60.0	45.0
65.0	0.25	0.75	0.00	40.0	65.0	30.0	0.00	0.50	0.50	40.0	65.0	40.0
70.0	0.00	1.00	0.00	35.0	70.0	30.0	0.00	1.00	0.00	35.0	70.0	30.0
Preference order $\Upsilon^{od} = 2\ 3\ 1$												
46.6	0.17	0.00	0.83	46.7	58.3	46.7	~	~	~	~	~	~
47.5	0.13	0.00	0.87	46.3	58.7	47.5	~	~	~	~	~	~
50.0	0.00	0.00	1.00	45.0	60.0	50.0	~	~	~	~	~	~
55.0	0.00	0.00	1.00	45.0	60.0	50.0	0.50	0.00	0.50	50.0	55.0	40.0
60.0	0.00	0.00	1.00	45.0	60.0	49.9	0.125	0.125	0.75	45.0	60.0	45.0
65.0	0.00	0.50	0.50	40.0	65.0	40.0	0.00	0.50	0.50	40.0	65.0	40.0
70.0	0.00	1.00	0.00	35.0	70.0	30.0	0.00	1.00	0.00	35.0	70.0	30.0
Preference order $\Upsilon^{od} = 3\ 1\ 2$												
46.6	0.17	0.00	0.83	46.7	58.3	46.7	0.17	0.00	0.83	46.7	58.3	46.6
47.0	0.15	0.00	0.85	46.5	58.5	47.0	0.15	0.00	0.85	46.5	58.5	47.0
48.0	0.10	0.00	0.90	46.0	59.0	48.0	0.10	0.00	0.90	46.0	59.0	48.0
49.0	0.05	0.00	0.95	45.5	59.5	49.0	0.05	0.00	0.95	45.5	59.5	49.0
50.0	0.00	0.00	1.00	45.0	60.0	50.0	0.00	0.00	1.00	45.0	60.0	50.0
Preference order $\Upsilon^{od} = 3\ 2\ 1$												
46.6	0.17	0.00	0.83	46.7	58.3	46.7	0.17	0.00	0.83	46.7	58.3	46.6
47.0	0.15	0.00	0.85	46.5	58.5	47.0	0.15	0.00	0.85	46.5	58.5	47.0
48.0	0.10	0.00	0.90	46.0	59.0	48.0	0.10	0.00	0.90	46.0	59.0	48.0
49.0	0.05	0.00	0.95	45.5	59.5	49.0	0.05	0.00	0.95	45.5	59.5	49.0
50.0	0.00	0.00	1.00	45.0	60.0	50.0	0.00	0.00	1.00	45.0	60.0	50.0

sub-preferred routes if and only if they are *satisficing*. Since route 2 is not *satisficing*, the remaining users choose route 3. Thus, both models yield similar route flows for these two values of AL^{od}. For $AL^{od} = 52.5$, route 2 becomes *satisficing* for our model. Thus the users will switch according to the strict preference order Υ^{od}. Thus, the users will first switch from route 3 to 2 and then from 2 to 1. In the case of the model of Zhao and Huang [19], the sub-UE solution corresponds to $U_2 > U_3$. Thus, no user chooses route 2. For $AL^{od} \geq 55$, all the users choose the most preferred route 1 for both models.

Consider the second strict preference order $\Upsilon^{od} = 1, 3, 2$. In this case, the users switch directly from route 3 to the most preferred route 1 as AL^{od} increases. In the case of our model, the users are assigned to the most preferred route until it is considered as *satisficing*. Then, the remaining users are assigned to the sub-preferred route 3. In the case of the

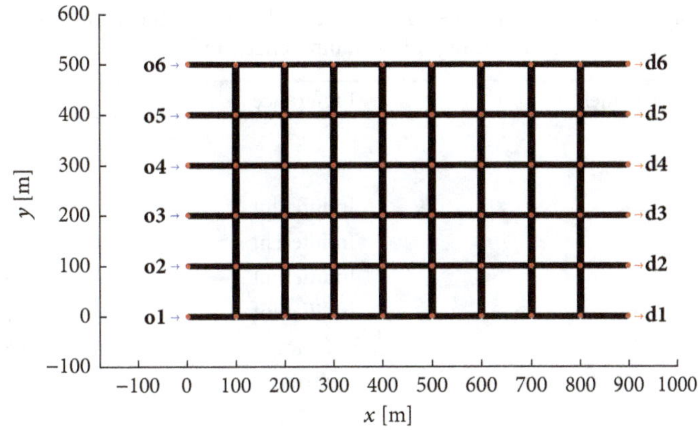

FIGURE 3: Manhattan network.

model of Zhao and Huang [19], the sub-UE solution also corresponds to $U_2 > U_3$ and thus no users choose route 2. So, for this strict preference order, both models give similar route flows as AL^{od} increases.

Consider the third strict preference order $\Upsilon^{od} = 2, 1, 3$. Given this preference order and for $AL^{od} \leq 50$, route 2 is not *satisficing* for our model. Therefore, the users will switch from route 3 (the less preferred) to the second most preferred route 1. The model of Zhao and Huang [19] cannot be applied for $AL^{od} < 55$, since $AL^{od} \neq U_2$. This would lead to a violation of the strict preference order assumption. We make this assumption more flexible. Although route 2 is the most preferred route, it is not considered as *satisficing* and the users choose other most preferred routes that are *satisficing*. This is the case of route 1. In the case of our model, for $AL^{od} \geq 55$, all the users switch from route 1 to route 2 (the most preferred one). In the case of the model of Zhao and Huang [19], the users will switch from route 1 to routes 2 and 3 (for $AL^{od} = 55$ and $AL^{od} = 60$) and then from 3 to 2 (for $AL^{od} = 65$ and $AL^{od} = 70$). Although route 3 is that least preferred, the users switch from route 1 to route 3. Note that we solve the following assignment problem for the model of Zhao and Huang [19]: $AL^{od} = U_2$ and $U_1 = U_3$.

Consider the fourth strict preference order $\Upsilon^{od} = 2, 3, 1$. In the case of our model, for $AL^{od} \leq 60$, route 2 is not *satisficing*. Therefore, all the users switch from route 1 to route 3 in accordance with the strict preference order. For $AL^{od} \geq 60$, route 2 becomes *satisficing* and the users switch from route 3 to route 2. In the case of the model of Zhao and Huang [19], the condition $AL^{od} = U_2$ is satisfied only for $AL^{od} \geq 55$. For $AL^{od} \geq 55$, we observe a route flow pattern similar to that in the previous strict preference order ($\Upsilon^{od} = 2, 1, 3$) for the same reasons discussed previously.

Consider the fifth and sixth strict preference orders $\Upsilon^{od} = 3, 1, 2$ and $\Upsilon^{od} = 3, 2, 1$. In the case of our model, route 2 is not *satisficing* for these two strict preference orders. Thus, users switch directly from route 1 to route 3, as AL^{od} increases. For $AL^{od} \geq 50$, all users choose the most preferred route 3. In the case of the model of Zhao and Huang [19], we solve the following assignment problem: $AL^{od} = U_3$ and $U_1 < U_2$ (sub-UE problem). Since $U_1 < U_2$ for all the listed values of AL^{od}, the remaining users that have not chosen route 3 choose route 1 for both strict preference orders. This is why we observe similar route flows for both models and both strict preference orders.

In summary, we validate our bounded rational framework considering a strict users preference search order, by comparing the route flows at equilibrium with the results obtained by the model of Zhao and Huang [19]. Moreover, our bounded rational framework is suitable for dynamic implementation with a traffic simulator and will be tested in the next section.

4. Dynamic Implementation on a Manhattan Network

In this section, we investigate the influence of different types of bounded rational user behavior on (i) individual route flows and (ii) network performance in terms of its internal, inflow, and outflow capacities. To do this, we consider the implementation of the bounded rational framework described in Algorithm 1. To determine the time-dependent cost paths that account for congestion, shock-waves, and spillback effects, we consider a mesoscopic LWR traffic simulator [21]. The tests are performed on a Manhattan network. We consider the indifferent and strict preference search orders and both definitions of the AL^{od}, as in (2) and (3).

4.1. Test Scenario Definition. For the dynamic implementation, we consider the Manhattan network composed of 134 links, as shown in Figure 3. All the links have the same length of 100 meters. Traffic lights regulate all the intersections. A green light duration of 45 seconds is considered for the traffic lights of the horizontal links whereas a duration of 15 seconds is considered for the traffic lights of the vertical links. Green times are set in the West-East and in the North-South directions. The offsets considered are of 10 and 20 seconds.

A triangular fundamental diagram is considered for each lane of the network, with the following parameters: $u = 15$ (m/s) for the free-flow speed; $w = 5$ (m/s) for the wave speed;

Table 5: The Gap value and the average travel times per route $\overline{TT_k}$ [in s] calculated from the distributions shown in Figure 6 are also listed. These values are listed for the DUE, SUE, and different settings of the indifference band.

Model	ID Model	Preference	Gap	$\overline{TT_k}$ [s]
DUE	1	~	0.3	315
SUE	2	~	0.3	295
IB ($\Delta^{od} = 0$)	3	Indifferent	0.3	294
IB ($\Delta^{od} = 100$)	4	Indifferent	7.3	337
IB ($\Delta^{od} = 500$)	5	Indifferent	14.7	332
IB [20]	6	Indifferent	16.2	350
IB ($\Delta^{od} = 0$)	7	Strict	68.0	312
IB ($\Delta^{od} = 100$)	8	Strict	68.8	297
IB ($\Delta^{od} = 500$)	9	Strict	68.8	297
IB [20]	10	Strict	69.1	297

and $k_{jam} = 0.2$ (veh/m/lane) for the jam density. The entry links (i.e., from O1 to 6) have two lanes. The total link flow is assigned equally on each lane.

The Manhattan network shown in Figure 3 has six entries (identified by O1 to O6 in Figure 3) and exits (identified by D1 to D6 in Figure 3). For each of the six entries, we consider a constant inflow (demand) of 0.5 (veh/s). There is no capacity restriction at the exits. There is a total of 36 possible od pairs. To define the choice set Ω^{od}, we consider 3 paths per od pair. These paths are calculated using a K-shortest path algorithm. This gives a total of 108 routes, considering the 36 possible od pairs.

For the dynamic tests and the bounded rational route choice model, we consider the endogenous definitions of the indifference band for AL^{od} as defined by (2) and (3) and two settings of the search order:

(i) an *indifferent* preference search order, where users randomly choose any of the *satisficing* alternatives; or, the least worst if there are no *satisficing* alternatives.

(ii) a *strict preference* search order, where users have a strict preference for the routes with the most reliable travel times. We consider the variances of the route travel times as the time reliability indicator. Then, the set of strict preference is built by ordering the routes from the lowest to the highest variance value for each od pair. This set of preferences is updated at every descent step of the MSA, based on the route travel time distributions of the previous simulation.

Considering this search order, users seek *satisficing* alternatives based on this set of strict preferences and on AL^{od}. Similarly, for the strict preference search order discussed in Section 3.3, the users choose only the most preferred route if it is perceived as *satisficing*, i.e., that conforms to (1). Then, if the most preferred route is not perceived as *satisficing*, the users consider the other most preferred routes until they find one that is *satisficing*. If none of the routes are *satisficing*, the users choose the route with the minimal travel time.

As a reference, we consider the DUE and SUE. To solve the SUE, we consider the Probit model with gamma distributed error terms and use Monte Carlo simulations [4]. For the indifference band defined by (2), we consider three exogenous values for Δ^{od}: 0; 100; and 500. We have a total of 10 simulation scenarios, considering both search orders defined above. The total simulation period is $T = 3000$ seconds. For the convergence, we set $tol = 10^{-2}$, $\Phi = 0$, and $N_{max} = 250$.

4.2. Analysis of the Individual Route Flows. In this section, we analyze the individual route flows for the 10 simulation scenarios. Each scenario is identified by one ID number, as listed in Table 5. We also list the Gap values in Table 5, that are calculated using (9). In Figures 4 and 5, we show the route flow distributions for each od pair of the network and all ten scenarios. In Figure 6, we show the distributions of the average route travel times for the 10 simulation scenarios. Note that the average travel time per route for each scenario (i.e., the average of these distributions) is also listed in Table 5.

We analyze the individual route flows shown in Figures 4 and 5. By setting $\Delta^{od} = 0$ (Model 3), we observe that for the indifferent preference search order, we obtain similar route flows compared to the SUE (Model 2). However, this is not observed for the strict preference search order, when comparing $\Delta^{od} = 0$ (Model 7) and the SUE (Model 2). This is due to the specific definition of the search order, where the routes with the most reliable travel times (i.e., with the lowest variances) may not correspond to the routes with the lowest travel times. This is also evidenced by the Gap values listed in Table 5, for the settings of the strict preference search order. Also note that in the case of the indifferent preference search order, setting $\Delta^{od} = 0$, only the lowest travel time route per od pair is considered as *satisficing* at each descent step of the MSA. For the indifferent preference search order, the users indifference increases as we increase Δ^{od} from 0 to 500. The route flows will then converge to 1/3 for all the od pairs (Model 5, in Figures 4 and 5). For $\Delta^{od} = 500$, the indifference band is sufficiently high with the result that all the routes in Ω^{od} for all od pairs are *satisficing*. Thus, the users can choose any of the routes. Since the users indifference increases, they will choose routes with higher travel times and consequently the

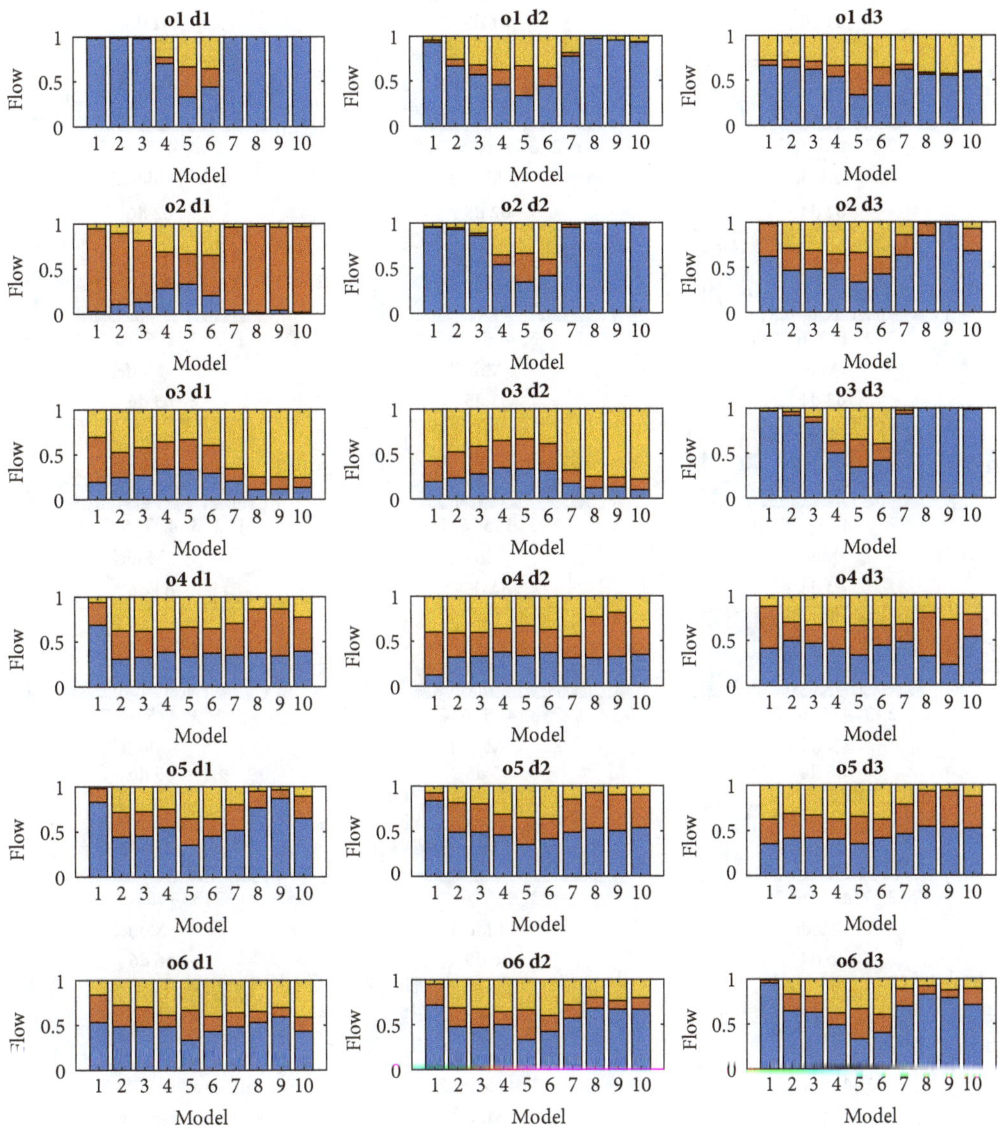

FIGURE 4: Route flow distributions for the 10 simulation scenarios and for the od pairs: $o = 1, \ldots, 6$ and $d = 1, 2, 3$. Each simulation scenario is identified by the Model ID equivalent to the ID values listed in Table 5.

Gap value also increases. Note that, here, the Gap indicates how far the simulation results are from the DUE; also it is calculated as in (9). On the other hand, the distributions of the average route travel times (Figure 6) also shift towards longer travel times due to an increase in user indifference. The average travel times per route also increase from 295 seconds for $\Delta^{od} = 0$ to 332 seconds for $\Delta^{od} = 500$. The strict preference search order reduces the variances of the distributions of the average route travel times compared to the indifferent preference search order.

4.3. Analysis of the Aggregated Traffic Stats of the Network. In this section, we analyze the network performance in terms of its inflow and outflow capacities and internal accumulation of vehicles, through the macroscopic fundamental diagram (MFD). We investigate the critical accumulation of vehicles n_c and the critical production P_c of the MFD obtained for the different settings of bounded rationality, compared against the values for the DUE and SUE, i.e., the reference models. To better highlight these differences, we define three criteria that represent

(i) the relative difference between the average TTD of the different PT settings ($\overline{TTD^*}$) and of the reference models ($\overline{TTD^{ref}}$):

$$\alpha_{TTD} = \frac{\overline{TTD^*} - \overline{TTD^{ref}}}{\overline{TTD^{ref}}} \times 100 \quad (12)$$

(ii) the relative difference between the average TTT of the different PT settings ($\overline{TTT^*}$) and of the reference models ($\overline{TTT^{ref}}$):

$$\alpha_{TTT} = \frac{\overline{TTT^*} - \overline{TTT^{ref}}}{\overline{TTT^{ref}}} \times 100 \quad (13)$$

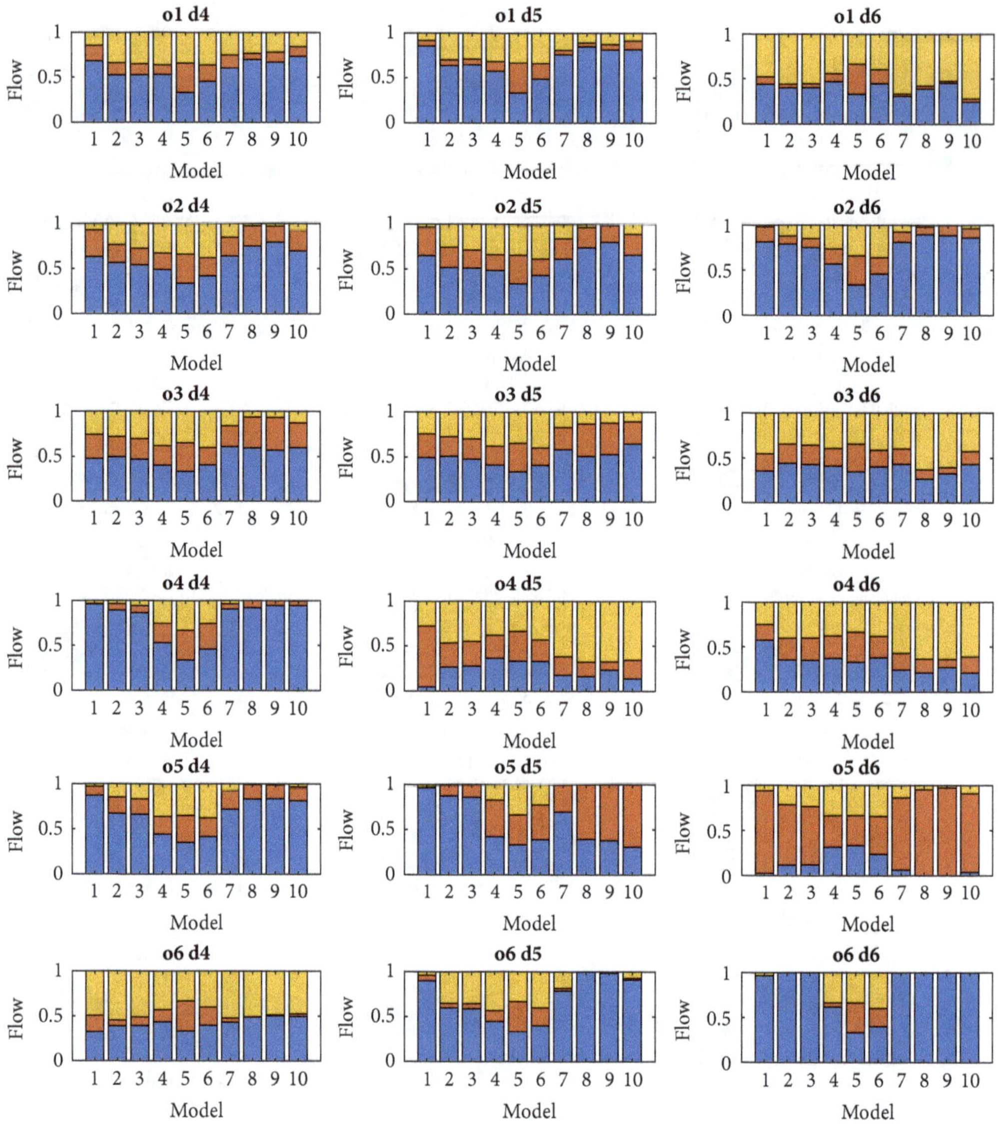

FIGURE 5: Same as in Figure 4, but for the od pairs: $o = 1, \ldots, 6$ and $d = 4, 5, 6$.

(iii) the relative difference between the average TTD of the different PT settings ($\overline{Q^*_{out}}$) and of the reference models ($\overline{Q^{ref}_{out}}$):

$$\alpha_{Q_{out}} = \frac{\overline{Q^*_{out}} - \overline{Q^{ref}_{out}}}{\overline{Q^{ref}_{out}}} \times 100 \tag{14}$$

The analysis of the three criteria is simple. If $\alpha_{TTD} < 0$, the network capacity decreases compared with the reference model. The vehicles accumulation inside the network is higher and congestion might spread backwards, increasing the average waiting times for vehicles to enter the network. The network inflow capacity decreases. Moreover, if the accumulation inside the network increases, the outflow performance of the network might also decrease. If $\alpha_{TTT} < 0$, the mean speed of vehicles inside the network is higher than the reference model. If $\alpha_{Q_{out}} > 0$, the outflow performance of the network is higher compared against the reference model.

We show the evolution of the total traveled distance (TTD) as well as the outgoing flow Q_{out} as a function of the total travel time (TTT) for the indifferent (Figure 7) and strict (Figure 8) preference orders. The average values for TTD, TTT, and Q_{out} are calculated for the simulation interval between 500 and 2500 seconds, for the 10 model settings. We then calculate α_{TTD}, α_{TTT}, and $\alpha_{Q_{out}}$ and estimate confidence intervals for these three criteria. The results are shown in Figure 7 for the indifferent preference order and in Figure 8 for the strict preference order. Our results show that the network capacity is higher for the strict preference search order case. This is observed by comparing the $\Delta^{od} = 500$ for both search orders, where the TTD is much lower for the indifferent preference search order compared to the strict preference search order. It can also be seen that the network capacity is approximately similar for the strict preference

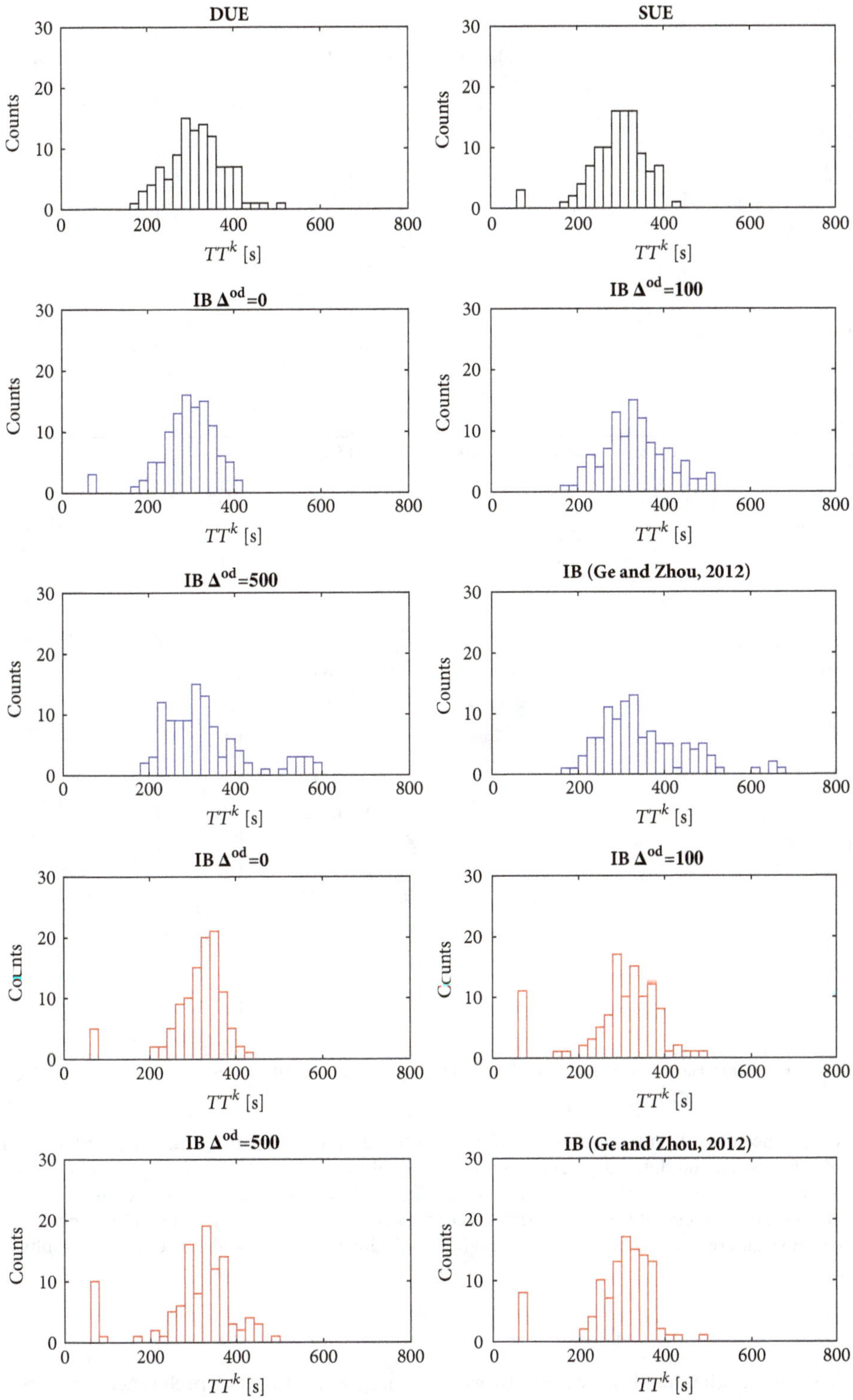

FIGURE 6: Average route travel time distributions for the DUE, SUE, and different settings of the indifference band.

search order and the different settings of the indifference band. While, for the indifferent preference search order, the network capacity decreases with an increase of the Δ^{od}. This is also evidenced in Figure 7 (ii) by the decrease of α_{TTD} as Δ^{od} increases. The average waiting time for the vehicles to enter the network also increases. The average waiting times per vehicle are 51 s for the DUE; 52 s for the SUE; 52 s for $\Delta^{od} = 0$; 54 s for $\Delta^{od} = 100$; 61 s for $\Delta^{od} = 500$; and 57 s for the setting of the indifference band defined by Ge and Zhou [20]. Note that these are the averaging waiting times for the indifferent

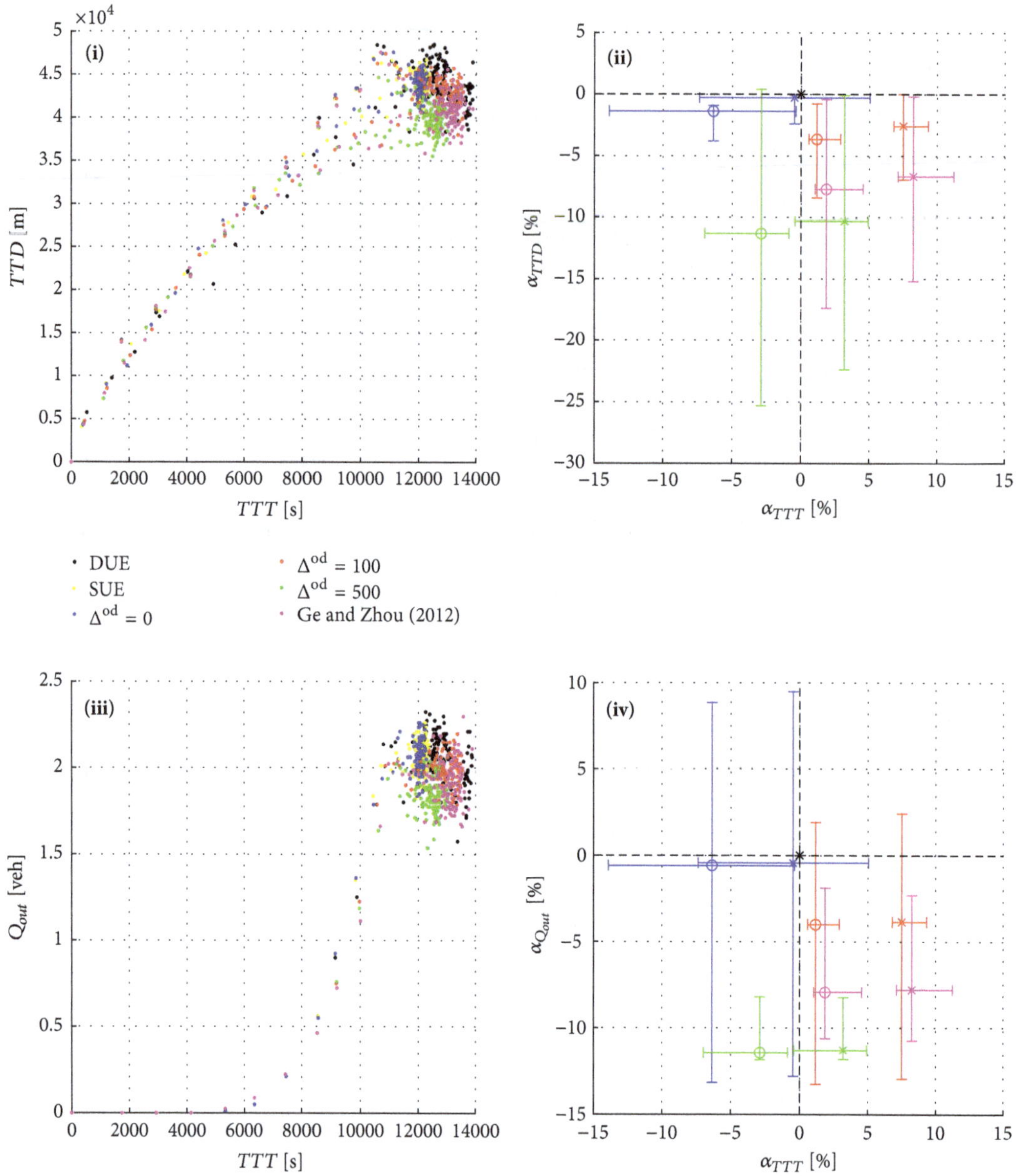

FIGURE 7: *(i)* Total travel distance (TTD) [in m] as function of the total travel time (TTT) [in s]. *(ii)* α_{TTD} versus α_{TTT} for the four settings of the indifference band. *(iii)* Vehicles outflow (Q_{out}) as function of the total travel time (TTT). *(iv)* $\alpha_{Q_{out}}$ versus α_{TTT} for the four settings of the indifference band. In subplots *(i)* and *(iii)*, the results are shown for the DUE, SUE, and the four settings of the indifference band. In subplots *(ii)* and *(iv)*, the circle dots represent the relative differences between the four settings of the indifference band and the DUE. The cross dots represent the relative differences between the four settings of the indifference band and the SUE. These results are for the indifferent preferences search order.

preference search order. On the other hand, since the network capacity is approximately similar for the strict preference search order and the different settings of Δ^{od}, the average waiting times per vehicle to enter the network are similar. The average waiting times for the strict preference search order are 74 s for $\Delta^{od} = 0$; 73 s for $\Delta^{od} = 100$; 75 s for $\Delta^{od} = 500$; and 74 s for the setting of the indifference band defined by Ge and Zhou [20]. From Figures 7 and 8, we can also observe a clear impact of the users search order on the total travel time spent on the network. For example, for $\Delta^{od} = 500$, the α_{TTT} is larger for the strict preference compared to the indifferent preference search order. This induces a lower vehicles mean speed and a lower internal network performance. Also, in the case of the indifferent preference search order, users will tend to choose routes with higher travel times as Δ^{od} increases. This leads to an increase of the accumulation of vehicles inside the network and consequently users spend more time to complete their trips. Also, the outflow Q_{out} of vehicles decreases as Δ^{od} increases (see Figures 7 and 8). Note that a lower outflow Q_{out} means lower system efficiency.

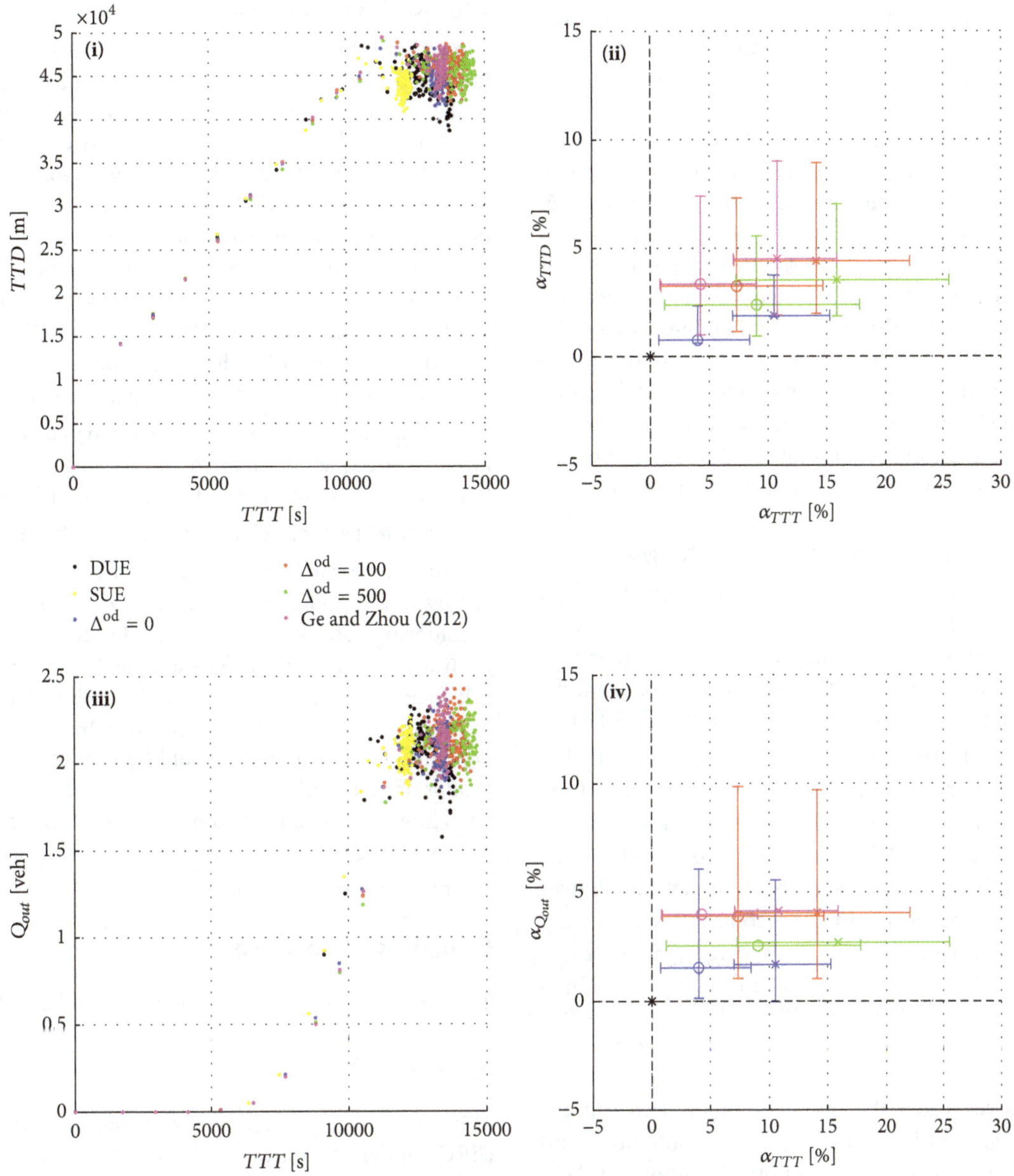

FIGURE 8: Same as in Figure 7, but for the strict preferences search order.

In summary, we show that different types of bounded user rationality have different impacts on the network performance. Considering the indifferent preference search order where users present an indifference behavior for all of the *satisficing* routes, as Δ^{od} increases, the internal and outflow capacities of the network decrease. However, when considering the strict user preference order, both the internal and outflow capacities of the network are approximately similar as Δ^{od} increases.

5. Conclusions

Users route choices determine the level of congestion on a road network. Thus understanding the effects of users behavior is important for transportation network planning policies. In this paper, we investigated the influence of two types of bounded rational behavior, considering users preferences for the search order (i.e., indifferent and strict preferences), on individual route flows and network performance. To do this, we considered a dynamic implementation of a bounded rational framework, using a mesoscopic LWR traffic simulator [21]. The route costs were time-dependent and accounted for congestion, shock-waves, and spillback effects. To model the bounded rationality behavior, we relaxed the definition of the search order of the DUE and SUE frameworks [4]. Thus, instead of using an all-or-nothing procedure to assign the users to the route(s) with the minimum travel time(s), they were assigned according to a more flexible definition of the search order according to user preferences. We also considered both definitions of the indifference band ((2) and

(3)) for AL^{od}. To account for the distributions of travel times, we used Monte Carlo simulations [4] and algorithm based on the Method of Successive Averages was presented to solve the network equilibria.

This work extended the framework of bounded rationality applied to dynamic traffic assignment modeling in some directions. First, it incorporates the stochasticity of route travel times that are treated through Monte Carlo simulations. Second, this framework accounts for indifferent and strict users' preferences for their route choices. Our framework is reduced to the DUE or SUE, if the search order is defined for users that are utility minimizers. Third, this framework extends the work of Zhao and Huang [19] to a dynamic context considering the setting of the indifference band. In our framework, we do not need to solve suboptimization problems to calibrate the aspiration levels of the sub-preferred routes.

To first assess and gain insight into the changes of route flows at equilibrium, AL^{od}, and for both user search orders, we considered a static implementation on the toy Braess network. The results obtained with the indifferent preference search order revealed that (i) the bounded rational model framework is able to reproduce both DUE and SUE; (ii) when AL^{od} is sufficiently large, the route flows converge to 1/3, showing the user indifference for the route choice; (iii) the algorithm discussed converges. Also, based on this simple numerical test, we showed that we converge towards the same solution of the BR-SUE calculated, based on averaging over all local BR-DUE problems. In the second test, considering the strict user preference order, we showed that the route flows calculated between our model and the model of Zhao and Huang [19] reach good agreement. This validated our methodology applied to determine the search order in a dynamic context.

We also investigated the influence of the two settings of the users search order on the individual route flows and network performance, considering both definitions of AL^{od} as defined in (2) and (3). These tests were performed in a dynamic context, using the mesoscopic LWR traffic simulator [21]. We first showed that we were able to reproduce the SUE, by setting $\Delta^{od} = 0$ for the indifferent preference search order. For the strict preference search order, we did not obtain route flows similar to the SUE when setting $\Delta^{od} = 0$. This is due to the fact that routes with the most reliable travel times did not necessarily have the minimal travel time per each od pair. We also showed that for the indifferent preference search order, the route flows also tended to 1/3 as we increased Δ^{od}.

We then showed that bounded user rationality had a significant impact on network performance. For the indifferent preference search orders, the network inflow capacity decreased as Δ^{od} increased; and the network performance decreased as Δ^{od} increased. The outflow Q_{out} also decreased as Δ^{od} increased. For the strict preference search order, the network capacity was approximately similar for the different settings of the indifference band. However, since users were allowed to choose routes with longer travel times as Δ^{od} increased, the TTT increased and the internal performance of the network decreased. In brief, we showed that different types of bounded rationality have clearly different influences on network performance. This is very important when guiding policy makers to decide the best measures to implement in order to increase network performance.

As future work, we can extend this work in many directions. We first plan to extend this work to the macroscopic fundamental diagram (MFD) simulation. We also plan to extend this framework to heterogeneous classes of users. The heterogeneity can either be included in the search order or on the setting of the AL^{od}. And, we also plan to extend this model to a day-to-day assignment, by considering a learning process (e.g., based on reinforcement learning models) either on the AL^{od} or the users search order definition. We also emphasize that further research in the setting of the AL^{od} is required. The setting of the preference orders in the search process allows considering heterogeneous classes of users, with a preference for transportation mode, for example. However, we note that this users heterogeneity can also be included in the setting of the AL instead of the search order. In this case, the total demand Q of the od pair should be split into homogeneous groups of users with the same preference. The AL should then be defined per od and class of users. For each class of users, they are assigned based on an all-or-nothing assignment to one of the routes listed in ω^{od}.

Conflicts of Interest

The authors declare that they have no conflicts of interest.

Acknowledgments

This project is supported by the European Research Council (ERC) under the European Union's Horizon 2020 research and innovation program (grant agreement No 646592-MAGnUM project). S. F. A. Batista also acknowledges funding support by the region Auvergne-Rhône-Alpes (ARC7 Research Program). Chuan-Lin Zhao acknowledges the funding supported by National Natural Science Foundation of China (71601012).

References

[1] J. G. Wardrop, "Some theoretical aspects of road traffic research," *Proceedings of the Institution of Civil Engineers*, vol. 1, no. 3, pp. 325–362, 1952.

[2] C. F. Daganzo and Y. Sheffi, "On stochastic models of traffic assignment," *Transportation Science*, vol. 11, no. 3, pp. 253–274, 1977.

[3] C. F. Daganzo, "Unconstrained extremal formulation of some transportation equilibrium problems," *Transportation Science*, vol. 16, no. 3, pp. 332–360, 1982.

[4] Y. Sheffi, *Urban Transportation Networks: Equilibrium Analysis with Mathematical Programming Methods*, Chapter 10 & 11, Prentice Hall Inc, Upper Saddle River, NJ, USA, 1985.

[5] S. Zhu, *The Roads Taken: Theory and Evidence on Route Choice in the Wake of the I-35w Mississippi River Bridge Collapse and Reconstruction [Ph.D. thesis]*, University of Minnesota, 2011.

[6] E. Avineri and J. N. Prashker, "Violations of expected utility theory in route-choice stated preferences: certainty effect and inflation of small probabilities," *Transportation Research Record*, vol. 1894, pp. 222–229, 2004.

[7] S. Zhu, D. Levinson, and Y. Ropert-Coudert, "Do People Use the Shortest Path? An Empirical Test of Wardrop's First Principle," *PLoS ONE*, vol. 10, pp. 1–18, 2015.

[8] D. Kahneman and A. Tversky, "Prospect theory: an analysis of decision under risk," *Econometrica*, vol. 47, no. 2, pp. 263–292, 1979.

[9] A. Tversky and D. Kahneman, "Advances in prospect theory: cumulative representation of uncertainty," *Journal of Risk and Uncertainty*, vol. 5, no. 4, pp. 297–323, 1992.

[10] E. Avineri, "The effect of reference point on stochastic network equilibrium," *Transportation Science*, vol. 4, no. 4, pp. 409–420, 2006.

[11] D. E. Bell, "Regret in decision making under uncertainty," *Operations Research*, vol. 30, pp. 961–981, 1982.

[12] G. Loomes and R. Sugden, "Regret theory: an alternative theory of rational choice under uncertainty," *The Economic Journal*, vol. 92, no. 368, pp. 805–824, 1982.

[13] C. Chorus, "Random Regret Minimization: An Overview of Model Properties and Empirical Evidence," *Transport Reviews*, vol. 32, no. 1, pp. 75–92, 2012.

[14] M. Li and H. Huang, "A regret theory-based route choice model," *Transportmetrica A: Transportation Science*, vol. 13, no. 3, pp. 250–272, 2016.

[15] H. A. Simon, *A Behavioral Model of Rational Choice*, Wiley, New York, NY, USA, 1957.

[16] H. A. Simon, *Theories of Decision-Making in Economics and Behavioural Science*, Palgrave Macmillan, London, UK, 1966.

[17] H. A. Simon, "A mechanism for social selection and successful altruism," *Science*, vol. 250, no. 4988, pp. 1665–1668, 1990.

[18] H. A. Simon, "Bounded rationality and organizational learning," *Organization Science*, vol. 2, no. 1, pp. 125–134, 1991.

[19] C.-L. Zhao and H.-J. Huang, "Experiment of boundedly rational route choice behavior and the model under satisficing rule," *Transportation Research Part C: Emerging Technologies*, vol. 68, pp. 22–37, 2016.

[20] Y. E. Ge and X. Zhou, "An alternative definition of dynamic user optimum on signalised road networks," *Networks and Spatial Economics*, vol. 46, no. 3, pp. 236–253, 2012.

[21] L. Leclercq and C. Becarie, "A meso lighthill-whitham and richards model designed for network applications," in *Proceedings of the 91st Transportation Research Board Annual Meeting (TRB)*, pp. 21–26, Washington, USA, 2012.

[22] H. S. Mahmassani and G.-L. Chang, "On boundedly rational user equilibrium in transportation systems," *Transportation Science*, vol. 21, no. 2, pp. 89–99, 1987.

[23] J. Vreeswijk, T. Thomas, E. Van Berkum, and B. Van Arem, "Drivers' perception of route alternatives as indicator for the indifference band," *Transportation Research Record*, vol. 2383, pp. 10–17, 2013.

[24] Y. Lou, Y. Yin, and S. Lawphongpanich, "Robust congestion pricing under boundedly rational user equilibrium," *Transportation Research Part B: Methodological*, vol. 44, no. 1, pp. 15–28, 2010.

[25] X. Di, H. X. Liu, J.-S. Pang, and X. J. Ban, "Boundedly rational user equilibria (BRUE): mathematical formulation and solution sets," *Transportation Research Part B: Methodological*, vol. 57, pp. 300–313, 2013.

[26] X. Di, X. He, X. Guo, and H. X. Liu, "Braess paradox under the boundedly rational user equilibria," *Transportation Research Part B: Methodological*, vol. 67, pp. 86–108, 2014.

[27] O. A. L. Eikenbroek, G. J. Still, E. C. van Berkum, and W. Kern, "The boundedly rational user equilibrium: a parametric analysis with application to the network design problem," *Transportation Research Part B: Methodological*, vol. 107, pp. 1–17, 2018.

[28] X. Di and H. X. Liu, "Boundedly rational route choice behavior: a review of models and methodologies," *Transportation Research Part B: Methodological*, vol. 85, pp. 142–179, 2014.

[29] H. Huang and W. H. K. Lam, "Modeling and solving the dynamic user equilibrium route and departure time choice problem in network with queues," *Transportation Research Part B: Methodological*, vol. 36, no. 3, pp. 253–273, 2002.

[30] W. Y. Szeto and H. K. Lo, "Dynamic traffic assignment: properties and extensions," *Transportmetrica*, vol. 2, no. 1, pp. 31–52, 2006.

[31] Y. E. Ge, B. R. Sun, H. M. Zhang, W. Y. Szeto, and X. Zhou, "A comparison of dynamic user optimal states with zero, fixed and variable tolerances," *Networks and Spatial Economics*, vol. 15, no. 3, pp. 583–598, 2015.

[32] V. H. Aguiar, M. J. Boccardi, and M. Dean, "Satisficing and stochastic choice," *Journal of Economic Theory*, vol. 166, pp. 445–482, 2016.

[33] K. Han, T. L. Friesz, W. Y. Szeto, and L. Hongcheng, "Elastic demand dynamic network user equilibrium: Formulation, existence and computation," *Transportation Research Part B: Methodological*, vol. 81, pp. 183–209, 2015.

[34] J. L. Gifford and C. Checherita, "Bounded rationality and transportation behavior: lessons for public policy," in *Proceedings of the 86th Transportation Research Board Annual Meeting (TRB)*, pp. 21–26, Washington, USA, 2007.

[35] G. Marsden, K. T. Frick, A. D. May, and E. Deakin, "Bounded rationality in policy learning amongst cities: lessons from the transport sector," *Environment and Planning A: Economy and Space*, vol. 44, no. 4, pp. 905–920, 2012.

[36] M. Sivak, "How common sense fails us on the road: contribution of bounded rationality to the annual worldwide toll of one million traffic fatalities," *Transportation Research Part F: Traffic Psychology and Behaviour*, vol. 5, no. 4, pp. 259–269, 2002.

[37] O. A. Nielsen, "On the distributions of the stochastic components in sue (stochastic user equilibrium) traffic assignment models," in *Proceedings of the Transportation Planning Methods: Seminar Held at The European Transport Forum Annual Meeting*, pp. 77–93, Brunel University, London, UK, 1997.

[38] B. T. Polyak, "A new method of stochastic approximation type," *Automation and Remote Control*, vol. 51, no. 7, pp. 937–964, 1990.

[39] H. X. Liu, X. He, and B. He, "Method of successive weighted averages (MSWA) and self-regulated averaging schemes for solving stochastic user equilibrium problem," *Networks and Spatial Economics*, vol. 9, no. 4, pp. 485–503, 2007.

[40] H. Taale, *Integrated Anticipatory Control of Road Networks a Game Theoretical Approach [Ph.D. thesis]*, Delft University of Technology, The Netherlands, 2008.

[41] A. Chen, X. Xu, S. Ryu, and Z. Zhou, "A self-adaptive Armijo stepsize strategy with application to traffic assignment models and algorithms," *Transportmetrica A: Transport Science*, vol. 9, no. 8, pp. 695–712, 2011.

[42] H. Sbayti, C. Lu, and H. S. Mahmassani, "Efficient Implementation of Method of Successive Averages in Simulation-Based Dynamic Traffic Assignment Models for Large-Scale Network Applications," *Transportation Research Record: Journal of the Transportation Research Board*, vol. 2029, no. 1, pp. 22–30, 2018.

13

Multidimensional Scaling-Based Data Dimension Reduction Method for Application in Short-Term Traffic Flow Prediction for Urban Road Network

Yi Zhao,[1] **Satish V. Ukkusuri**,[2] **and Jian Lu**[3]

[1]*College of Automobile and Traffic Engineering, Nanjing Forestry University, Longpan Road #159, Nanjing, 210037, China*
[2]*Lyles School of Civil Engineering, Purdue University, 550 Stadium Mall Drive, West Lafayette, IN, 47907, USA*
[3]*School of Transportation ,Southeast University, Si Pai Lou #2, Nanjing, 210096, China*

Correspondence should be addressed to Yi Zhao; zhaoyi207@126.com

Academic Editor: David F. Llorca

This study develops a multidimensional scaling- (MDS-) based data dimension reduction method. The method is applied to short-term traffic flow prediction in urban road networks. The data dimension reduction method can be divided into three steps. The first is data selection based on qualitative analysis, the second is data grouping using the MDS method, and the last is data dimension reduction based on a correlation coefficient. Backpropagation neural network (BPNN) and multiple linear regression (MLR) models are employed in four kinds of urban traffic environments to test whether the proposed method improves the prediction accuracy of traffic flow. The results show that prediction models using traffic data after dimension reduction outperform the same prediction models using other datasets. The proposed method provides an alternative to existing models for urban traffic prediction.

1. Introduction

The success of ITS technology is heavily dependent on the availability of timely and accurate estimates of prevailing and emerging traffic conditions. Hence, there is a strong need for traffic prediction. The method used for prediction should be able to utilize advanced traffic models to analyze data, especially real-time traffic data, from different sources to estimate and predict traffic conditions. Accurate prediction of traffic conditions makes it possible to implement proactive ATMS and ATIS strategies in advance to meet various traffic control, management, and operation objectives.

Urban road systems have a greater need for traffic prediction than freeway and intercity highway systems. Increases in urban populations and vehicle ownership in the last few decades have resulted in severe traffic congestion, with serious economic and social consequences [1]. Therefore, an advanced short-term traffic flow prediction method is required to support the implementation of real-time traffic management measures, such as a dynamic signal control schemes, traffic information release, and real-time traffic induction.

1.1. Literature Review. Short-term traffic flow prediction has been an important subject in the field of intelligent transportation research since the 1970s. Up to 200 kinds of short-term traffic flow prediction models have been developed in the past 50 years.

The first prediction model relies on statistical theory and assumes that traffic flow follows a linear system. Prediction models mainly include the time series model [2], historical average model [3], the Kalman filter [4], and linear regression model [5]. However, traffic flow is very uncertain and nonlinear [6]. Hence, nonlinear theoretical prediction models based on classification theory [7], chaos theory [8], wavelet theory [9], mutation theory [10], and the nonparametric regression method [11] have been developed to address this issue. Over the last decade, studies on traffic prediction models have emphasized the use of the artificial intelligence (AI) model rather than the traditional statistical model. The AI model has no strict mathematical relationship between input and output data but tries to obtain knowledge through training. The most popular AI models are the support vector machine-based model [12, 13] and the neural network-based model [14, 15]. Certain traffic flow theory-based simulation models have also been used for traffic flow prediction. In the simulation models, a series of mathematical models, such as the motion wave model [16], cellular automata model [17], and three-phase traffic flow theory [18], have been used to capture the dynamic characteristics of traffic.

Among the aforementioned prediction models, each has its own advantages and disadvantages [19]. Therefore, researchers have tried to combine different prediction models to improve the stability and accuracy of a single model. Bates et al. [20] first combined two separate sets of forecasts of airline passenger data to form a composite set of forecasts, which showed the superiority and potential of a combined model. Fuzzy logic [21], Bayesian theory [22], and Grey relation analysis [23] are all commonly employed tools used to combine different models. The idea of combining different models has confirmed the need to keep improving the stability and accuracy of single prediction models.

1.2. Challenges in Short-Term Traffic Prediction. In 2014, Vlahogianni [24] pointed out that recent developments in technology and the widespread use of powerful computers and mathematical models give researchers an unprecedented opportunity to expand their horizons and direct work in 10 challenging, yet relatively underresearched, directions. The main motivations of this paper lie in overcoming two of these challenges.

The first challenge involves the fact that most short-term traffic forecasting algorithms were built to function at a freeway, arterial, or corridor level. Short-term traffic forecasting for urban arterials is more complex than that for freeways due to constraints such as signalization. This was pointed out by Kirby [25] 20 years ago, but there has not yet been any change. The limited research on urban traffic focuses on either arterial streets or small local areas. Yin et al. [26] used a fuzzy-neural approach to predict urban traffic flow, but only two adjacent junctions were adopted as examples. Stathopoulos and Karlaftis [27] presented a multivariate state space approach to predict traffic flow on urban arterial streets near a downtown. In almost all existing research on urban traffic prediction, the input data for prediction models only consist of historical data from one or several road sections. Moreover, only arterial streets or suburban roads, rather than central streets, are studied. Generally speaking, the characteristics of urban road traffic flow have not been suitably considered in existing prediction models. Hence, it is likely that the prediction results will be unstable and inaccurate if existing prediction models are directly applied to urban road networks.

The second challenge is that identifying spatial and temporal flow patterns has been an important consideration in short-term traffic forecasting research. The most striking feature of urban traffic is that the source of road flow is very complex and the flow is affected by signal control. In order to obtain accurate prediction results, a large amount of traffic flow data should be collected and the relationships between these data should be analyzed. Traffic flow from different road sections is relevant in terms of both time and space. In other words, a spatial-temporal correlativity analysis is required in urban traffic prediction. Some attempts at such an analysis have been made, such as Cheng's [28] study of spatiotemporal autocorrelation and Wang's [29] study of bipattern recognition. The principal methods for spatial-temporal correlativity analysis include the spatiotemporal random effects model [30], aggregation method [31], and state space approach [27]. The main purpose of this study is to combine spatial-temporal correlativity analysis and existing prediction models and to test whether the combined method can reach a better prediction result.

1.3. Motivations and Contributions of the Paper. Short-term traffic prediction models have been developed for more than half a century. However, their application to urban traffic is limited. With increasing vehicle ownership and network complexity, there are increasing challenges in short-term traffic prediction for urban traffic. In addition, developments in detection and communication techniques can provide more comprehensive information about traffic conditions for prediction. Hence, the aim of the study is urban traffic flow prediction.

Although large amount of data of urban road traffic can be collected, there are two factors that prevent these data from been directly used in urban road traffic prediction models. The first is computation speed. As the study focuses on short-term prediction, large amounts of data may reduce the computation speed of the models, decreasing the value of the short-term predictions. The second is the accuracy of the prediction results. Traffic data from adjacent or homogeneous road sections are highly similar. In other words, large amounts of data contain duplicate and redundant data, which affect the prediction results. Therefore, a method for reducing data dimensions should be applied to the prediction model. To the best of our knowledge, multidimensional scaling (MDS) has not been applied to urban traffic prediction. The advantage MDS-based data dimension reduction is its ability to visualize the level of similarity of a traffic flow data set. Hence, the proposed data dimension reduction method makes it possible to use less data to represent a whole data set, resulting in improved prediction accuracy.

To summarize, the research goals are as follows:

(a) To propose an MDS-based data dimension reduction method that can be used to conduct spatial-temporal correlativity analysis of urban traffic data and divide large amount of traffic flow data into smaller groups according to the level of similarity.

(b) To demonstrate that the proposed method can be combined with existing prediction models. The proposed method will be used to generate a smaller data set to represent all the traffic flow data. The accuracy of the prediction result using the small data set will then be evaluated against those obtained using other data sets.

(c) To illustrate that the proposed method can be combined with different kinds of prediction models and that the combined methods can be adapted to different traffic environments in urban road networks.

2. Data Collection

As previously stated, the main characteristic of urban traffic is its complex direction and signal control management. Compared to that in a single freeway or intercity highway, the traffic direction in urban road networks is more fluid, and signal control results in poor continuity in traffic flow. To collect enough traffic flow data for the study, a virtual network was built in VISSIM to run a simulation, as shown in Figure 1. To reflect the complexity of an urban network, the virtual network is comprised of five main streets and six collector streets. The distance between adjacent streets is randomly set. It is assumed that the structure of the network has no effect on the prediction process; hence, other networks with different structures are not discussed. At each cross-section, lanes with the same driving direction are denoted as one road section. There are 142 road sections in total, and each section has a detector for collecting traffic flow information.

The traffic flow is input into the network through link origins, and the traffic flow subjects to Poisson distribution. In order to test the effectiveness of the proposed method under different environments, four kinds of OD input data are employed: OD_1, OD_2, OD_3, and OD_4. OD_1, OD_2, and OD_3 have the same flow pattern, but OD_4 has a different pattern. The same flow pattern means the same number of traffic origins and destinations, and the same main flow direction. The first three kinds of OD differ in the traffic loading level, with a low loading for OD_1, medium loading for OD_2, and high loading for OD_3, while OD_4 has a medium traffic loading. A fixed signal timing scheme is adopted, and the scheme is optimized using built-in modules in VISSIM.

Based on existing research [32], common prediction intervals for short-term prediction include 5 min, 10 min, 15 min, and 20 min. Considering the signal cycle at intersections is ~2 min, using a prediction interval of 5 min may lead to instability of prediction result. Alternatively, a long prediction interval reduces the significance of the short-term traffic prediction. Hence, an interval of 10 min was used. Traffic data are collected on different days, and data should be collected in the same time period to obtain enough historical data. Hence, random seeds are renewed every hour while running the simulation. Thirty-nine groups of effective traffic flow data can be collected from each road section for simulation.

3. Application of MDS-Based Data Dimension Reduction Method

3.1. Basic Idea of Proposed Data Dimension Reduction Method. The proposed data dimension reduction method is composed of three steps, as follows.

Step 1 (selection of historical data based on qualitative analysis). The road section for which predictions are going to be made is denoted as the target section. All the road sections of the network are connected, and the relationship between them is either strong or weak. It is not necessary to consider the influence of all road sections on the target section. Two principles should be observed when selecting appropriate road sections that have an obvious influence on the target section. Firstly, road sections can be included if vehicles on the section can reach the target section in the upcoming prediction interval. Secondly, road sections where traffic is moving away from the target section are excluded, as there is a low possibility of traffic reaching the target section.

Only the historical data of sections that are supposed to exert an obvious influence on the target section can be used for prediction.

Step 2 (data grouping using MDS method). There are many road sections that remain that must be considered after Step 1 is completed. Hence, the next step is to divide the sections into several groups, making it easier to reduce their dimensions. Similarity is the best criterion for grouping; however, as each section has 39 groups of traffic flow data, it is difficult to directly identify the level of similarity between the different sections. Hence, the MDS method is used here (a detailed explanation of it will be given in a later section). However, the MDS method only provides a visual representation of the patterns of similarities or differences. The final grouping result should involve grouping of road sections on the same street based on the MDS analysis results.

Step 3 (data dimension reduction based on correlation coefficients). After the data grouping in Step 2, it can be assumed that road sections in the same group have affect the target section in similar ways. Hence, the data dimensions can be reduced by choosing one representative road section from each group. The data of the selected road sections will serve as the input data for the prediction section to which the BPNN model and MLR model will be applied. The BPNN model requires a low correlation of input data to enhance its robustness, while the MLR model requires a high correlation of input data. Considering that the historical data of the target section will also serve as the input data, the road sections with highest and lowest correlation with the target section in each group will be extracted. The input data for prediction can then be determined according to the requirements of the prediction model. The Pearson product-moment correlation coefficient, denoted as ρ_{ij}, will be employed to analyze the degree of correlation between the target section and the other road sections.

$$\rho_{ij} = \frac{E(\mathbf{ij}) - E(\mathbf{i})E(\mathbf{j})}{\sqrt{E(\mathbf{i}^2) - E(\mathbf{i})^2}\sqrt{E(\mathbf{j}^2) - E(\mathbf{j})^2}} \quad (1)$$

where ρ_{ij} is the Pearson product-moment correlation coefficient between the ith and the jth sections (i and j are section numbers) and $E(\mathbf{i})$, $E(\mathbf{j})$, $E(\mathbf{ij})$, $E(\mathbf{i}^2)$, and $E(\mathbf{j}^2)$ are the expectations for $\mathbf{i}, \mathbf{j}, \mathbf{ij}, \mathbf{i}^2, \mathbf{j}^2$, respectively (**i** and **j** are traffic data for different road sections).

3.2. MDS Method. MDS is a means of visualizing the level of similarity between individual cases in a dataset. It is a form

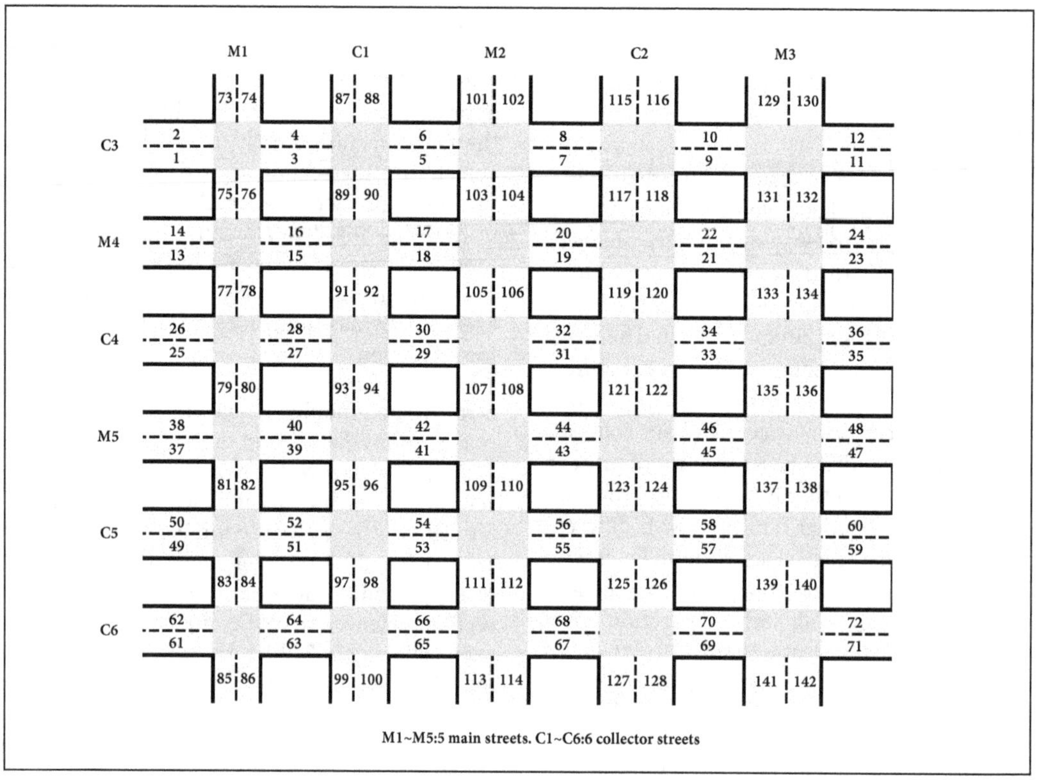

FIGURE 1: Structure of virtual network.

of nonlinear dimensionality reduction. The MDS algorithm places each object in an N-dimensional space, such that the between-object distances are preserved as well as possible.

If the traffic data set used for prediction is obtained from n road sections and the number of data values for each section is m, the data to be analyzed can be expressed in an $n \times m$ matrix $\mathbf{X}$. As traffic flow data are of the nonmetric multidimensional type, $\mathbf{X}$ should be normalized before the Euclidean distance is calculated. The Euclidean distance will be employed to describe the distance and dissimilarity between objects. The Euclidean distance δ_{ij} from $\mathbf{i}$ to $\mathbf{j}$ is given by the Pythagorean formula. The Euclidean distances between each road section can be expressed using the distance matrix $\mathrm{D}(\mathbf{X})$.

$$\mathbf{X} = \begin{bmatrix} x_{11} & x_{12} & \cdots & x_{1m} \\ x_{21} & x_{22} & \cdots & x_{2m} \\ \vdots & \vdots & \ddots & \vdots \\ x_{n1} & x_{n2} & \cdots & x_{nm} \end{bmatrix} \tag{2}$$

$$\delta_{ij} = \sqrt{\sum_{h=1}^{m} \left(x_{ih} - x_{jh}\right)^2} \tag{3}$$

$$\mathrm{D}(\mathbf{X}) := \begin{bmatrix} \delta_{11} & \delta_{12} & \cdots & \delta_{1n} \\ \delta_{21} & \delta_{22} & \cdots & \delta_{2n} \\ \vdots & \vdots & \ddots & \vdots \\ \delta_{n1} & \delta_{n2} & \cdots & \delta_{nn} \end{bmatrix} \tag{4}$$

where x_{ij} is the jth traffic flow data of road section i ($i \in [1, n]$ and $j \in [1, m]$) and δ_{ij} is the Euclidean distance from $\mathbf{i}$ to $\mathbf{j}$ (which are data from the ith and jth sections, respectively).

The goal of the MDS method is to find an embedding from n section vectors into $\mathbb{R}^{\mathrm{N}}$, an N-dimensional vector space, such that the distances are preserved. If N is 2 or 3, n section vectors should be plotted to visualize the similarities. Supposing n vectors can be found in $\mathbb{R}^{\mathrm{k}}$, whose Euclidean distance matrix $\widehat{\mathrm{D}}(\mathbf{Y})$ is close to $\mathrm{D}(\mathbf{X})$, then matrix $\mathbf{Y}$, which is composed of the n vectors, is called the fitting chart of $\mathrm{D}(\mathbf{X})$.

$$\mathbf{Y} = \left(\mathbf{y}_1; \mathbf{y}_2; \cdots \mathbf{y}_n\right) = \begin{bmatrix} y_{11} & y_{12} & \cdots & y_{1k} \\ y_{21} & y_{22} & \cdots & y_{2k} \\ \vdots & \vdots & \ddots & \vdots \\ y_{n1} & y_{n2} & \cdots & y_{nk} \end{bmatrix} \tag{5}$$

$$\widehat{\delta}_{ij} = \left\|\mathbf{y}_i - \mathbf{y}_j\right\| = \sqrt{\sum_{s=1}^{k} \left(y_{is} - y_{js}\right)^2} \tag{6}$$

$$\widehat{\mathrm{D}}(\mathbf{Y}) = \begin{bmatrix} \widehat{\delta}_{11} & \widehat{\delta}_{12} & \cdots & \widehat{\delta}_{1n} \\ \widehat{\delta}_{21} & \widehat{\delta}_{22} & \cdots & \widehat{\delta}_{2n} \\ \vdots & \vdots & \ddots & \vdots \\ \widehat{\delta}_{n1} & \widehat{\delta}_{n2} & \cdots & \widehat{\delta}_{nn} \end{bmatrix} \tag{7}$$

where $\mathbf{Y}$ is a matrix composed of n vectors in $\mathbb{R}^k$; $\mathbf{y}_i$ is the ith vector in $\mathbb{R}^k$, denoted as $[y_{i1}, y_{i2}, \ldots, y_{ik}]$; and $\widehat{\delta}_{ij}$ is the Euclidean distance between $\mathbf{y}_i$ and $\mathbf{y}_j$.

There are various approaches to determine $\mathbf{Y}$. Usually, MDS is formulated as an optimization problem, where $\mathbf{Y}$ is a minimizer of the cost function $\sigma(\mathbf{Y})$.

$$\sigma(\mathbf{Y}) = \min\left(\sum_{i=1}^{n}\sum_{j=1}^{n}\left(\delta_{ij} - \widehat{\delta}_{ij}\right)^2\right) \tag{8}$$

The most common approach used to determine $\mathbf{Y}$ is an iterative process commonly referred to as the Shepard–Kruskal algorithm. However, the algorithm is not a major focus of this study. SPASS can be employed to solve the algorithm problem. The goodness-of-fit statistic *Stress*, which is based on the differences between the actual distances and their predicted values, is adopted to express how well $\mathbf{X}$ is represented by $\mathbf{Y}$. The smaller the *Stress* value, the better the fit. The goodness-of-fit is excellent when the *Stress* is less than 0.025, good when it is 0.025–0.050, fair when it is 0.050–0.100, and poor when it is 0.100–0.200.

$$Stress = \sqrt{\frac{\sum_{i=1}^{I-1}\sum_{j=i+1}^{I}\left(\delta_{ij} - \widehat{\delta}_{ij}\right)^2}{\sum_{i=1}^{I-1}\sum_{j=i+1}^{I}\delta_{ij}^{2}}} \tag{9}$$

3.3. Dimension Reduction of Test Data Using Proposed Method. Road section 104 in the virtual network in Figure 1 is used as a case study. The proposed data dimension reduction method is applied to reduce the dimensions of the collected data. According to the principles in Step 1, only 56 road sections should be considered, including the target section. All 56 road sections are renumbered as shown in Figure 2, where the target section is No. 42.

The historical data for 56 road sections can be expressed in a 56×39 $\mathbf{X}$ matrix. According to Step 2 of the method, $\mathbf{X}$ matrices under four kinds of ODs are analyzed using the MDS method, and the results are shown in Figure 3. The *Stress* values for all the OD types are under 0.050, which indicates a good goodness-of-fit. Based on the similarity analysis, road sections on the same street are grouped with a priority. The 56 road sections can be divided into four groups for each OD. For each kind of OD, there are several road sections that are not within the group, such as section 6 for OD_2 and OD_3, and section 18 for OD_4. Apart from these exceptions, the first three ODs have the same grouping result, but OD_4 has a different grouping result. Figure 4 illustrates the grouping results more clearly, where each color represents a group.

Based on the grouping result, the sections with the lowest and highest correlation with the target section are selected by employing the Pearson product-moment correlation coefficient. The results are shown in Table 1.

4. Prediction Model Construction and Prediction Result Analysis

The BPNN and MLR models were employed to test the effectiveness of the proposed method for data dimension reduction. The BPNN model is a widely accepted knowledge discovery model, and its main feature is its ability to learn and adapt to any uncertain system with complex non-linear relationships. The MLR model relies on linear relationships. Ideally, the input data should have a low correlation for the BPNN model and a high correlation for the MLR model. This difference makes it impossible to test the effectiveness of proposed data dimension reduction method on different prediction models.

4.1. Evaluation Index Selection. To test the prediction performance of the two prediction methods, two measures, the mean absolute percentage error (MAPE) and the root-mean-square deviation (RMSE), are employed. They are calculated as follows:

$$MAPE = \frac{\sum_{t=1}^{S}\left(\left(Y(t) - \widehat{Y}(t)\right)/Y(t)\right)}{S} \tag{10}$$

$$RMSE = \sqrt{\frac{1}{S}\sum_{t=1}^{S}\left(\frac{Y(t) - \widehat{Y}(t)}{Y(t)}\right)^2} \tag{11}$$

where $Y(t)$ is the observed value, $\widehat{Y}(t)$ is the predicted value, and S is the number of observed values.

4.2. Construction of BPNN Model. The BPNN model employed contains three layers: input layer, hidden layer, and output layer. The other components are set as follows.

Input data: Three groups of input data for each OD are used for prediction: group AX, group SX, and group TY. Group AX contains all 56 road sections selected via qualitative analysis. Group SX contains the four sections with the lowest correlation and the target section, i.e., five sections in total. Group TY contains only the target section.

Temporal correlation: Based on the concept of the time series prediction model, for all sections, the prediction input data are composed of the historical data of the last three intervals before the prediction period. In other words, if the goal is to predict the traffic flow of the target section in time period t, the input data for section X include X(t-1), X(t-2), and X(t-3).

Data assignment for each section: There are 39 data values for each section. Eighteen are used for training, eight are used for testing, and ten are used for prediction.

Number of neurons: The range of the number of neurons is $[\sqrt{N(\text{in}) + N(\text{out})}, \sqrt{N(\text{in}) + N(\text{out})} + 10]$, where $N(\text{in})$ and $N(\text{out})$ are the number of nodes of the input and output layer, respectively. The final number of neurons is determined via training and testing.

Function setting: The functions that prove to be most effective based on several attempts of testing are selected and assigned to the BPNN model. All the BPNN models have the same functions. The transmission function for the hidden layer is 'tansig' and that for the output layer is 'purelin'. The training function is 'trainlm' and the weight learning function is 'learndm'.

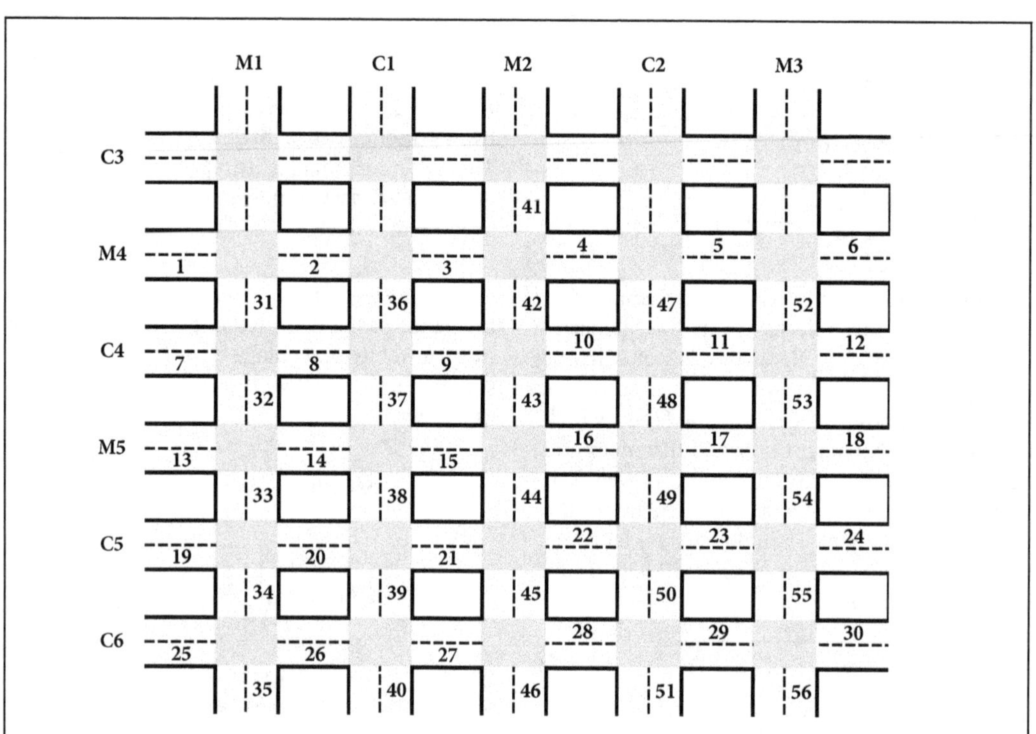

FIGURE 2: Road section selection based on qualitative analysis.

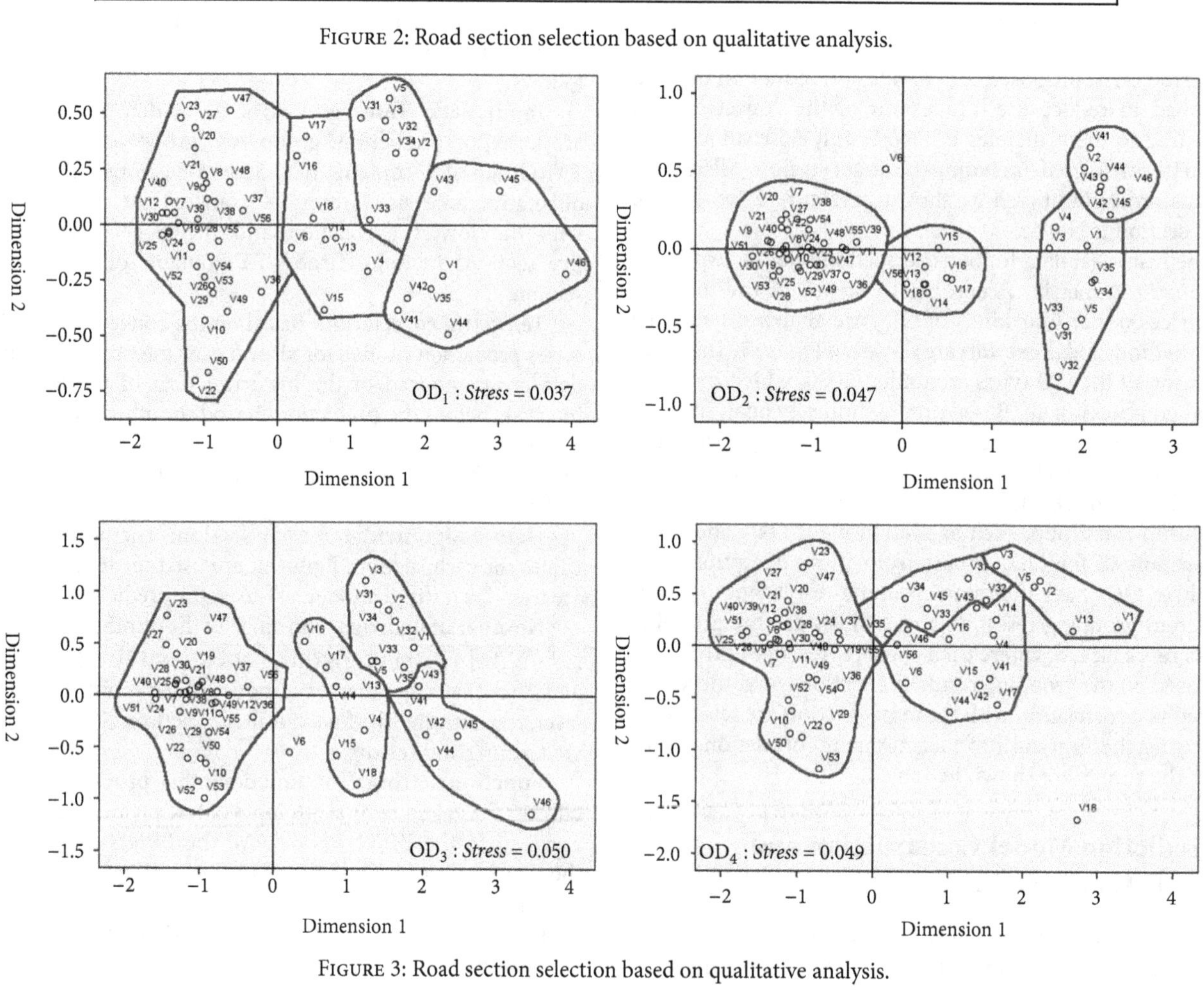

FIGURE 3: Road section selection based on qualitative analysis.

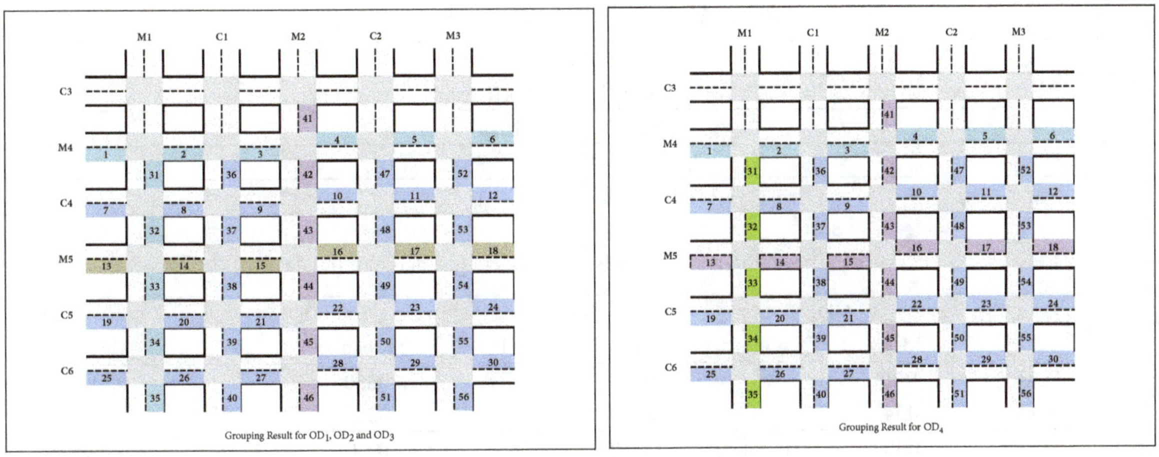

FIGURE 4: Grouping results for different kinds of ODs.

TABLE 1: Correlation analysis based on grouping result.

Input data type	Sections with lowest correlation in each group	Sections with highest correlation in each group
OD_1	27\32\12\45	39\2\15\42
OD_2	46\32\13\53	43\39\15\4
OD_3	48\35\17\45	26\4\15\42
OD_4	5\31\14\56	4\32\39\42

4.3. Construction of MLR Model. The dependent variable in the analysis is the traffic of the target section. The independent variables are divided into three groups for each OD, i.e., group AX, group SX, and group TY. ANOVA is used to select sections with a significant difference, owing to the large error that occurs when 56 road sections are used for MLR analysis. The ANOVA results are shown in Table 2. Group AX contains all the sections after implementing ANOVA. Group SX contains the four sections with the highest correlation and the target section. Group TY contains only the target section.

As with the BPNN model, temporal correlation is also considered in the MLR model. For each section that is used as an independent variable, the input data are composed of the historical data of the last three time intervals before the prediction period. Only the data from the last 10 intervals of the target section are used for prediction, while the others are used for fitting.

4.4. Assessment of Prediction Results. Because the initial weights in the BPNN model are random, the prediction result varies when the model is rerun. The prediction process is repeated 30 times using the BPNN model. The 10 best results are selected, and the mean values of the evaluation indices are calculated, as shown in Table 3. The prediction result obtained using the MLR model is shown in Table 4.

In most existing research, only the historical data of the target section are used in the prediction model. Therefore, group TY is treated as the representative of regular prediction method. Group AX uses the data of all the related sections for prediction, without considering the spatial-temporal correlativity of the sections. Group SX uses the selected sections by employing the proposed data dimension reduction method, which takes the spatial-temporal correlativity of the sections into consideration. The improvements in the results for methods using group SX and group AX are compared against the method that uses group TY. The degree of improvement is denoted as P in Tables 3 and 4.

The following conclusions can be made from the comparison of the prediction results:

(a) According to the MAPE index, the prediction accuracy of the BPNN model is more than 85%, while that of the MLR model is ~80%. The BPNN model is better than the MLR model when a single prediction method is employed.

(b) For the two kinds of OD flow patterns, both the method using group AX and that using group SX result in greater improvements in the prediction accuracy than the method using group TY.

(c) For the three different levels of traffic loading, both the method using group AX and that using group SX result in greater improvements in the prediction accuracy than the method using group TY.

(d) When using the BPNN model for prediction, the prediction accuracy of the method using group AX and that of the method using group SX are very similar. However, because of the difference in the size

TABLE 2: ANOVA results for 56 road sections.

OD type	ANOVA result	Amount
OD_1	6, 15, 20, 24, 35, 37, 41, 45, 46	9
OD_2	1, 4, 5, 11, 16, 18, 22, 41, 51	9
OD_3	2, 6, 12, 15, 40, 41, 45, 46	8
OD_4	1, 5, 13, 16, 35, 40, 41, 44, 47	9

TABLE 3: Prediction result using BPNN model for different input data.

Input Type		OD_1		OD_2		OD_3		OD_4	
		Mean	P	Mean	P	Mean	P	Mean	P
Group TY	MAPE	0.118		0.113		0.130		0.109	
	RMSE	0.142		0.145		0.162		0.143	
Group SX	MAPE	0.100	15.4%	0.092	18.9%	0.110	15.1%	0.093	14.6%
	RMSE	0.122	14.1%	0.116	19.7%	0.135	16.6%	0.131	8.9%
Group AX	MAPE	0.102	13.8%	0.089	22.8%	0.115	11.2%	0.092	15.6%
	RMSE	0.132	7.1%	0.114	22.6%	0.150	7.3%	0.127	11.6%

TABLE 4: Prediction result using MLR model for different input data.

Input Type		OD_1		OD_2		OD_3		OD_4	
		Result	P	Result	P	Result	P	Result	P
Group TY	MAPE	0.213		0.089		0.194		0.161	
	RMSE	0.244		0.106		0.252		0.191	
Group SX	MAPE	0.138	35.3%	0.077	14.4%	0.152	22.0%	0.108	33.3%
	RMSE	0.151	38.0%	0.098	7.1%	0.173	31.6%	0.134	29.7%
Group AX	MAPE	0.175	17.8%	0.110	-22.8%	0.179	7.7%	0.151	6.5%
	RMSE	0.203	16.9%	0.142	-34.2%	0.204	19.3%	0.173	9.3%

of the input data, the method using group AX requires more time for prediction than that using group SX.

(e) When using the MLR model for prediction, the prediction accuracy of the method using group SX is better than that of the method using group AX.

(f) Based on the above, it can be concluded that the proposed data dimension reduction method can improve the prediction accuracy for different conditions, using different prediction models and conditions under different traffic environments.

4.5. Application in Urban Traffic Management. The most common problem in urban traffic management is traffic jams. When a traffic jam occurs, administrators have a limited amount of time to stop it from spreading and causing other problems. More time for management can be obtained if traffic jams can be prejudged. In addition, if the traffic flow on roads that lead to an intersection can be predicted, the occurrence of congestion at the intersection can be determined in advance. The proposed short-term traffic flow prediction method can play an important role in predicting traffic flow towards intersections and identifying possible traffic jams. Then, based on the prediction results, the management measures can be appropriately adjusted.

The proposed method enables the application of existing prediction models to the urban traffic environment, as it overcomes the complexity and instability limitations of existing prediction models. Large amounts of historical traffic data from different road sections can be collected for each road leading to an intersection. Then, the data dimension can be reduced using the proposed MDS-based data dimension reduction method and a smaller dataset can be obtained to represent the larger dataset. By using a small dataset as the input data in existing prediction models, such as the BPNN and MLR models, a better prediction result is likely to be reached. The result is the basis for predicting traffic jams. The proposed MDS-based data dimension reduction method can also be combined with other prediction models.

5. Concluding Remarks

This study presents an MDS-based data dimension reduction method and applies it to short-term traffic flow prediction in urban road networks. A virtual network is built to generate traffic flow data, and the proposed method is used to reduce the data dimensions. The test results show significant advantages of applying the proposed method to short-term traffic flow prediction over the regular prediction method, which only considers the target section's historical data.

Additionally, the proposed prediction method is better at considering all the related sections, in terms of both efficiency and accuracy.

The key contributions of the study are as follows:

First, an MDS-based data dimension reduction method was developed and successfully applied to short-term traffic flow prediction in an urban road network.

Second, this research has demonstrated that the proposed data dimension reduction method can be combined with existing prediction models and adapted to different traffic environments, regardless of the flow patterns or traffic loading levels. Furthermore, the proposed method is compatible with different types of prediction models, both nonlinear and linear.

Third, an alternative method to apply various prediction methods to urban traffic prediction is presented. In urban traffic systems, the network is more complex and analysis of the spatial-temporal correlativity is necessary.

During the course of this study, several elements for future research were identified, which are as follows. (1) The results reported here are from a hypothetical network. Hence, results from real-world implementation would make the conclusions stronger. (2) Although different ODs are tested in the research, the same network is used. Networks with different structures should be analyzed to prove the effectiveness of the method. (3) The proposed method can also be applied to other prediction models, such as combined prediction models or models that can predict more sections at one time.

Disclosure

An earlier version of this paper has been presented at "Transportation Research Board 97th Annual Meeting" as a presentation.

Conflicts of Interest

The authors declare that they have no conflicts of interest.

References

[1] I. Thomson and A. Bull, "Urban traffic congestion: its economic and social causes and consequences," *Cepal Review*, pp. 105–116, 2002.

[2] M. Zare and M. Shirazi, "Arima Model for Network Traffic Prediction and Anomaly Detection," in *Proceedings of the International Symposium on Information Technology*, vol. 4, pp. 1–6, 2008.

[3] M. Billy and P. K. D. D. Williams, "Urban Freeway Traffic Flow Prediction," *Transportation Research Record*, vol. 1644, pp. 132–141, 2014.

[4] I. Okutani and Y. J. Stephanedes, "Dynamic prediction of traffic volume through Kalman filtering theory," *Transportation Research Part B: Methodological*, vol. 18, no. 1, pp. 1–11, 1984.

[5] H. Sun, H. X. Liu, H. Xiao, R. R. He, and B. Ran, "Use of Local Linear Regression Model for Short-Term Traffic Forecasting," *Transportation Research Record*, vol. 1836, pp. 143–150, 2014.

[6] W. Hui, "Vehicle speed prediction based on clustering and adaptive neuro-fuzzy inference," *Computer Engineering and Applications*, vol. 44, no. 6, pp. 240–242, 2008.

[7] A. S. Nair, J. C. Liu, L. Rilett, and S. Gupta, "Non-Linear Analysis of Traffic Flow," in *Proceedings of the IEEE Conference on Intelligent Transportation Systems*, pp. 681–685, Oakland, CA, USA, 2001.

[8] S. Y. Chen and W. Wang, "Chaos Forecasting for Traffic Flow based on Lyapunov Exponets," *China Civil Engineering Journal*, vol. 37, no. 9, pp. 96–99, 2004.

[9] G. He, S. F. Ma, and Y. Li, "Study on the Short-term Forecasting for Traffic Flow Based on Wavelet Analysis," *Systems Engineeering—Theory & Pratictice*, vol. 22, no. 9, pp. 101–106, 2002.

[10] X.-J. Cheng, J. Liu, and M.-S. Ma, "Algorithm of short-term traffic flow forecasting using fractal theory," *Journal of Transportation Systems Engineering and Information Technology*, vol. 10, no. 4, pp. 106–110, 2010.

[11] G. A. Davis and N. L. Nihan, "Nonparametric Regression and Short-Term Freeway Traffic Forecasting," *Journal of Transportation Engineering*, vol. 117, no. 2, pp. 178–188, 1991.

[12] V. N. Vapnik, *The Nature of Statistical Learning Theory*, Springer, NY, USA, 1995.

[13] Q. H. Xu and R. Yang, "Traffic Flow Prediction Using Support Vector Machine Based Method," *Journal of Highway and Transportation Research and Development*, vol. 22, no. 12, pp. 131–134, 2005.

[14] J. Hua and A. Faghri, "AppHcations of Artificial Neural Networks to Intelligent Vehicle-Highway Systems," *Transportation Research Record*, vol. 1453, pp. 83–90, 1994.

[15] B. L. Smith and M. J. Demetsky, "Short-term traffic flow prediction models-a comparison of neural network and non-parametric regression approaches," in *Proceedings of the IEEE International Conference on Systems, Man and Cybernetics*, pp. 1706–1709, San Antonio, TX, USA, 1994.

[16] G. F. Newell, "A simplified theory of kinematic waves in highway traffic, Part I: general theory," *Transportation Research Part B: Methodological*, vol. 27, no. 4, pp. 281–287, 1993.

[17] K. Nagel and M. Schreckenberg, "A cellular automaton model for freeway traffic," *Journal de Physique I*, vol. 2, no. 12, pp. 2221–2229, 1992.

[18] B. S. Kerner, *The physics of traffic*, Springer-Verlag, 2004.

[19] E. I. Vlahogianni, J. C. Golias, and M. G. Karlaftis, "Short-term traffic forecasting: overview of objectives and methods," *Transport Reviews*, vol. 24, no. 5, pp. 533–557, 2004.

[20] J. M. Bates and C. W. J. Granger, "The combination of forecasts," *Operational Research Quarterly*, vol. 20, no. 4, pp. 451–468, 1969.

[21] J. E. Jarrett and J. S. Plouffe, "The fuzzy logic method for simpler forecasting," *International Journal of Engineering Business Management*, vol. 3, no. 3, pp. 18–52, 2011.

[22] J. Wang, W. Deng, and Y. Guo, "New Bayesian combination method for short-term traffic flow forecasting," *Transportation Research Part C: Emerging Technologies*, vol. 43, pp. 79–94, 2014.

[23] H. Y. Chen, J. B. Zhao, and C. L. Liu, "Properties of combination forecasting model based on degree of grey incidence," *Journal of Southeast University (Natural Science Edition)*, vol. 34, no. 1, pp. 130–134, 2004.

[24] E. I. Vlahogianni, M. G. Karlaftis, and J. C. Golias, "Short-term traffic forecasting: where we are and where we're going," *Transportation Research Part C: Emerging Technologies*, vol. 43, pp. 3–19, 2014.

[25] H. R. Kirby, S. M. Watson, and M. S. Dougherty, "Should we use neural networks or statistical models for short-term motorway traffic forecasting?" *International Journal of Forecasting*, vol. 13, no. 1, pp. 43–50, 1997.

[26] H. B. Yin, S. C. Wong, J. Xu, and C. K. Wong, "Urban traffic flow prediction using a fuzzy-neural approach," *Transportation Research Part C: Emerging Technologies*, vol. 10, no. 2, pp. 85–98, 2002.

[27] A. Stathopoulos and M. G. Karlaftis, "A multivariate state space approach for urban traffic flow modeling and prediction," *Transportation Research Part C: Emerging Technologies*, vol. 11, no. 2, pp. 121–135, 2003.

[28] T. Cheng, J. Haworth, and J. Wang, "Spatio-temporal autocorrelation of road network data," *Journal of Geographical Systems*, vol. 14, no. 4, pp. 389–413, 2012.

[29] J. Wang, P. Shang, and X. Zhao, "A new traffic speed forecasting method based on bi-pattern recognition," *Fluctuation and Noise Letters*, vol. 10, no. 1, pp. 59–75, 2011.

[30] Y. Wu, F. Chen, C. Lu, and S. Yang, "Urban Traffic Flow Prediction Using a Spatio- Temporal Random Effects Model," *Journal of Intelligent Transportation Systems*, vol. 20, no. 3, pp. 282–293, 2015.

[31] R. Giacomini and C. W. Granger, "Aggregation of space-time processes," *Journal of Econometrics*, vol. 118, no. 1-2, pp. 7–26, 2004.

[32] S. Oh, Y.-J. Byon, K. Jang, and H. Yeo, "Short-term travel-time prediction on highway: a review of the data-driven approach," *Transport Reviews*, vol. 35, no. 1, pp. 4–32, 2015.

14

Enhancing Freeway Safety through Intervening in Traffic Flow Dynamics based on Variable Speed Limit Control

Jinming You,[1] Shouen Fang,[1] Lanfang Zhang,[1] John Taplin,[2] and Jingqiu Guo[1]

[1] *Key Laboratory of Road and Traffic Engineering of The Ministry of Education, Tongji University, Shanghai 201804, China*
[2] *Business School, The University of Western Australia, WA, 6009, Australia*

Correspondence should be addressed to Jingqiu Guo; guojingqiu@hotmail.com

Guest Editor: Zhiyuan Liu

New technologies and traffic data sources provide great potential to extend advanced strategies in freeway safety research. The High Definition Monitoring System (HDMS) data contribute comprehensive and precise individual vehicle information. This paper proposes an innovative Variable Speed Limit (VSL) based approach to manage crash risks by intervening in traffic flow dynamics on freeways using HDMS data. We first conducted an empirical analysis on real-time crash risk estimation using a binary logistic regression model. Then, intensive microscopic simulations based on AIMSUN were carried out to explore the effects of various intervention strategies with respect to a 3-lane freeway stretch in China. Different speed limits with distinct compliance rates under specified traffic conditions have been simulated. By taking into account the trade-off between safety benefits and delay in travel time, the speed limit strategies were optimized under various traffic conditions and the model with gradient feedback produces more satisfactory performance in controlling real-time crash risks. Last, the results were integrated into lane management strategies. This research can provide new ideas and methods to reveal the freeway crash risk evolution and active traffic management.

1. Introduction

There is a growing body of evidence confirming a positive relationship between the road safety benefits and vehicle speed enforcement, especially on freeways. In China, for example, Shanghai and Jiangsu with intensive freeway networks are actively employing intelligent technology systems for coordinating traffic flow and improving road safety. Previous studies have highlighted the higher vehicle speed on freeways associated with increased crash risk and injury severity [1, 2]. Meanwhile, speed variation among vehicles can disturb traffic flow and create more conflict situations [3]. Active traffic management (ATM) has been emerging in recent years aiming to provide traffic control to improve traffic flow and reduce congestion on freeways. Proper traffic control can significantly reduce delays and improve traffic distribution at a bottleneck, especially under congested [4] and work zone conditions [5, 6]. As a key application of ATM, Variable Speed Limit (VSL) systems aim to dynamically regulate freeway speeds based on real-time traffic flow information.

In the last decades, VSL has been intensively investigated on two main directions: traffic enhancement and safety improvement [7]. For instance, Hegyi et al. [8] proposed the macroscopic traffic flow model METANET with coordinated control of ramp metering and VSL to minimize the total time vehicles spent on the road; the method significantly reduced congestion. Naïve and Empirical Bayes are used to evaluate the effects of the VSL system and results indicate that VSL reduces crashes by 4.5% to 8% [9]. However, instead of mandatory, the advisory VSL does not show significant impact on traffic flow [10]. Especially under low speed limits, some drivers tend to violate the limits in pursuit of their personal benefits. Hence low speed limits may widen the range of flows under homogeneous traffic and contribute to a raise in lane changes [11]. Recent studies have shown that lane-changing maneuvers are a major source of traffic disturbance on a multilane freeway. Therefore, a proper setting is essential in VSL strategies. Instead of just using VSL before or during periods of high congestion, it can be applicable during off-peak periods as well [12]. So far, limited studies have been focusing on enhancing freeway safety by intervening in traffic flow dynamics based on VSL.

This study aims to apply real-time crash prediction in traffic control management. Previous studies on crash precursors have employed different kinds of traffic data such as loop detectors [1, 2, 13–15], automatic vehicle identification [16, 17], traffic counter data [18], and weather [19, 20] as well as road geometry data [21]. Different data mining and detection methods have been utilized to fully investigate the interrelationship between crash risk and traffic operation data. Abdel-Aty et al. [1] developed a neural network-based classifier to evaluate rear-end crash risk with traffic parameters from five stations: it can identify 75% of the crashes, with false positives of 34%. Ahmed et al. [16] exploited automatic vehicle identification (AVI) data and the model achieves an accuracy of 75.93% and 72.92% for rear-end and all crashes, respectively. As large numbers of false positives might affect drivers' compliance with the system and reduce its effectiveness, various refining approaches have been employed to optimize the evaluation algorithm. Traffic data with high resolution and multiple sources are needed for a better evaluation model with higher predictive accuracy and robustness. Ahmed et al. [17] enhance the AVI data with remote traffic microwave sensor data and their model can successfully identify 88.9% of the crashes with a false positive rate of only 6.5%. Kwak et al. [22] found that traffic flow characteristics leading to crashes differ by segment type and traffic flow state.

However, with respect to data type and resolution, the detectors are often limited and traffic data from continuous detectors cannot be collected or the collected data do not meet the requirements of the models. For instance, in China, detectors are installed far apart on freeways and most segments have not been equipped with detectors or surveillance devices. Regarding to method, generalized linear models could provide direct evidence of the traffic parameters' impacts on crash risk. When dealing with highly nonlinear relationships between traffic flow and crash, it requires more computational, flexible, and nonlinear models [23]. Recent studies found that nonlinear models are capable of achieving higher crash prediction accuracy with less false positives. However, the limitations of the available nonlinear models include heavy computations to reveal deeper connections between the traffic parameters as well as for model calibration. Meanwhile, few studies have thoroughly investigated the application of real-time crash risk prediction. Based on an accurately quantified crash risk evaluation, proper traffic management strategies can be applied and therefore improve road safety. For example, Yu et al. [7] proposed a VSL control algorithm in mountainous freeways and the result indicated a positive outcome in crash risk control. However, most VSL control studies are not safety-oriented and so the only parameter utilized in the crash risk model is speed. Meanwhile, generally it is hard to obtain accurate speed variation data, and the effect of speed dispersion on traffic safety has not been intensively investigated [3].

New technologies and traffic data sources provide great potential to extend advanced strategies in freeway safety research. For example, the High Definition Monitoring System (HDMS) data contribute comprehensive and precise individual vehicle information, including vehicle type, speed, lane number, and plate number, as well as high quality photos captured by advanced vehicle license plate recognition systems. In China, HDMS have been installed on major freeways for public security management. The major contributions of this paper consist of the following aspects (Figure 1): (1) to employ HDMS data with individual vehicle information to study the crash mechanism on freeways; (2) to develop a Logistic Regression model for real-time crash risk estimation; (3) to evaluate safety benefits of the optimized VSL based on enhanced AIMSUN simulations on a 100 km freeway stretch; (4) to investigate sensitivities of VSL impacts on driver compliance.

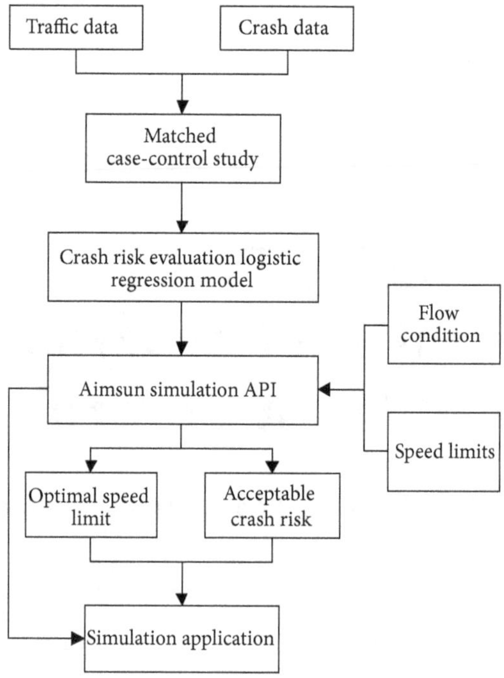

FIGURE 1: Architecture of this study.

2. Crash Risk Model

2.1. Data Preparation. The study area is G15 Freeway in Nantong, Jiangsu Province, China, with a total length of approximately 100 km, from Sutong Bridge to Fuan Toll (as shown in Figure 2).

Data are obtained from the Public Security Traffic Managing System of the Traffic Management Research Institute, Ministry of Public Security. The freeway is a 6-lane one (3 lanes in each direction). The primary dataset includes all crash data and HDMS data from January to October 2016. The extracted HDMS data cover the lane number, direction, vehicle type and speed, recorded time of vehicle passing, etc. The study area includes five pairs of HD cameras.

The raw crash dataset includes 5924 crashes. However, the majority of crashes are not recorded with detailed location or direction information. 88% of the crashes are involved with multivehicles. Among them, 96% of the crashes are recorded with causes of hitting the fixed objects such as the guardrails

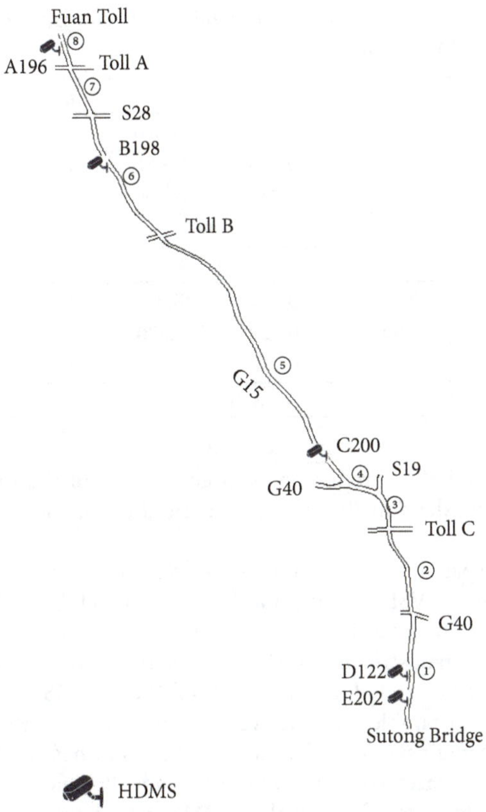

FIGURE 2: The study area and subsegment division.

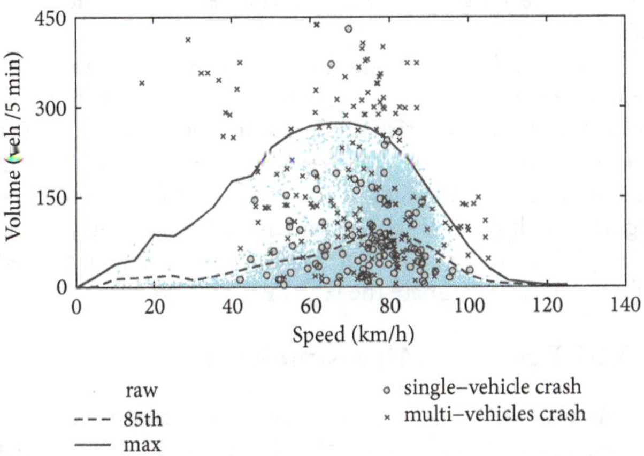

FIGURE 3: The traffic status 5-10 minutes prior to crashes.

and the medians, or hitting the unfixed objects such as the crash barriers. In order to investigate the impact of traffic dynamics on crashes, the traffic status prior to crashes has been examined as Figure 3. In Figure 3, the raw HDMS data are aggregated with the interval of 5 minutes. The datasets from April 1st to May 31st in 2016 have been utilized to show the relationship between traffic volume and speed. The solid curve "max" shows the margin of the data during this period and the dash curve shows the 85th margin of the data during this period. The dash curve demonstrates that, under most conditions, the speed/volume distribution is within the curve. The speed/volume data 5-10 minutes prior to each crash have

TABLE 1: Summary statistics of variables.

Variable	Mean	S.D.	First Quartile	Third Quartile
Q (veh)	103.59	84.80	43.00	138.00
V (km/h)	80.37	12.83	75.04	87.93
DV (km/h)	22.28	5.40	18.73	26.08

been plotted in Figure 3 as well. The symbol "o" refers to the crashes involved with single vehicle and the symbol "x" refers to the crashes involved with multivehicles.

First, the "max" curve shows the similar trend with that of the traditional capacity/speed curve. However, as under most conditions, the traffic state is normal and it is difficult to obtain the saturated flow state with different speeds. Hence, the "max" curve mainly reflects the nonfree flow state, in which the volume is approaching the maximum capacity. The area beyond the "max" curve reflects the chaos flow state or congestion state. Second, comparing the state of single-vehicle crashes and the state of multivehicles we could find that the single-vehicle crashes are more likely to occur within the 85th curve; i.e., the crashes are likely to occur under free flow conditions. This is consistent with several existing studies [24]. Another finding is that the number of multivehicle crashes is in majority in the area beyond the "max" curve. It indicates that when the traffic state approaches congestion state, the crashes are likely to involve multivehicles; especially when the speed is below 60 km/h, only multivehicle crashes are detected. Meanwhile, under all conditions, when the speed is below 40km/h, no single crashes have been recorded.

Additionally, as single-vehicle crashes are usually caused by random effects, such as driving distraction and breaking down, only multivehicles crashes with detailed temporal and spatial information are utilized in this study to investigate the relationship within traffic dynamics and crash risk. The data 5-10 minutes prior to the crashes are utilized to represent the traffic status prior to crashes. The method is commonly used in existing studies [15] as well.

A matched case-control method is utilized to extract the related samples for each sampled crash. A 4:1 control-case ratio is used, as recommended in several previous studies [2]. Each crash case and noncrash case is matched with corresponding traffic data on the same road segment. The four control samples are selected from 14 days before the recorded crash time, 7 days before, 7 days after, and 14 days after. Considering the transferability of the model, the HDMS data 5-10 minutes prior to the samples are aggregated as traffic flow, mean speed, and speed dispersion, which are labeled as Q, V, and DV, respectively.

Some filtering rules are also applied to select the available samples. Due to occasional HDMS system failure, some samples would be matched with invalid HDMS data or missing HDMS data. Noise and outliers are removed from the final dataset. Finally the crash dataset contains 171 samples and the control dataset has 618 non-crash samples. The summary statistics of variables are listed in Table 1.

2.2. Binary Logistic Regression Model. Logistic regression analysis is commonly used to quantify the crash risk in

Table 2: Crash risk evaluation model for the whole segment.

Variables	B.	S.E.	Wals.	Df.	Sig.
Q	0.012	0.001	101.806	1	0.000
DV	0.120	0.020	37.882	1	0.000
Constant	-5.487	0.527	108.473	1	0.000
AUC	0.755	***Nagelkerke*** R^2		0.288	

Table 3: Comparison of the performance of the three models.

Data type	Detector density	Modeling technique	Prediction accuracy
AVI	3 detectors per 2.35 km	Bayesian	0.759
Loop detector	6 detectors per 5.46 km	Genetic Programming	0.608
HDMS	Single detector per 20 km	Logistic Regression	0.755

real-time crash analysis. The traffic condition can be divided into two parts, crash cases (y_i =1) and noncrash cases (y_i =0) with respective probability p_i and 1-p_i. The probability of a crash occurrence is estimated by

$$P(y_i = 1 \mid x_i) = p_i = \frac{e^{\beta_0+\sum \beta_k x_{ki}}}{1 + e^{\alpha+\sum \beta_k x_{ki}}} \quad (1)$$

The odds of crash occurrence $g(x)$ can be calculated as

$$g(x) = \frac{p_i}{1 - p_i} = e^{\beta_0+\sum \beta_k x_{ki}} \quad (2)$$

where x_{ki} denotes input variables series; β_0 denotes the constant in the logistic model; β_k denotes coefficients for the independent input variables. Here β_k can be estimated by solving the log-likelihood function in

$$\ln L(\beta) = \sum_{i=1}^{n} \left[\beta_0 + \beta_1 x_{1i} + \cdots + \beta_k x_{ki} - \ln\left(1 + e^{\beta_0+\sum \beta_k x_{ki}}\right)\right] \quad (3)$$

The crash risk model is estimated with the binary logistic regression procedure in SPSS 19. Backward LR (likelihood ratio) variables selection is applied to select the significant parameters in the proposed model. As shown in Table 2, results indicate that flow and speed dispersion are significant variables for crash risk estimation. The larger values of *Q* and *DV* indicate a higher crash risk. The *AUC* (area under the receiver operating characteristic curve, which illustrates the performance of the classifier) value indicates that the logistic regression model could successfully classify most of the crash and noncrash cases.

In order to compare the predicting ability of the HDMS data with the ability of the other kinds of traffic data, such as loop detector data [22] or Automatic Vehicle Identification (AVI) data [16], several existing studies are listed in Table 3. Table 3 reveals that the HDMS data provide better evidence for crash risk estimation and the corresponding model has a relatively better prediction accuracy despite of the simple form of model and the discrete distribution of the HDMS devices.

Table 4: Crash risk evaluation models for downstream and upstream of the HDMS.

Variables	B	S.E.	Wals	df	Sig.
Q	0.012	0.001	64.270	1	0.000
DV	0.139	0.026	27.961	1	0.000
Constant	-5.912	0.714	68.650	1	0.000
Downstream ***AUC***		0.759			
Variables	B	S.E.	Wals	df	Sig.
Q	0.013	0.002	38.924	1	0.000
DV	0.093	0.029	10.320	1	0.001
Constant	-4.948	0.776	40.608	1	0.000
Upstream ***AUC***		0.755			

The spatial issue should be addressed for the implementation of VSL. Hence, another two models have been formulated to investigate the spatial effect, a downstream model and an upstream model. As shown in Figure 4, samples with crashes located downstream of the HDMS are the downstream samples and vice versa. Finally the downstream dataset contains 105 crashes and 360 noncrashes, and the upstream dataset contains 68 crashes and 266 noncrashes (the samples located just at the HDMS station are classified into both downstream and upstream samples).

As before, binary logistic regression has been used to estimate the crash risk models. The results are shown in Table 4.

Results indicate that the performance of the crash risk models considering the spatial effects is similar to the performance of the crash risk model for the whole segment. The reason for this is that the crash risk is stable on each segment and the traffic parameters of adjacent locations on the same segment are highly correlated, which has been shown by Fang et al. [25]. Hence, the model for the whole segment is used subsequently to estimate the real-time crash risk.

3. VSL Based on Microsimulation

3.1. Aimsun API. In order to verify the method based on dynamic VSL control of crash risk, a sub-segment of the G15 Freeway segment utilized in Section 2 is employed in the simulation (Segment 1 in Figure 2). The segment extends from "Nantong Toll Interchange" to "Sutong Bridge" and the total length is approximately 6 km.

Aimsun API (Application Programming Interface) can be a helpful platform to evaluate certain traffic management strategies. We can obtain the necessary real-time traffic data (flow, speed, occupancy, etc.) with required aggregation levels or detailed vehicle information. The project is built with Visual C++ 6.0 based on Visual Studio 2005. Using Aimsun API functions, the detectors, VMS, and traffic control plans are modeled and the attributes are defined in our in-depth simulations.

To simulate the real HDMS data, AIMSUN API function is used to gather the real-time vehicle information. The

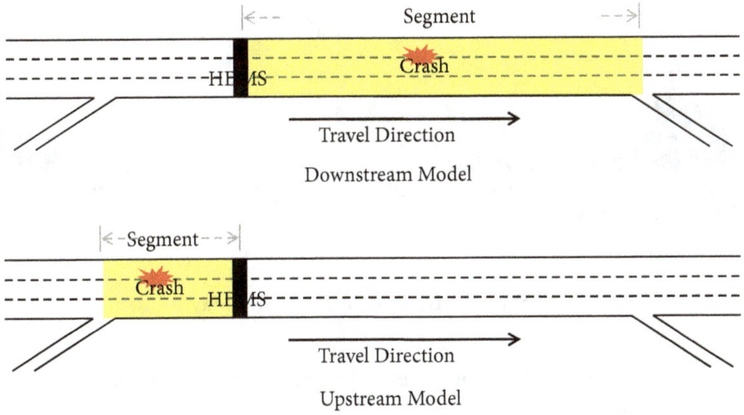

FIGURE 4: Spatial classification for downstream and upstream models.

scenarios are tested on the G15 Freeway with a design speed of 120km/h. In order to code the freeway segment in the simulation, Baidu Map GIS data source is utilized to build the freeway network. The drivers are assumed to comply with the speed limit, with a certain compliance rate, when the VSL starts to function on the segment.

The step size is 1 second. The Aimsun software development kit has been utilized to develop a module to extract the parameters for the crash risk evaluation model during the simulation process. The values for Q_i, DV_i, with a 5 minutes aggregation are recorded every minute. Thus the crash risk could be estimated every minute and exported as a report for final analysis as well as the average delay of all vehicles in the whole simulation process.

The real-time crash risk probability (p_i) for time period (i) is estimated with (4) using the parameters in Table 2:

$$p_i = \frac{e^{-5.487+0.012*Q_i+0.12*DV_i}}{1+e^{-5.487+0.012*Q_i+0.12*DV_i}} \quad (4)$$

The average delay (AD) for each simulation application can be calculated with

$$AD = \frac{\sum_{i=1}^{n}(T_i - T_{ei})}{n} \quad (5)$$

where T_i denotes the real time vehicle i spent on the freeway, T_{ei} denotes the expected time vehicle i spent on the freeway, and n denotes the number of vehicles passing by.

3.2. Simulation Calibration

3.2.1. Speed Distribution and Compliance Calibration. The speed distribution and compliance level are calibrated before the simulations. The original speed limit for the G15 Freeway is 120 km/h and the proportion of vehicles with speed above 120 km/h is set as the non-compliance level. Drivers tend to speeding on the freeway as the freeway is designed with better alignments especially long stretch of straight line. The traffic on the freeway mainly comprises private cars and trucks. All the speed data for May 2016 is used to calibrate the parameters. Figure 5 present the speed distributions for different vehicle types. The calibration includes the speed distribution and the compliance rate. The compliance rate for cars and trucks is 88.37% and 79.05% respectively. Other parameters such as lane width, lane number and road type are also calibrated. Part of the final scenario settings are listed in Table 5.

TABLE 5: Simulation scenario settings in Aimsun.

Scenario Parameters	Settings
Road type	Freeway
Lane width	3.75 meter
Maximum speed limit	120 km/h
Detection cycle	1 second
Car following model	Minimum Headway+ deceleration estimation
Global arrivals	Exponential

3.2.2. Calibration of Traffic Temporal Distribution. Traffic spatial distribution should also be addressed to validate the simulations.

In existing studies, aggregate statistics have been validated such as the GEH statistics by FHWA [7]. In this study, HDMS data have been utilized to calibrate the simulations. 1-hour HDMS data starting at 2016-05-05, 13:50:00, are used to validate the simulation. The distribution diagram of time headway in Figure 6 indicates that the temporal distribution of the real data is similar to the temporal distribution of the simulation data, especially when the value of time headway exceeds 10 seconds. The significant difference appears on the distribution when the time headway is less than 4 seconds. The reason is that in real world some drivers may drive aggressively with a relatively low following distance especially under congestions, or try to change the lane with a relatively low transverse distance, leading to the relatively high proportion of low time headway. While in the simulation scenarios, the vehicles follow the car-following and lane-changing rules and they would never exceed the limit values. The result of an additional Pearson test shows that the coefficient value of the two curves is 0.75, which suggests moderate fidelity of the simulation.

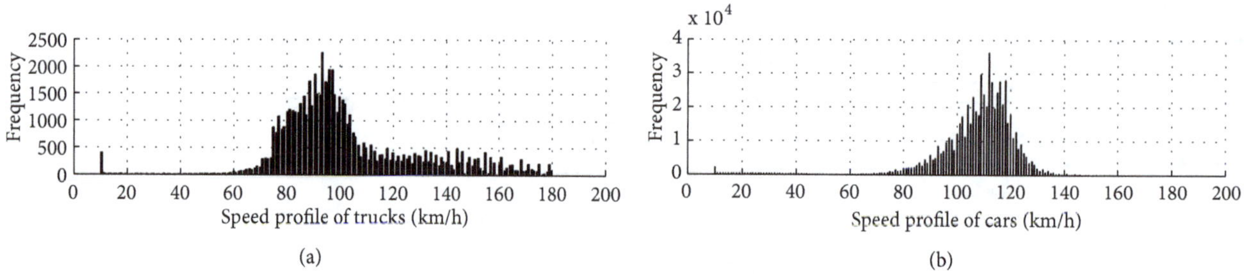

FIGURE 5: Speed distributions of vehicles (May 2016).

TABLE 6: Speed limit impacts under different flow conditions.

Flow condition (veh/h)	Speed Limit (VSL) (km/h)	Acceptable crash risk	Average Delay (s)	Crash risk benefit (%)	Average Delay Loss (%)
5000	60	0.23	153.24	29.81	258.98
4000	70	0.27	158.92	39.73	167.51
3000	80	0.16	44.64	25.47	35.89
2000	120	0.09	17.94	0	0

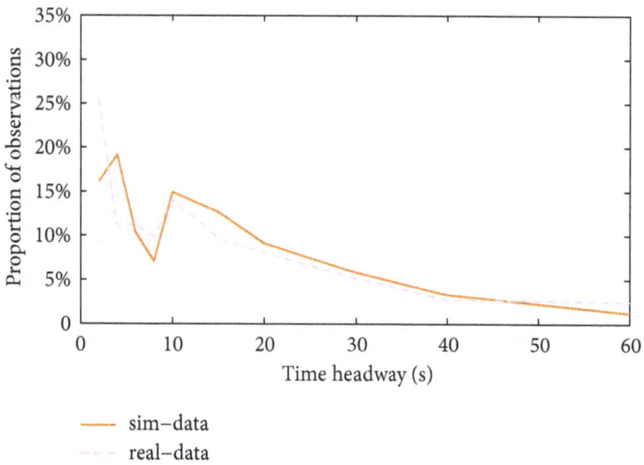

FIGURE 6: Comparison of the distributions of time headway.

3.3. VSL Performance Assessment under Different Flow Conditions. A set of six speed limits has been tested to evaluate the VSL performance under different flow conditions, namely 90, 80, 70, 60, 50, and 40 km/h. The traffic demand ranges from 2,000 to 5,000 veh/h. The Aimsun simulation results depend on the random seeds, reflecting the impact of random factors, and simulations were replicated five times to account for the variability. Each replication has 20 minutes to warm up with the traffic demand 2,000 veh/h and 60 minutes more to simulate the whole process with a different flow.

As the traffic is dynamic, the real traffic flow varies over time, as does the crash risk. Thus the objective is to keep the crash risk within an acceptable limit. In this study, the commonly used 85th percentile index in traffic safety is selected as the crash risk threshold for each replication; i.e., the traffic is evaluated as safe below that threshold. Once the crash risk exceeds the limit, proper strategies should be implemented to minimize the risk. As inappropriate speed limits would decrease the capacity and increase traffic delay, a comprehensive analysis should be made to achieve an optimal cost benefit ratio. Figure 7 shows plots of crash risk (the possibility of crash occurrence) and average delay in relation to speed limit under different flow conditions.

Figure 7 indicates that a lower speed limit would achieve lower crash risk; however the benefits vary at different flow levels. The average delay increases more and more as the speed limit is set lower. Hence, the crash risk benefit and average delay loss under different speed limits can be related to the original crash risk and the average delay of the replication without VSL, so that trade-offs can be made as suggested in Table 6. The results above are based on the hypothesis that all drivers would comply with the VSL; i.e., compliance rate equals 100%. However, when the VSL is advisory and also inappropriate, most of the drivers would not comply with it; thus the VSL strategy does not have any significant impact on traffic conditions [10]. Hence, simulations with a lower compliance rate are also examined, the rate being set to 50%. The results are shown in Figure 8. Figure 8 indicates that at flows of 2,000 to 3,000 veh/h and speed limits below 70 km/h, with a compliance rate of 50% the crash risk increases as the speed limit decreases. The compliance rate of 50% is low, with half of the drivers driving at the VSL and the other half driving as they prefer. At low flow levels, drivers would pursue higher driving speeds, so that speed dispersion would increase and make accidents more likely.

4. Application of VSL Strategies in Simulation

4.1. VSL Control Strategy. The objective of the VSL control strategy is to manage the traffic within an acceptable crash risk level and feedback is needed to adapt the strategy to the real-time traffic condition. Two kinds of strategies have been implemented in the simulations. The first is implementing

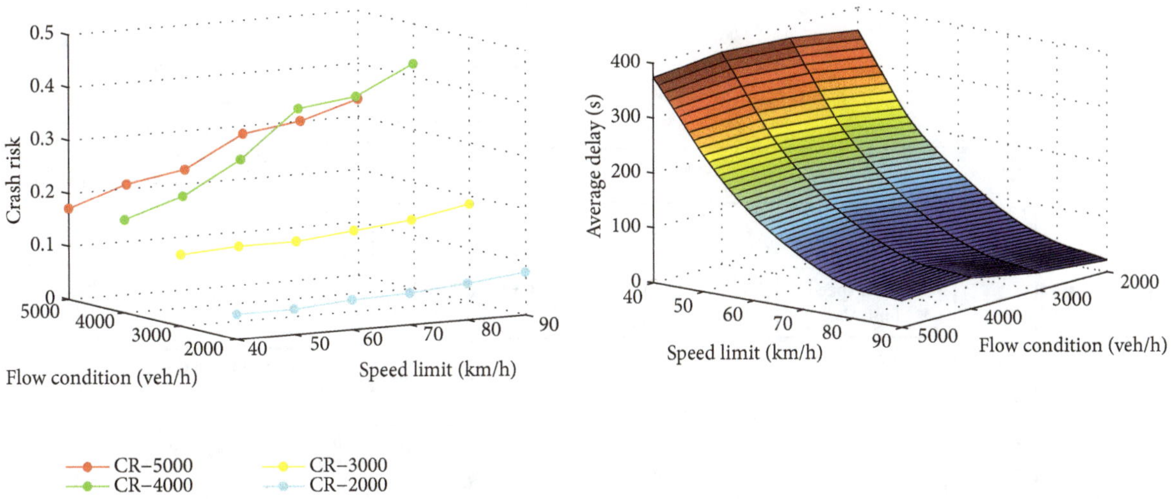

(a) Crash risk versus speed limits and flow conditions

(b) Average delay versus speed limits and flow conditions

FIGURE 7: Average delay and crash risk versus speed limit under different flow conditions.

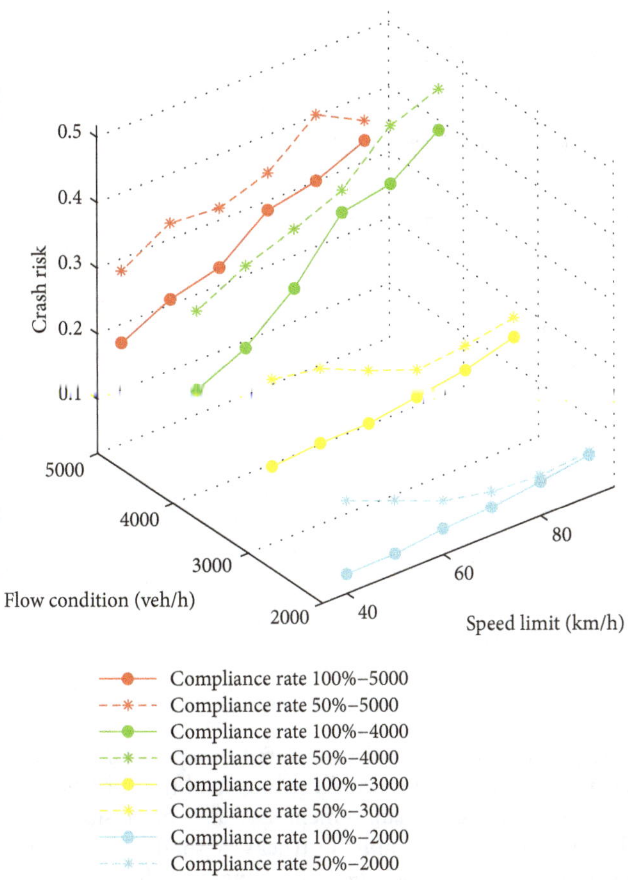

FIGURE 8: Comparison of crash risk versus speed limit for alternative compliance rates under different flow conditions.

and withdrawing the optimal VSL gradually (Strategy A) and the other is implementing and withdrawing the optimal VSL rapidly (Strategy B). Strategy A can be described as in Figure 9. When the real-time crash risk exceeds the acceptable risk threshold τ), the corresponding target VSL V_τ is set and the VSL value is decreased at the gradient of 5 km/h every minute towards the target VSL V_τ. In Strategy B, when the crash risk exceeds τ, the VSL is set to V_τ immediately. The HDMS data are extracted from the HDMS database starting at 2016-05-02, 13:50:00. The corresponding traffic demand and speed distribution in the simulation are calibrated again to fit the real HDMS data. The compliance rates of the simulations is set at 88.37% for cars and 79.05% for trucks, being the statistical results in Section 3.2. It took 15 minutes to warm up and the whole simulation time is 75 minutes. The simulations, and the comparison with raw data, are labeled as "raw," "Strategy A," and "Strategy B." The simulations are started by the same random seed.

4.2. Application Results. The simulations results with different strategies are shown in Figure 10. The crash risk curve of the raw HDMS data displays two crash risk peaks after 40 minutes, exceeding the crash risk threshold significantly. Compared with the raw curve, Strategy B could decrease the average crash risk by 10.15%, but the risk is still high after the simulation time of 50 minutes. The crash risk curve indicates that there are no significant crash risk peaks under the control of Strategy A. The average crash risk has been decreased by 22.63% by Strategy A compared with the Raw strategy curve. The crash risk remains at a low level. Strategy A outperforms Strategy B in which the speed limit is set to the target speed immediately. When drivers pass by the speed limit sign, they have to decelerate rapidly to comply with the speed limit. As a result, the speed dispersion increases rapidly and the crash risk increases as well.

Average delays are 56.31s, 24.87s, and 91.84s for Raw, Strategy A, and Strategy B, respectively. Thus Strategy A generates the shortest travel time and this strategy could control the traffic condition efficiently and steadily, whereas improper speed limit implementation may lead to unexpected traffic congestions.

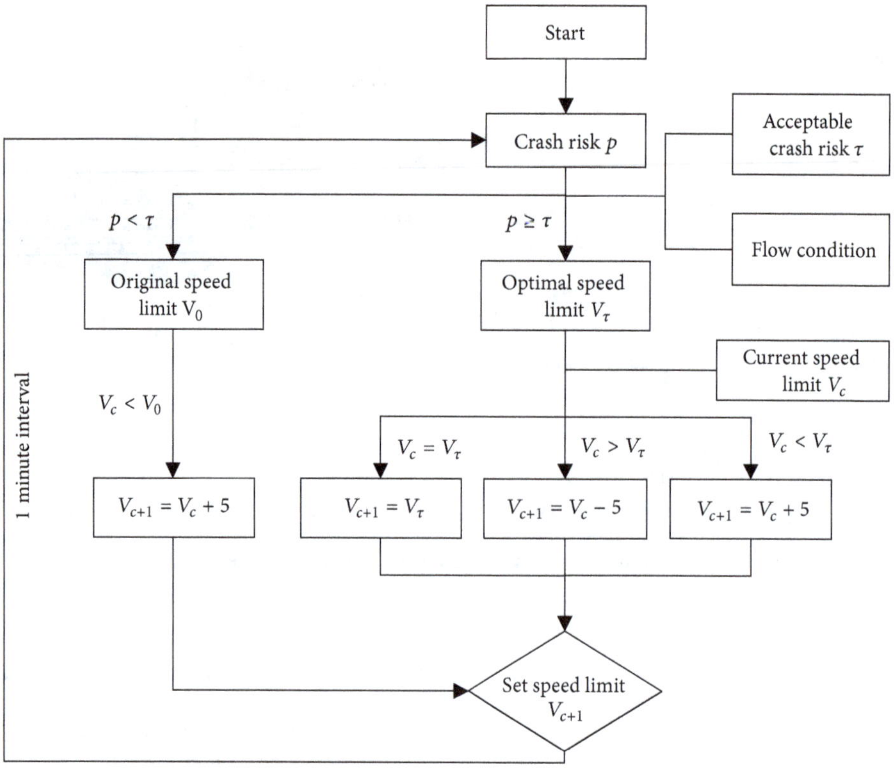

FIGURE 9: Strategy A: the VSL strategy with gradient feedback control.

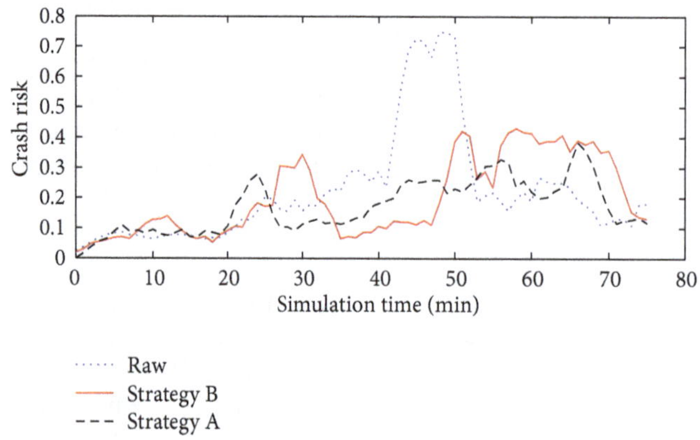

FIGURE 10: Simulation results under three conditions.

5. Discussion

Results of this study demonstrate that the proposed VSL method could improve traffic safety, but more developments are required to produce integrated control strategies that are efficient and also applicable in real-time to large-scale networks [26]. Integrated control with ramp metering and VSL has been used to improve traffic flow efficiency based on an optimal VSL rate [27]. Lane management (LM) method has been employed in most countries, such as US, Europe, and China. Lord et al. [28] indicated that truck-free freeways would have a better safety record than mixed traffic and separating truck traffic from passenger cars improves safety. Toll lanes have been suggested to separate *Lights* and *Heavies* and the method could reduce total travel costs [29]. In this study, lane management has also been simulated. To separate the cars and the trucks, a solid line rule has been applied to restrict the lane-changing behavior on the mainline. The truck can only access the outer lane while the cars can access the other two lanes. Each condition with lane management has been simulated with different VSL under different traffic conditions. The results are shown in Figure 11.

Figure 11 indicates that, under the traffic conditions of 2,000 veh/h and 3,000 veh/h, lane management has little impact on the crash risk. However, under traffic conditions of

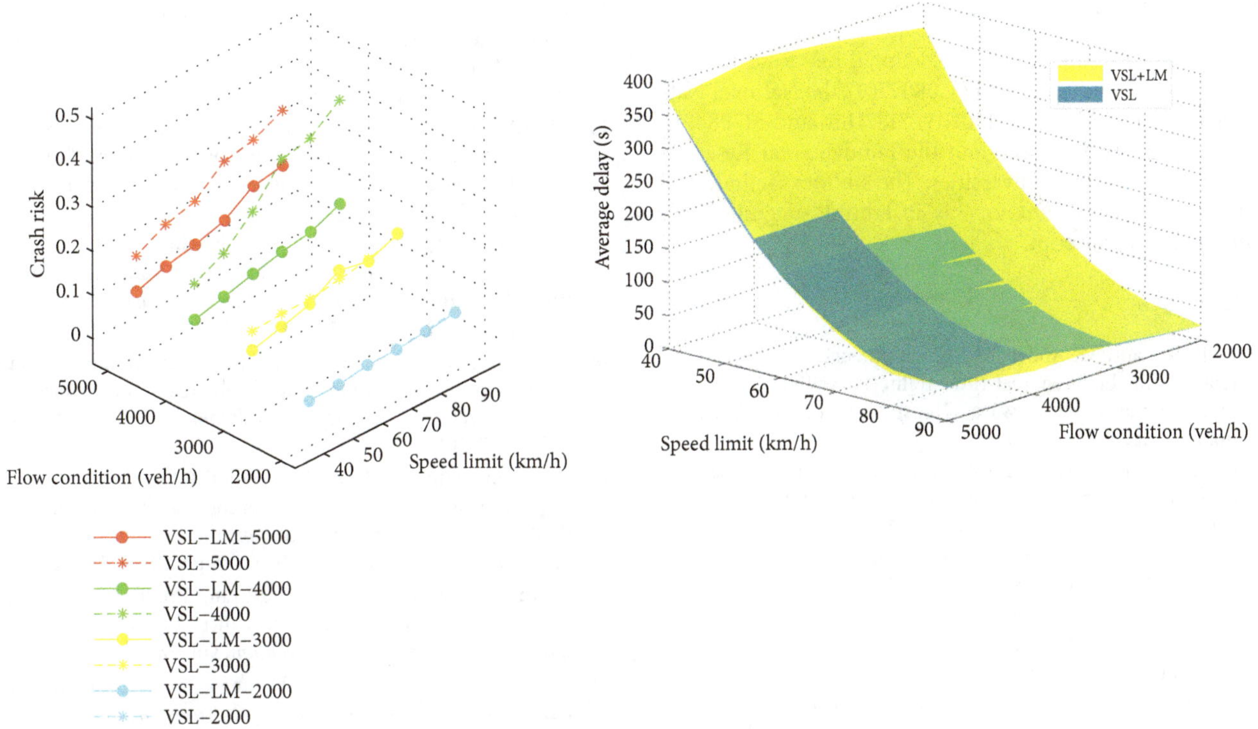

(a) Crash risk versus flow conditions and speed limits

(b) Average delay versus flow conditions and speed limits

FIGURE 11: Impact of lane management on crash risk and average delay under different traffic conditions.

4,000 veh/h or 5,000 veh/h, lane management has significant impacts on ameliorating the crash risk. By implementing lane management, the crash risk could be substantially controlled despite high speed limits.

As shown in Figure 11(b), lane management at lower traffic levels would have no impact on the average delay. Thus, by implementing lane management properly with VSL, traffic management administration would get significant improvement in crash risk without affecting the level of service of the freeway.

6. Conclusions

The study proposes an innovative dynamic variable speed approach through intervening in traffic flow dynamics. A binary logistic regression model based on HDMS data is built to estimate crash risk. HDMS data provide detailed vehicle information instead of aggregated data from loop detectors or other detectors. They provide better evidence on the crash mechanism. Microsimulations have been conducted with the AIMSUN simulation software. AIMSUN API is utilized to extract the detailed real-time vehicle information to calculate the crash risk. Different speed limits with several compliance rates under certain traffic conditions have been simulated. Considering the trade-off between safety benefits and travel time delay, we aim to optimize speed limit strategies under various traffic conditions.

Two kinds of VSL strategies have been applied to control the real-time crash risk in the simulated conditions of real traffic accidents; the strategy of implementing and withdrawing the optimal VSL gradually (gradient control) could provide better control effects and keep the crash risk at a lower level. Furthermore, lane management control has also been assessed. Results indicate that such integrated control could significantly reduce the crash risk without increasing average traffic delay. The trends in optimal integrated traffic control to reduce real-time crash risk prove to be promising.

Several potential directions are open for future exploration. For example, further work is being conducted to study the performance of applying the strategies into various road types or road network. In future studies, as more and more surveillance devices and vehicle on-board devices are installed, real-time data such as weather condition as well as driving behavior could be obtained. Meanwhile, with the continuous spatial distribution of surveillance devices and detectors, the aggregated traffic control of multiple segments could be investigated to achieve balanced traffic conditions in the road network as more driver-friendly integrated control strategies are developed to fit the new era of ITS.

Conflicts of Interest

The authors declare that they have no conflicts of interest.

Acknowledgments

The research is supported by the National Key Research and Development Program (2016YFC0802701), the National Natural Science Foundation (71301119, 71871161, and 51708421), Shanghai Pujiang Program, and the Fundamental Research Funds for the Central Universities. The authors are indebted to Professor Xiaobo Qu for his insightful suggestions to improve this manuscript.

References

[1] A. Pande and M. Abdel-Aty, "Comprehensive analysis of the relationship between real-time traffic surveillance data and rear-end crashes on freeways," *Transportation Research Record*, no. 1953, pp. 31–40, 2006.

[2] C. Xu, P. Liu, W. Wang, and Z. Li, "Evaluation of the impacts of traffic states on crash risks on freeways," *Accident Analysis Prevention*, vol. 47, no. 1, pp. 162–171, 2012.

[3] A. Theofilatos and G. Yannis, "A review of the effect of traffic and weather characteristics on road safety," *Accident Analysis & Prevention*, vol. 72, pp. 244–256, 2014.

[4] X. Yang, Y. Lin, Y. Lu, and N. Zou, "Optimal variable speed limit control for real-time freeway congestions," *Procedia: Social and Behavioral Sciences*, vol. 96, pp. 2362–2372, 2013.

[5] W. Zhizhou, L. Hao, and Y. Hongsu, "Variable speed limit control on highway work zone considering large vehicle mix rates," *Procedia - Social and Behavioral Sciences*, vol. 96, pp. 1784–1791, 2013.

[6] A. Tympakianaki, A. Spiliopoulou, A. Kouvelas, I. Papamichail, M. Papageorgiou, and Y. Wang, "Real-time merging traffic control for throughput maximization at motorway work zones," *Transportation Research Part C: Emerging Technologies*, vol. 44, pp. 242–252, 2014.

[7] R. Yu and M. Abdel-Aty, "An optimal variable speed limits system to ameliorate traffic safety risk," *Transportation Research Part C: Emerging Technologies*, vol. 46, pp. 235–246, 2014.

[8] A. Hegyi, B. De Schutter, and H. Hellendoorn, "Model predictive control for optimal coordination of ramp metering and variable speed limits," *Transportation Research Part C: Emerging Technologies*, vol. 13, no. 3, pp. 185–209, 2005.

[9] S. Long, L. Gentry, and G. H. Bham, "Driver perceptions and sources of user dissatisfaction in the implementation of variable speed limit systems," *Transport Policy*, vol. 23, pp. 1–7, 2012.

[10] A. Nissan and H. N. Koutsopoulosb, "Evaluation of the impact of advisory variable speed limits on motorway capacity and level of service," *Procedia - Social and Behavioral Sciences*, vol. 16, pp. 100–109, 2011.

[11] F. Soriguera, I. Martínez, M. Sala, and M. Menéndez, "Effects of low speed limits on freeway traffic flow," *Transportation Research Part C: Emerging Technologies*, vol. 77, pp. 257–274, 2017.

[12] M. Abdel-Aty, J. Dilmore, and A. Dhindsa, "Evaluation of variable speed limits for real-time freeway safety improvement," *Accident Analysis & Prevention*, vol. 38, no. 2, pp. 335–345, 2006.

[13] X. Qu, S. Wang, and J. Zhang, "On the fundamental diagram for freeway traffic: a novel calibration approach for single-regime models," *Transportation Research Part B: Methodological*, vol. 73, pp. 91–102, 2015.

[14] Y. Kuang, X. Qu, and S. Wang, "A tree-structured crash surrogate measure for freeways," *Accident Analysis & Prevention*, vol. 77, pp. 137–148, 2015.

[15] Z. Li, S. Ahn, K. Chung, D. R. Ragland, W. Wang, and J. W. Yu, "Surrogate safety measure for evaluating rear-end collision risk related to kinematic waves near freeway recurrent bottlenecks," *Accident Analysis & Prevention*, vol. 64, pp. 52–61, 2014.

[16] M. Ahmed, M. Abdel-Aty, and R. Yu, "Bayesian updating approach for real-time safety evaluation with automatic vehicle identification data," *Transportation Research Record*, no. 2280, pp. 60–67, 2012.

[17] J. Zhang, X. Qu, and S. Wang, "Reproducible generation of experimental data sample for calibrating traffic flow fundamental diagram," *Transportation Research Part A: Policy and Practice*, vol. 111, pp. 41–52, 2018.

[18] H. D. Son, Y.-J. Kweon, and B. B. Park, "Development of crash prediction models with individual vehicular data," *Transportation Research Part C: Emerging Technologies*, vol. 19, no. 6, pp. 1353–1363, 2011.

[19] E. I. Vlahogianni, M. G. Karlaftis, and F. P. Orfanou, "Modeling the effects of weather and traffic on the risk of secondary incidents," *Journal of Intelligent Transportation Systems: Technology, Planning, and Operations*, vol. 16, no. 3, pp. 109–117, 2012.

[20] C. Xu, A. P. Tarko, W. Wang, and P. Liu, "Predicting crash likelihood and severity on freeways with real-time loop detector data," *Accident Analysis & Prevention*, vol. 57, pp. 30–39, 2013.

[21] C. Lee, F. Saccomanno, and B. Hellinga, "Analysis of crash precursors on instrumented freeways," *Transportation Research Record*, no. 1784, pp. 1–8, 2002.

[22] H.-C. Kwak and S. Kho, "Predicting crash risk and identifying crash precursors on Korean expressways using loop detector data," *Accident Analysis & Prevention*, vol. 88, pp. 9–19, 2016.

[23] Y. Lao, G. Zhang, Y. Wang, and J. Milton, "Generalized nonlinear models for rear-end crash risk analysis," *Accident Analysis & Prevention*, vol. 62, pp. 9–16, 2014.

[24] Z. Christoforou, S. Cohen, and M. G. Karlaftis, "Identifying crash type propensity using real-time traffic data on freeways," *Journal of Safety Research*, vol. 42, no. 1, pp. 43–50, 2011.

[25] S. Fang, W. Xie, J. Wang, and D. R. Ragland, "Utilizing the eigenvectors of freeway loop data spatiotemporal schematic for real time crash prediction," *Accident Analysis & Prevention*, vol. 94, pp. 59–64, 2016.

[26] M. Papageorgiou, C. Diakaki, V. Dinopoulou, A. Kotsialos, and Y. Wang, "Review of road traffic control strategies," *Proceedings of the IEEE*, vol. 91, no. 12, pp. 2043–2065, 2003.

[27] R. C. Carlson, I. Papamichail, M. Papageorgiou, and A. Messmer, "Optimal motorway traffic flow control involving variable speed limits and ramp metering," *Transportation Science*, vol. 44, no. 2, pp. 238–253, 2010.

[28] D. Lord, D. Middleton, and J. Whitacre, "Does separating trucks from other traffic improve overall safety?" *Transportation Research Record*, no. 1922, pp. 156–166, 2005.

[29] A. de Palma, M. Kilani, and R. Lindsey, "The merits of separating cars and trucks," *Journal of Urban Economics*, vol. 64, no. 2, pp. 340–361, 2008.

15

An Innovative Approach for Modelling Urban Road Traffic using Timed Automata and Formal Methods

Eduardo Valente,[1,2] Camelia Avram,[1] José Machado ,[2] and Adina Astilean[1]

[1] *Technical University of Cluj-Napoca, Faculty of Automation and Computer Science, Str. G. Baritiu, nr. 26-28, 400027 Cluj-Napoca, Romania*
[2] *University of Minho, MEtRICs Research Center, Campus of Azurém, 4800-058 Guimarães, Portugal*

Correspondence should be addressed to José Machado; jmachado@dem.uminho.pt

Academic Editor: Abdelaziz Bensrhair

Modelling and implementing adequate controllers for urban road traffic control constitute a huge challenge nowadays because of the complexity of systems, as well as possible scenarios and configurations, in each road in a city. A series of issues related to modelling these behaviours are common to arise when using formalisms, tools, and computation machines to perform complex calculations and limitations. This paper presents a formal, flexible, and adaptable approach, with no limitations, from the scientific point of view. For this purpose, modelling formalisms (cellular automata and timed automata) and analysis techniques (simulation and formal verification) are proposed to reach the main goals of modelling complex and adaptable behaviours in urban road traffic with multiple over time changeable configurations. A case study is presented, in order to illustrate the approach and demonstrate in detail the unlimited application of the presented approach.

1. Introduction

The continuous increasing of the number of vehicles in parallel with a slight possibility of building new roads and the corresponding infrastructure are only two of the main reasons that are permanently leading to the search for new solutions, capable of preventing congestion and improving road safety.

Numerous authors highlighted the necessity of a thorough analysis of various traffic conditions, having the role to reveal the most important specific characteristics. They found their expression in the three main categories of models, developed as a result of macro, meso, and micro approaches, which were modified, extended, and improved in a multitude of studies [1–3].

Based on different categories of information, these models are used in a lot of scenarios and for various purposes. While macroscopic model simulations determine the traffic flow or average velocities for different vehicle densities, where vehicles are considered moving entities in the traffic infrastructures, micromodels need a large amount of data to reproduce the dynamic behaviour of individual vehicles in different traffic conditions [4, 5]. A lot of research has been dedicated to developing microscopic car-following models applicable in situations in which collisions can occur, with the offered information being useful for operational analysis, too [6]. Considering the importance and the complexity of lane-changing operations as components of various traffic simulation tools [7], different approaches and algorithms were proposed for macro and especially micro modelling applications [8].

Practical reasons generated by inherent disadvantages of each category of the basic models, such as computational cost and some unsatisfactory evaluation results regarding the capabilities of different groups to adapt to real situations, lead to the necessity to consider appropriate hybrid approaches, respectively, the possibility of combining models to work together in a common framework [9]. It must be mentioned that different model combinations may lead to network representation consistency problems, and new specific aspects related to this approach must be taken into account [6].

Traffic simulation systems, as dedicated tools used for the analysis of traffic congestions, played a crucial role in the development of traffic models. In existing research directions, the event driven, or time driven, evolution of the traffic was considered along with the deterministic or stochastic nature of processes. Under particular conditions for many traffic models, two possible approaches were focused upon in an attempt to realize efficient models within a unified framework; the theory of traffic modelling focused on two major directions, attempting to realize efficient models and a unified framework.

A considerable number of road traffic simulators were proposed, and a particular importance has been given to microsimulation models [10]. VISSIM [11–14] TRANSMIS [15], TransModeler [16], AIMSUN [17], MITSIM [18, 19], CORSIM [20], and PARAMICS [21] are only representative examples, some of the most used simulation tools for road traffic monitoring, management, control, prediction, and other related applications.

Comparative aspects regarding the characteristics and capabilities of traffic simulators constituted standalone subjects of many studies, taking into account different criteria by the authors. While comparisons referring to traditional traffic [22–30] only consider basic criteria, others aim to evaluate the actual simulators as useful tools for the analysis of intelligent transportation systems. The most common characteristics that have been taken into account in the first case are the flow model category, modelled objects/entities and phenomena, realized functionalities, the facilities that the model can analyse, implemented restrictions, the system complexity, the simulation speed, and the dimension of the simulated network. In the second case, adaptability, flexibility, extensibility, real-time data integration and analysis, computational cost, visualization facilities, and possibilities to use wireless sensors and to integrate Geographic Information Systems (GIS) as components used in the decision process form some elements of the comparison basis.

Recent researches show that the potential of traffic simulators must be increased, in terms of pedestrian safety traffic [31–35], new algorithms [36], and calibration and validation [37–39], and the growing complexity of the next generation of transportation systems requires automated verification techniques.

Focusing especially on the necessity of validation and verification of microsimulation traffic models and based on the theory of timed automata [38], UPPAAL model-checker [40–42], and the corresponding associated language, this paper presents a unitary framework to modelling, simulation, and verification of urban traffic. A flexible traffic micromodel, easily changed and adaptable to various local traffic conditions, was designed. Having an increased integration capacity, its use can be extended to the scale of an entire city. An adaptive and systematic approach to modelling traffic behaviour considers different configurations of the several actors (vehicles, lanes, road, crossovers, and pedestrians) that could intervene in this process. It must be specified that, in adopting these solutions, a unitary approach is possible because UPPAAL is a tool designed to check systems that can be modelled as networks of timed automata, extended with integer variables, structured data types, user defined functions, and channel synchronisation [40, 41].

The rest of the paper is organised as follows: Section 2 is devoted to a discussion on a few work hypotheses related to formalisms and tools used in the developed study; Section 3 presents the problem and a systematic methodology; Section 4 details a case study for illustration and the obtained results are analysed. Section 5 presents some conclusions and future work.

2. Considered Hypotheses

The key elements of the urban road traffic modelling approach to be considered are presented in this section. A careful choice of the theoretical and implementation support was needed to ensure a systematic, flexible, and adaptive solution (Section 4). Relevant characteristics of the main components, their corresponding role in the system, and their interactions are highlighted.

2.1. Cellular Automata. Cellular automata were used for the first time in the 40s by Ulam and von Neumann in order to analyse possible behaviours of complex systems. Later, due to a considerable number of studies in the field, the theory of cellular automata was continuously developed, and many derived structures have been used to simulate a large variety of behaviours.

The basic components of cellular automata are cells that are organised in different configured n-dimensional grids. The system evolution can be described by a variety of sets of rules that are defined, taking into account the states of neighbouring cells, and are applied in a preestablished number of phases. A complete formal definition of this theoretical concept is presented in [29, 35]. In a simplified manner, it can be defined as follows:

Cellular automaton *CA* is a quintuple:

$CA = \langle S, s_0, G, d, F \rangle$;

S is a finite set of states;

s_0 is the initial state, s_0 in *S*;

G is the cellular neighbourhood;

$G = \{i, i + r1, i + r2, \ldots, i + rn\}$, where n specifies the neighbourhood size;

d is the dimension of C;

$f: S^n \longrightarrow S$ is the local transition rule, or the local cellular interaction rule;

CA(t) is the configuration at time *t*;

$CA(t) = (s_0(t), s_1(t), \ldots, s_N(t))$, where *N* is finite size of *CA* and $s_i(t)$ is state of cell i at time *t*;

F is the global mapping, $F: C(t) \longrightarrow C(t+1)$.

The utility of cellular automata was proved in many categories of applications, with the urban road traffic being one of the most important [45–47]. Taking into account the above considerations, in the first stage of this work, cellular automata based formalism was chosen for modelling purposes. It was considered a good option not only because it was suitable to model specific characteristics of the application, but also because it could constitute a possible basis to further implement discrete events and real-time modelling features [48]. Their subsequent integration played a crucial role in the process of obtaining a realistic simulation and a correct formal verification, with both being two important final goals of the presented application.

2.2. Timed Automata. Timed Automata were developed as a consequence of the need to model the timing behaviour of certain categories of systems that was not possible, using discrete automata. In order to include time variables [40], the theory of the finite automata was extended. In this context, the time was related to a set of real-valued variables modelling clocks [41]. There is a direct relation between the restrictions on the behaviour of timed automatons and the restrictions imposed by clock variables.

Consequently, this formalism is well adapted for modelling processes in which the time must be taken into account such as in urban road traffic modelling. In this case, time is important in order to calculate the speed or the acceleration, or to model other various aspects.

A complete formal description of timed automata is presented in [40], while a brief, formal definition of timed automata is given below [49]:

A timed automaton is a six-tuple $A = (L, L^0, C, A, E, I)$, where

$\boldsymbol{L}$ is a set of locations;

$\boldsymbol{L^0} \in \boldsymbol{L}$ is the set of initial locations;

$\boldsymbol{C}$ is the finite set of clocks;

$\boldsymbol{A}$ is a set of actions and coactions and the internal τ-action;

$\boldsymbol{E} \subseteq L \times A \times B(C) \times \boldsymbol{2}^C \times L$ represents the set of transitions. An arc (l, l', a, λ, b) represents a transition from the location l to the location l' when entering the symbol a. The set $\lambda \subseteq \mathbf{C}$ represents the clocks to be reset in this transition, and b is a time restriction over C.

$\boldsymbol{I} : L \longrightarrow B(C)$ assigns invariants to locations.

To define the semantics of a timed automaton, a clock valuation is a function $\boldsymbol{u} : C \longrightarrow R \geq \mathbf{0}$ from the set of clocks to the nonnegative real values. Let R^C be the set of all clock valuations. Let $u_0(x) = 0$ for all $x \in C$. It is considered that the notation means that guards and invariants are sets of clock valuations; thus writing $u \in I(l)$ means that u satisfies $I(l)$.

During a run of a timed automaton, all clock values increase with the same speed [50].

The clocks can be reset to zero (independently of each other) in the transition of the automaton, keeping the time elapsed since the last reset [51].

Based on the above considerations, the application presented in the paper was developed starting with a model based on cellular automata, which corresponds to the requirements to be a dynamic, discrete space and discrete time formalism and then translating the resulting cellular automata to a discrete automata [45]. Furthermore, temporal aspects were taken into account.

It must be pointed out that even though a large number of formalisms could be used to model timed systems and in particular the proposed traffic system, timed automata were adopted here especially because the behavioural analysis of the considered models was realized using UPPAAL model-checker. This choice has the advantage that the necessity to convert models when the simulation and verification processes are performed is eliminated.

2.3. UPPAAL. There are several tools used to analyse timed automata and extensions, including UPPAAL, Kronos, and TIMES. These are becoming more and more mature, but they are all exclusively academic research tools.

UPPAAL is a model-checker [38], adapted to verify real-time system behaviours. The considered models are built in a network of processes, with each process being considered an automaton. The entire model is composed of three parts: its global and local declarations, the automata templates, and the system definition. Because the use of templates is possible in UPPAAL, it is suitable for modelling complex systems to use modules corresponding to the component parts of the system that is being modelled. This characteristic allows the instantiation of models of the same kind and the reusing of models for systems of the same structure, having similar component parts. Also, the utilization of modelling modules allows more simplicity in terms of the modelling complex systems behaviour.

Besides the mentioned advantages, UPPAAL allows obtaining results for simulation and formal verification techniques, using the same environment [40, 48]. This tool is also well adapted to simulate and verify networked complex mechatronic systems [41].

Due to the above specified reasons, UPPAAL was chosen for simulation and formal verification purposes contributing to a more complete understanding of the presented application behaviour.

2.4. Timed Computation Tree Logic. In order to formalize the desired behaviour of the system modelled with timed automata, it is important to choose a formal logic compatible with the verification tool. Consequently, this will be a timed version of Computational Tree Logic (CTL), a simplified version of Timed Computation Tree Logic (TCTL) [52], used by UPAAL.

CTL [43] is a propositional, branching-time temporal logic that enables expressing queries over the possible (desired and/or undesired) behaviours of a model. In CTL, the behaviour of a system is seen as a tree, representing behaviour possibilities (Figure 1). CTL formulae consists of path formulae and state formulae. State formulae are evaluated in individual states. Path formulae quantify over paths in the behaviour of the model and over the states in those paths.

3. Modelling of Urban Road Traffic

This section is divided into two main parts: firstly, a series of issues are dealt with in relation to modelling urban road traffic; secondly, a flexible, adaptive, and systematic approach for solving these issues is presented.

3.1. Aspects of Modelling Urban Road Traffic. In order to correctly represent urban road traffic, aspects to be considered include modelling the following:

(i) modelling different vehicles (cars, buses, trams, trucks, etc.)

(ii) modelling the driver's behaviour;

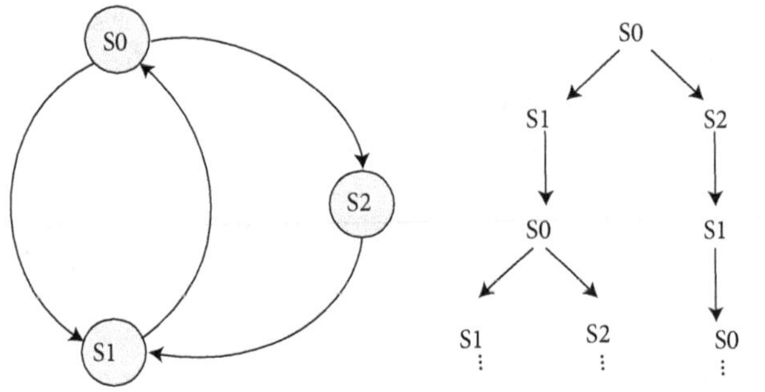

FIGURE 1: A state machine and its computation tree. adapted from [43].

(iii) modelling road and respective traffic-lanes, considering different sizes for road and lanes;

(iv) modelling crossroads and upcoming road;

(v) modelling car parks, stops, and changing traffic-lanes for different kinds of vehicles (except trams);

(vi) modelling the velocity and acceleration of vehicles;

(vii) modelling the space occupied by each vehicle taking into account the size of the vehicle and the speed corresponding to each road (dynamic cell length).

When considering modelling tasks, there is also an interaction between the actors of the system.

The automobiles can come across the following implemented traffic-elements:

(i) pedestrian crossings for pedestrians;

(ii) car parks placed transversely/alongside on the right side of the traffic-lane.

The buses can come across the following implemented traffic-elements:

(i) pedestrian crossings;

(ii) bus-stops inside or outside the traffic-lane.

The trams can come across the following implemented traffic-elements:

(i) pedestrian crossings;

(ii) tram-stops inside the traffic-lane.

For the purposes mentioned above, a modular modelling approach is adopted, using timed automata. This approach considers a modular model for each vehicle, road, lane, crossroad, upcoming road, and physical component.

In fact, this approach seems to be efficient, but it is neither flexible nor adaptive, despite being systematic. The reuse and readaptation of modules are not simple and when, for instance, the size of one road or one lane changes, the modules must be built again, with new specific conditions for each model, for each actor of the system. Another important issue is the fact that a huge number of models (composed of different modules) pose issues when it comes to extending the reachable state space of the overall system model.

In order to deal with this complexity, it was decided to create a model to representing accurately by a matrix the movement of the vehicles and the traffic-elements and being easily extendable allowing addition of more rows or columns. Matrices were used for modelling all parts of these complex systems.

This model is the first model with relevant traffic road rules included along with a significant level of detail. This methodology features the following main characteristics:

(i) The physical environment is a one-dimensional grid of rectangular cells, all equal in size (7.5 meters of length).

(ii) It is a single cell model because each automobile will occupy only one road's cell for each time iteration, and the cells can have only two possible states: occupied by an automobile or empty.

(iii) The size of the neighbourhood is the same for each cell; the model is anisotropic because the automobiles only respond to stimuli in front of them.

(iv) This model is a dynamic system with a limited number of automobiles, buses, trams, and roads that evolve and change in time and space depending on the same rules; only if the cell in front is free, can the different vehicles move along the lane.

(v) The time is a stochastic feature and its choice is completely nondeterministic. In an instant t the automobile can circulate in a cell at a velocity of 50 km/h, and in the instant t+1 it can move at 2 km/h. This model considers, for vehicles, the possibility of severe deceleration and acceleration even if the road is completely free in order to model the individual driver's behaviour.

(vi) This model can be easily extended. The number of roads and automobiles can be easily changed; the only requirements are to define in global declaration those variables and the matrix for extending the map, to define the inputs in terms of the traffic flow used in the simulation, to define the upcoming road, and to set up the variable for each automobile.

By using this approach, vehicles can circulate irrespectively of the different traffic-elements and interact with them. For example, a pedestrian crossing will influence the behaviour of the automobiles, buses, and trams when pedestrians cross the traffic-lane. If there is also a bus or a tram stop on the next lane, this will not affect an automobile on the road because it does not need to change its behaviour; rather buses and the trams will have to stop. These different behaviours are given by the functions and channels implemented in the road declarations explained in the next subsection.

3.2. Modelling Approach. Different structures are defined in order to represent the complexity of the urban road traffic. Two types of urban road traffic structures are taken into consideration: infrastructure (road, crossroads, traffic signs, parking lots, and pedestrian crossings) and moving vehicles (automobiles, trams, and buses). The formalism adopted

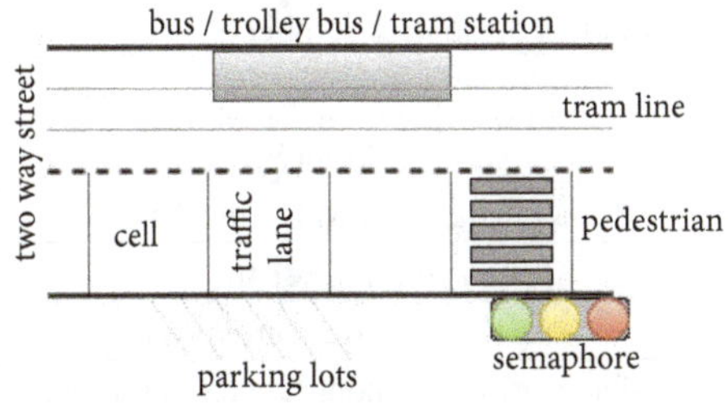

FIGURE 2: First layer of the structure.

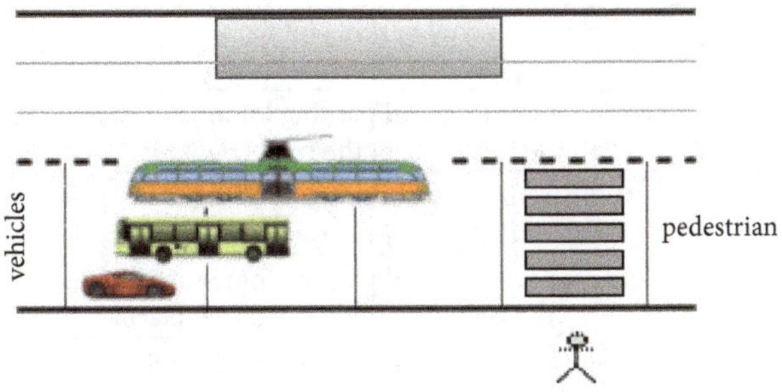

FIGURE 3: Second layer of the structure.

to model the considered system is timed automata (TA). UPPAAL environment will be used for performing the simulation and formal verification analysis techniques.

To model and analyse the two types of structures, a two-layered model based on timed automata is considered; one represents the components of the infrastructure and the relations between them (Figure 2) and the second one represents the mobile components of the urban road traffic (Figure 3). The interactions between layers are defined into a set of rules which are implemented in "C" programming language.

Several equations are considered to represent the evolution of each component. Arrays and matrices are built to memorize the components (as arrays) and their (current and future) behaviour and their interactions (as a matrix).

The circulation of the individual vehicles in a traffic flow is described by a set of rules that reflect the movements of cars and the lane-changing behaviour, evolving in time and space.

The characteristics of the defined structures include:

(i) *traffic-lanes*: name, length, number of cells (determined for each road separately taking into consideration the length and the safety distance), bus/tram stop, and parking lot;

(ii) *road*: number of lanes;

(iii) *crossroad*: number of roads which form the crossroad, name of the road;

(iv) *traffic signs*: (stop, give way, traffic-lights, etc.) defining the rule that needs to be observed;

(v) *car parks/parking lots*: located inside a road or aside from the road and can be parallel, perpendicular, or placed at an angle of 45° from the direction of travel;

(vi) *pedestrian crossings*: located inside a road;

(vii) *automobile*: number of automobiles; they occupy a single cell at a given time and can travel anywhere or have a predefined route;

(viii) *tram*: number of trams; they occupy three cells and have a predefined route and predefined stops;

(ix) *bus*: number of buses; they occupy two cells and have a predefined route and predefined stops.

The automobiles can come across the following implemented road-elements:

(i) pedestrian crossings;

(ii) car parks placed transversely on the right side of the traffic-lane;

(iii) car parks placed on the right side of the traffic-lane;

(iv) (fixed or mobile) obstacles that can be overtaken if the traffic allows it.

The buses can come across the following implemented road-elements:

(i) pedestrian crossings;

(ii) bus-stops inside the traffic-lane;

(iii) bus-stops outside the traffic-lane, on the right side of the traffic-lane;

(iv) obstacles that can be overtaken if the traffic allows it.

The trams can come across the following implemented road-elements:

(i) pedestrian crossings;

(ii) tram-stops inside the traffic-lane;

(iii) overtaking obstacles is not possible.

Using this formalism to represent urban road traffic and the interactions within it allows us both a general representation and the ability to model complex structures. This way, the model allows any modification of the simulated map without changing the UPPAAL model, only by changing the initialization variables. Since the length of road is different and the matrix cannot be defined with variable number of columns, the value -1 is introduced.

Global declarations include the number of automobiles, buses, trams, and roads, using the function "typedef". The maximum number of cells that can be defined for a road (the longest road) has been declared by using the variable "maxnoCells". With this variable and the variable number of road (noS) the matrix "idexSC" was created. This matrix is a map of road cell coordinates. Each line corresponds to a road ID—the ID of the first line is 0 and the ID of the last line is the last road's ID. The column number one corresponds to the number of cells that each road covers and the following columns are filled with -1 which means that a road's cell is empty. The first column will be important as it allows the limitation of the road size.

Due to UPPAAL's limitations it is not possible to create arrays with different sizes, and some of the values of "-1" present in the matrix have no significance. Therefore, it is easy to extend the map and to implement other features in subsequent models.

The choice of the next road in an intersection was implemented with a similar application. First, a new variable "maxNextRoad" (maximum next road) was declared. This variable is equal to three because a crossroad features a choice of no more than three roads usually. In case of a crossroad with more than three streets the problem is managed in a similar manner. Using the variable "maxNextRoad" and with the variable number of the road ("noS"), another matrix "indexMAP" was created. Each line corresponds to a road ID whose first line ID is 0 and whose last line ID is the last road's ID. The first column contains the number of possible next road choices at the end of the current road, and the other columns have the road's IDs of the next road.

To implement these road-elements it was necessary to instantiate the matrix "indexSC" for new negative numbers which will have a different meaning from an empty cell, affecting the behaviour of the vehicles. The elements present inside the road (pedestrian crossings, bus-stops inside the road, and tram-stops inside the road) were coded in the respective road ID with a negative number.

However, to create the road-elements outside the road, it was necessary to define adjacent roads to implement a turning movement that a vehicle will execute to go to an adjacent road where "*i*" the road-element "*s*" is located. The implementation rule is the following: if a traffic-lane has to its right an exterior road-element (Parking lots across the road, Parking lots alongside, and bus-stops outside the road), the road ID-1 will have the corresponding negative digit of the road-element, and in the road ID-2 will be necessary to create the border of the road that will give information that the vehicle cannot move to another road ID on the right.

In summary, the different negative digits that a matrix "indexSC" can have will give the instructions for an interaction between a vehicle and a road-element. The meaning (the road-element) of the negative digits is as in Table 1.

With the "indexSC" variable and the variable number of the roads (noS), three matrices "indexMapAUTOMOBILE", "indexMapBUS", and "indexMapTRAM" are created. Each line of those matrices corresponds to a road ID, whose first line ID is 0 and whose last line ID is the last road's ID.

To define the physical environment several variables were defined: "maxnoCells" (represents the maximum length of the road considered, maximum number of cells); "indexSC" matrix (size of noS x maxnoCell+1: this matrix is the description of the physical infrastructure; each line corresponds to a road ID, whose first line ID is 0 and whose last line ID is the last road's ID and the column number one corresponds to the number of cells that each road has and the following columns are filled with -1 which means that a road's cell is empty); "maxNextRoad" (maximum number of next roads); "indexMAP" matrix (each line corresponds to a road ID, whose first line ID is 0 and whose last line ID is the last road's ID; the first column contains the number of possible next roads at the end of the road, and the other columns have the road's IDs of the next road.); and "*currentRoadA*" (the current road for each automobile was defined; an array is created and presented as the initial road's ID where an automobile will start the simulation). This array contains the inputs of the traffic flow used in the simulation. The value of "current Road" will change every time an automobile leaves a road and continues to move along another road of the map. The current road for each automobile is defined, and the value defined in the array "current Road" represents the initial road's ID where an automobile will start the simulation. Figure 4 represents the interactions between the modelling layers.

Table 1: Description of the matrix number of cells per different road (with different road-elements).

-1	Empty cell
-2	Bus station inside the traffic-lane
-3	Tram station inside the traffic-lane
-4	Pedestrian crossings
-5	Possible parking lot alongside for automobiles on the right side of the road
-55	Border of a parking lot alongside for automobiles on the right side of the road
-6	Possible transversal parking lot for automobiles on the right side of the road (right place)
-66	Border of a transversal parking lot for automobiles on the right side of the road (right place)
-7	Possible transversal parking lot for automobiles on the right side of the road (left place)
-77	Border of a transversal parking lot for automobiles on the right side of the road (left place)
-22	Bus station outside the right side of the road
-222	Border of a bus station outside the right side of the road
-8	Road's end and cells with no meaning for the simulation

The average speed of the vehicles along a traffic-lane can be expressed by the following equation.

$$\overline{v} = \frac{\sum_{i=1}^{n} v_i}{n}, \quad [m/s] \tag{1}$$

The traffic density is calculated as follows.

$$\rho = \frac{n}{L}, \quad [vehicles/m] \tag{2}$$

Based on these values, the traffic flow is:

$$J = \frac{\sum_{i=1}^{n} v_i}{L}, \quad [vehicles/s]\,. \tag{3}$$

The average speed of the vehicles that cross the lane section is expressed by:

$$\overline{v} = \frac{\sum_{j=1}^{n} v_j}{m}, \quad [m/s]\,. \tag{4}$$

Two types of traffic behaviours were implemented: the "car-by-car" model and the free traffic model (this involves that when a car is driving on the road with no constraints and an obstacle emerges, the traffic allows the driver to overtake it).

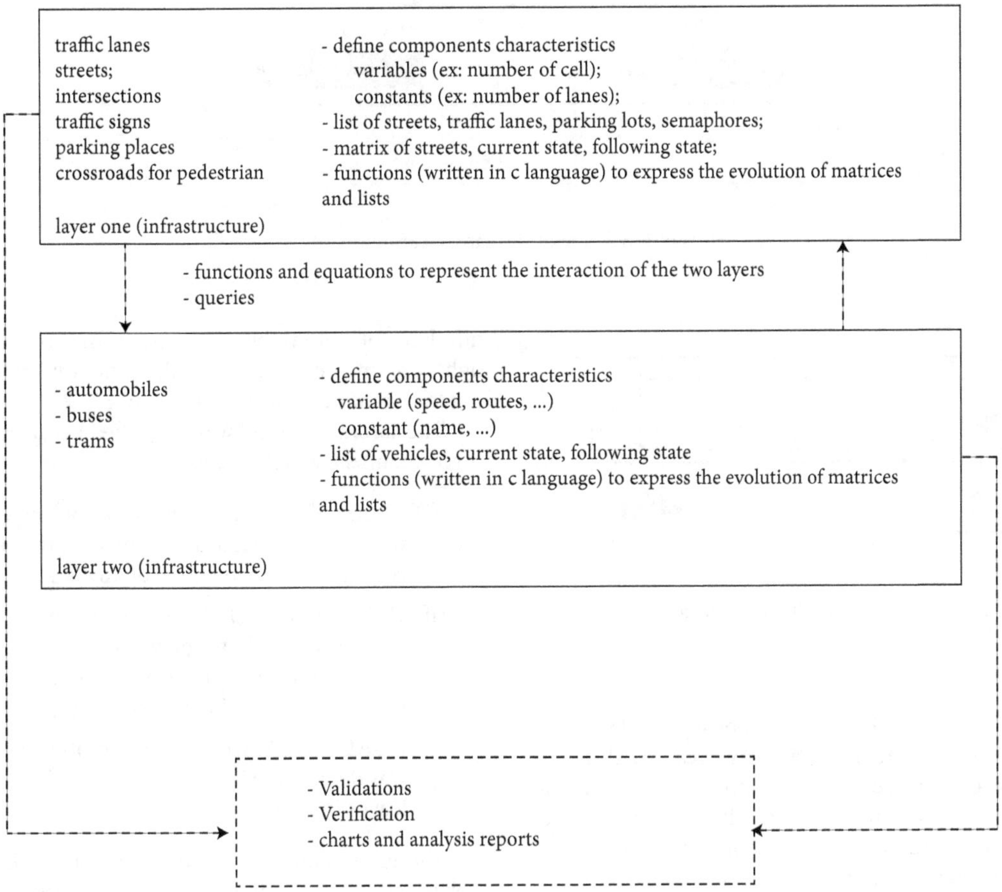

FIGURE 4: Interactions between layers.

The "car-by-car" model used in this model is presented in Figure 5.

In general, the "car-by-car" models are centred on the following relation.

$$[Response]_n \propto [Stimulus]_{n+1} \tag{5}$$

The vehicle $\mathbf{n} = \mathbf{1}, \mathbf{2}, \ldots$ can only accelerate or decelerate as a response to the different flow conditions. The equations of "car-by-car" models consider the vehicle's analysed speed ($\mathbf{v_n}$), the speed difference between it and the leader vehicle ($\mathbf{\Delta v_n}$), and the distance between these vehicles ($\mathbf{d_n}$) among others. In general, it has the following form [53].

$$\overline{x}_n(t) = f(v_n(t), d_n(t), \Delta v_n(t)) \tag{6}$$

The stimulus function, f, has different interpretations for each "car-by-car" model. The simplest model is a follow-the-leader one which only uses the speed difference between the vehicle's leader and its tracker, represented by the following equation.

$$\overline{x}_n(t) = \frac{1}{\tau(\dot{x}_{n+1}(t) - \dot{x}_n(t))} \tag{7}$$

where τ is a setting parameter related to the time scale; $1/\tau$ can be interpreted as the sensitivity of the tracker vehicle's speed to the leader vehicle's speed variations.

Changing the traffic-lanes in order to overtake an obstacle or not is presented in Figure 6 and is done by using the following steps: First, three cells (in front) on both traffic-lanes need to be free (current road lane and neighbour road lane); the checking is done in cell X [t or t+1, i]; if yes, secondly, the vehicle changes the traffic-lane. This works only to prove the concept because in real life the traffic is congested or at least there is more than one car on the streets. To be more accurate a third step (or check) is also introduced: simultaneously, in front of the car one cell should be free (cell i+1) and two cells on the next traffic-lane should be free (cells i and i+1), too.

4. Application of the Methodology: Case Study

After presenting the developed and tested approach, we include a case study in order to allow illustrating the application of the approach and to allow the extrapolation of the obtained results for other similar systems.

4.1. Presentation of Different Situations. One of the key aspects of creating good traffic road models is to define in a precise and unambiguous way the road that will be studied with all relevant information for a correct vehicular circulation.

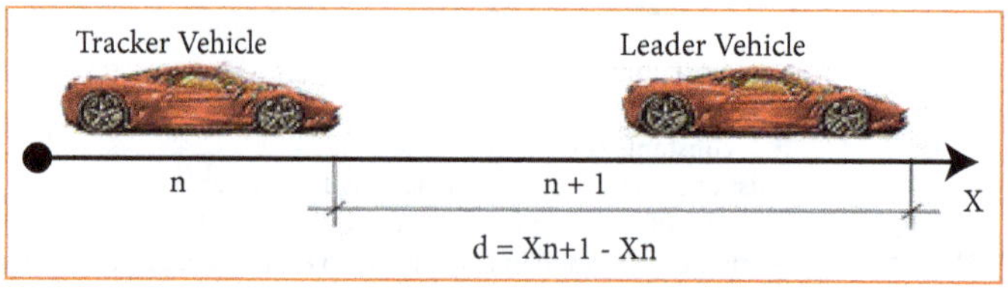

FIGURE 5: Scheme of the car-by-car model [44].

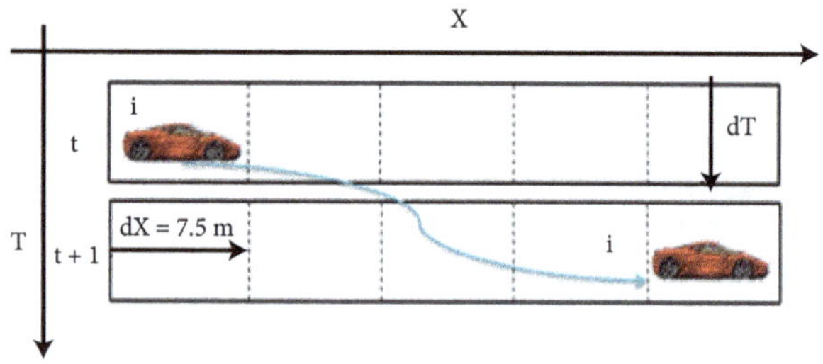

FIGURE 6: Changing traffic-lanes model.

In order to test and validate the proposed model a case study is considered, and the simulation involves a small group of roads in the city centre of Cluj-Napoca, containing the maximum road-elements that a road can have in order to test and verify the proposed model, as presented in Figure 7.

This section details the physical environment of the simulation, describing the road (number, direction, and length of the traffic-lanes), the location of the pedestrian crossings, bus stations and tram stations, tram railways, the intersections between roads, and the way traffic-lights for road, rail, and pedestrian traffic operate.

The total network length (Figure 7) is 3735 meters and this simulation contains 12 pedestrian crossings, 32 traffic-lanes for automobiles, 22 lines of buses, 2 tram railways, 8 traffic-lights for drivers and other 8 traffic-lights for pedestrians, 4 bus-stops, and 2 tram-stops, as well as 2 parking spaces.

The simulated map contains the maximum number of elements that can be simultaneously simulated due to software constraints. Observing a large number of elements, tracking their status over time, is also costly.

In order to analyse some specific road traffic problems from the simulated network, some small parts are simulated and tested separately.

The image in Figure 7 is a 2D model of the entire studied traffic environment. For each road various pieces of information were collected and structured in Table 2.

The time measurements were taken at midday and at 6 pm, when roads are the busiest. The mode of operation of the traffic-lights taken into consideration in the simulation is the same at both times.

4.2. Illustration of Methodology Application. The model is based on templates for roads, traffic-lights, and vehicles (automobiles, buses, and trams, because three different types of vehicles are considered). In the global declaration, the numbers of automobiles, buses, trams, roads, and traffic-lights are first declared, using the function "typedef" according to the model described in Section 3.2.

The automaton road was simplified because there are only three channels, which means

(i) ROAD_STARTS: when an automobile, a bus, or a tram was detected by a sensor at the beginning of a road and the road is free of any automobile;

(ii) UPDATING_CELLS: when an automobile, bus, or tram or more than one vehicle on the road are moving; several functions were created in order to model the driving behaviour on the road;

(iii) ROAD_ENDS: when an automobile, a bus, or a tram was detected by a sensor at the end of the road and it leaves the road.

The automaton road is the "brain" of the simulation, thus storing all the information regarding traffic conditions, the set of rules, and the subsequent transition rules.

The automaton automobile has three places:

(i) OUT_of_the_ROAD_START_MOVING: when an automobile is moving towards the beginning of the road;

(ii) MOVING_INSIDE_the_ROAD: when an automobile was detected by a sensor and is moving on the road;

(iii) OUT_of_the_MAP: when an automobile was detected by a sensor and there is no possible next road and it exits the considered map.

The bus and tram automatons have structures similar to that of an automobile, but they also feature a predefined path on the map.

The traffic light automaton has three states defining the three colours (red, green, and yellow). Yellow can be skipped, consequently reducing the loop to only two colours (red and green).

In the road considered in the case study, traffic behaviour is not exclusively influenced by the state of the road cells, as it was considered in the previous model. The vehicles interact with other road-elements present in the road such as pedestrian crossings, bus stations, tram stations, and traffic-lights. In the studied model, the interaction between vehicles and the road-elements contained in the group of the roads was taken into consideration.

The automobiles can come across the following implemented road-elements:

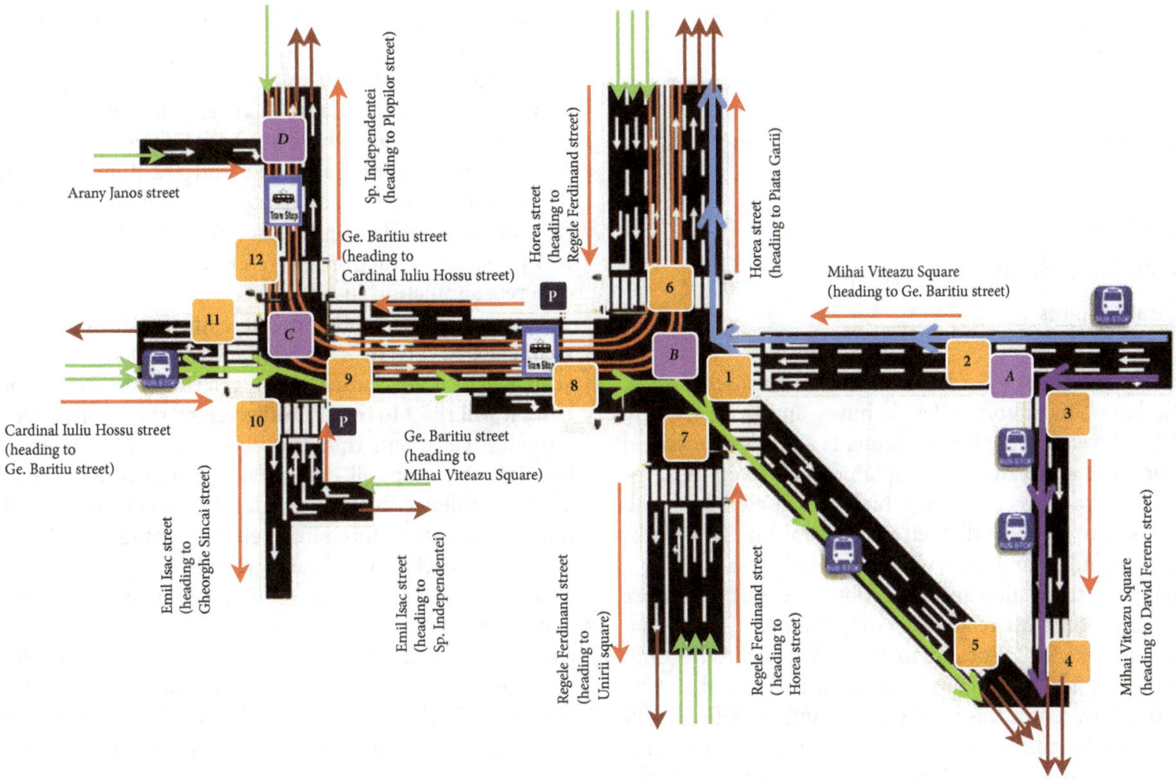

FIGURE 7: Representation of the entire road network considered in this simulation.

TABLE 2: Structured information for each modelled and simulated road.

Length considered	Length of the traffic lane (number of cells)				
Number and direction of traffic-lanes	Number of traffic lanes - direction				
Possible next road in the intersection A	Next road to follow				
Possible next road in the intersection B	Possible next road to follow (directions)				
Traffic flow control at the intersection B for each traffic-lane	Traffic-lane	Right		Left and central	Time
	Red:	Time	Time	Time	seconds
	Yellow:	Time	Time	Time	seconds
	Green:	Time	Time	Time	seconds
Pedestrians flow control in the pedestrian crossing number one	Traffic-light: Green: time [s] Red: time [s]				
Features of the Bus stop	Location	coordinates			
	Length	Length of the bus station [m]			
	Buses	Buses no. lanes			
Pedestrian crossing number two	Pedestrians' desire				
Maximum velocity	Max. speed allowed on this sector km/h				
Types of vehicles allowed	Automobiles, Motorbikes and Buses				

(i) pedestrian crossings;

(ii) parking lots placed transversely on the right side of the traffic-lane;

(iii) parking lots placed alongside the right side of the traffic-lane;

(iv) traffic-lights.

The buses can come across the following implemented road-elements:

(i) pedestrian crossings;

(ii) bus-stops inside the traffic-lane;

(iii) bus-stops outside the traffic-lane, on the right side of the traffic-lane;

(iv) traffic-lights.

The trams can come across the following implemented road-elements:

(i) pedestrian crossings;

(ii) tram-stops inside the traffic-lane;

(iii) traffic-lights.

4.3. Simulation Results. To implement the considered map, 74 traffic-lanes, 480 automobiles, 30 buses, and 18 trams were declared. Afterwards, the traffic-elements and the predefined routes for buses and trams were implemented. Given the large volume of interactions between vehicles, road-elements, and roads, the entire map was divided into several small networks in order to simulate and analyse each modelled feature.

During the simulation and validation process, the authors faced several problems related to computational power that UPPAAL needed. The allocated memory that UPPAAL could use during the simulation was restricted and could not be extended. To overcome this problem, the authors defined several scenarios to validate all the considered traffic-elements. First the roads from the original case study were considered (74 traffic-lanes, 90 automobiles, 10 buses, and 1 tram) in order to verify and test all the already presented interactions. In that map, the road-elements were not considered because of the limitations of UPPAAL (e.g., pedestrians, car parking, drivers' behaviour, and dynamic cell length). Another simulation was performed in order to study the interaction between traffic-elements. This simulation included a road which contained all the different road-elements (pedestrian crossing, bus stop, and tram stop cells within the road and dynamic cell length).

Figure 8 presents possible time interactions for an automobile travelling inside the map (route: intersection D ⟶ C ⟶ B ⟶ A ⟶ road 4; see Figure 6) with this sequence of road-elements being presented. Basically, an automobile will only react to the stimuli given by road cells and pedestrian crossing cells, and when it is passing through a bus stop or a tram stop it will react with the same behaviour as for a road cell. These stimuli will influence the time needed to move through each different cell, with each red line being the minimum time needed to travel each cell and each green line being the maximum time needed to travel each cell. The sums of the minimum and maximum times needed to travel each cell of the road, by an automobile, are 4 and 239 time units, respectively. In this figure, two concrete ways of interaction between road-elements and an automobile are presented. The values presented in these two possible interactions were extracted from the option "concrete simulation" in UPPAAL.

Figure 9 shows the possible time interactions at the moment that a bus is travelling on the road with the same sequence of road-elements. When a bus is travelling on this road, it will react to the stimuli given by road cells, pedestrian crossing cells, and bus stop cells, and when it is passing through a tram stop cell, it will react as for a road cell. These stimuli will influence the time needed to travel through each different cell, with each red line being the minimum time needed to travel each cell and each green line being the maximum time needed to travel each cell. The sums of the minimum and maximum times needed to travel each cell of the road, by an automobile, are 33.5 and 272 time units, respectively. In this figure two concrete ways of interaction between road-elements and a bus are also presented. The values for the two possible interactions were extracted from the separator concrete simulation in UPPAAL.

Figure 10 shows the possible time interactions in the moment that a tram is travelling on the road with the same sequence of road-elements. When a tram is travelling on this road, it will react to the stimuli given by road cells, pedestrian crossings cells, and tram stop cells, and when it is passing through a bus stop cell, it will react like a road cell. These stimuli will influence the time needed to travel through each different cell, with each red line being the minimum time needed to travel each cell and each green line being the maximum time needed to travel each cell. The sum of the minimum time needed to travel through each cell of the road by an automobile is 272 time units. In this figure two concrete ways of interaction between road-elements and a bus are also presented. The values for the two possible interactions were extracted from the separator concrete simulation in UPPAAL.

4.4. Formal Verification. Formal verification is used in this work as a complementary technique to simulation. In fact, simulation considers some possible scenarios of evolution of the developed models, but formal verification checks all possible behaviours of those models. Due to this reason, the validation of the presented approach seems important by using this exhaustive technique.

For formal verification of the model it is enough to know that, in the UPPAAL version, A is the universal quantifier on paths (for any path...), E is the specific quantifier on paths (there is a path....), [ ] is the universal quantifier over states in a path (for any state...), and <> is the specific quantifier over states in a path (there is a state ...).

In UPPAAL, in the option Verifier, several queries to check the correct behaviour of this model were implemented. The initial values and variables taken into consideration for the formal verification of the map in Figure 6 are the following:

(i) number of automobiles = 90;

(ii) number of buses = 10;

(iii) number of trams = 1;

(iv) number of roads = 9;

(v) number of road cells per road = [30, 30, 8, 22, 14, 14, 14, 14, 14];

(vi) indexSC[noS][maxnoCells +1]: status (free or occupied) of each cell in each road;

(vii) novis[noS]: number of vehicles on the road.

Cellular automata:

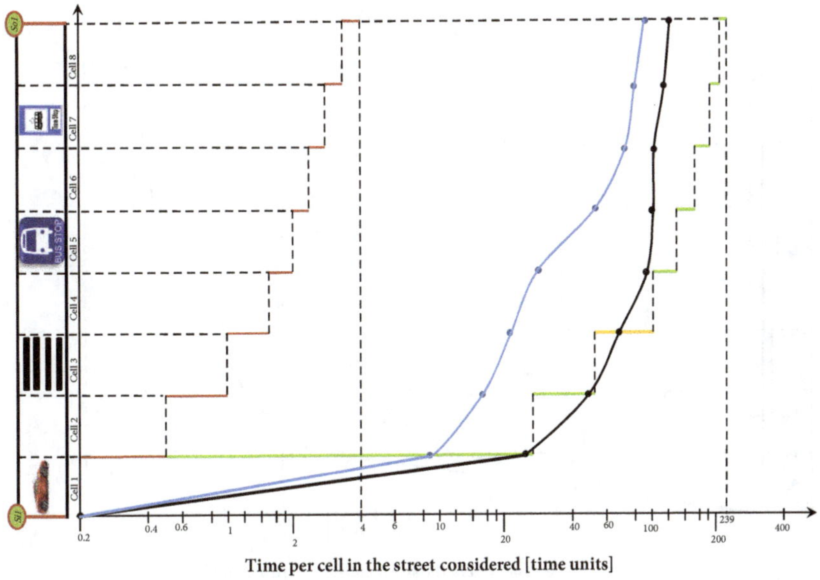

FIGURE 8: Time intervals needed for an automobile to go through a sequence of cells.

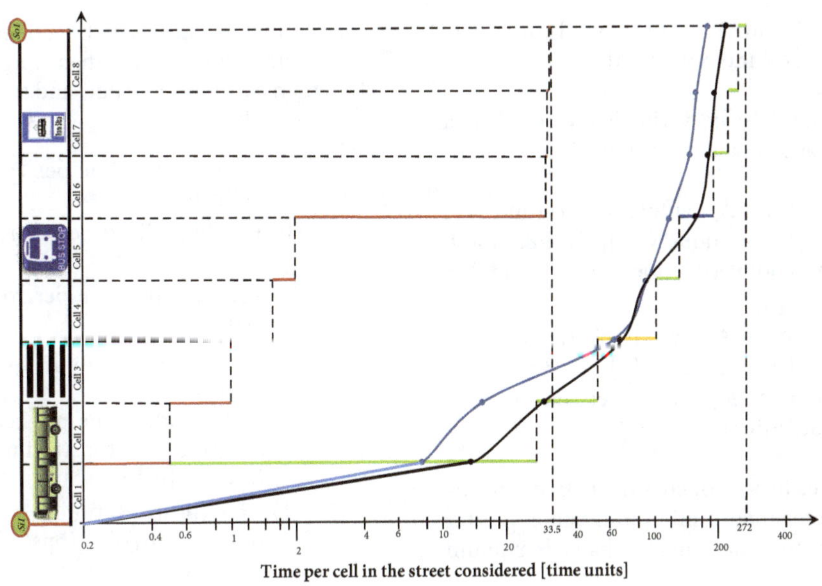

FIGURE 9: Time intervals needed for a bus to go through this sequence of cells.

(i) automobile (idA, time_per_road, time_per_cellA);

(ii) bus(idB, time_per_road, time_per_cellB);

(iii) tram(idT, time_per_road, time_per_cellT);

(iv) road (idS, GLOBAL, novis);

(v) traffic-lights (idSem, tR, tY, tG).

A list of queries was generated in order to verify various properties. Some of the queries were built in order to double check some important traffic behaviours. The response of the query could be either "yes" or "no". Based on this response, a report of a traffic simulation scenario can be generated.

The following properties can be verified and validated:

(i) Properties that need to be verified: greater than a minimum value

(a) E[] forall (i : idA) automobile(i).time_per_road(idS) > 0.5 ∗ indexSC[p][q]

(b) E<>forall (i : idA) automobile(i).time_per_road(idS) > 0.5 ∗ indexSC[p][q]

(ii) Properties that need to be verified: smaller than a maximum value

(a) E[] forall (i : idA) automobile(i).time_per_road(idS) < 27 ∗ indexSC[p][q]

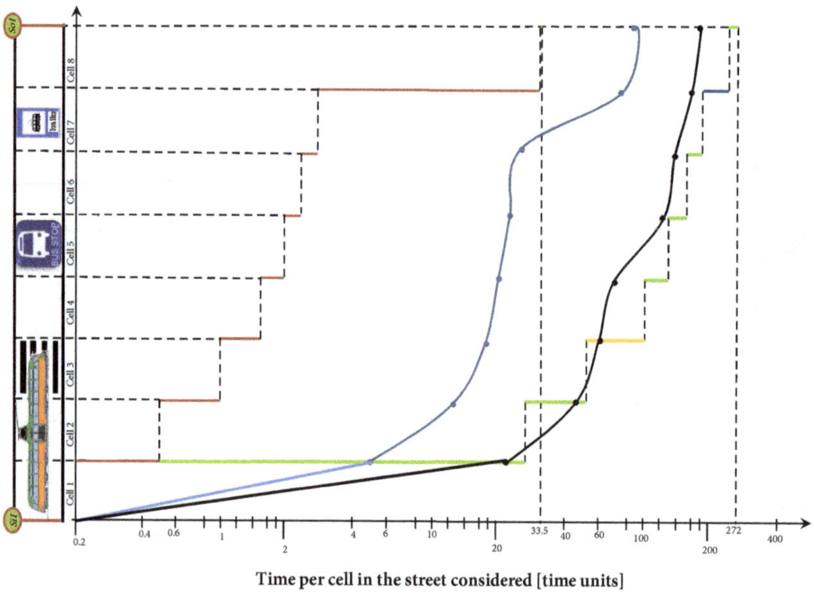

FIGURE 10: Time intervals needed for a tram to go through this sequence of cells.

(b) E<>forall (i : idA) automobile(i).time_per_road(idS) < 27 ∗ indexSC[p][q]

(iii) Properties that need to be verified: greater than a minimum value and smaller than a maximum value

(a) E[] (forall (i : idA) automobile(i).time_per_road(idS) > 0.5 ∗ indexSC[p][q]) && (forall (i : idA) automobile(i). time_per_road (idS) < 27 ∗ indexSC[p][q])

(b) E<> (forall (i : idA) automobile(i).time_per_road(idS) > 0.5 ∗ indexSC[p][q]) && (forall (i : idA) automobile(i). time_per_road (idS) < 27 ∗ indexSC[p][q])

(iv) Properties that need to be verified: smaller than a maximum value, greater than a minimum value, greater than a minimum value, and smaller than a maximum value (automobile)

(a) E[] automobile(idA).time_per_road(idS) > 0.5 ∗ indexSC[p][q]

(b) E<> automobile(idA).time_per_road(idS) > 0.5 ∗ indexSC[p][q]

(c) E[]automobile(idA).time_per_road(idS) < 27 ∗ indexSC[p][q]

(d) E<> automobile(idA).time_per_road(idS) < 27 ∗ indexSC[p][q]

(e) E[](automobile(idA).time_per_road(idS) > 0.5 ∗ indexSC[p][q]) && (automobile(i).time_per_road(idS) < 27 ∗ indexSC[p][q])

(f) E<> (automobile(idA).time_per_road(idS) > 0.5 ∗ indexSC[p][q]) && (automobile(idA).time_per_road(idS) < 27 ∗ indexSC[p][q])

(v) Properties that need to be verified: smaller than a maximum value, greater than a minimum value, greater than a minimum value, and smaller than a maximum value (bus)

(a) E[] bus(idB).time_per_road(idS) > 0.5 ∗ indexSC[p][q]

(b) E<> bus(idB).time_per_road(idS) > 0.5 ∗ indexSC[p][q]

(c) E[] bus(idB).time_per_road(idS) < 27 ∗ indexSC[p][q]

(d) E<> bus(idB).time_per_road(idS) < 27 ∗ indexSC[p][q]

(e) E[](bus(idB).time_per_road(idS) > 0.5 ∗ indexSC[p][q]) && (bus(i).time_per_road(idS) < 27 ∗ indexSC[p][q])

(f) E<> (bus(idB).time_per_road(idS) > 0.5 ∗ indexSC[p][q]) && (bus(idB).time_per_road(idS) < 27 ∗ indexSC[p][q])

(vi) Properties that need to be verified: smaller than a maximum value, greater than a minimum value, greater than a minimum value, and smaller than a maximum value (tram)

(a) E[] tram(idT).time_per_road(idS) > 0.5 ∗ indexSC[p][q]

(b) E<> tram(idT).time_per_road(idS) > 0.5 ∗ indexSC[p][q]

(c) E[] tram(idT).time_per_road(idS) < 27 ∗ indexSC[p][q]

(d) E<> tram(idT).time_per_road(idS) < 27 ∗ indexSC[p][q]

(e) E[](tram(idT).time_per_road(idS) > 0.5 ∗ indexSC[p][q]) && (tram(i).time_per_road(idS) < 27 ∗ indexSC[p][q])

(f) E<> (tram(idT).time_per_road(idS) > 0.5 * indexSC[p][q]) && (tram(idT).time_per_road(idS) < 27 * indexSC[p][q])

(vii) Properties that need to be verified: if the vehicles leave the road

(a) E<> forall (i : idA) automobile(i).response = -1

(b) E<>forall (i : idB) bus(i).response = -1

(c) E<> forall (i : idT) tram(i).response = -1

(viii) Properties that need to be verified: if the time needed to travel through a cell is small or greater than the predefined values, or if it is in a fixed interval

(a) E[] automobile(idA).time_per_cellA[m] < 27

(b) E[] automobile(idA).time_per_cellA[m] > 1/2

(c) E[] ((automobile(idA).time_per_cellA[m] >1/2) && (automobile(idA). time_per_cellA[m] < 27))

(ix) Properties that need to be verified: if the number of vehicles does not exceed the size of each road

(a) E[] novis[idS] >= 0

(b) E[] novis[idS] <= indexSC[p][q]

(c) E[] ((novis[idS] >= 0) && (novis[idS] <= indexSC[p][q]))

(d) E[] ((novis[idS] >= 0) && (novis[idS] <= indexSC[p][q]/3))

(e) E[] ((novis[idS] >indexSC[p][q]/3) && (novis[idS] <= indexSC[p][q]*2/3))

(f) E[] ((novis[idS]>indexSC[p][q]*2/3) && (novis[idS] <= indexSC[p][q]))

(g) ...

(x) Properties that need to be verified: determine the load of each road during the simulation process

(a) E<> ((GLOBAL > 0 && GLOBAL <=10) && ((novis[idS] >= 0) && (novis[idS] <= indexSC[p][q]/3))

(b) E[] ((GLOBAL > 0 && GLOBAL <=10) && ((novis[idS] >= 0) && (novis[idS] <= indexSC[p][q]/3))

(c) E<> ((GLOBAL > 0 && GLOBAL <=10) && ((novis[idS] >= 0) && (novis[idS] >indexSC[p][q]/3) && (novis[idS] <= indexSC[p][q]* 2/3))

(d) E[] ((GLOBAL > 0 && GLOBAL <=10) && ((novis[idS] >= 0) && (novis[idS] >indexSC[p][q]/3) && (novis[idS] <= indexSC[p][q]* 2/3))

(e) E<> ((GLOBAL > 0 && GLOBAL <=10) && ((novis[idS] >= 0) && (novis[idS]>indexSC[p][q]*2/3) && (novis[idS] <= indexSC[p][q]))

(f) E[] ((GLOBAL > 0 && GLOBAL <=10) && ((novis[idS] >= 0) && (novis[idS]>indexSC[p][q]*2/3) && (novis[idS] <= indexSC[p][q]))

(g) ...

In these queries inserted in UPPAAL, maximum values for validation variables taken into consideration are the values presented at the beginning of the simulation scenario. The minimum time is the multiplication of 0.5 time units by the number of cells contained in the first column of the indexSC[noS][maxnoCells +1] Matrix. The maximum time is the multiplication of 27 time units by the number of cells, which is contained in the first column in the indexSC[noS][maxnoCells +1] matrix.

This structuring modality of the presented simulation scenarios was chosen to prove the capacity of the system to simulate in detail the behaviour of each traffic component in a realistic context. During different verification processes, depending on the reachable states which are taken into account for investigation purposes, the element types of the simulation scenario have to be changed in a corresponding manner. At the same time, because not all the existing traffic-elements influence the considered evolution of the system on the interest state, the number of implied traffic components can decrease significantly, extending the urban area in which the verification process is applicable from the point of view of UPAAL limitations.

4.5. Extrapolation of Results for Similar Situations. Cellular automata allow the observation of different phenomena, managing to break down components into individual variables. They allow the understanding of how local changes affect the whole grid of cells. The formalism of timed automata, due to their elementary structure, is appropriate for a modular approach, and the resolution (level of detail) and system size obtained (the network size that needs to be covered) are appropriate to the model proposed. In the context of urban traffic, cellular automata based on microscopic models have the capacity to simulate in detail all the elements presented in this environment. The quantity of traffic-elements implemented generates a model containing a large number of evolutionary rules and interactions.

This model has a high flexibility in accordance with the environment where it will be applied, due to the fact that it is implemented on a large group of road-elements and possible interactions. The stimuli created for each vehicle will depend on the road-elements contained in the road and the traffic conditions, due to the modular approach used.

In the context of formal verification, this model presents all the possible scenarios that can occur, even the ones that are physically impossible, and the presented results allow the maximum level flexibility.

It was demonstrated that UPPAAL software is capable of dealing with low/medium complexity models, but for the implementation of high complexity models it is limited, due to the computational power that UPPAAL can accesses and cannot be extended. Taking its limits into consideration, the application proved to be a success in this context.

5. Conclusions and Future Work

The new road traffic simulation approach presented in this paper is based on the theory of timed automata and UPAAL model-checker. It was able to overcome the disadvantages regarding the lack of adaptability and flexibility of some previous urban road traffic models and to respond to the increasing necessity to verify the behaviour of traffic simulation models.

The behavioural characteristics of a representative and complex urban road traffic system are rendered in a realistic manner, considering the interactions between the infrastructure component and the moving vehicles. A two-layered timed automata based model was developed for this purpose. Unlike existing systems, this approach offers real extension possibilities, easily applied in practice due to the modular components based on matrix structures. They can be extended by adding a corresponding number of rows or columns. All the modifications of the simulated map could be performed simply by changing the initialization variables, without the necessity of modifying the UPAAL model. A case study highlighted some limitations of the UPAAL model-checker, more suitable for low/medium complexity models.

Features of this work can be developed and explored more deeply, providing room for new perspectives in this domain. Taking into consideration the scenario in which this approach will be implemented, some improvements could be made in terms of the specific problem and of the application itself.

Conflicts of Interest

The authors declare that there are no conflicts of interest.

Acknowledgments

This research was carried out under European funding, Structural Projects: POSDRU/89/1.5/S/52603-4D-POSTDOC.

References

[1] R. Mardiati, N. Ismail, and A. Faroqi, "Review of microscopic model for traffic flow," *ARPN Journal of Engineering and Applied Sciences*, vol. 9, no. 10, pp. 1794–1800, 2014.

[2] K.-C. Chu, L. Yang, R. Saigal, and K. Saitou, "Validation of stochastic traffic flow model with microscopic traffic simulation," in *Proceedings of the 2011 7th IEEE International Conference on Automation Science and Engineering, CASE 2011*, pp. 672–677, Italy, August 2011.

[3] P. J. Gundaliya, T. V. Mathew, and S. L. Dhingra, "Heterogeneous traffic flow modelling for an arterial using grid based approach," *Journal of Advanced Transportation*, vol. 42, no. 4, pp. 467–491, 2008.

[4] M. Treiber and A. Kesting, "Microscopic calibration and validation of car-following models – a systematic approach," *Procedia - Social and Behavioral Sciences*, vol. 80, pp. 922–939, 2013.

[5] E. Amini, M. Tabibi, E. R. Khansari, and M. Abhari, "A vehicle type-based approach to model car following behaviors in simulation programs (case study: Car-motorcycle following behavior)," *IATSS Research*, 2018.

[6] J. Casas, J. Perarnau, and A. Torday, "The need to combine different traffic modelling levels for effectively tackling large-scale projects adding a hybrid meso/micro approach," in *Proceedings of the 14th Meeting of the Euro Working Group on Transp. - In Quest for Adv. Models, Tools, and Methods for Transp. and Logist., EWGT, 26th Mini-EURO Conf. - Intelligent Decis. Making in Transp. and Logist., MEC, 1st Eur. Sci. Conf. on Air Transp. - RH*, pp. 251–262, Poland, September 2011.

[7] M. Rahman, M. Chowdhury, Y. Xie, and Y. He, "Review of microscopic lane-changing models and future research opportunities," *IEEE Transactions on Intelligent Transportation Systems*, vol. 14, no. 4, pp. 1942–1956, 2013.

[8] V. L. Knoop, S. P. Hoogendoorn, Y. Shiomi, and C. Buisson, "Quantifying the number of lane changes in traffic: An empirical analysis," in *Proceedings of the 91st Annual Meeting Transportation Research Board*, Washington, DC, USA, 2012.

[9] J. Barcelo, D. García, and J. Perarnau, "Methodological notes on combining macro, meso, and micro models for transportation analysis," in *Proceedings of the Workshop on Modeling and Simulation*, 2005.

[10] E. Bernauer, L. Breheret, S. Algers et al., "Review of micro-simulation models, smartest/D3," in *The "SMARTEST" Project, Simulation Modelling Applied to Road Transport European Scheme Tests*, Institute for Transport Studies, University of Leeds, 1998.

[11] *VISSIM 5.2 User Manual, PTV Planung Transport Verkehr AG*, Germany, 2010.

[12] D. Lin, X. Yang, and C. Gao, "VISSIM-based simulation analysis on road network of CBD in Beijing, China," *Procedia - Social and Behavioral Sciences*, vol. 96, pp. 461–472, 2013.

[13] M. Fellendorf and P. Vortisch, "Microscopic traffic flow simulator VISSIM," in *Fundamentals of Traffic Simulation*, J. Barceló, Ed., pp. 61–93, Springer, New York, NY, USA, 2010.

[14] "Surrogate "safety measures from traffic simulation models"," FHWA-RD-03-050, Office of Research, Development, and Technology, Office of Safety, RDT, 2012.

[15] B. W. Bush, *An Algorithmic Overview of TRANSIMS*, Los Alamos National Laboratories, 2010.

[16] *TransModeler Users Guide, Traffic Simulation Software.Caliper*, MA, 2008.

[17] J. Barceló and J. Casas, "Dynamic network simulation with AIMSUN, simulation approaches in transportation analysis recent advances and challenges," in *MasoKuwahara Operations Research/Computer Science Interfaces Series*, R. Kitamura, Ed., pp. 57–98, 31 edition, 2005.

[18] MITSIMLab, *User's Guide for MITSIMLab and Road Network Editor (RNE)*, Massachusetts Institute of Technology, Intelligent Transportation Systems Program, 2002.

[19] T. Toledo, C. F. Choudhury, and M. E. Ben-Akiva, "Lane-changing model with explicit target lane choice," *Transportation Research Record*, no. 1934, pp. 157–165, 2005.

[20] CORSIM, *Advanced CORSIM Training Manual*, Minnesota Department of Transportation, 2004.

[21] P. Quadstone, "QuadstoneParamics V4.2 analyser reference manual," in *Quadstone Limited*, 2003.

[22] C. R. Long, T. Yanzhi, and L. Der-Horng, "Comparison of PARAMICS and GETRAM/AIMSUN microscopic traffic simulation," in *Proceedings of the 83rd Annual Meeting of the Transportation Research Board*, 2003.

[23] J. Olstam and A. Tapini, "Comparison of Car-Following Models," in *VTI Publication M960A.2004*, Swedish National Road and Transport Research Institute, Linköping, Sweden, 2004.

[24] S. Panwai and H. Dia, "Comparative evaluation of microscopic car-following behavior," *IEEE Transactions on Intelligent Transportation Systems*, vol. 6, no. 3, pp. 314–325, 2005.

[25] C. Krogscheepers and K. Kacir, "Latest trends in micro simulation: an application of the PARAMICS model," in *Proceedings of the The 20th Annual South African Transport Conference*, South Africa, 2001.

[26] H. Sbayti and D. Roden, *Best Practices in the Use of Micro Simulation Models Prepared for: American Association of State Highway and Transportation Officials (AASHTO)", repaired as part of NCHRP Project -36, Task 90*, National Cooperative Highway Research Program, Transportation Research Board, 2010.

[27] E. Ugnenko, E. Uzhvieva, and Y. Voronova, "Simulation of traffic flows on the road network of urban area," *Procedia Engineering 134*, pp. 153–156, 2016.

[28] J. Shen and X. Jin, "Detailed traffic animation for urban road networks," *Graphical Models*, vol. 74, no. 5, pp. 265–282, 2012.

[29] L. Li, "Research on traffic flow mathematical model in urban traffic," *International Journal of Smart Home*, vol. 9, no. 11, pp. 123–132, 2015.

[30] Z. Tian, L. Jia, H. Dong, F. Su, and Z. Zhang, "Analysis of urban road traffic network based on complex network," *Procedia Engineering*, vol. 137, pp. 537–546, 2016.

[31] M. M. Ishaque and R. B. Noland, "Behavioural issues in pedestrian speed choice and street crossing behaviour: A review," *Transport Reviews*, vol. 28, no. 1, pp. 61–85, 2008.

[32] G. Waizman, S. Shoval, and I. Benenson, "Micro-simulation model for assessing the risk of vehicle–pedestrian road accidents," *Journal of Intelligent Transportation Systems: Technology, Planning, and Operations*, vol. 19, no. 1, pp. 63–77, 2013.

[33] V. Astarita, G. Guido, A. Vitale, and V. P. Giofré, "A new microsimulation model for the evaluation of traffic safety performances," *Transport*, no. 51, pp. 1–16, 2012.

[34] F. Cunto and F. F. Saccomanno, "Microlevel traffic simulation method for assessing crash potential at intersections," in *Proceedings of the 86th Annual Meeting of the Transportation Research Board*, Washington, DC, USA, 2007.

[35] L. P. Hurd, "Formal language characterizations of cellular automaton limit sets," *Complex Systems*, vol. 1, no. 1, pp. 69–80, 1987.

[36] H. Zhang, X. Liang, and K. Xu, "VISSIM network simulation under actuated signal control," *Journal of Computational Information Systems*, vol. 10, no. 4, pp. 1573–1580, 2014.

[37] R. Dowling, A. Skabardonis, J. Halkias, G. McHale, and G. Zammit, "Guidelines for calibration of microsimulation models: framework and applications," in *Proceedings of the 83rd Annual Meeting of the Transportation Research Board*, Washington, DC, USA, 2004.

[38] B. P. Byungkyu and W. Jongsun, *Microscopic Simulation Model Calibration and Validation Handbook*, Final Contract Report, Traffic Operations Laboratory, Center for Transportation Studies, University of Virginia, 2006.

[39] Q. Chao, Z. Deng, and X. Jin, "Vehicle-pedestrian interaction for mixed traffic simulation," *Computer Animation and Virtual Worlds*, vol. 26, no. 3-4, pp. 405–412, 2015.

[40] R. Alur and D. L. Dill, "A theory of timed automata," *Theoretical Computer Science*, vol. 126, no. 2, pp. 183–235, 1994.

[41] G. Behrmann, A. David, and K. G. Larsen, "A tutorial on uppaal," in *Formal Methods for the Design of Real-Time Systems*, vol. 3185 of *Lecture Notes in Computer Science*, pp. 200–236, Springer, Berlin, Germany, 2004.

[42] C. Avram, J. Machado, and A. Aştilean, "Modeling and formal analysis of urban road traffic," in *Proceedings of the 11th International Conference of Numerical Analysis and Applied Mathematics 2013, ICNAAM 2013*, pp. 1408–1411, Greece, September 2013.

[43] E. M. Clarke, E. A. Emerson, and A. P. Sistla, "Automatic verification of finite-state concurrent systems using temporal logic specifications," *ACM Transactions on Programming Languages and Systems (TOPLAS)*, vol. 8, no. 2, pp. 244–263.

[44] E. B. Lima, *Modelos microscópicos para simulação do tráfego baseados em autômatos celulares*, Dissertação de Mestrado em Computação, Universidade Federal Fluminense, Niterói, Brazil, 2014, http://www2.ic.uff.br/PosGraduacao/Dissertacoes/361.pdf.

[45] J. Vasic and H. J. Ruskin, "A discrete flow simulation model for urban road networks, with application to combined car and single-file bicycle traffic," *Lecture Notes in Computer Science (including subseries Lecture Notes in Artificial Intelligence and Lecture Notes in Bioinformatics): Preface*, vol. 6782, no. 1, pp. 602–614, 2011.

[46] M. Tiriolo, L. Adacher, and E. Cipriani, "An urban traffic flow model to capture complex flow interactions among lane groups for signalized intersections," *Procedia - Social and Behavioral Sciences*, vol. 111, pp. 839–848, 2014.

[47] J. Shi, L. Cheng, J. Long, and Y. Liu, "A new cellular automaton model for urban two-way road networks," *Computational Intelligence and Neuroscience*, vol. 2014, Article ID 685047, 6 pages, 2014.

[48] G. A. Wainer, "Cellular modeling with cell-DEVS: a discrete-event cellular automata formalism," in *ACRI 2014, LNCS 8751*, J. Was, G. C. Sirakoulis, and S. Bandini, Eds., pp. 6–15, 2014.

[49] J. Bengtsson and Y. Wang, "Timed automata: semantics, algorithms and tools," in *Lectures on Concurrency and Petri Nets*, vol. 3098 of *Lecture Notes in Computer Science*, pp. 87–124, Springer, Berlin, Germany, 2004.

[50] M. U. Iftikhar, J. Lundberg, and D. Weyns, "A model interpreter for timed automata," in *ISoLA*, pp. 243–258, 2016.

[51] J. Wiedermann and P. Lukas, "Communicating mobile nanomachines and their computational power," in *Proceedings of NanoBet Conference*, vol. 125, 2008.

[52] G. Kunz, E. Perondi, and J. MacHado, "Modeling and simulating the controller behavior of an Automated People Mover using IEC 61850 communication requirements," in *Proceedings of the 2011 9th IEEE International Conference on Industrial Informatics, INDIN 2011*, pp. 603–608, Portugal, July 2011.

[53] L. Evans and R. Rothery, *Experimental Measurements of Perceptual Thresholds in Car following*, Highway Research Record 454, Transportation Research Board, Washington, DC, USA, 1973.

16

Evaluation and Application of Urban Traffic Signal Optimizing Control Strategy based on Reinforcement Learning

Yizhe Wang,[1,2] Xiaoguang Yang,[1,2] Yangdong Liu,[1,3] and Hailun Liang[1,2]

[1]*Key Laboratory of Road and Traffic Engineering of the Ministry of Education, Tongji University, 4800 Cao'an Road, Shanghai 201804, China*
[2]*Intelligent Transportation System Research Center of Tongji University, 4801 Cao'an Road, Shanghai 201804, China*
[3]*Hangzhou Hikvision Digital Technology Co., Ltd., No. 555 Qianmo Road, Binjiang District, Hangzhou 310052, China*

Correspondence should be addressed to Xiaoguang Yang; yangxg@tongji.edu.cn

Guest Editor: Hamzeh Khazaei

Reinforcement learning method has a self-learning ability in complex multidimensional space because it does not need accurate mathematical model and due to the low requirement for prior knowledge of the environment. The single intersection, arterial lines, and regional road network of a group of multiple intersections are taken as the research object on the paper. Based on the three key parameters of cycle, arterial coordination offset, and green split, a set of hierarchical control algorithms based on reinforcement learning is constructed to optimize and improve the current signal timing scheme. However, the traffic signal optimization strategy based on reinforcement learning is suitable for complex traffic environments (high flows and multiple intersections), and the effects of which are better than the current optimization methods in the conditions of high flows in single intersections, arteries, and regional multi-intersection. In a word, the problem of insufficient traffic signal control capability is studied, and the hierarchical control algorithm based on reinforcement learning is applied to traffic signal control, so as to provide new ideas and methods for traffic signal control theory.

1. Introduction

Traffic congestion has become a world-concerned problem all over the world. With the increasing number of vehicles, traffic congestion has deeply affected people's daily life and the development of social economy. Traffic control is one of the most important technological means of regulating traffic flow, improving obstruction, and improving its safety and even energy conservation and emission reduction. At present, traffic signal control problem not only has a long-time congestion phenomenon at peak time, but also has obvious ability of grooming in peak time. In order to ease the traffic pressure, rational analysis and control are considered as an important tool. Its progress and development are always keeping pace with the times, accompanied by information technology, computer technology, and system science.

According to the system's ability to adapt to the environment and the level of intelligent decision-making, Gartner proposed the evolution of urban transport control system development level in 1996 [1]. The first-generation self-adaptive control system adopts the multi-time timing control of fine division of period, or completely isolated self-adaptive control, to realize the simple regulation of traffic flow. The second-generation traffic signal control system dynamically adjusts the parameters of the signal timing scheme (cycle length, split, offset). Typical second-generation control systems include SCATS [2] and SCOOT [3]. The UK Transport Research Laboratory had a worldwide reputation for contributions to the field of traffic signal control, especially as originators of the TRANSYT and SCOOT signal coordination methods [4]. The third-generation control system uses similar idea to the second-generation to dynamically adjust the signal timing parameter in response to the fluctuation of the time-varying traffic flow at the intersection. HK Lo and HF Chow investigated the relationship of finer resolutions and larger errors in adaptive traffic control system through an extensive simulation of scenarios in Hong Kong with a recently developed dynamic traffic control model, DISCO

[5]. Aboudolas K and Papageorgiou M tested a preliminary simulation-based investigation of the signal control problem for a large-scale urban road network using store-and-forward modeling demonstrating the comparative efficiency and real-time feasibility of the developed signal control methods [6]. The fourth-generation traffic signal control system is an integrated traffic management and control system. Meneguzzer C presented two alternative deterministic, discrete-time DP models of the interaction between signal control and route choice, which are proposed and compared with the conventional iterative optimization and assignment (IOA) method for network traffic signal setting [7]. The fifth-generation traffic signal control system is based on the abilities of artificial intelligence and self-learning.

2. Literature Review

2.1. Traffic Big Data Environment. Digitized and informational infrastructure of urban road traffic and constructions of related systems have developed rapidly in the past ten years, and urban traffic control is developing from the "data poverty" times to the "data rich" times. Meanwhile, the appearance of ICV (intelligent connected vehicle) and autonomous vehicles will construct the future traffic environment jointly, which significantly differs from conventional manual driving vehicles in terms of individual information acquisition, perception ability, reaction time, interactive behavior, etc. New requirements of traffic control have formed a high-level demand for the next generation of traffic control [8]. The research on the next generation of traffic signal control for regional transportation under the "data rich" environment is on the agenda. Ma D et al. proposed the lane-based saturation degree estimation for signalized intersections and maximum queue length estimation for traffic lane groups, which enriched the way to obtain traffic parameters and increased the precision of estimation method. For example, the results show that the new method of maximum queue length estimation has a higher precision compared to the existing method based on a similar concept, with maximum and average deviations of 39.36% and 12.25%, respectively, over twenty cycles [9, 10].

Under the conditions of limited cross-section traffic flow data, many existing adaptive traffic control systems have adopted traffic models to actively predict the evolution of network traffic flows and then adopted the aggregative indicator method to optimize and solve timing parameters. However, the real-time detection of the spatiotemporal data based on urban road network traffic status can provide rich and high-quality basic data and fine-grained assessment of control effects for traffic control. In the face of the main defects encountered in the existing self-adaptive traffic control system, a closed-loop feedback self-adaptive control system with better uncertainty response capability and higher intelligent decision-making level is an inevitable result of the objective needs of the development and application technologies [11]. Ma D proposed a calculation method for the occupancy per cycle under different traffic conditions presented, based on the relationship between the three basic traffic flow parameters, speed, traffic flow, and density [12]. The results show that the precision of this method was affected by the detector location and bus ratio insignificantly [13].

2.2. Reinforcement Learning Traffic Control. According to real-time collection of states, rewards, and punishments, the single intersection's signal control of reinforcement learning can find an optimization strategy of traffic signal control suitable for traffic flow characteristics through the interaction. In recent years, more and more domestic scholars have studied principles of reinforcement learning and discussed the applications of reinforcement learning algorithms in traffic control. Reinforcement learning has developed rapidly in the optimizing control [14, 15].

Scholars have done a lot of research on reinforcement learning theory, algorithms, and applications and have obtained many famous research results. Ma D proposed a new control method that brings significant and positive effects to the bottleneck link itself and to the entire test area [16]. Yang W concluded that critical issues in developing agent-based traffic control systems for integrated network were addressed as interoperability, adaptability, and extendibility [17]. Zhang L drew a conclusion that extensive simulation results for the designed Shanghai simulation scenarios indicate that most of the observed counts match quite well with the traffic simulation volumes and demonstrate the potential of MATSIM for large-scale dynamic transport simulation [18]. Aslani M developed adaptive traffic signal controllers based on continuous residual reinforcement learning (CRL-TSC) that was more stable, and the best setup of the CRL-TSC leads to saving average travel time by 15% in comparison to an optimized fixed-time controller [19].

Reinforcement learning control has the advantages of real-time online and feedback control, which especially accords with the control thoughts of signal adaptive control in urban intersections. However, there is a question as to whether the traffic signal optimization strategy based on the reinforcement learning is applicable to all the traffic environments.

3. Traffic Signal Control Strategy Based on Reinforcement Learning

Reinforcement learning is a typical data-driven control method. In this paper, the method of signal control scheme improvement is proposed. According to the different traffic flow characteristics, the subregions are divided. Based on the three key parameters of cycle length, arterial coordination signal offset, and green split, a set of hierarchical control algorithms based on reinforcement learning is constructed to optimize and improve the current signal timing scheme.

3.1. Control Subregions Division and Cycle Optimization. As for the regional coordination control, the primary content is the division of the coordination subregions. In the signal control road network, each intersection has its influence range, and the intersection and section within this range are greatly affected by it. To quantify the impact and define the scope of influence, literature defines direct relevance to describe the relationship between adjacent intersections, finding that when the upstream node traffic flows into the downstream node, it is close to or greater than the downstream node's import capacity. It is found that the path correlation is mainly affected by the traffic network topology and OD distribution between the two intersections. The more the OD paths through two nodes at the same time, the stronger the correlation between nodes. The higher the flow rate of OD path passing through both nodes at the same time, the stronger the correlation between nodes. The more the OD paths that pass through both nodes at the same time are unique, the stronger the correlation between nodes will be.

The optimization range is region-level road network optimization. The control subregions are divided by characteristic parameters such as average travel time; vehicle OD amount between intersections and traffic coordination control subregions are finally determined.

The signal cycle refers to the time required for the signal color to display one cycle in the set phase order, that is, the sum of the steps of each control step in one cycle. The signal cycle is the key control parameter that determines the effectiveness of traffic signal control. If the signal cycle is too short, it is difficult to ensure that the vehicles in all directions can pass through the intersection smoothly, resulting in frequent stops at the intersection and a decline in the utilization rate of the intersection. If the signal cycle is too long, it will cause the driver to wait for too long, greatly increasing the delay time of the vehicle. The cycle in the green wave control is taken as the common cycle by the maximum signal cycle of the key intersection of the arterial, and the signal cycle of the remaining intersections is reallocated to each phase according to the traffic flow ratio.

According to different evaluation indexes, the optimal cycle is obtained by using model-based algorithm. Regarding the evaluation indicators of traffic efficiency at intersections, traffic capacity, saturation, service level, travel time, number of stops, and queue length are commonly used at home and abroad. The delay is mainly due to the travel time loss caused by traffic friction and traffic control. It is closely related to the cycle duration, green split, and saturation. It is an important indicator for evaluating the traffic service level and operational efficiency of signalized intersections, including queue delay, parking delay, control delay, and lane approach delay.

3.2. Offset Optimization Based on Bayesian Optimization Algorithm. The phase offset is also called the time offset or the green time offset. The phase offset includes the absolute phase offset and the relative phase offset. Absolute phase offset refers to the offset between the starting or ending point of the signal green light (red light) in the coordinated direction of the arterial at each intersection and the starting or ending point of the signal green light (red light) in the coordinated direction of the arterial at a certain intersection (generally a key intersection). Relative phase refers to the time offset between the starting or ending points of the green light (red light) signal in the coordinated direction of the arterial at adjacent intersections. The relative phase offset is equal to the difference value between the absolute phase offset of two intersections, which is determined by the actual vehicle speed.

According to the coordination effect between the intersections, it is divided into several control subregions, and internal coordination control is implemented for its traffic characteristics. The basic principles of control subregions division are as follows:

(1) The distance between adjacent intersections is less than 600 meters and control subregion contains no more than 10 intersections.

(2) The optimal period length of each intersection is an integer multiple relationship.

The following lines with inconsistent coordination effects should not be included in a subregional coordination:

(1) An excessively long connection, and the traffic flow along the connection is highly discrete.

(2) There are traffic production sources or attraction sources (such as large parking lots and shopping malls) and very frequent pedestrian activities along both sides of certain lines, which seriously interfere with traffic flow.

The Bayesian optimization algorithm belongs to the sequential model-based optimization (SMBO) algorithm. This algorithm determines the value of the next (optimal) sample set by analyzing historical observations of a loss function$f$. Since the Bayesian optimization algorithm was proposed around 2010, it has been used to optimize the hyperparameters of machine learning models in the field of machine learning in recent years. The so-called superparameter is the model parameter that needs to be set artificially. In this competition, due to the large number of timing parameters that need to be optimized, which includes the signal split and phase offset of multiple different intersections, the solution space dimension is relatively high and the optimization is quite difficult. The overall idea of the Bayesian optimization algorithm is as follows:

Calculate the posterior expectation of the loss function f using the observed sample set $\mathbf{X}_{1:n}$.

Generate a new set of samples $\mathbf{X}_{new}$ to sample the loss function f, which can maximize the expectation of f in the value range of independent variables.

Repeat the above steps until the preset convergence condition is reached. End the optimization process.

The algorithm will be described in detail below and the process will be summarized.

To calculate the posterior expectation of the loss function f, the likelihood model of the sample and the prior probability model of f should be obtained in advance. In the Bayesian optimization process, we can assume that the sample obeys the multivariate Gaussian distribution and obtain the Gaussian likelihood function:

$$y = f(\mathbf{X}) + \epsilon, \quad \epsilon \sim N\left(0, \sigma_\epsilon^2\right) \tag{1}$$

For the prior distribution, we assume that the loss function f can be described by a Gaussian process (GP). The essence of the Gaussian process is the generalization of the multivariate Gaussian distribution to the function distribution. Therefore, just as the Gaussian distribution is determined by its expectation and variance, the Gaussian process is completely determined by its expectation function $m(\mathbf{X})$ and the covariance function $k(\mathbf{X},\mathbf{X}')$. The Gaussian process is widely used in the application of all probabilistic models because its description of the posterior distribution of the loss function is easier for us to analyze and calculate.

One of the most widely used acquisition functions is the expected improvement (EI) function. The EI function is defined as

$$EI(\mathbf{X}) = \mathbf{E}\left[\max\left\{0, f(\mathbf{X}) - f\left(\widehat{\mathbf{X}}\right)\right\}\right] \quad (2)$$

where $\widehat{\mathbf{X}}$ is the current optimal sample set, and this function gives a new sample set that can best enhance the expectation of the loss function. Moreover, the expected lifting function can be calculated based on the Gaussian process model, namely,

$$EI(\mathbf{X}) = \begin{cases} \left(\mu(\mathbf{X}) - f\left(\widehat{\mathbf{X}}\right)\right)\Phi(Z) + \sigma(\mathbf{X})\emptyset(Z) & \text{if } \sigma(\mathbf{X}) > 0 \\ 0 & \text{if } \sigma(\mathbf{X}) = 0 \end{cases} \quad (3)$$

$$Z = \frac{\mu(\mathbf{X}) - f\left(\widehat{\mathbf{X}}\right)}{\sigma(\mathbf{X})} \quad (4)$$

where $\mathbf{\Phi}(Z)$ and $\emptyset(Z)$ are the cumulative distribution function and probability density distribution function of the multivariate standard Gaussian distribution, respectively. When the posterior expectation $\mu(\mathrm{X})$ is higher than the current loss function optimal value $f(\widehat{\mathbf{X}})$, EI will get a larger value. When the uncertainty $\sigma(\mathrm{X})$ of X is high, EI will get a larger value.

After the above analysis and introduction, the whole principle and process of Bayesian optimization can be summarized to form a Bayesian optimization algorithm:

Given the observed value $f(\mathbf{X})$ of the loss function, the posterior expectation of the loss function f is updated based on the Gaussian model.

Solve the expected lifting function (EI function) to find the new best sample set: $\mathbf{X}_{new} = \arg\max EI(\mathbf{X})$.

Calculate the value of the loss function at $\mathbf{X}_{new}$.

Repeat the above steps until the preset number of repetitions (i.e., the number of iterations) is reached or the convergence condition is met.

In (2) of the above steps, we can use the gradient-based solution method to optimize the EI function to get $\mathbf{X}_{new}$.

On the basis that the parameters such as the optimal cycle length are determined, the phase offset of the intersections after deduplication $\mathbf{X} = \{x_1, \ldots, x_n\}$ can be regarded as input loss function sample set. The sample $\mathbf{X}$ of the function, which is returned by the online feedback, can be iterated multiple times based on the Bayesian optimization algorithm.

3.3. Split Optimization Based on Q-Learning Algorithm. In the urban transportation system, the traffic flow, vehicle speed, and traffic density are the most intuitive reflections of traffic conditions. They are the three characteristic parameters of traffic flow and the research focus and foundation of traffic flow theory. Among them, the traffic flow refers to the number of vehicles passing through per unit time; the vehicle speed refers to the distance that the vehicle passes per unit time; and the traffic density refers to the number of vehicles on the section per unit length. The traffic flow theory is the basis for the establishment of urban traffic signal control system.

The traffic model uses a discrete-time difference equation or a continuous time subdivision tool to introduce a dynamic relationship between the concepts of traffic volume Q, vehicle speed V, and traffic density K, which summarizes the physical quantities of the traffic network and is used to describe the collective average behavior of a large number of vehicles. In the free flow, the interaction between vehicles can be neglected, and the traffic flow increases linearly with the vehicle density. The wide moving jam flow is usually characterized by stop-go-stop traffic, that is, a series of jams. The density of vehicles in the region is high and the average speed and flow of vehicles are small. The average velocity of the synchronized flow is significantly lower than that of the free flow.

At present, Q-learning algorithm is one of the most frequently used methods in the fields of reinforcement learning, proposed by Watkins in 1989 [20]. Q-learning algorithm is widely used in the fields of control, depending on the update mode of its special value function.

In Q-learning, the solution formula of the mainstream value function is as follows.

$$\begin{aligned} Q(s_t, a_t) &\longleftarrow Q(s_t, a_t) \\ &\quad + \alpha\left[\left(r_{t+1} + \gamma \max_{a\in A} Q(s_{t+1}, a) - Q(s_t, a_t)\right)\right] \end{aligned} \quad (5)$$

According to the formula, at the moment of t, the state of Q-learning is s_t. If the taken action is a_t, the corresponding value function will be Q(s_t, a_t). The update of the value function is determined by three factors. The first is the current value of the action state value function, Q(s_t, a_t), that needs to be updated. The second is to control the corresponding maximum value of all Q-values of actions in the postexecution state of s(t+1), and the third is the immediate return, r(t+1), after the action. Besides, there are also two model parameters, learning rate $\alpha \in [0,1]$ and discount factor $\gamma \in (0,1]$. The former is used to balance the relationship between the learning and utilization of the algorithm. When a⟶1, the controller tends to explore new knowledge; otherwise it will use the existing knowledge. The latter is used to coordinate the present relationship with the future. When $\gamma \longrightarrow 1$, the controller tends to consider the future return, and when $\gamma \longrightarrow 0$, the controller mainly considers immediate return [21].

Whether in theoretical research or in engineering practice analysis, road traffic density is an effective indicator for measuring the degree of traffic congestion. The operation of

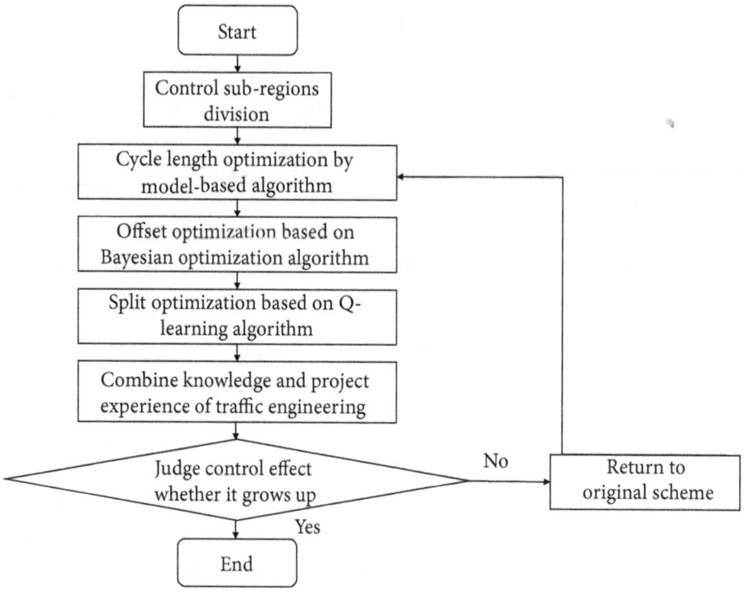

FIGURE 1: The flow chart of the hierarchical control algorithm based on reinforcement learning.

the traffic on the section is affected by the signal control of the upstream and downstream intersections. The release signal at the upstream intersection directly changes the density of the section, which indirectly affects the traffic capacity and saturation at the section of the stop line and indirectly affects the density of the queue section. The mutual influence of the two is especially noticeable in the supersaturated state. Since the penetration rate of the connected vehicles in different sections is unknown, it is impossible to visually reflect the actual flow of the road through the number of discrete connected vehicles. Even by expanding the sample, it is difficult to guarantee accuracy, but it can clearly reflect the speed of the overall traffic flow. Therefore, this paper uses traffic density as the core parameter to provide a basis for green split optimization.

3.4. The Flow of the Control Algorithm. Firstly, according to different evaluation indexes, the optimal cycle is obtained by using model-based algorithm. Using the combination of the average travel time of vehicles and the Bayesian optimization method based on the Gaussian process, which is commonly used in the optimization of machine learning algorithms, the arterial coordination control is set. The phase offset is optimized by the two-way flow ratio of the upstream and downstream roads and the reasonable setting of the pedestrian crossing phase. Then set different green wave bandwidths to match the upstream and downstream traffic of the morning rush hour and the tidal phenomenon with uneven travel speed. The intelligent algorithm such as Q-learning is used to optimize the green split of each intersection by using key traffic flow parameters at each intersection.

In conclusion, the flow of the traffic signal control strategy based on reinforcement learning is as shown in Figure 1.

4. Verification of Traffic Signal Control Strategy Based on Reinforcement Learning

4.1. Verification of Single Intersection Signal Control Strategy Based on Reinforcement Learning. The intelligent algorithm such as Q-learning is used to optimize the green split of single intersection by using key traffic flow parameters at the intersection. We have compared the Q-learning control method adopted in this paper with the traditional timing signal control and the adaptive control method in second-generation traffic signal control system. The delay of intersection means the average delay for all vehicles passing through all of lane groups at the intersection in the same cycle. The results are shown in Figure 2.

Having compared the traditional timing control with the traffic signal control method based on Q-learning algorithm applied in this paper, the study has shown that the application of Q-learning control method has achieved good performance. In terms of effectiveness, compared with the traditional timing control, the optimization effects of traffic signal control based on Q-learning have, respectively, reached 31.68%, 30.10%, 37.59%, 38.07%, 40.69%, and 43.89%, which has shown that compared with traditional timing control, the traffic signal control based on Q-learning can achieve better optimization effects. However, compared with the existing traffic control strategy, the optimization effects of traffic signal control based on Q-learning have, respectively, reached -4.21%, -5.28%, 3.14%, 6.23%, 13.11%, and 9.72%. The optimization effects of traffic signal control based on reinforcement learning are inferior under low flow conditions, while they are better under medium and high flow conditions.

4.2. Verification of Arterial Intersection Signal Control Strategy Based on Reinforcement Learning. The green wave coordinated control has three important parameter conditions: the

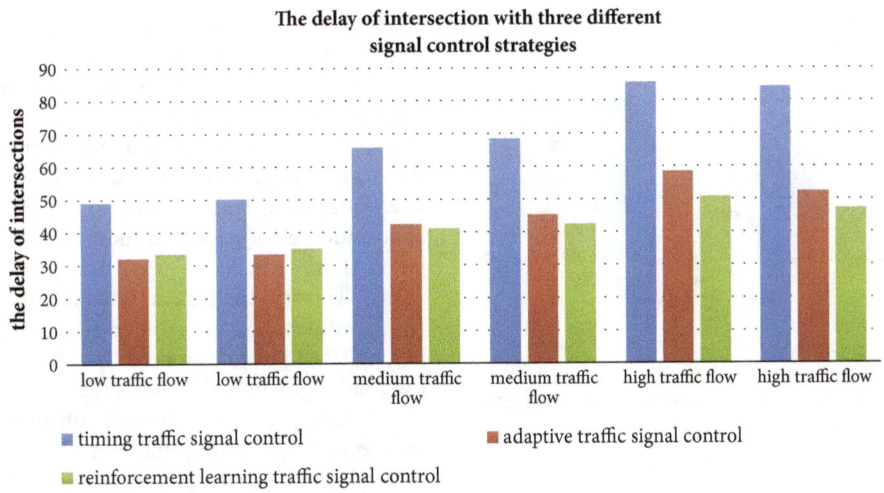

FIGURE 2: Comparisons of traffic control methods of single intersections.

signal clocks of each intersection should be synchronized; the signal cycle should be the same and have phase offset (the travel time calculated by the adjacent intersection based on the actual average speed). Only with these three conditions can the validity of green wave be guaranteed.

Discrete connected vehicle trajectory data cannot directly obtain the data required for signal distribution and intersection channelization scheme under traditional conditions but can obtain more detailed and complete trajectory-level data. The complete physical trajectory of the vehicle during driving can not only reflect the driving path of the vehicle on the road network, but also reflect the changing characteristics of the vehicle speed with time and space. It is the most comprehensive and complete expression form of the traffic flow operating state, containing a wealth of traffic flow information which is key parameter for offset optimization (for example, travel speed, queue length, delay, and stop times).

For the coordinated control problem of the supersaturation state during morning rush hour, the above research basically follows the idea based on the strong mathematical hypothesis model, but the control system deviates from the original trajectory due to the inaccuracy based on the strong mathematical hypothesis model and the interference from the outside of the control system. In response to these shortcomings, some scholars have further proposed the idea of predictive control, which enables the system to correct the trajectory deviation in real time and achieve the purpose of optimal control. However, the establishment of the optimal control model is still a centralized processing idea. In the application of intersection control problems, it focuses on the control problem of single-point intersections. From the perspective of the structure of the control algorithm, the hierarchical control structure can integrate more control personnel's design ideas, which is of great help to solve the problem of complex control state of the road network. With the development of intelligent control technology, Bayesian optimization methods based on Gaussian process, fuzzy control, reinforcement learning calculation, and neural network have been also widely used in traffic control. However, these applications present a similar feature that is loosely integrated with the actual traffic condition. The computational speed of the online control system is still a big obstacle, and it is more difficult to push it to practical applications. Therefore, traffic control at the network level for rush hours should be based on offline large-scale optimization calculation based on traffic model (based on travel time, then obtaining the relative phase offset between adjacent intersections) and intelligent algorithm (Bayesian optimization methods based on Gaussian process), seeking to achieve a system-optimized phase offset timing scheme.

On the other hand, in the traditional arterial coordinated control scheme, the green wave velocity, the forward green wave bandwidth, and the reverse green wave bandwidth between the starting point and the ending point are always the same or almost the same, and no or less consideration is given to the individualization of the velocity distribution between the sections and the tidal of the traffic flow. Therefore, we adopted different green wave speed optimization methods for different road sections, combined with the unique tidal phenomenon. In the direction of large traffic flow, based on the calculated green wave bandwidth, the bandwidth of the reverse green wave is appropriately increased to match the traffic demand of rush hour. At the same time, the vehicle traffic phase is, respectively, covered to the forward and reverse two-way green wave at the pedestrian crossing, which minimizes the probability that the vehicle stops at the signal control pedestrian crossing.

As is shown in Figure 3, the comparison of original signal control scheme and hierarchical traffic signal control method in different traffic flows has been given. Compared with the original signal control scheme, the arterial traffic signal optimization control method based on hierarchical traffic signal control performs better in medium and high traffic flow. The average delay per vehicle of the original signal control scheme is, respectively, 28.47 seconds, 40.34 seconds, and 61.38 seconds. While with regard to the arterial traffic signal optimization control method based on hierarchical traffic

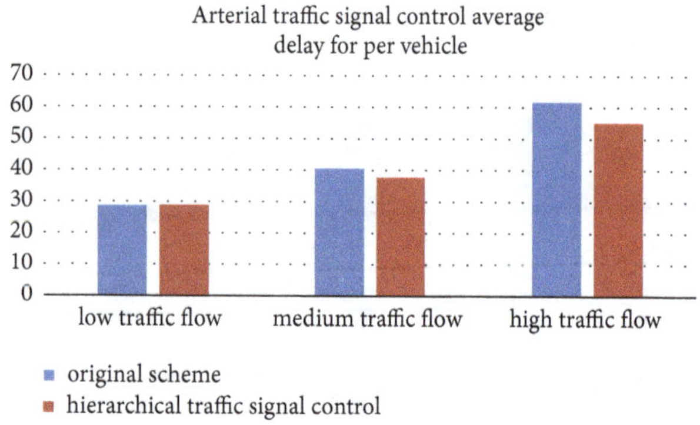

FIGURE 3: Evaluation of the traffic control method of arterial multiple intersections.

signal control, the average delay per vehicle is, respectively, 28.77 seconds, 37.56 seconds, and 54.79 seconds. Then, the optimization percentage is, respectively, -1.05%, 6.89%, and 10.74%. However, due to the randomness of traffic flows, although there is some increasing average delay in low traffic flow, the general trend has similar conclusions.

4.3. Verification of Regional Traffic Signal Control Strategy Based on Reinforcement Learning. The concept of regional signal control can be divided into a narrow sense and a broad sense. In a narrow sense, regional signal control is a signal control method that unifies several intersections with strong correlation and carries out mutual coordination, namely, the so-called regional signal coordination control. In a broad sense, regional signal control refers to the monitoring of all intersections within the region under the management of a command and control center. It is a comprehensive signal control for single isolated intersection, multiple intersections of the arterial and the highly connected intersection group. It can be classified according to control strategy (timed offline control system, adaptive online control system), control mode (scheme selection, solution generation), and control structure (centralized, distributed).

The purpose of vehicle path feature extraction is to obtain the information of the nodes (i.e., intersections and OD points) that each vehicle passes through, so as to be able to calculate other dynamic features of the sections and road networks (such as traffic, average speed, and road network OD matrix). However, the trajectory data does not contain information such as when the vehicle passed through which node, and we only know the coordinate points of the vehicle trajectory. Then, can we extract the vehicle trajectory by using the trajectory coordinate points and the node information?

In the beginning, we tried clustering based on clustering methods, trying to cluster the coordinate points according to the separated road segments. However, after experimenting with various mainstream clustering methods, it is found that clustering cannot solve the tagging problem of coordinate points. How do we mark the vehicle coordinate points with the tag of the node? After further experimentation, we thought that the vehicle path extraction can be carried out by using the function inpolygon that comes with MATLAB. The core idea is as follows:

(1) Taking each node as the center, the regular polygons that can cover a certain section range are constructed, respectively. These regular polygons represent the in-out range of the node, which is referred to as the node regular polygon. If a coordinate point is within a node regular polygon, it is considered to belong to the node; otherwise, it does not belong to the node.

(2) Traverse the trajectory data and find out which node the coordinate points belong to. If a coordinate point does not belong to all nodes, it means that the coordinate point is on the road segment and special labeling is carried out.

(3) Compress the trajectory information of each vehicle, and record the starting line number of each vehicle passing through the node and the number of coordinate points at each node.

(4) Continue to compress the trajectory information of each vehicle, record all the nodes that each vehicle passes through (in the form of a string and a vector), and derive the OD characteristics of the vehicle.

In this paper, the method of signal control scheme improvement is proposed. According to the different traffic flow characteristics, the subregions are divided. Based on the three key parameters of cycle, arterial coordination signal offset, and green split, a set of hierarchical control algorithms based on reinforcement learning is constructed to optimize and improve the current signal timing scheme. Firstly, according to different evaluation indexes, the optimal period is obtained by using model-based algorithm. Using the combination of the average travel time of vehicles and the Bayesian optimization method based on the Gaussian process, which is commonly used in the optimization of machine learning algorithms, the arterial coordination control is set. The phase offset is optimized by the two-way flow ratio of the upstream and downstream roads and the reasonable setting of the pedestrian crossing phase. Then set different green wave bandwidths to match the upstream and downstream traffic of the rush hour and the tidal phenomenon with uneven travel speed. The intelligent algorithm such as Q-learning is used to optimize the green split of each intersection by using key traffic flow parameters such as traffic flow, density, and speed at each intersection. In the end, this paper uses the hierarchical traffic signal control algorithm based on reinforcement learning and combines the relevant knowledge of traffic engineering and engineering project experience to fine-tune the phase offset and green split to solve the problem of green wave bottleneck point of the arterial and signal interference caused by the right-turning vehicle and then obtain the optimal solution.

There are four key evaluation indexes included the number of vehicles leaving the road network at the end of the simulation, and the total delay time, the total travel time, and the total stopping number have been reduced to varying degrees shown in Figure 4.

For the different traffic flow states, the optimization effect caused by the Q-learning signal control method proposed in this paper in the high flow conditions (32.73%) is better than that in the medium flow and low flow conditions (22.32%,

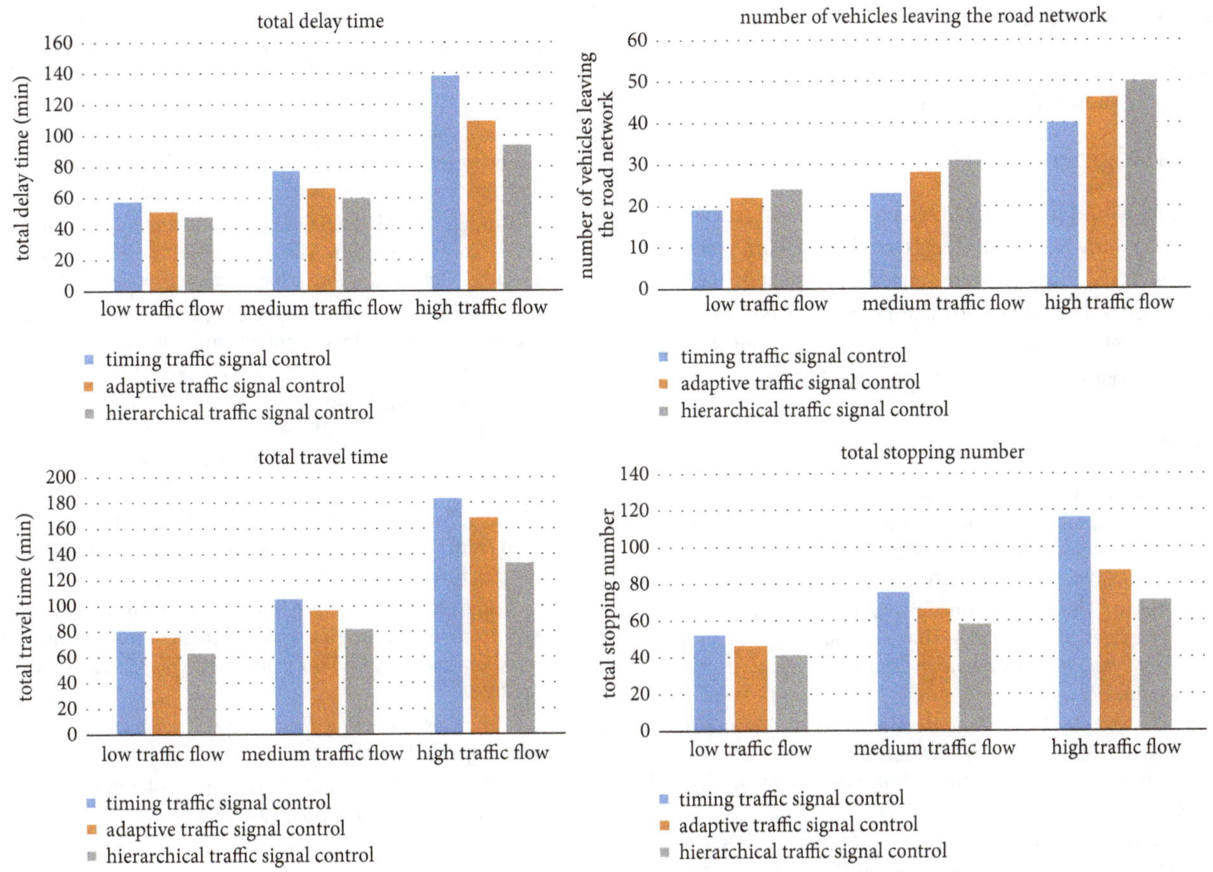

FIGURE 4: Evaluation of the traffic control method of regional road network.

17.11%). This also proves that the traffic control strategy based on reinforcement learning is more suitable for complex traffic environment (medium and high traffic flow, multi-intersection).

4.4. Analysis of Simulation Results. Through the above comparisons and analysis, it can be concluded that the traffic signal optimization strategy based on reinforcement learning is not applicable for all traffic environments. As to single intersections and arteries, their control effects are inferior to the current adaptive traffic signal control strategy in the low flow conditions. However, the traffic signal optimization strategy based on reinforcement learning is suitable for complex traffic environments (high flows and multiple intersections), and the effects of which are better than the current optimization methods in the conditions of high flows, as to both single intersections and arteries. In the future, we will focus on the network, continue to study the network traffic signal optimization method based on reinforcement learning, and then compare the effects with traditional optimization algorithms.

5. Conclusion and Discussion

This paper uses the hierarchical traffic signal control algorithm based on reinforcement learning and combines the relevant knowledge of traffic engineering and engineering project experience to fine-tune the phase offset and green split to solve the problem of green wave bottleneck point of the arterial and signal interference caused by the right-turning vehicle and then obtain the optimal solution.

In terms of the temporal dynamics of traffic control, reinforcement learning does not have complex optimizing modules and instant decisions can be made to respond to the uncertainty of time-varying traffic flow according to the characteristics of traffic flow observed in real time, which also corresponds with the actual conditions. Therefore, this paper focuses on the application of reinforcement learning in the field of traffic control and concludes that the traffic control method based on reinforcement learning has a better applicability in the complex traffic environment (high flows and multiple intersections), but it is not applicable to all traffic conditions. Furthermore, different from single intersection signal control, facing the integrated control for mainline and networked level traffic, it is still necessary to make further analysis in the aspects of data models and samples, coordination optimization techniques and multi-agent strategies, and mechanism analysis of the interaction between heuristic guidance and higher-level optimization mechanisms such as the pure stochastic optimization and hierarchical algorithm.

Conflicts of Interest

The authors declare that there are no conflicts of interest regarding the publication of this paper.

Acknowledgments

The authors would like to acknowledge the Intelligent Transportation System Research Center of Tongji University for data support. The research is supported by Project of National Natural Science Foundation of China (Project No. 61773293 and No. 61773288) and Key Project of National Natural Science Foundation of China (Project No. 51238008).

References

[1] N. H. Gartner, C. Stamatiadis, and P. J. Tarnoff, "Development of advanced traffic signal control strategies for intelligent transportation systems: multilevel design," *Transportation Research Record*, no. 1494, pp. 98–105, 1995.

[2] A. G. Sims, "The Sydney coordinated adaptive traffic system," in *Proceedings of the Engineering Foundation Conference on Research Directions in Computer Control of Urban Traffic Systems*, Calif, USA, 1979.

[3] Y.-T. Wu and C.-H. Ho, "The development of Taiwan arterial traffic-adaptive signal control system and its field test: A Taiwan experience," *Journal of Advanced Transportation*, vol. 43, no. 4, pp. 455–480, 2009.

[4] R. Vincent, "Safer signalized junction design and self-optimizing control," *Journal of Advanced Transportation*, vol. 28, no. 3, pp. 217–226, 2010.

[5] H. K. Lo and H. F. Chow, "Adaptive traffic control system: Control strategy, prediction, resolution, and accuracy," *Journal of Advanced Transportation*, vol. 36, no. 3, pp. 323–347, 2010.

[6] C. Meneguzzer, "Dynamic process models of combined traffic assignment and control with different signal updating strategies," *Journal of Advanced Transportation*, vol. 46, no. 4, pp. 351–365, 2012.

[7] K. Aboudolas, M. Papageorgiou, and E. Kosmatopoulos, "Store-and-forward based methods for the signal control problem in large-scale congested urban road networks," *Transportation Research Part C: Emerging Technologies*, vol. 17, no. 2, pp. 163–174, 2009.

[8] J. Hao, *Studies of the application of data-driven control method in traffic control*, Beijing Jiaotong University, Beijing, China, 2013.

[9] D. Ma, X. Luo, S. Jin, W. Guo, and D. Wang, "Estimating Maximum Queue Length for Traffic Lane Groups Using Travel Times from Video-Imaging Data," *IEEE Intelligent Transportation Systems Magazine*, vol. 10, no. 3, pp. 123–134, 2018.

[10] D. Ma, X. Luo, S. Jin, D. Wang, W. Guo, and F. Wang, "Lane-based saturation degree estimation for signalized intersections using travel time data," *IEEE Intelligent Transportation Systems Magazine*, vol. 9, no. 3, pp. 136–148, 2017.

[11] Y. Wang, X. Yang, H. Liang, and Y. Liu, "A Review of the Self-Adaptive Traffic Signal Control System Based on Future Traffic Environment," *Journal of Advanced Transportation*, vol. 2018, Article ID 1096123, 12 pages, 2018.

[12] D.-F. Ma, D.-H. Wang, F. Sun, Y.-M. Bie, and S. Jin, "Method of spillover identification in urban street networks using loop detector outputs," *Journal of Central South University*, vol. 20, no. 2, pp. 572–578, 2013.

[13] D. Ma, D. Wang, Y. Bie, S. Jin, and Z. Mei, "Identification of spillovers in urban street networks based on upstream fixed traffic data," *KSCE Journal of Civil Engineering*, vol. 18, no. 5, pp. 1539–1547, 2014.

[14] Littman M. L., "Markov games as a framework for multi-agent reinforcement learning," in *Proceedings of The 11th International Conference on Machine Learning*, pp. 157–163, 1994.

[15] S. J. Bradtke and Michael O. D., "Reinforcement learning methods for continuous-time Markov decision problems," *Advances in Neural Information Processing Systems*, pp. 393–400, 1995.

[16] D. Ma, F. Fu, S. Jin et al., "Gating control for a single bottleneck link based on traffic load equilibrium," *International Journal of Civil Engineering*, vol. 14, no. 5, pp. 281–293, 2016.

[17] W. Yang, L. Zhang, Y. Shi, and M. Zhang, "Applications of Agent Technology in Urban Traffic Signal Control Systems: A Survey," *Journal of Wuhai University of Technology (Transportation Science Engineering)*, vol. 38, no. 4, pp. 709–718, 2014.

[18] L. Zhang, W. Yang, J. Wang, and Q. Rao, "Large-scale agent-based transport simulation in shanghai, china," *Transportation Research Record*, vol. 2399, no. 1, pp. 34–43, 2013.

[19] M. Aslani, S. Seipel, and M. Wiering, "Continuous residual reinforcement learning for traffic signal control optimization," *Canadian Journal of Civil Engineering*, vol. 45, no. 8, pp. 690–702, 2018.

[20] C. J. Watkins, *Learning from delayed rewards*, University of Cambridge, UK, 1989.

[21] C. J. C. H. Watkins and P. Dayan, "Q-learning," *Machine Learning*, vol. 8, no. 3-4, pp. 279–292, 1992.

Permissions

All chapters in this book were first published in JAT, by Hindawi Publishing Corporation; hereby published with permission under the Creative Commons Attribution License or equivalent. Every chapter published in this book has been scrutinized by our experts. Their significance has been extensively debated. The topics covered herein carry significant findings which will fuel the growth of the discipline. They may even be implemented as practical applications or may be referred to as a beginning point for another development.

The contributors of this book come from diverse backgrounds, making this book a truly international effort. This book will bring forth new frontiers with its revolutionizing research information and detailed analysis of the nascent developments around the world.

We would like to thank all the contributing authors for lending their expertise to make the book truly unique. They have played a crucial role in the development of this book. Without their invaluable contributions this book wouldn't have been possible. They have made vital efforts to compile up to date information on the varied aspects of this subject to make this book a valuable addition to the collection of many professionals and students.

This book was conceptualized with the vision of imparting up-to-date information and advanced data in this field. To ensure the same, a matchless editorial board was set up. Every individual on the board went through rigorous rounds of assessment to prove their worth. After which they invested a large part of their time researching and compiling the most relevant data for our readers.

The editorial board has been involved in producing this book since its inception. They have spent rigorous hours researching and exploring the diverse topics which have resulted in the successful publishing of this book. They have passed on their knowledge of decades through this book. To expedite this challenging task, the publisher supported the team at every step. A small team of assistant editors was also appointed to further simplify the editing procedure and attain best results for the readers.

Apart from the editorial board, the designing team has also invested a significant amount of their time in understanding the subject and creating the most relevant covers. They scrutinized every image to scout for the most suitable representation of the subject and create an appropriate cover for the book.

The publishing team has been an ardent support to the editorial, designing and production team. Their endless efforts to recruit the best for this project, has resulted in the accomplishment of this book. They are a veteran in the field of academics and their pool of knowledge is as vast as their experience in printing. Their expertise and guidance has proved useful at every step. Their uncompromising quality standards have made this book an exceptional effort. Their encouragement from time to time has been an inspiration for everyone.

The publisher and the editorial board hope that this book will prove to be a valuable piece of knowledge for researchers, students, practitioners and scholars across the globe.

List of Contributors

Xiaomeng Shi and Zhirui Ye
Jiangsu Key Laboratory of Urban ITS, Southeast University, China
Jiangsu Province Collaborative Innovation Center of Modern Urban Traffic Technologies, China
School of Transportation, Southeast University, 2 Dongnandaxue Rd, Nanjing, Jiangsu 211189, China

Nirajan Shiwakoti
School of Engineering, RMIT University, Carlton, Melbourne, VIC 3053, Australia

Offer Grembek
Safe Transportation Research & Education Center, Institute of Transportation Studies, UC Berkeley, 2614 Dwight Way, Berkeley, CA 94720-7374, USA

Mohammad Maghrour Zefreh and Adam Torok
Budapest University of Technology and Economics, Department of Transport Technology and Economics, Stoczek St. 2, Building ST, 1111 Budapest, Hungary

Chao Sun, Xiaohong Chen and H. Michael Zhang
The Key Laboratory of Road and Traffic Engineering, Ministry of Education, Tongji University, Shanghai 201804, China

Chao Sun and Ze Huang
Shenzhen Urban Transport Planning Center Co., Ltd., Shenzhen Key Laboratory for Traffic Information and Traffic Engineering, Shenzhen 518021, China

H. Michael Zhang
Department of Civil and Environmental Engineering, University of California Davis, California, CA 95616, USA

Ellen F. Grumert and Andreas Tapani
Swedish National Road and Transport Research Institute (VTI), 581 95 Linköping, Sweden

Ellen F. Grumert
Department of Science and Technology, Linköping University, 601 74 Norrköping, Sweden

Lei Zhao, Hongzhi Guan, Junze Zhu and Yunfeng Wei
College of Architecture and Civil Engineering, Beijing University of Technology, Beijing 100124, China
Beijing Key Laboratory of Transportation Engineering, Beijing 100124, China

Manuel Contreras-Jara, Tomás Echaveguren and José Vargas Baecheler
Laboratory of Transportation Systems Management (GESITRAN), Departamento de Ingeniería Civil, Facultad de Ingeniería, Universidad de Concepción, Edmundo Larenas 219, Concepción, Chile

Tomás Echaveguren and Alondra Chamorro Giné
National Research Center for Integrated Natural Disaster Management (CIGIDEN), CONICYT/ FONDAP/15110017, Avda. Vicuña Mackenna 4860, Santiago, Chile

Alondra Chamorro Giné and Hernán de Solminihac Tampier
Department of Construction Engineering and Management, School of Engineering, Pontificia Universidad Católica de Chile, Avda. Vicuña Mackenna 4860, Santiago, Chile

Hernán de Solminihac Tampier
Latin American Center of Economic and Social Policies (CLAPES UC), Avda. Libertador Bernardo O'Higgins 440, Piso 13, Santiago, Chile

Yuwei Bie, Mudasser Seraj, Can Zhang and Tony Z. Qiu
Department of Civil and Environmental Engineering, University of Alberta, Edmonton, Canada

Tony Z. Qiu
Intelligent Transportation Systems Research Center, Wuhan University of Technology, Wuhan, China

Yanqun Yang
College of Civil Engineering, Fuzhou University, Fuzhou, Fujian 350108, China

Said M. Easa
Department of Civil Engineering, Ryerson University, Toronto, Ontario, Canada M5B 2K3
College of Civil Engineering, Fuzhou University, Fujian 350108, China

Zhibin Lin
Durham University Business School, Mill Hill Lane, Durham, DH1 3LB, UK

Xinyi Zheng
Faculty of Humanities and Social Sciences, Fuzhou University, Fuzhou, Fujian 350108, China

Volker Lücken, Nils Voss, Julien Schreier, Thomas Baag, Michael Gehring, Matthias Raschen, Christian Lanius, Rainer Leupers and Gerd Ascheid
Institute for Communication Technologies and Embedded Systems (ICE), RWTH Aachen University, Aachen, Germany

Chunjiao Dong, Chunfu Shao, Juan Li and Zhihua Xiong
MOE Key Laboratory for Urban Transportation Complex Systems Theory and Technology, Beijing Jiaotong University, Beijing 100044, China

Chunjiao Dong and Chunfu Shao
Key Laboratory of Industry of Big Data Application Technology for Comprehensive Transport, Ministry of Transport, Beijing Jiaotong University, Beijing 100044, China

Chunjiao Dong
Center for Transportation Research, Tickle College of Engineering, University of Tennessee, 600 Henley Street, Knoxville 37996, Tennessee, USA

Yizhe Wang, Xiaoguang Yang, Hailun Liang and Yangdong Liu
Key Laboratory of Road and Traffic Engineering of the Ministry of Education, Tongji University, 4800 Cao'an Road, Shanghai 201804, China

S. F. A. Batista and Ludovic Leclercq
Univ. Lyon, ENTPE, IFSTTAR, LICIT, F-69518, Lyon, France

Chuan-Lin Zhao
School of Civil and Transportation Engineering, Beijing University of Civil Engineering and Architecture, Beijing 100044, China

Yi Zhao
College of Automobile and Traffic Engineering, Nanjing Forestry University, Longpan Road #159, Nanjing, 210037, China

Satish V. Ukkusuri
Lyles School of Civil Engineering, Purdue University, 550 Stadium Mall Drive, West Lafayette, IN, 47907, USA

Jian Lu
School of Transportation, Southeast University, Si Pai Lou #2, Nanjing, 210096, China

Jinming You, Shouen Fang, Lanfang Zhang and Jingqiu Guo
Key Laboratory of Road and Traffic Engineering of The Ministry of Education, Tongji University, Shanghai 201804, China

John Taplin
Business School, The University of Western Australia, WA, 6009, Australia

Eduardo Valente, Camelia Avram and Adina Astilean
Technical University of Cluj-Napoca, Faculty of Automation and Computer Science, Str. G. Baritiu, nr. 26-28, 400027 Cluj-Napoca, Romania

Eduardo Valente and José Machado
University of Minho, MEtRICs Research Center, Campus of Azurém, 4800-058 Guimarães, Portugal

Yizhe Wang, Xiaoguang Yang, Yangdong Liu and Hailun Liang
Key Laboratory of Road and Traffic Engineering of the Ministry of Education, Tongji University, 4800 Cao'an Road, Shanghai 201804, China

Yizhe Wang, Xiaoguang Yang and Hailun Liang
Intelligent Transportation System Research Center of Tongji University, 4801 Cao'an Road, Shanghai 201804, China

Yangdong Liu
Hangzhou Hikvision Digital Technology Co., Ltd., No. 555 Qianmo Road, Binjiang District, Hangzhou 310052, China

Index

www.ingramcontent.com/pod-product-compliance
Lightning Source LLC
LaVergne TN
LVHW080734210926
848485LV00003B/556